1800
99BB

THE STRUCTURE
AND FUNCTION OF MUSCLE

SECOND EDITION

VOLUME II

Structure Part 2

CONTRIBUTORS

J. R. BENDALL

G. H. BOURNE

R. COUTEAUX

ARPAD I. CSAPO

E. J. FIELD

HARRY A. FOZZARD

CLARA FRANZINI-ARMSTRONG

R. P. GOULD

P. HUDGSON

SOHAN L. MANOCHA

ZBIGNIEW OLKOWSKI

ERNEST PAGE

E. VIVIER

THE STRUCTURE

AND FUNCTION OF MUSCLE

Second Edition

VOLUME II

Structure Part 2

Edited by

Geoffrey H. Bourne

Yerkes Primate Research Center
Emory University
Atlanta, Georgia

ACADEMIC PRESS   1973   New York and London
A Subsidiary of Harcourt Brace Jovanovich, Publishers

ACADEMIC PRESS, INC.
111 Fifth Avenue, New York, New York 10003

United Kingdom Edition published by
ACADEMIC PRESS, INC. (LONDON) LTD.
24/28 Oval Road, London NW1

LIBRARY OF CONGRESS CATALOG CARD NUMBER: 72-154373

PRINTED IN THE UNITED STATES OF AMERICA

CONTENTS

1. The Uterus—Model Experiments and Clinical Trials

Arpad I. Csapo

2. Capacitative, Resistive, and Syncytial Properties of Heart Muscle— Ultrastructural and Physiological Considerations

Ernest Page and Harry A. Fozzard

3. Contractile Structures in Some Protozoa (Ciliates and Gregarines)

E. Vivier

4. The Microanatomy of Muscle

R. P. Gould

5. Postmortem Changes in Muscle

J. R. Bendall

6. Regeneration of Muscle

P. Hudgson and E. J. Field

7. Muscle Spindle

Zbigniew Olkowski and Sohan L. Manocha

8. Motor End Plate Structure

R. Couteaux
Revised by G. H. Bourne

9. Membranous Systems in Muscle Fibers

Clara Franzini-Armstrong

LIST OF CONTRIBUTORS

Numbers in parentheses indicate the pages on which the authors' contributions begin.

J. R. BENDALL, *Agricultural Research Center Meat Research Institute, Langford Bristol, England* (243)

G. H. BOURNE, *Yerkes Primate Research Center, Emory University, Atlanta, Georgia* (483)

R. COUTEAUX, *Faculté des Sciences, Université de Paris, Paris, France* (483)

ARPAD I. CSAPO, *Department of Obstetrics and Gynecology, Washington University School of Medicine, St. Louis, Missouri* (1)

E. J. FIELD, *Medical Research Council Demyelinating Diseases Unit, Newcastle General Hospital, Newcastle upon Tyne, England* (311)

HARRY A. FOZZARD, *Department of Medicine and Physiology, Pritzker School of Medicine, The University of Chicago, Chicago, Illinois* (91)

CLARA FRANZINI-ARMSTRONG, *Department of Physiology, University of Rochester, School of Medicine and Dentistry, Rochester, New York* (531)

R. P. GOULD, *Department of Anatomy, The Middlesex Hospital, Medical School, London, England* (185)

P. HUDGSON, *The Muscular Dystrophy Research Laboratories and the Medical Research Council Demyelinating Diseases Research Unit, Newcastle General Hospital, Newcastle upon Tyne, England* (311)

SOHAN L. MANOCHA, *Yerkes Primate Research Center, Emory University, Atlanta, Georgia* (365)

ZBIGNIEW OLKOWSKI, *Department of Radiation Therapy, Winship Clinic, Emory University Medical School, Atlanta, Georgia* (365)

ERNEST PAGE, *Department of Medicine and Physiology, The University of Chicago, Chicago, Illinois* (91)

E. VIVIER, *Department of Animal Biology, Science Faculty, Lille, and Laboratory of Electron Microscopy, Pasteur Institute, Paris, France* (159)

PREFACE

In the years that have elapsed since the first edition of this work was published in 1960, studies on muscle have advanced to such a degree that a second edition has long been overdue. Although the original three volumes have grown to four, we have covered only a fraction of the new developments that have taken place since that time. It is not surprising that these advances have not been uniform, and in this new edition not only have earlier chapters been updated but also areas in which there was only limited knowledge before have been added. Examples are the development of our knowledge of crustacean muscle (172 of 213 references in the reference list for this chapter are dated since the first edition appeared) and arthropod muscle (205 of 233 references are dated since the last edition). Obliquely striated muscle, described in 1869, had to wait until the electron microscope was focused on it in the 1960's before it began to yield the secrets of its structure, and 33 of 43 references dated after 1960 in this chapter show that the findings described are the result of recent research. There has also been a great increase in knowledge in some areas in which considerable advances had been made by the time the first edition appeared. As an example, in Dr. Hugh Huxley's chapter on "Molecular Basis of Contraction in Cross-Striated Muscles," 76 of his 126 references are dated after 1960.

The first volume of this new edition deals primarily with structure and considers muscles from the macroscopic, embryonic, histological, and molecular points of view. The other volumes deal with further aspects of structure, with the physiology and biochemistry of muscle, and with some aspects of muscle disease.

We have been fortunate in that many of our original authors agreed to revise their chapters from the first edition, and it has also been our good fortune to find other distinguished authors to write the new chapters included in this second edition.

To all authors I must express my indebtedness for their hard work and patience, and to the staff of Academic Press I can only renew my confidence in their handling of this publication.

Geoffrey H. Bourne

PREFACE
TO THE FIRST EDITION

Muscle is unique among tissues in demonstrating to the eye even of the lay person the convertibility of chemical into kinetic energy.

The precise manner in which this is done is a problem, the solution of which has been pursued for many years by workers in many different disciplines; yet only in the last 15 or 20 years have the critical findings been obtained which have enabled us to build up some sort of general picture of the way in which this transformation of energy may take place. In some cases the studies which produced such rich results were carried out directly on muscle tissue. In others, collateral studies on other tissues were shown to have direct application to the study of muscular contraction.

Prior to 1930 our knowledge of muscle was largely restricted to the macroscopical appearance and distribution of various muscles in different animals, to their microscopical structure, to the classic studies of the electro- and other physiologists and to some basic chemical and biochemical properties. Some of the latter studies go back a number of years and might perhaps be considered to have started with the classic researches of Fletcher and Hopkins in 1907, who demonstrated the accumulation of lactic acid in contracting frog muscle. This led very shortly afterward to the demonstration by Meyerhof that the lactic acid so formed is derived from glycogen under anaerobic conditions. The lactic acid formed is quantitatively related to the glycogen hydrolyzed. However, it took until nearly 1930 before it was established that the energy required for the contraction of a muscle was derived from the transformation of glycogen to lactic acid.

This was followed by the isolation of creatine phosphate and its establishment as an energy source of contraction. The isolation of ADP and ATP and their relation with creatine phosphate as expressed in the Lohmann reaction were studies carried out in the thirties. What might be described as a spectacular claim was made by Engelhart and Lubimova, who in the 1940's said that the myosin of the muscle fiber had ATPase activity. The identification of actin and the relationship of actin and myosin to muscular contraction and the advent of the election microscope and its application with other physical techniques to the study of the general morphology and ultrastructure of the muscle fibers were events in the 1940's which greatly developed our knowledge of this complex and most mobile of tissues.

In the 1950's the technique of differential centrifugation extended the knowledge obtained during previous years of observation by muscle cytologists and electron microscopists to show the differential localization of metabolic activity in the muscle fiber. The Krebs cycle and the rest of the complex of aerobic metabolism was shown to be present in the sarcosomes—the muscle mitochondria.

This is only a minute fraction of the story of muscle in the last 50 years. Many types of disciplines have contributed to it. The secret of the muscle fiber has been probed by biochemists, physiologists, histologists and cytologists, electron microscopists and biophysicists, pathologists, and clinicians. Pharmacologists have insulted skeletal, heart, and smooth muscle with a variety of drugs, *in vitro, in vivo,* and *in extenso;* nutritionists have peered at the muscle fiber after vitamin and other nutritional deficiencies; endocrinologists have eyed the metabolic process through hormonal glasses. Even the humble histochemist has had the temerity to apply his techniques to the muscle fiber and describe results which were interesting but not as yet very illuminating—but who knows where knowledge will lead. Such a ferment of interest (a statement probably felicitously applied to muscle) in this unique tissue has produced thousands of papers and many distinguished workers, many of whom we are honored to have as authors in this compendium.

Originally we thought, the publishers and I, to have *a book* on muscle which would contain a fairly comprehensive account of various aspects of modern research. As we began to consider the subjects to be treated it became obvious that two volumes would be required. This rapidly grew to three volumes, and even so we have dealt lightly or not at all with many important aspects of muscle research. Nevertheless, we feel that we have brought together a considerable wealth of material which was hitherto available only in widely scattered publications. As with all treatises of this type, there is some overlap, and it is perhaps unnecessary to mention that to a certain extent this is desirable. It is, however, necessary to point out that most of the overlap was planned, and that which was not planned was thought to be worthwhile and was thus not deleted.

We believe that a comprehensive work of this nature will find favor with all those who work with muscle, whatever their disciplines, and that although the division of subject matter is such that various categories of workers may need only to buy the volume which is especially apposite to their specialty, they will nevertheless feel a need to have the other volumes as well.

The Editor wishes to express his special appreciation of the willing collaboration of the international group of distinguished persons who made this treatise possible. To them and to the publishers his heartfelt thanks are due for their help, their patience, and their understanding.

Emory University, Atlanta, Georgia   GEOFFREY H. BOURNE
October 1, 1959

CONTENTS OF OTHER VOLUMES

1

THE UTERUS—MODEL EXPERIMENTS AND CLINICAL TRIALS

ARPAD I. CSAPO

I. Introduction

In the previous edition of this treatise, I referred self-gibingly to the uterus as "headache" muscle. This term has been invented by investigators for whom "muscle" was the frog sartorius, and the smooth muscles, continuously changing under vitally important regulatory effects, were nuisances. Few students of muscle recognized the broad biological significance and basic nature of these regulatory changes and the opportunity to characterize them by studying excitability and contractility. At a strategic conference, the question "What is the effect of hormones on the excitable membranes?" was unhesitantly answered by the spokesman on excitation, "None whatsoever." Apparently, a public debate over the blessings and apprehensions experienced during the massive exposure of millions to pre- and postconceptional therapy was needed for altering this posture, for interest in smooth muscles increased recently.

However, a few investigators did recognize the challenges of these cyclically changing "mini" cells decades ago. By persistent redesign of techniques and procedures, they penetrated some of the secrets of smooth muscles. As a result, we learned that the progestational hormone protects pregnancy by disrupting "communication" among the billions of cells of the myometrial community. We also learned that clinically effective uterine activity demands the synchronic function of this community, and that it is this synchrony which the progesterone block prohibits by suppressing the conduction of electric train discharges over the organ (Csapo, 1955). Recent studies revealed the nature of the block by showing that progesterone increases tissue resistance by decreasing electrical coupling between myometrial cells (Ichikawa and Bortoff, 1970). When population control strategists and neonatologists recognize the broad regulatory significance of this block, reproduction and obstetrics will enter a new era.

To promote this recognition we shall survey our facts so as to reconstruct the basic mechanism of function and regulation of smooth muscles. The comparative aspects of these organs and the functional characterisation of a variety of smooth muscles, including the uterus, are described in comprehensive writings.* This chapter will focus on the limited subject of how basic studies in uterine function and regulation promoted developments in reproduction, obstetrics, and regulatory biology, and the reader is referred to recent reviews (Csapo, 1969a, 1970, 1971a,b; Csapo and Wood, 1968) for further information.

II. Why Smooth Muscles?

Hollow chambers and tubes in mammalian viscera are coated with varying layers of smooth muscle tissue. By their contractility, these multi-

---

* Reynolds, 1949; Goto and Tamai, 1960; Csapo, 1960, 1961a, 1962; Bulbring, 1962, 1964; Bulbring *et al.*, 1970; Kuriyama, 1961; Bozler, 1962; Marshall, 1962, 1964; Barr, 1963; Burnstock *et al.*, 1963; Schofield, 1963; Jung, 1964; Lepage and Sureau, 1964; Sureau *et al.*, 1965; Suzuki, 1966; Needham and Schoenberg, 1967.

billion cell communities aid hollow organs in the performance of a variety of functions. Considering the final outcome, these functions are as different from one another as the capturing of the ovum by the fimbriae; the precisely controlled transport of the blastocyst through the ampullary–isthmic junction and the isthmus toward the uterine cavity; the protection of the embryo during prenatal life and the delivery of a newborn at term by the uterus; the promotion of blood flow in the blood vessels; the passage of urine through the urethra, bladder, and ureter; the emptying of the gastrointestinal tract; etc.

However, different as they are in structure and biological function, these smooth-muscle-coated organs have certain common operational principles. Within physiological limits they all are capable of distension, accommodating the volume increase of their luminal contents; they promote the directional passage of these contents, and do so (with few exceptions) without voluntary control. Accordingly, smooth muscles are not specialized for speed, but for slow movements demanded by their physiological functions. The organs coated by them are capable of simultaneous internal shortening and lengthening, as required by the different forms of synchronic and peristaltic contractile functions.

Smooth muscle organs are built from spindle-shaped contractile cells embedded in connective tissue network. These cells, the complete units of contractile function, are of considerable practical interest to man; indeed far greater than the structurally more perfect single fibers of skeletal muscle. Man's everyday existence depends on their precisely controlled performance, starting as early in life as during ovum transport, nidation, intrauterine development, and parturition. All smooth muscles are subject to disease, frequently resulting from regulatory (rather than structural) failures. Uterine function alone controls the fate of 331,000 newborns every day (400 millions every year) during their hazardous first journey.

Since the physiological function and regulation of smooth muscle organs are broadly different, they cannot be discussed together in depth, within the limits of this review. Thus, the present treatment is limited to the uterus. This smooth muscle organ has been thoroughly examined for its fine structure, chemical composition, energetics, ionic profiles, excitability, mechanical activity, pharmacological reactivity, and regulation. These studies were conducted at various levels of organization, in animals as well as in man, including the most complex level of all, the parturient patient, whose uterine function determines the fate of two lives. Because of these multidisciplinary studies, the uterus offers a useful model for the examination of the relationship between structure, function, and regulation of mammalian smooth muscles.

In their basic contractile function, there is a similarity not only among smooth muscles, but also among smooth and cross-striated muscle organs. However, as important as is this basic similarity are the superimposed differences among the various muscles. For, in addition to a common functional "backbone," muscles are also equipped with a variety of structural, functional, and regulatory features, which allow them to perform specific assignments. The examination of the uterus at various levels of organization, ranging from subcellular constituents to the delivery of patient, promises to be a rewarding experience. More than that, it might illustrate the point that, incomplete as it is, current knowledge about the mechanism of muscle function significantly promoted academic and therapeutic advances in medicine.

Muscle has been split to angstrom dimensions and reconstituted from isolated constituents to be examined with all available tools of modern analytical biology. This extensive research was focused upon one single problem, the molecular mechanism of contraction. No final answer has been obtained, but man's desire to know was satisfied by progress. Our concept of life greatly depends on our degree of understanding the mechanism of function in a single cell. Which one we choose to study may not be a crucial decision in basic research, for the organization of function in living matter is probably accomplished according to common biological principles, shared by a variety of cells. Focusing upon the mechanism of function at the molecular level may well reassure us that it makes little difference whether the cell contracts, digests food, secretes hormones, or transmits messages. Studies at the molecular level must be complemented by additional experiments, conducted at the cellular and organ levels, so as to approximate physiological reality.

Muscle has been examined preferentially to other tissues because its function can be measured more easily and with greater precision than that of other cells. Among the various contractile organs, skeletal muscle has been preferentially selected for the analysis of the mechanism of function. This choice is justified by the unique structural and functional precision of striated muscle. Fully functional single cells, even subcellular units, were prepared successfully from this organ. Mammalian smooth muscles are lesser suited for analytical studies of this type. Their structural organization is complex, their spindle-shaped cells are small and are embedded in an abundant connective tissue network. It is virtually impossible, therefore, to isolate single cells, without sacrificing their structural integrity. The tissue culture technique may overcome this limitation.

A community of over 200 billion smooth muscle cells must work in concert to guarantee the safe delivery of the newborn. Any regulatory error which upsets the synchronic and directional contractility of this community, jeopardizes normal parturition and exposes the fetus to grave hazards. Preventive obstetrics, an ever increasing reality in the management of labor, evolved from analytical uterine physiology. Progress was slow and silent, yielding no spectacular cures. However, being firmly anchored to classic and molecular muscle physiology, as well as to regulatory biology, this analytical approach is the one to stay. If medical developments are measures of progress, it would appear that those students of the uterus who examined its function comparatively, step by step, at different levels of organization, in various species, and using a variety of techniques, contributed to tangible accomplishments.

III. Structure

The myometrial cell is a contractile system surrounded by an excitable membrane (Csapo, 1962). As described in earlier reviews (Csapo, 1955, 1962), these spindle-shaped units of contractile function (Fig. 1) are not only small (5–10 μ wide and 50–300 μ long), but their dimensions are altered by regulatory conditions. For, unlike the single fibers of cross-striated muscle, the myometrial cells are continuously exposed to structural and functional changes from puberty till menopause, during the menstrual cycles, pregnancy, and postpartum involution. The individual cells form fiber bundles of about 100 μ in diameter and are embedded in a connective tissue network.

The fiber bundles unite in an interlacing longitudinal and circular structure, with vascular zones in between them. Together they create a contractile coat around the endometrial lining of the uterine cavity. Mammalian contractile tubes and chambers all have this type of basic structure (Bulbring *et al.*, 1970), enabling them to exert pressure on their contents in the process of emptying themselves.

Electron microscopy revealed myofibrils of about 0.5–1 μ in diameter inside the myometrial cells (Csapo, 1955). These myofibrils run parallel to the long axis of the cell. The space between myofibrils is occupied by cellular elements, the nucleus, and mitochondria, spaced like beads in a row (Csapo, 1955) (Fig. 2). Thus, the sites of energy production and expenditure are in close vicinity.

Unlike in cross-striated muscle (Huxley, 1956), in the myometrium the detailed structural arrangement of the contractile proteins (myosin and actin), the bridges between these filaments, and the position of other supportive proteins have not been resolved. However, smooth muscles are being examined (Needham and Schoenberg, 1967) for these fine structural details. Some of the current accomplishments and unresolved problems will be discussed presently.

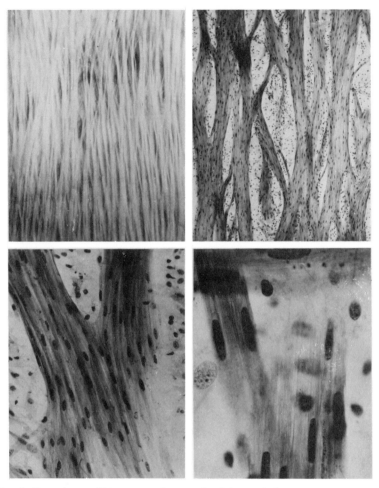

Fig. 1. Parturient rabbit uterus. Longitudinal layer, light microscopy. Note the fiber bundles, the spindle-shaped cells and the myofibrils, as magnification is increased. (Csapo, 1955).

IV. The Contractile System

The analytical study of the mechanism of myometrial function began with the isolation of the protein complex actomyosin, the high energy phosphate compound ATP, and the *in vitro* examination of their interaction (Csapo, 1955, 1962). In contrast to cross-striated muscle, which contains about 50 mg actomyosin per gram wet weight, the nonpregnant uterus has only 2–6 mg and the pregnant uterus 7–13 mg of this protein complex (Fig. 3). The concentrations measured depend on extraction procedures and the methods of quantitation.

Actin-free myosin has been prepared from uterine actomyosin (Csapo, 1959) by treatment with dissociating agents (potassium iodide or potas-

Fig. 2. Electron micrograph of the midportion of one myometrial cell. Note the myofilaments within and the mitochondria between myofibrils (×6750). (Csapo, 1955.)

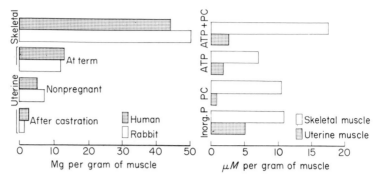

Fig. 3. The concentrations of actomyosin and high-energy phosphates in skeletal and uterine muscles. Note the concentration changes induced by alterations in regulatory conditions. (Csapo, 1955.)

sium chloride solutions containing ATP and magnesium chloride) and subsequent separation in the centrifuge. Myosin-free actin has also been obtained from the uterus by separating it electrophoretically from tropomyosin (Needham and Schoenberg, 1967). Uterine actin and myosin, when crossed with the proteins of skeletal muscle, readily formed actomyosin. On precipitation by dilution, uterine actomyosin lost its ability to yield "contractile threads." Contractility was restored, however, by a thermolabile, nondialyzable (aqueous) uterine extract (Csapo, 1959). This uterine (X) factor may serve a function similar to that of the troponin (Ebashi *et al.*, 1967) of cross-striated muscle. Troponin is required for the calcium sensitivity and the activation of ATPase, when actin and myosin are brought together to form actomyosin.

The ATPase activity of uterine actomyosin is low (0.54–1.2 μM of phosphorus set free per milligram of protein nitrogen per minute); the enzyme is activated by calcium and inhibited by magnesium (Csapo, 1955, 1959; Needham and Schoenberg, 1967). Like in cross-striated muscle, tripsin digestion splits uterine myosin to meromyosins and in so doing increases the ATPase activity of the H-meromyosin fraction (Csapo, 1959). The biochemistry of the myometrium and other smooth muscles is discussed in detail elsewhere (Needham and Schoenberg, 1967).

V. Comparison between Cross-Striated and Uterine Muscles

The balanced opinion (Csapo, 1955, 1962) that the mechanism of function may be similar in the structurally highly organized cross-striated and the lesser precise uterine muscle is not based entirely on the observed similarities in their biochemical composition. Similarities were also observed between these tissues in energetic, excitability, and mechanical properties.

When the classic experiment of Lundsgaard (originally performed on the frog sartorius) was repeated with the uterus (Csapo, 1955), the significance of ATP and CP in the energetics of uterine contractility became apparent (Fig. 4). When the aerobic, anaerobic, or both path-

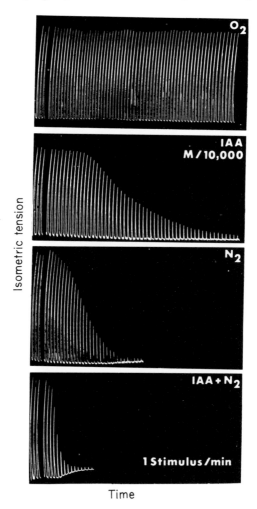

Fig. 4. The effect of reduced and suspended energy supply on uterine tension. Rabbit in natural estrus. Note the gradual loss of tension when resynthesis of high energy phosphates is suppressed at different degrees. (Csapo, 1955.)

ways of resynthesis of high energy phosphates were blocked, tension (in the electrically tentanized uterus) declined as a function of the disappearance of CP. When, after the depletion of CP, ATP also disappeared the uterus developed sustained contracture.

The threshold–membrane potential relation states (Jennerick and Gerard, 1953) that threshold is lowered by a gradual decrease in membrane potential. At a critical membrane potential value, threshold becomes zero, promoting spontaneous activity, while mechanical rest is maintained by an excess membrane potential over this critical value. The uterus obeys this relationship (Csapo, 1962) (Fig. 5), as does the frog sartorius. Spontaneous activity in the uterus is abolished by hyperpolarization, while it is increased by slight depolarization. At the critical potential, spontaneous activity is maximal, and further depolarization leads to sustained contracture.

The strength–duration relation states (Rushton, 1930) that within limits the product of the strength and duration of the electric stimulus determines the threshold mechanical response. The uterus obeys this relationship (Csapo, 1955), as does the frog sartorius (Fig. 6).

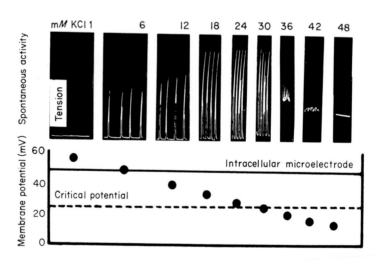

Fig. 5. Spontaneous uterine activity as a function of the membrane potential. Parturient rabbit uterus. Note the decrease in membrane potential and increase in spontaneous activity when the potassium gradient across the cell membrane is reduced. Note the maximum activity when the membrane is depolarized to the critical level, the display of contracture when depolarization exceeds the critical value, and the suppression of spontaneous activity by hyperpolarization. (Goto and Csapo, 1959.)

It is proposed in the sliding model that the contractile process results from a molecular interreaction followed by a displacement between the thick myosin and thin actin filaments, arranged with spatial precision in the sarcomere. An outstanding feature of this model is that it is consistent with the mechanical characteristics of cross-striated muscle, for example with the length–tension and load–shortening velocity relationships. If these mechanical features are expressions of fine structure and chemical composition, as is currently believed, then one would expect that smooth muscles, which share chemical and mechanical similarities, also share structural similarities with cross-striated muscles. In the previous chapters evidence was presented that the uterus is such a smooth muscle.

And yet, in spite of extensive efforts (Needham and Schoenberg, 1967), the structural basis of the sliding model could not be demonstrated in the uterus and in other mammalian smooth muscles. The thick and thin filaments were only observed regularly in some invertebrates, but not in vertebrates. In vertebrates, only one type, the 40–80 Å wide filaments were resolved and the thick filaments could not be found regularly, not even by the X-ray diffraction technique.

The possibility has been considered (Needham and Schoenberg, 1967), therefore, that these thin filaments are made up of F-actin, while myosin is dispersed in the resting smooth muscle cells, aggregating into thick filaments only during activity. This hypothesis was based on the certainty that actomyosin is present in the uterus, but is soluble at low ionic strength in the form of tonoactomyosin, and that the solubility of uterin myosin is different from that of cross-striated muscle. The alternative has been considered also, that in living myometrial cells the thick myosin filaments are present, but being labile, they dissolve during the tortuous processes of isolation and electron microscopic preparation.

Accordingly, smooth muscle filaments have been suspended in the presence of ATP, calcium, and magnesium, or glycerinated and intact fiber bundles have been exposed to ATP prior to processing them for electron microscopy (Needham and Schoenberg, 1967). These procedures did bring out thick, 130 Å wide filaments. However, further experiments showed (Garamvolgyi et al., 1971) that during preparation the crucial step is stretch, for stretch alone did bring out the thick filaments, irrespective of ATP pretreatment.

At first these findings were interpreted to mean that the thick filaments only appear in smooth muscles during contractile activity. However, it has been shown that this is not the case, for thick filaments did appear regularly in electron micrographs of stretched smooth muscles, even if their activity has been prohibited (Garamvolgyi et al., 1971). It would appear, therefore, that intact smooth muscles do contain thick filaments, presumably consisting of myosin, for a slight stretch, which preserves them during preparatory procedures, is nearly always present under physiological conditions *in vivo* and is revealed by the resting tension or resting pressure of the organ. If so, then the sliding model would seem to be applicable to smooth muscle, and there is no need for proposing alternative contraction mechanisms to explain their function.

VII. Working Capacity

The working capacity of the uterus is determined by the concentration of actomyosin and high energy phosphates in the myofibrils (Csapo, 1955, 1962). The synthesis of this contractile system is dependent on estrogen stimulation. After ovariectomy, the actomyosin concentration of the uterus and with it the maximum tension decrease. These changes are prevented by estrogen treatment. Estrogen deficiency and substitution therapy decrease and increase respectively the working capacity of the uterus, but the ratio tension/actomyosin concentration remains constant (Fig. 9) (Csapo, 1955, 1962).

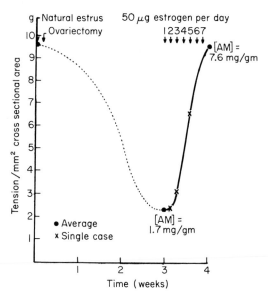

Fig. 9. The effect of estrogen on working capacity. Rabbit uterus. Note the relationship between estrogen, actomyosin concentration ([*AM*]) and tension (*T*) and the constant ratio: *T*/[*AM*]. ● = average; x = single case. (Csapo, 1955.)

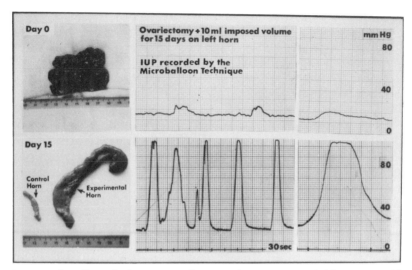

Fig. 10. The effect of chronic stretch on working capacity. Rabbit uterus. After ovariectomy, one horn is distended for 15 days by the imposition of 10 ml intrauterine volume. Note the increased mass and IUP of this horn, and compare it with the castrate atrophy of the contralateral (undistended) horn. (DeMattos *et al.*, 1967.)

However, estrogen is not the only regulatory factor which increases the working capacity of the uterus. Chronic stretch can accomplish the same effect (Csapo, 1970). This physical regulatory factor also induces a marked myometrial hypertrophy (Reynolds, 1949) and an increase in the working capacity of the uterus (Fig. 10), even in ovariectomized rabbits, which characteristically show castrate atrophy in the undistended horn (DeMattos *et al.*, 1967). This effect of chronic stretch is significant. During normal menstrual cycles, endometrial proliferation imposes upon the uterus a volume increase of about 5 ml, while pregnancy imposes an over 4000 ml volume increase. In addition to these effects on working capacity, stretch also facilitates excitability and spontaneous activity in the uterus (Csapo, 1962, 1970; Csapo and Wood, 1968). Through these multiple actions, stretch sustains the uterus in a functional readiness without the support of endogenous "stimulants." However, stimulants can further promote uterine activity.

VIII. Excitation

The contractile system of the myometrium, as that of cross-striated muscle, is activated through the excitation process, generated by the cell membrane (Bozler, 1962). Nearly unsurmountable technical difficulties had to be overcome before myometrial excitation could be examined meaningfully. These difficulties resulted from the small size of the myometrical cells, the inhomogeneity of the tissue and the changing regulatory conditions of the organ. Single cells and excised strips were studied *in vitro* and the intact organ *in vivo* using a variety of techniques. These variations in experimental conduct promoted the electrophysiological characterization of the myometrium. The agreement reached is impressive and is well reflected in the writings of those who pioneered in uterine electrophysiology.*

Of course, not all investigators applied the intracellular microelectrode technique meaningfully to the delicate myometrial cells. Tissue injuries and prolonged *in vitro* perfusion destroyed the regulatory identity of the uterus and led to technical errors (reflected by the quality of tracings) and opinionated conclusions (Kao, 1967).

The contractility of the electrically tetanized uterus is stable and maximal for several hours. This is so because in an electric field all myometrial cells are simultaneously activated and the working capacity is not (only excitability is) seriously affected by prolonged perfusion (Fig. 11A). Figure 11B illustrates that after 10 min *in vitro* exposure, the pregnant uterus (examined by suction electrodes) displays no appreciable electric activity, while after 150 min perfusion it shows distinct discharges. Also, the strength–duration curve (Fig. 11C) reflects high threshold after 10 min exposure and low threshold after 150 min exposure of the organ to *in vitro* conditions. Evidently, after prolonged perfusion, the progesterone block weakens and the regulatory character of the uterus is altered. Tissue damage accelerates the deterioration of the myometrium.

When the excised uterus is stimulated in an electric field, it displays its maximum contractile capacity in isometric tension, or isotonic work (under appropriate load). Activity is maximal because all cells are simultaneously activated. In contrast, spontaneous activity is submaximal as a rule (Fig. 12). The degree of participation in activity of the various uterine regions is determined by regulatory conditions.

The contractile system of the uterus is continuously changing under regulatory influences. However, this control over contractility appears to be of lesser medical significance than that exerted over excitability,

---

* Goto and Tamai, 1960; Csapo, 1960, 1961a, 1962, 1969a, 1970, 1971a,b; Csapo and Wood, 1968; Bulbring, 1962, 1964; Bulbring *et al.*, 1970; Kuriyama, 1961; Bozler, 1962; Marshall, 1962, 1964; Barr, 1963; Burnstock *et al.*, 1963; Schofield, 1963; Jung, 1964; Lepage and Sureau, 1964; Sureau *et al.*, 1965; Suzuki, 1966; Needham and Schoenberg, 1967.

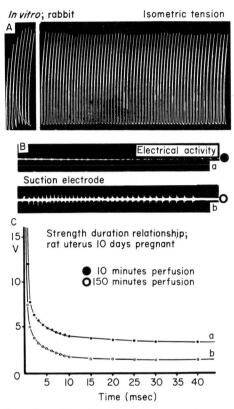

Fig. 11. The effect of prolonged perfusion on excitability. Note that prolonged perfusion does not affect electrically induced maximum tension appreciably (A), but it does affect the spontaneous train discharges (B) and threshold (C). (Csapo, 1971b.)

for clinical abnormalities are more readily associated with undesirable changes in excitability than with contractility. Excitation (the initial step in uterine function) was found highly sensitive to regulatory influences, functional disorders, and therapeutic measures. Of course, when excitation is suppressed, contractility need not be, because without excitation there is no contraction.

These general considerations advise caution in the interpretation of the *in vitro* data. Certain information can only be obtained *in vitro* by analytical experiments. However, the physiological and clinical significance of these studies have to be verified by experiments at higher levels of organization, e.g., by a step-by-step synthesis until the level of the intact animal or patient is reached.

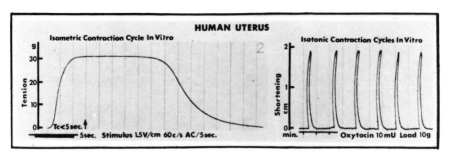

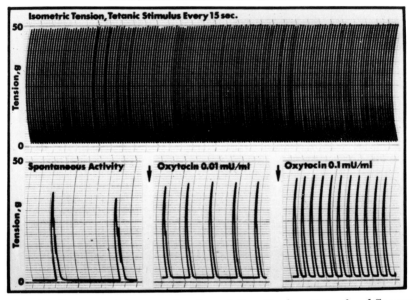

Fig. 12. Uterine activity as a function of the stimulus. Note the differences in tension under electric, spontaneous, and pharmacological stimulation. (Csapo, 1961a.)

Resting and action potentials were recorded from myometrial cells of 5–10 μ in diameter. When these cells were penetrated (with microelectrodes of less than 0.5 μ in diameter), several attempts failed before a steady and sustained deflection revealed the true value of the resting potential (Fig. 13) (Goto and Csapo, 1959; Goto *et al.*, 1959). Depending on regulatory conditions, (inside negative) potential differences of 35–60 mV have been recorded between the interior and the exterior of the myometrical cells. In a variety of excitable tissues, the magnitude of the resting potential was related to the concentration gradients of potassium and sodium across the cell membrane. In case of the myometrium, this demonstration has been hampered by difficulties in the precise

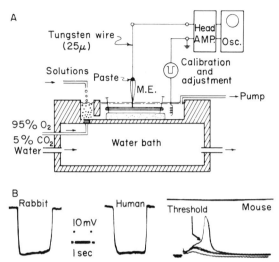

Fig. 13. (A) Diagram of the microelectrode, technique.   (B) Original tracings of the resting and action potentials. Note the recorded potential difference across the membrane when the microelectrode penetrates the myometrial cell (Goto and Csapo, 1959). Note also the action potential when electrically induced depolarization reaches the threshold value. (Kuriyama, 1961.)

measurements of the extracellular space (values of 30–40% are commonly quoted). However, it was observed (Fig. 5) that the resting potential is lowered by a stepwise increase in the $[K]_0$, although a less than 58 mV slope was recorded for a tenfold change in $[K]_0$. This difference between the theoretical and measured values remained unexplained.

The ionic theory of excitation is based on the observation that changes in the permeability of the cell membrane provide the underlying mechanism for the changes in membrane potential. However, the uterine action potential, unlike those of nerve and cross-striated muscle, could not be fully accounted for by this theory (Fig. 14) (Kuriyama and Csapo, 1961). Uterine spike discharges were not extinguished in sodium-deficient Krebs solution (80% of the sodium being replaced by choline or sucrose). Therefore, an inward flow of sodium does not account fully for the uterine action potentials. Bates' advice, "treasure our exceptions," seems applicable. It appears mandatory, therefore, to quantitate the ionic changes in the uterus reliably (under a variety of endocrine conditions) before attempting to define the regulatory modifications of myometrial excitation in ionic terms.

When the membrane potential of the myometrial cell is lowered by stepwise increasing current, no spike is generated until depolarization reaches a critical threshold value (Fig. 13) (Goto *et al.,* 1959). At this

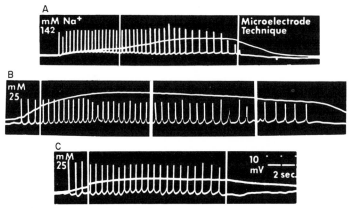

Fig. 14. The effect of sodium deficiency on the train discharge. Rat uterus post partum; (A) normal Krebs solution, (B) choline chloride solution (120 min), (C) sucrose Krebs solution (120 min). Note that 120 minutes exposure to a modified Krebs solution, in which 80% of the sodium is replaced by choline chloride or sucrose, does not abolish the spontaneous train discharges. (Kuriyama and Csapo, 1961.)

level, a self-generated process completes depolarization; in fact, the inside of the cell may become positive to the outside (overshoot). Subsequently, the membrane is repolarized and the status quo is restored. The recorded signal is the action (or spike) potential. Repetitive spikes form train discharges (Fig. 15) (Kuriyama and Csapo, 1961).

A stepwise increase in the $[K]_0$ gradually lowers the resting potential of the uterus (Fig. 5). As depolarization is increased, the required stimulus for eliciting a (threshold) mechanical response decreases. In the unstimulated uterus, this decrease in threshold increases spontaneous activity, and at the critical potential, maximum activity is recorded. Further depolarization induces a sustained contracture. In contrast, hyperpolarization suspends spontaneous activity (Fig. 5).

These changes in mechanical activity were accounted for by the effects of depolarization and hyperpolarization, respectively, on the train discharges. Depolarization (by excess potassium) increased spike frequency, the total number of spikes in the train, and the repetition frequency of subsequent trains (Fig. 16) (Kuriyama and Csapo, 1961). Extreme depolarization suspended the spike discharges. Apparently, uterine activity is altered within wide limits by subtle changes in membrane characteristics, and regulatory agents seem to take advantage of this relatively simple method of control.

Spontaneous activity is a significant functional feature of the uterus. In comparison with cross-striated and cardiac muscles, the resting poten-

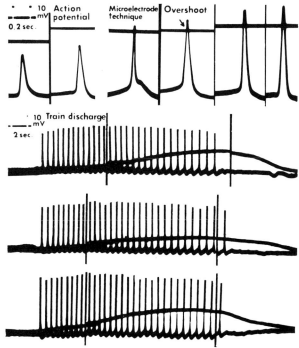

Fig. 15. Single action potentials and repetitive train discharges. Rat uteri, post partum. Note the differences in the shape and magnitude of the action potentials without or with overshoot. Note also the stability of the membrane and action potentials during a series of train discharges recorded sequentially from the same cell. (Kuriyama and Csapo, 1961.)

tial of the myometrium is low under all regulatory conditions. The membrane potential being close to the critical value (where threshold becomes zero), the excess potential (ensuring rest) is small. The low setting of the resting potential facilitates autorhythmicity. This conclusion is substantiated by the demonstration that when cross-striated muscle is rendered calcium-deficient (by exposure to calcium-free Krebs solution), the normally stable resting fibers develop spontaneous activity, as their excess potential is reduced (previous edition of this treatise). However, it should be noted that under these conditions the membrane potential changes were accompanied by a calcium imbalance, which may account by itself for spontaneous activity.

In the uterus, calcium deficiency develops rapidly (Fig. 17) (Csapo, 1961a; Kuriyama, 1961). The rapid response of the uterus to changes in ionic environment is the evidence of a quick penetration of ions

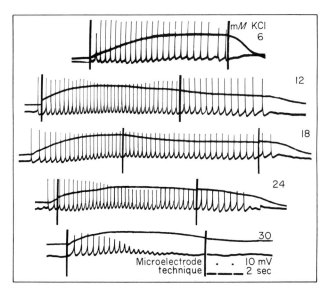

Fig. 16. The effect of stepwise depolarization on the train discharge. Rat uterus, post partum. Note that as the membrane is depolarized (step by step) by increasing the [K]ₒ, the train discharge is facilitated. However, when depolarization exceeds the critical potential, the train discharge is abolished. (Kuriyama and Csapo, 1961.)

into its extracellular space. This finding strengthens the significance of sustained spike discharges after 120 min exposure of the uterus to sodium-deficient Krebs solution. It also reveals the extreme sensitivity of this autorhythmic organ to calcium deficiency, providing a clue for mechanism of control. Calcium deficiency results in a rapid depolarization of the myometrial cell membrane, an extinction of spike discharges, an increase in threshold, a loss of spontaneous activity, electric excitability and pharmacological reactivity (Fig. 17). However, contractility is not altered significantly, for the calcium-deficient uterus develops close to maximum tension when stimulated in a strong electric field.

The recovery of the calcium-deficient uterus in normal Krebs solution is also rapid, indicating that changes in membrane calcium alone may bring about these changes in excitability. Excitability is restored in a graded manner by a gradual increase in the $[Ca]_0$. The significance of these early observations (Fig. 17) was promoted recently, when calcium was considered by the students of cross-striated muscle as the key ion in the activation of the contractile system. The further clarification of the role of calcium in the activation of the myometrium is mandatory, for spontaneous activity may well be regulated through this ion. In addition to temperature, autorhythmicity is markedly affected by

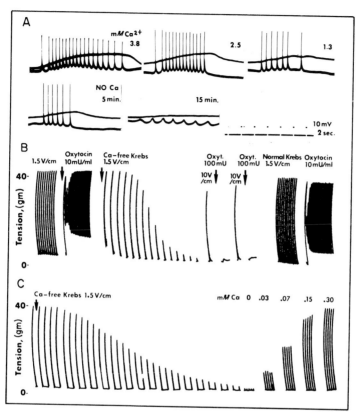

Fig. 17. The effect of calcium deficiency on the train discharge and electrical and pharmacological excitability. (A) Train discharge (parturients rat uterus—microelectrode). Note that calcium deficiency reduces and eventually suspends the spike discharge. (B) Isometric tension (parturient rabbit uterus). Note that calcium deficiency also suspends normal electrical excitability and pharmacological reactivity, but not contractility (induced by massive electric stimulation). (C) Isometric tension (25-day-pregnant rabbit uterus). Note that excitability is restored in a graded manner by the gradual restitution of normal $[Ca]_o$. (Csapo, 1961.)

$[K]_o$, $[Ca]_o$, stretch, metabolic inhibitors, and of course, by a variety of hormonal regulatory agents.

Variations in the action potentials of the myometrium are considerable (Fig. 18) (Kuriyama and Csapo, 1961). Bursts of spikes (rather than one) are usually recorded in the spontaneously contracting uterus. The character and discharge frequency of the spikes are different, depending on the effect of neighboring cells on the recorded cell. Further variations are caused by regulatory conditions and by the methods of recording.

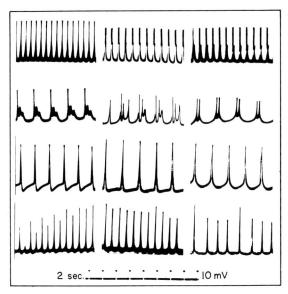

Fig. 18. Variations in the train discharge. Rat uterus post partum, microelectrode technique. Different uteri, or different regions of the same uterus, discharge action potentials of different character, due to the influence of surrounding regions reaching the recorded cell. (Kuriyama and Csapo, 1961.)

The amplitude of the uncomplicated single spike (recorded by intracellular electrodes) may vary between 30 and 70 mV, depending on endocrine conditions. The duration of the signal is about 100 msec. The average rate of rise and fall is 13 V/sec and 5 V/sec, respectively. Uncomplicated spikes, discharged at regular frequency in regular trains, are rare. They are usually seen shortly before, during, and after parturition. In other regulatory conditions, the shape of the potentials is complex, as a rule, demanding explanation. These complex signals probably represent asynchronic discharges of the region surrounding the impaled cell. The uncomplicated spikes (forming regular trains) signal electric synchrony. An endogenous hormone, which imposes upon the uterus electric (and mechanical) asynchrony is progesterone, as we will discuss presently.

Regular spikes of large amplitude and uncomplicated shape do not, by themselves, guarantee advanced mechanical activity. Several spikes ought to be discharged, at frequencies of about 1–2/sec, for the train to be effective. The significance of the number and repetition frequency of subsequent spikes is illustrated by Fig. 19 (Kuriyama and Csapo, 1961). One spike triggers a twitch. A series of spikes of low frequency triggers an "incomplete" tetanus. Only a train of high spike frequency

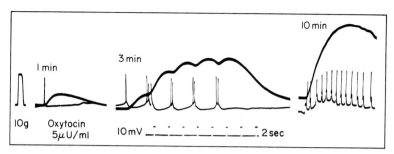

Fig. 19. The relationship between electrical and mechanical activity. Parturient rabbit uterus, microelectrode technique. Note the intimate relationship (as in cross-striated muscle) between the frequency of the train discharge and tension; the twitch and the incomplete and the complete tetanus. (Kuriyama and Csapo, 1961.)

induces a smooth, "complete," tetanus. Apparently, regulatory agents utilize this mechanism of control. By altering the reptition frequency of the action potentials, they can sustain a mechanical activity at any desired level.

No structurally distinct pacemaker has been demonstrated in the uterus. However, train discharges characteristic of pacemaker activity have been recorded from various uterine regions. Therefore, it is generally believed that all myometrial cells are potential pacemakers and the regions of low threshold initiate "slow" depolarization (of about 5–15 mV) in repeated waves. These waves lower the membrane potential to the critical (threshold) value, triggering the train discharges (Fig. 20A) (Kuriyama and Csapo, 1961). Pacemaker regions are distinguished by these cyclically generated slow depolarization waves preceding the trains of spikes. If the train discharges are recorded at a distance from the pacemaker, the slow depolarization waves are absent (Fig. 20B) (Kuriyama and Csapo, 1961). Thus, the distinction between pacemaker and conducted train discharges is made by the presence or absence, respectively, of the slow depolarization waves. Local, or electronic, potentials are distinguished from slow waves by their magnitude, shape, and frequency (Fig. 20C) (Kuriyama and Csapo, 1961). These membrane potential oscillations do not generate train discharges, unless the resting potential is lowered significantly. The biological significance of these oscillatory potentials is uncertain. They may be related to the transmission mechanism.

Conduction between neighboring cells is thought to be accomplished by electrotonic spread following low-resistance pathways. The distance reached by the same uterine action potential, before becoming decremental (and thus demanding attenuation) has not been determined in different endocrine conditions.

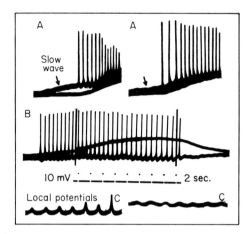

Fig. 20. The slow wave, propagated spikes, and local potentials. Parturient rat uterus, microelectrode technique. Note the slow waves, manifesting in cyclic depolarizations preceding the train discharges at pacemaker regions (A); the propagating train discharges not preceded by slow waves (B); and the local potentials manifesting in slight depolarizations unaccompanied by spikes (C). (Kuriyama and Csapo, 1961.)

The conduction velocity in the same fiber bundle is a function of distance between the stimulating and recording electrodes. In the parturient uterus, values of 5–100 cm/sec were measured with microelectrodes at 5 mm and 50 μ distances, respectively (Kuriyama, 1961). With suction electrodes, a velocity of 15 cm/sec was recorded at 45 mm distance (Talo and Csapo, 1970). The direction of spread affects conduction velocity (Fig. 21) (Goto *et al.*, 1959). In the transverse direction (Figure 21B), it is considerably lower than in the longitudinal direction (Figure

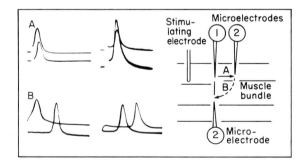

Fig. 21. Variations in conduction velocity. Note the difference in conduction velocity (indicated by the time lag between two action potentials) when it is measured in the same (A) and in different (B) fiber bundles. (Kuriyama, 1961.)

21A, as indicated by the distance between the spikes). The velocity and extent of conduction in the uterus are of considerable medical significance. Therefore, these problems will be discussed together with the progesterone block, and with the mechanism of intrauterine pressure development.

IX. Regulation

Function in cross-striated muscle is stable, in the uterus it is continuously changing. The cyclic alterations of the physical and endocrine conditions of the uterus complicate the study of the basic mechanism of its function. However, they present an equally challenging problem, the mechanism of regulation. Some investigators approached this problem by examining the synthesis, release, distribution, and metabolism of hormones and other biologically active compounds. These studies were rewarding, for they revealed the chemical character of the controlling factors and their everchanging profile during sexual maturation, the menstrual cycle, pregnancy, menopause, etc. However, it is the *interreaction* between these factors and target cells which constitute the final regulatory event itself, for the functional changes which result from these interreactions are the fulfillments of regulation.

The examination of these interreactions in target organs is a relatively recent venture. Physiological and electrophysiological techniques quantitated excitatory and contractile changes in the uterus long before the chemical methods became sensitive and specific enough for measuring reliably the tissue concentration of hormones which bring them about. Recent advances in analytical steroid chemistry are significant because, allied with physiology, they permit meaningful examination of the mechanism of regulation.

The maternal and fetal endocrine systems and hydrodynamic changes impose upon the pregnant uterus a complex regulatory hierarchy. However, many of these regulatory factors only affect uterine function indirectly. We shall only consider here those five endogenous regulatory factors that control myometrial function directly: uterine volume, estrogen, progesterone, oxytocin, and prostaglandins. A full treatment of this subject is found in earlier reviews of the regulatory physiology of the uterus.[*]

[*] Reynolds, 1949; Gato and Csapo, 1959; Gato *et al.*, 1959; Kuriyama, 1961; Kuriyama and Csapo, 1961; Csapo, 1961a, 1962, 1969a, 1970, 1971a,b; Csapo and Wood, 1968; Marshall, 1962, 1964; Schofield, 1963; Jung, 1964; Lepage and Sureau, 1964; Sureau *et al.*, 1965; Suzuki, 1966; Needham and Schoenberg, 1967; Talo and Csapo, 1970.

Regulatory processes have been examined at different levels of organization. Their physiological or medical significances were determined by model experiments in intact animals and subsequent clinical trials. The meaningful interpretation of the clinical evidence has been promoted by judiciously conducted *in vitro* experiments serving as background information. No single method offered the desired information, and among reliable techniques, none was more informative than the other, while together they were complementary. Thus, some investigators changed their techniques and the level of organization at which myometrial function and regulation has been examined.

X. Uterine Volume

The cyclically increasing endometrial mass during menstrual cycles and the growing uterine contents during pregnancy, impose sustained stretch upon the uterus. This physical change affects uterine function at multiple points and so profoundly, that the principal regulatory problem is not how the myometrium is activated, but how the activity is suppressed.

Stretch-induced myometrial hypertrophy (Fig. 10) and the effect of cell length on tension (Fig. 7) explain how the working capacity of the uterus is promoted by increasing uterine volume. These effects of stretch on contractility are complemented by additional actions on excitability. The microelectrodes revealed (Fig. 22) (Kuriyama and Csapo, 1961) that stretch lowers the membrane potential and threshold, increases the frequency and total number of spikes in the train, increases the repetition frequency of the subsequent train discharges, and increases conduction velocity. No stimulants are known which have more profound effects on uterine function.

These effects are readily manifest when the volume–pressure relationship of the uterus is examined *in vivo*. Figure 22 also illustrates how critically the intrauterine pressure (IUP) is controlled by uterine volume through changes in the character of the train discharges and the velocity of conduction (Csapo *et al.*, 1963). In the slack (unstretched) uterus, conduction is slow, as indicated by the delayed appearance of train discharges at distant uterine regions. The unstretched uterus is an asynchronic organ; it develops pressure cycles of complex shape and low magnitude. A mere increase in volume (moderate stretch) increases the conduction velocity, improves synchrony, and in so doing triggers IUP cycles of uncomplicated shape and high magnitude. This change

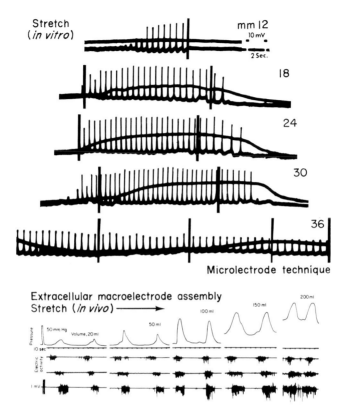

Fig. 22. The effect of stretch on electrical and mechanical activity. Note that increasing stretch of the parturient rat uterus *in vitro* increases the discharge frequency and the number of spikes in the train, as well as tension, until the optimum length is reached. Note that this effect of stretch also manifests *in vivo* (in the parturient rabbit uterus) and that conduction velocity controls the rate of rise, magnitude, and shape of the pressure cycles. (Kuriyama and Csapo, 1961; Csapo *et al.*, 1963.)

from asynchronic to synchronic activity is all that is desired for the initiation of labor. Any regulatory agent that can bring about this change becomes an inducer of labor (or abortion).

These demonstrations provide the basis for the concept that the growing uterine contents facilitate their own expulsion. However, as will be discussed presently, this effect of uterine volume on IUP is balanced by an opposing action, the progesterone block. When the block is complete, the stretch effect cannot manifest. The only limitation of the electrical and mechanical advantages of stretch is optimum length. Stretching the uterus beyond this optimum length induces sustained electric activity, it lowers tension, and only increases the resting tension (Fig. 22).

Investigators who overlooked the fact that during the entire duration of pregnancy the potential of the uterus to deliver its contents is granted by its own growing volume were looking unsuccessfully for a hormonal inducer of labor. This research may not have been completely sterile, for after the evolution of IUP triggered the onset of labor, endogenous stimulants promote its progress. This promotion is of considerable practical significance.

The Laws of Laplace and Pascal

In a closed spheroid, pressure (P) is generated by the wall tension (T) according to the Law of Laplace: $P = (2w/R)T$ (where w = wall thickness and R = radius). This relationship states explicitly that P and T are in balance at all times, and that at a given T the magnitude of P becomes a function of R. It is also implicit in the Laplace Theorem that the generation of the same P, at increasing R, demands an increase in T.

Good agreement was found between theoretical predictions and experimental reality when single pressure cycles were considered and no time was allowed for alterations in the characteristics of the uterine wall induced by regulatory changes. For when the physical and functional character of the uterine wall changes, the numerical values of P and T are modified. For example, the Laplace Theorem predicts that P increases when R decreases (the smaller the uterine volume, the greater the IUP). However, this prediction only holds when T and the elasticity of the uterine wall remain constant, for it is not considered in the Laplace Theorem (describing the hydrodynamic character of nonliving matter) that in contractile muscle spheroids, an increase in R can increase T (and thus P) through a stretch effect (Csapo, 1970).

Figures 7 and 22 illustrate the length-tension and volume-pressure relationships of the uterus. It is apparent that within limits (well defined by the resting T and P) T increases with increased cell length and thus P increases with increasing R. This effect is explained by increased: (1) excitability, (2) conductivity, and (3) contractility, resulting from moderate stretch (Csapo, 1970).

However, the effect of stretch on the resting T and P is not apparent during pregnancy, in spite of the marked increase in R (from 1 to 70 mm). Apparently, a compensatory increase in uterine elasticity balances the effect of stretch on the resting T and P. Only extreme volume increases (hydramnios) manifest in increased resting P, probably because they exhaust this compensatory mechanism. Apparently, the basic laws of hydrodynamics are applicable to the uterus. However, regulatory influences, which change the physical character of the organ, must be carefully examined in hydrodynamic considerations (Csapo, 1970).

In the normal pregnant human uterus (a closed fluid-filled organ), P is equal and uniform at all points, as predicted by the law of Pascal. However, the applicability of this law is also limited. When the presenting part becomes engaged, Pascal's law no longer applies, because the direct contact between two semisolids terminates hydrodynamic criteria.

XI. Estrogen

A withdrawal syndrome of estrogen, uterine atrophy, was already noted centuries ago by sow gelders. Ovariectomy deprives the uterus of its contractile proteins and high-energy phosphates, and thus limits its working capacity (Fig. 9). Normal excitability and pharmacological reactivity also deteriorate. The membrane potential drops to about 35 mV. The spike discharges disappear, as a rule, and with them the spontaneous mechanical activity is reduced to a low level-oscillation.

However, ovariectomy does not induce atrophy if the uterine volume of the castrated animal is increased. In fact, hypertrophy can occur in castrates if the stretch effect is appropriate (Fig. 10). Apparently the physical stimulus of chronic stretch closely simulates the estrogen effect and in so doing provides a challenging model for the study of the molecular mechanism that controls uterine growth and function.

Substitution therapy in estrogen-deficient animals induces marked effects on the uterus. It restores the contractile system and working capacity (Fig. 9). It increases the membrane potential to 45–55 mV, brings about spike discharges, and in so doing, restores spontaneous activity and pharmacological reactivity. However, the functional climax of the uterus is only reached during parturition. Apparently several regulatory factors must work in concert to bring about this functional climax of the uterus.

The regulatory significance of estrogen in the delivery mechanism was not clarified until recently. It has been shown in the rat that the estrogen concentration of the ovarian vein blood sharply increases 24 hrs before term. When the ovaries were removed 48 hrs before term, only a small fraction of the litter was delivered (Fig. 23) (Csapo, 1969a). The characteristic uterine symptoms of estrogen deficiency (ex-

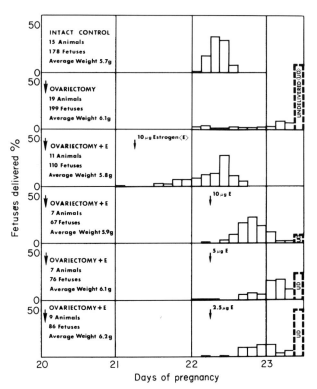

Fig. 23. The effects of ovariectomy and estrogen substitution therapy on the delivery mechanism. Pregnant rats. Note the precise timing of normal delivery, the failure in delivery after ovariectomy, and the correction of this failure by estrogen substitution therapy. (Csapo, 1969a.)

amined in the excised uterus) explained this failure in delivery. Mechanical activity decreased and electrical stimulation of limited uterine regions did not induce activity at distant portions. However, this short-term (48 hrs) estrogen deficiency did not reduce the contractile capacity of the uterus (probably because the uterine volume was sustained by the undelivered litter). Apparently, the failure in delivery was due to a diminution of normal excitability. The evolution of IUP *in vivo* did not show the terminal acceleration characteristic of the normal animal (Fig. 24) (Salgo, 1971).

A single dose of 10 μg estrogen fully corrected this regulatory error, in less than 24 hrs (Fig. 23). The abnormalities of uterine function disappeared (both *in vitro* and *in vivo*) and the animals delivered normally. These experiments provide a challenging model for further studies of the delivery mechanism.

XII. Progesterone

The regulatory significance of this ovarian steroid was recognized when it was discovered that ovariectomy in pregnant rabbits terminated pregnancy unless it was protected by progesterone replacement therapy (Corner and Allen, 1930), which "inactivated" the uterus (Reynolds, 1949). This myometrial effect of progesterone, promoting pregnancy and prenatal life, was later explained in terms of a blocking action (Csapo, 1955).

However, subsequent research showed that ovariectomy, carried out at different times in the same or in different species, had different consequences. Progesterone withdrawal was not observed before parturition and replacement therapy was not effective in all species. Some observers, therefore, did not assign a key role to progesterone in the control of pregnancy and parturition. They preferred the view (Caldeyro-Barcia, 1964) that in different species basically different regulatory mechanisms control the same physiological functions, the maintenance of pregnancy, and the initiation of labor.

The controversy which developed paralyzed academic and clinical developments, for profoundly different obstetric concepts evolve when progesterone is considered a key regulatory factor of the pregnant uterus and when it is not. Recent comparative studies conducted with advanced methods in laboratory mammals and man are restoring the broad regulatory significance of progesterone. However, since under physiological conditions several controlling factors simultaneously affect the uterus, the action of a given one has limited meaning.

This point is of critical importance in myometrial regulation and therefore demands illustration. When increasing rather than decreasing progesterone levels were observed (Short, 1961) in near term patients, investigators did not question the reliability of the method, nor did they look for an imbalance, a relative change between the activity promoting and suppressing regulatory factors, but rather dismissed the progestational hormone as a controlling agent of the pregnant uterus. In this rush judgment, the powerful activity-promoting effect of increasing uterine volume (through stretch) was ignored. For, when this effect is considered, it becomes apparent that a regulatory imbalance promoting the evolution of uterine activity can readily develop at any time during pregnancy when the progesterone levels fail to increase in parallel with the uterine volume. Thus, the regulatory conditions and the function of the uterus are not described by the absolute progesterone levels, but by the balance between uterine volume and progesterone (V/P) (Csapo *et al.*, 1970), and of course by the superimposed actions of

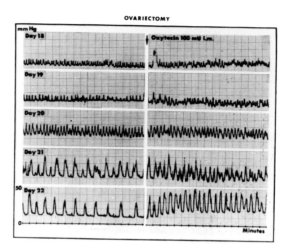

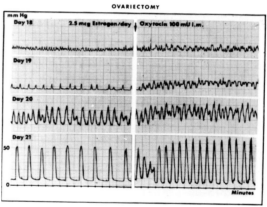

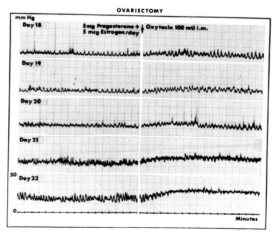

the other regulatory factors. The significance of this V/P ratio will be discussed presently. The point of importance here is that the regulatory factors work in concert, and therefore, uterine function is determined by the balance among them. The investigation of one selected factor has only demonstrative meaning.

Comparative studies showed that the myometrial effect of progesterone has the same blocking action in laboratory animals and man (Csapo, 1969a, 1970, 1971a,b; Csapo and Wood, 1968). However, superimposed species differences are encountered in the topography of the block, in the character, timing and extent of progesterone withdrawal, and in the effectiveness of replacement therapy. The differences result from anatomical variations in the site of progesteronegenesis. In some species, the extrauterine corpora lutea dominate uterine function throughout pregnancy, by a systemic progesterone effect. In others, progesteronegenesis shifts partly, or entirely, to the intrauterine placenta, resulting in a local progesterone action. This luteoplacental shift, occurring at different degrees and at different times in gestation, creates various mixtures of systemic and local progesterone effects, resulting in functional inhomogeneity in the uterus (Csapo, 1969b).

The recognition of the regulatory importance of the luteoplacental shift is a relatively recent achievement. Documentation is advanced, but still incomplete. Since the myometrial block cannot be discussed meaningfully without considering the various combinations of the systemic and local progesterone effects and the resulting relative endocrine and functional asymmetry of the uterus, the regulatory variations have to be considered.

When the myometrium is controlled by circulating (systemic) progesterone, the uterus tends to be endocrinologically and functionally symmetric. The level of the systemic block is controlled by the plasma concentration of progesterone by the binding of the hormone in the plasma and the uterus and the metabolism of this steroid. Therefore, plasma progesterone withdrawal and exogenous substitution therapy profoundly affect the systemic block. In contrast, placental progesterone is thought to be distributed locally, resulting in an endocrine and functional asymmetry of the uterus (Csapo, 1961a). This concept is supported by the recent finding that uterine regions near the placenta have a higher progesterone concentration than distant regions (Kumar and Barnes,

---

Fig. 24. The effects of ovariectomy, estrogen, and estrogen plus progesterone therapy. An *in vivo* study in pregnant rats. Note the accelerated evolution of IUP after ovariectomy; the failure of terminal acceleration, the correction of this failure by estrogen therapy, and the suspension of the evolution of IUP by progesterone treatment. (Salgo, 1971.)

1965; Zander *et al.*, 1969). Thus, this local mechanism is relatively independent of the circulating progesterone levels, although the plasma concentration does reflect the biosynthetic activity of the placenta and, therefore, indirectly the local placental effect. Thus, in case of local placental control, plasma progesterone withdrawal is only an indirect signal of myometrial progesterone deficiency, and the success of exogenous therapy should depend on the effectiveness of the systemic transport mechanism. These considerations are helpful in the interpretation of experimental data.

The uterus blocked by progesterone is in a unique functional condition. Maximum working potential is retained, but is not displayed except under experimental conditions, when all myometrial cells are simultaneously activated in an electric field (Figs. 11 and 17). The block only suppresses intrinsic excitation, conduction, and pharmacological reactivity, but not contractility. This condition resembles the procaine block of the cross-striated muscle and is not described accurately by the term "inactivation."

In classic muscle physiology, the impairment of conduction is termed "block." If, therefore, it is shown that conduction is impaired when the progesterone effect dominates the uterus, the block is demonstrated. This demonstration has been accomplished (Fig. 25) (Csapo, 1969a), justifying the use of the term. Thus, the progesterone block is no longer

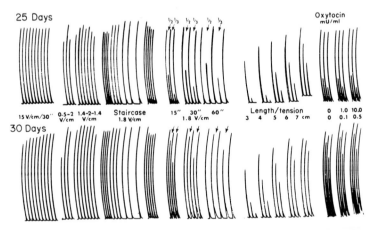

Fig. 25. The mechanical manifestations of nonpropagating and propagating activity. Pregnant and parturient rabbit uteri. The progesterone block is examined by recording the mechanical response (tension) of the excised uterus *in vitro*, to partial, and full-length electrical stimulation. Note the different tension responses of the pregnant and term uteri and the effects of the repetition frequency of stimulation, stretch, and oxytocin. (Csapo, 1969a.)

a concept, but an experimental reality. The mechanism of progesterone effect has also been explained recently when it was shown that this hormone increases tissue resistance by decreasing electric coupling between myometrial cells (Ichikawa and Bortoff, 1970). These demonstrations conclude an important phase in reproductive and regulatory biology, but only to open up more detailed examinations of the mechanism of progesterone action. To promote this new venture, the major steps of the early work will be briefly considered.

The progesterone block has been discovered by comparing the maximum isometric tension of pregnant and parturient (or hormone-treated) rabbit uteri. The experimental conditions were such that the portion of the uterine strip directly exposed to the stimulating electric field could be varied (Fig. 25). The parturient uterus developed close to maximum tension, even if only one-third of its length was tetanized, while tension of the pregnant (or progesterone-treated) uterus was approximately proportional to the fractional length of the stimulated region. Apparently, the unstimulated region of the blocked uterus remained inactive, and by elastic yielding, lowered the tension developed by the active region. It was concluded that the parturient uterus conducts the electrical activity to a considerable degree, while the pregnant (or progesterone-treated) uterus does not (Csapo, 1955, 1961a).

These experiments (Fig. 25) were recently extended to include observations of the transient preparturient uterus and of the effects of surgically induced progesterone withdrawal (ovariectomy), progesterone substitution therapy, stretch, and oxytocin. These studies confirmed the reality of the progesterone block, the decremental character of myometrial conduction, the gradual recovery of the uterus from the block prior to delivery, and the promotion of conduction by stretch and oxytocin. A significant point, brought out by these experiments, was the observation that neither stretch nor oxytocin could promote conduction when progesterone completely blocked the uterus. However, during recovery from the block these stimulants markedly promoted conduction and therefore increased uterine function. The medical significance of these findings is considerable.

However, in the experiments described, the block was demonstrated by recording mechanical rather than electrical activity. Therefore, the two extreme conditions—block and propagating activity (as well as the transient)—have been reexamined *in vivo* by an extracellular microelectrode assembly in pregnant and parturient (or hormone-treated) rabbits. These experiments documented the block by electrophysiological evidence (Csapo and Takeda, 1965), as will be described presently.

However, the classic method used in cross-striated muscle for the

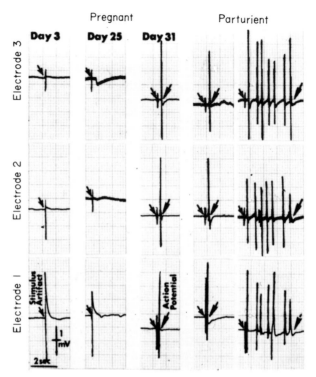

Fig. 26. The progesterone block. Suction electrode recording technique, rabbit uteri at different days of pregnancy. The uterine strip is stimulated *in vitro* with single D.C. pulses between a pair of electrodes 5 mm apart. The action potentials are recorded at 15, 30, and 45 mm distances from the stimulated region. Note the stimulus artifacts (signaling the time of stimulation) and the absence of action potentials (generated by the stimulus) until the animal approaches term. Note the appearance of conducted action potentials in the near term and parturient uteri and the conduction velocity (about 15 cm/sec). Note also that the parturient uterus may respond to a single stimulus with a train discharge rather than a single action potential. (Talo and Csapo, 1970.)

demonstration of conduction blocks was to excite a limited region (with a pair of stimulating electrodes) and show that no action potentials appeared beyond the stimulated region. When the muscle recovered from the block, the action potentials reappeared. The technical difficulties of adapting this classic method in studies of myometrial conduction were recently overcome. Using the suction electrode technique (Fig. 26) (Talo and Csapo, 1970), it has been shown that while the pregnant (and progesterone-treated) uterus is not conductive, the parturient uterus *is*. When discrete regions were stimulated electrically, only stimulus artifacts were recorded from the pregnant uterus. No action potentials appeared shortly after stimulation, not even near the stimulating electrodes, indicating that if occasionally the electrodes revealed local activity, this activity was not coming from a distance and was not directly induced by the stimulus. In contrast, the parturient uterus discharged distinct action potentials within a few milliseconds after stimulation. These spikes appeared at different times, depending on the distance of the recorded region from the stimulating electrodes. The time lag between stimulus artifact and action potential indicated the conduction time. Occasionally, the parturient uterus responded to a single D.C. stimulus by a train discharge, rather than by a single spike. These studies confirmed the reality of the intermediate condition of limited propagation, the effect of progesterone in suppressing conduction, and the effect of stretch and oxytocin in promoting conduction.

The fact that the block and propagating activity are controlled by the imposition and withdrawal of progesterone (and are not *in vitro* artifacts) was demonstrated by experiments *in vivo* (Figs. 27 and 28) (Csapo and Takeda, 1965). An extracellular microelectrode assembly has been implanted into the pregnant rabbit uterus. The evolution of electric activity and IUP were recorded during pregnancy in intact ovariectomized and progesterone-treated animals. These studies revealed (Fig. 27) that during normal pregnancy only occasional, irregular discharges are present before term. The number of these discharges increased as term approached, but they did not form regular trains and were not conducted from one uterine region to the other. Therefore, the one-to-one relationship between train discharge and pressure cycle was absent, and this local, nonpropagating, asynchronic electric activity only induced low-level oscillations in IUP.

As pregnancy approached term (Fig. 27), at first irregular train discharges and pressure cycles of limited magnitude appeared. With the laps of time, the train discharges became more regular, but were recorded with delay at distant electrodes, indicating slow conduction. Eventually regular trains appeared, with short delay at distant electrodes, signaling improved spike generation and conduction. As a result of this improved electrical activity the IUP had a rapid (quadratic) rise, it reached high values and showed regular shape. There was a one to one relationship between train discharge and pressure cycle.

Ovariectomy (Fig. 28) brought out the evolution of electrical and mechanical activity in 24 hr during late pregnancy and oxytocin promoted activity about 12 hrs after surgery. Progesterone substitution therapy disallowed this accelerated evolution (Fig. 28) (Csapo and Takeda, 1965). These studies documented (1) the progesterone block *in vivo*, (2) the condition of electrical and mechanical asynchrony and synchrony, (3) limited conduction and undisturbed propagation, (4) the intimate relationship between electrical and mechanical activity, and

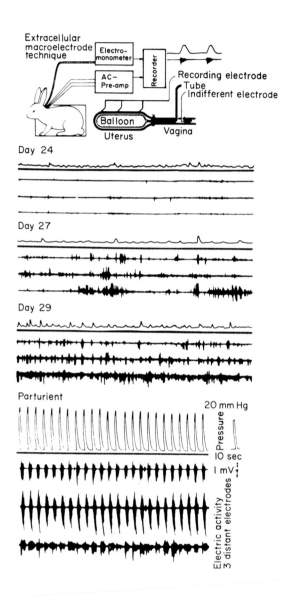

Fig. 27. The evolution of electrical and mechanical activity. Pregnant rabbit uterus, *in vivo*. Note the gradual evolution of electrical activity and the low-level oscillation of IUP when train formation is poor and when distant uterine regions are not in synchrony. Note also the rapid and quadratic rise of the high and regular IUP, when the train discharges become synchronic. (Csapo and Takeda, 1965.)

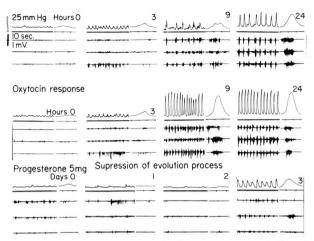

Fig. 28. The accelerated evolution and the suppression of the electrical and mechanical activity. Pregnant rabbit uterus, *in vivo*. The evolution process is induced on the twenty-fifth day of pregnancy by ovariectomy and placental dislocation. Recording was by extracellular macroelectrodes and an intrauterine balloon. Note the rapid evolution of electric and mechanical activity, the train formation, and the appearance of the oxytocin response at 9 hours after surgery. Note also the suppression of the evolution process by progesterone. (Csapo and Takeda, 1965.)

(5) that the myometrial block is a progesterone effect (Csapo and Takeda, 1965).

However, all these *in vitro* or *in vivo* experiments were conducted in rabbits, a species in which pregnancy is sustained by the corpora lutea and, therefore, the progesterone effect is systemic. In rabbits, placental progesterone support is insignificant, and the ovaries are indispensable during pregnancy unless the placentas are challenged into compensatory hypertrophy. Thus, the medical significance of these studies remained uncertain, for extrapolations to the regulatory conditions of the pregnant human uterus were limited by indications that in patients a local, placental-controlled progesterone block operates.

Since it is difficult to conduct penetrating studies in patients, an animal model was needed for the initial examination of the local effect of placental progesterone. Studies in rats revealed that, while in this species the corpora lutea dominate the pregnant uterus, there is a significant placental progesterone effect during late pregnancy (Csapo and Wiest, 1969). The systemic component of progesterone support was suspended by ovariectomy. As a result, the plasma progesterone level dropped below the parturient value. However, those animals that responded to ovariectomy by compensatory hypertrophy sustained a preparturient

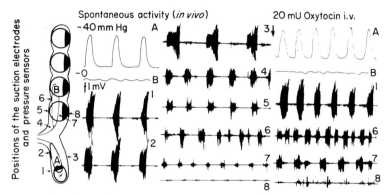

Fig. 29. The electrical asymmetry and the asymmetric oxytocin response of the pregnant rat uterus. Rat 18 days pregnant, 2 days after bilateral ovariectomy and unilateral placental dislocation. Suction electrode and extraovular microballoon techniques. Note that the placental dislocated horn aborted all conceptus sacs but one, while the contralateral horn is intact pregnant. The aborted horn has synchronic, regular, and advanced electrical activity (1, 2, and 3) and high and regular IUP (A). In contrast, the electrical activity of the intact pregnant horn is only moderately regular near the cervix (4 and 5), while it is irregular at higher antiplacental regions (6) and is suppressed near (7) or over (8) the placental bed. This asymmetric and asynchronic electrical activity only induces low level oscillations in IUP (B), due to the elastic yielding of the inactive uterine regions. Note also that oxytocin only improves the electrical (1) and mechanical (A) activity of the placental dislocated horn. Since the antiplacental region of the intact horn (6) is only moderately affected and the placental bed (8) is not, the IUP (B) remains unaffected. (Csapo, 1969b.)

uterine progesterone level and pregnancy (in spite of a drastic decrease in plasma progesterone). When in such animals one set of placentas was dislocated in one horn of the bicornuate uterus, the uterine progesterone level decreased in this horn and most of the conceptus sacs were discharged after 24 hrs. In contrast, the contralateral intact horn of higher uterine progesterone levels remained normally pregnant. Thus, the endocrine and functional asymmetry of the uterus has been demonstrated.

The electrophysiological comparison of the inact and placental-dislocated horns (of these bicornuate uteri) revealed marked differences in electrical and mechanical activity (Fig. 29) (Csapo, 1969b). The placental-dislocated horn (examined with suction electrode during abortion) showed regular, synchronic train discharges and high and regular IUP, as characteristic of a propagating uterus. The intact pregnant horn of the same animal was also active, but it only discharged irregular electric activity at antiplacental regions and insignificant electric activity over the placental beds. Since only part of this horn was electrically active, the limited and asynchronic activity only induced low-level IUP. Thus, the placental-controlled local block has been demonstrated by electrophysiological evidence in a uterus, whose contralateral horn displayed advanced electrical activity and IUP (Csapo, 1969b).

Oxytocin treatment of these bilaterally ovariectomized, unilaterally placental dislocated animals revealed that when the block is local and uterine function is asymmetric, the oxytocin effect is also asymmetric. Only the placental-dislocated horn and the antiplacental regions of the intact pregnant horn responded to oxytocin; the placental regions remained refractory. Evidently, the myometrial cells of the same uterine horn may or may not respond to oxytocin, depending on their regulatory and electrophysiological character. The proportion of responding and refractory cells seem to determine the effect of oxytocin on the IUP, for if the inactive region is too large, it lowers the IUP by its elastic yielding. Thus, the oxytocin response of the uterus may not be controlled by a sensitivity of the uterus to this polypeptide but by the number of active (and inactive) cells. This brings us back to the classic view that muscle function at the organ level is controlled by the recruitment of fibers. However, once the weakening of the block permits the oxytocin response of a sizable number of cells, the IUP does become dependent on the dose of oxytocin administered, because induced activity generates more activity by affecting the defense mechanism of pregnancy. As will be discussed presently, this autocatalytic nature of the evolution of uterine activity is a significant factor in the abortifacient action of prostaglandins.

It was also demonstrated that this local placental effect, on electrical and mechanical activity, is mediated by progesterone. Progesterone substitution therapy in ovariectomized, unilaterally placental-dislocated rats increased the plasma and uterine progesterone levels and prevented abortion (Csapo and Wiest, 1969). These studies were complemented by examination of the IUP under a variety of endocrine conditions (Fig. 30) (Salgo, 1971).

These collaborative experiments of physiologists and steroid chemists yielded (Csapo and Wiest, 1969) the first quantitative correlation between ovarian, uterine, placental, and plasma progesterone; electrical and mechanical activity of the uterus; and abortion, premature, and term delivery. They demonstrated that in the intact pregnant rat the gradual cessation of corpus luteum function results in a sharp decrease in ovarian vein, peripheral plasma, and uterine (but not in placental) progesterone concentrations. Delivery begins when the uterine progesterone level drops below 2 μg/100 gm, for it is at this time that the

INTACT PREGNANT

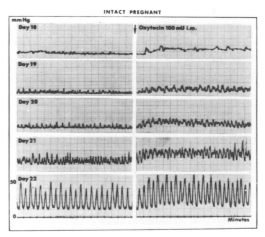

OVARIECTOMY + UNILATERAL PLACENTAL DISLOCATION

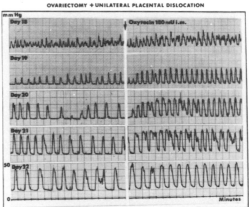

OVARIECTOMY + UNILATERAL PLACENTAL DISLOCATION

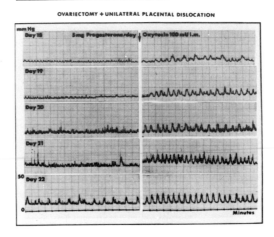

evolution of IUP accelerates. Therefore, this value is considered the blocking concentration in this animal under the conditions of normal term pregnancy (term uterine volume, endogenous stimulants, etc.). Ovariectomy precipitates a rapid and drastic plasma, but a lesser uterine progesterone withdrawal, if the hypertrophied placentas partly compensate for the loss of luteal function. Progesterone treatment increases plasma and uterine (but not placental) concentrations, suppresses the evolution of IUP, and prevents premature delivery. If intact rats are given progesterone near term (in daily doses of 2.0–4.0 mg), plasma and uterine progesterone concentrations are sustained at higher than parturient levels and the animals cannot deliver at term. Doses as low as 0.25 mg/day are effective in delaying delivery until an excessive increase in uterine volume overcomes this minimal progesterone effect.

The distinction between systemic and local progesterone effects has considerable biological significance. For example, it accounts for the species differences in the timing of the evolution of IUP and oxytocin response (OR) during pregnancy (Fig. 31) (Csapo, 1969a). Rapid evolution before delivery (characteristic of luteal-dominated species) is explained by a rapidly developing imbalance in V/P, resulting from a marked systemic progesterone withdrawal. Gradual and prolonged evolution already distinct during third trimester pregnancy is accounted for by the gradually increasing imbalance in the V/P ratio, due to the gradual failure of the lesser complete local (placental) progesterone block. Apparently, the evolution of uterine activity is a reflection of the timing and extent of the luteoplacental shift and of the consequent changes in the character of the myometrial block. In both instances the evolution of uterine activity begins with the withdrawal of the luteal effect. Its subsequent course depends on the gradual failure of the placental effect and on the increase in the number of active cells controlled by the residual placental effect (Csapo, 1969a, 1970, 1971a,b; Csapo and Wood, 1968).

A persistent theoretical and medical problem is the ineffectiveness of progesterone therapy when the uterus is placental controlled. This failure of exogenous progesterone in restoring the weakening progesterone block exposes current ignorance about progesterone transport. However, advances can be expected now since (1) in the guinea pig a suitable animal was found for model experiments (Porter, 1969), (2)

Fig. 30. The normal and accelerated evolution of intrauterine pressure and its suppression in pregnant rats. Extraovular microballoon technique. Note the gradual evolution of IUP in intact pregnant rats, the accelerated evolution in ovariectomized and placental dislocated animals, and the suppression of evolution by progesterone treatment. (Salgo, 1971.)

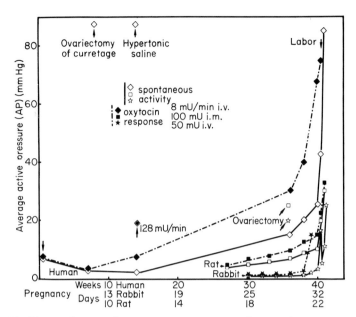

Fig. 31. The evolution of intrauterine pressure and oxytocin response of the pregnant human, rat, and rabbit uteri. Extraovular microballoon technique. Note the difference in the timing of the evolution process and in the maximum pressure reached during labor in the different species. Note also the accelerated evolution of IUP after ovariectomy and after intraamniotic hypertonic saline treatment. (Csapo, 1969a.)

reliable progesterone assay methods became available for the quantitation of this hormone (not only in the plasma, but also in the uterus) (Csapo and Wiest, 1969; Csapo *et al.*, 1969; Zauder *et al.*, 1969), and (3) considerable interest is focused upon the isolation and characterization of progesterone-binding proteins in plasma and target tissues. When advanced, these studies can explain the direction of movement of endogenous or exogenous progesterone and its myometrial concentration in different species under a variety of regulatory conditions.

The mechanism of action has not been fully resolved, as yet, for any hormone and progesterone is no exception. Investigators recorded a higher membrane potential in myometrial cells when the uterus has been exposed to progesterone and the possibility has been considered that this increase in membrane potential is one underlying reason of reduced excitability and conduction. However, hyperpolarization provides no adequate explanation, for when it is accomplished by lowering the $[K]_0$, the uterus does not acquire the properties of the progesterone-blocked organ. Spontaneous activity is suspended, but oxytocin response remains (Csapo, 1962).

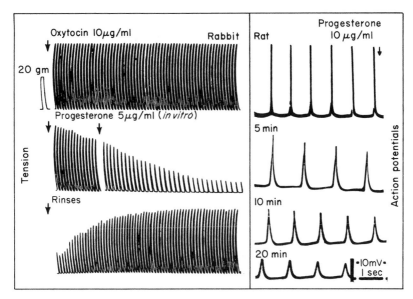

Fig. 32. The effect of *in vitro* progesterone treatment. Parturient rabbit and rat uterus. Note the decrease in the rate of rise and the magnitude of the action potentials and tension after *in vitro* treatment with 10 μg/ml progesterone. (Csapo, 1961b.)

Microelectrode studies revealed pronounced, but not consistent, irregularities in the train discharges. It is difficult to interpret these results, for they were carried out (due to inherent technical difficulties) during observation periods of several hours. Since *in vivo* only occasional and irregular spikes were recorded (without train formation) when the block was completed (Fig. 27), it is reasonable to assume that the train discharges recorded were *in vitro* artifacts, due to the disappearance of progesterone during perfusion. This concern is supported by the demonstration that the threshold changes markedly during prolonged *in vitro* observations (Fig. 11).

Investigators attempted to overcome this serious limitation of the *in vitro* study conditions by adding progesterone to the Krebs solution. As illustrated by Fig. 32, dramatic blocking effects were observed within minutes. The rate of rise of the spikes was gradually reduced. At sufficiently high progesterone concentration, the spike discharges disappeared and only cyclic membrane potential changes of small amplitude remained. Accordingly, the mechanical activity (tension) also decreased. However, the biological significance of these observations is uncertain, for the *in vitro* and *in vivo* effects of progesterone are only similar but not analogous and the *in vitro* effect can be induced by a whole family of steroids which are ineffective *in vivo* (Csapo, 1961a).

The possibility has been repeatedly considered that the blocking effect of progesterone, its influence on endogenous excitability and pharmacological reactivity, is mediated by a control over a small calcium fraction, which is instrumental in activation. It has been speculated that under progesterone domination this calcium fraction is more strongly bound in the cell membrane than when the block is withdrawn. Furthermore, that since spontaneous activity demands a relative instability of this calcium, the stabilizing effect of progesterone (manifesting in hyperpolarization, in the suppression of excitation, the rate of rise of the action potentials, conduction, etc.) is due to the stabilization of this calcium. However, these contentions have to be promoted by future, decisive experiments (Csapo, 1961a, 1962). Thus, the molecular mechanism of the progesterone effect is yet to be discovered.

XIII. Oxytocin

The myometrial effect of this pituitary polypeptide hormone was discovered at the turn of the century. However, critical experiments were only conducted when synthetic oxytocin became accessible. A systematic study (Csapo, 1961b) revealed that oxytocin is not, strictly speaking, a myoplastic stimulant. A fully contractile uterus that is rendered inexcitable (by calcium deficiency, or by treatement with excess potassium) is not activated by it (Fig. 17) (Csapo, 1961a,b). The oxytocin effect demands a myometrial condition in which trains of action potentials are generated and (at least partly) conducted, or this potential is present in a latent form, but cannot manifest itself because of the high threshold. Electrical activity and oxytocin response are suspended in the uterus by calcium deficiency, but are restored by 0.3 mM $[Ca]_o$ (Fig. 17). Evidently, there is no oxytocin effect when at least a fraction of the calcium required for excitation and excitation–contraction coupling is bound or is removed (Csapo, 1961b).

In the parturient or postpartum uterus (the most suitable preparation for the study of stimulatory effects), oxytocin lowers threshold, increases the rate of rise, increases the number and frequency of the spike discharges, increases the duration and repetition frequency of individual trains, and consequently increases mechanical activity (Fig. 33) (Csapo, 1961b, 1962; Csapo and Takeda, 1965; Marshall, 1962; Jung, 1964). It also improves conduction markedly and through this effect it corrects asynchrony. It is the imposition of electrical and mechanical synchrony on a previously asynchronic uterus that distinguishes oxytocin as a powerful therapeutic agent (Csapo, 1961a,b, 1962, 1969b).

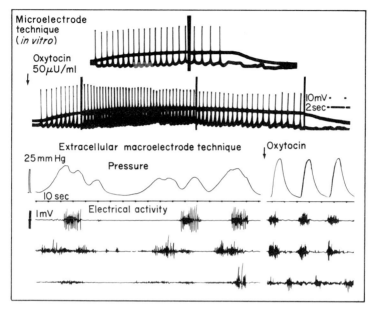

Fig. 33. The effect of oxytocin on the train discharge, tension, and intrauterine pressure. Parturient rabbit uterus *in vitro* (upper) and *in vivo* (lower). Note the oxytocin-induced increase in the frequency and number of spikes in the train discharges and increase in tension. Note the *in vivo* asynchrony of electrical activity and the irregular shape of the IUP cycles corrected by oxytocin. (Kuriyama and Csapo, 1961; Csapo and Takeda, 1965.)

However, as stated earlier, oxytocin does not improve upon myometrial function in all, but only under limited regulatory conditions. The differences observed in the oxytocin response of the uterus in different species are due to variations in the regulatory character of the uterus at various stages of nonpregnancy and pregnancy. In those animals which during pregnancy are largely under the systemic (luteal) effect of progesterone, the evolution of IUP and OR only manifests when the progesterone effect is largely withdrawn shortly before delivery (Fig. 31) (Csapo, 1969a). In contrast, in those species in which the uterus is controlled by a local (placental) progesterone effect, the evolution of IUP and OR are apparent earlier in pregnancy, because the block weakens gradually as the nonplacental region increases (Fig. 31). Model experiments (Fig. 29) (Csapo, 1969b) demonstrated that in case of a local control, the oxytocin effect is restricted to the nonplacental regions of the uterus. These studies also revealed that the effect of oxytocin on the IUP is controlled by the number of cells participating in activity, and thus the ratio active/inactive regions. Apparently, the classic concept, the recruitment of muscle fibers, is more meaningful than that of "sensitivity."

The mechanism of oxytocin effect is not known. Since a critical relationship is documented between oxytocin and calcium (Fig. 17) (Csapo, 1961a,b), it is feasible to examine in precise quantitative terms the calcium influence in spontaneously contracting and oxytocin stimulated uteri.

The precise role of endogenous oxytocin in myometrical regulation is not known due to technical difficulties in quantitating this labile polypeptide in plasma and tissue. The first indirect evidence for oxytocin release during delivery was obtained in rabbits (Csapo, 1961b). However, direct evidence is still lacking, in spite of repeated efforts at demonstration, including studies in patients (Caldeyro-Barcia, 1964). This uncertainty does not affect, however, broad therapeutic use. As will be discussed presently, the judicious application of oxytocin represents a major step toward preventive obstetrics.

XIV. Prostaglandins

The hormonelike actions of prostaglandins (PG) have been observed in a variety of organs, including smooth muscles. The effects of this family of C_{20} fatty acids have been repeatedly reviewed (Bergstrom et al., 1968, 1971; Ramwell and Shaw, 1970; Lancet, 1970; Karim and Filshie, 1970; New York Academy of Sciences, 1970). This present discussion is limited, therefore, to the myometrial actions of PG. Like the ovarian steroids, PG have been known to researchers and clinicians for over 40 years, when the human uterus was found to respond to semen. However, in spite of the early isolation and characterization of six crystalline PG (of the E and F series) and the evidence that these compounds are normally present and exert a variety of effects in a number of tissues, studies with PG only produced slow advances in regulatory biology and reproduction.

The academic significance and therapeutic potential of PG became fully recognized when it was reported that these compounds are effective abortifacients (Karim and Filshie, 1970; New York Academy of Sciences, 1970; Bergstrom et al., 1971). Bold clinical trials revealed (Karim and Filshie, 1970; New York Academy of Sciences, 1970; Bergstrom et al., 1971) that the massive and prolonged intravenous infusion of PG F2α and E2 may terminate pregnancy in complete abortion. What these studies did not resolve as yet is the frequent failure of PG to induce abortion and the high incidence of incomplete abortions (Table I). Incomplete abortions demand surgical completion, which this priority

TABLE I

Clinical Outcome of PG F2α-Induced Abortions

Institution	Gestation (weeks)	Complete	Incomplete	Failure
Yale University	7–16	1	4	5
Columbia University	12–16	3	6	1
Washington University	12–17	3	5	2
Total	7–17	7 (23%)	15 (50%)	8 (27%)

program was set up to avoid, for the broad therapeutic success of PG rests on the validity of the premise that they are nonsurgical abortifacients, effective at all stages of gestation. It is reassuring, therefore, that local (intravaginal or intrauterine) application reduced the effective dose and the most common side effects of PG (nausea, vomiting, diarrhea) (New York Academy of Sciences, 1970; Bergstrom *et al.*, 1971). Considering that these clinical results have been achieved in less than 2 years, one may look upon the many unresolved problems optimistically. However, in seeking therapeutic resolutions, investigators must realize the paramount importance of basic research with PG, specifically the clarification of the mechanism of their myometrial effect.

The therapeutic value of endogenous, nonsurgical abortifacients is that they can spare millions of women from chronic exposure to massive steroidal and other exogenous contraceptives. Since the majority of fertile women do not become pregnant during a given menstrual cycle (even if unprotected from conception), the chronic exposure of this majority to prophylactic contraceptive agents is unnecessary. However, since it is uncertain which individual patient will conceive in a given cycle and which will fail to do so, this consideration is valid only if safe and effective postconceptional therapy provides predictable retrospective measures in nonsurgical fertility control.

Steroid contraceptives load patients chronically with highly potent synthetic drugs. In contrast, abortifacients such as PG may unload women acutely from the actions of endogenous regulatory factors, vitally important for pregnancy maintenance. This difference is of considerable importance in population control strategy, demanding systematic studies of the premise that PG do in fact unload both the animal and human uteri from the effect of regulatory factors vitally important for pregnancy maintenance. These studies are in progress (Csapo, 1971c; Csapo *et al.*, 1971a,b) and will be discussed presently.

Recent clinical trials at three medical centers (Csapo *et al.*, 1971a; G. G. Anderson *et al.*, 1971; Jewelewicz *et al.*, 1971) revealed that out of 30 patients infused for 12 hrs (and reinfused if necessary) with up to 200 μg/min PG F2α, only 23% aborted completely, 50% incompletely and 27% failed to abort. Thus, 77% of the study patients required surgical completion of abortion. The uniformity of these results, obtained independently at three centers, makes the reported (Karim and Filshie, 1970) high success rate of PG induced abortions doubtful. More important than this realistic assessment of the efficacy of intravenous PG F2α treatment, however, is the relationship which these studies established between IUP, plasma hormone levels, and clinical progress in PG induced abortions. This relationship is readily brought out by a method of analysis which rests on hydrodynamic laws and basic studies in uterine physiology.

The key to correct analysis is the recognition (Coren and Csapo, 1963) that it is uterine wall tension (wT), rather than IUP, which promotes clinical progress and is directly controlled by progesterone (Csapo, 1955, 1961a). Investigators measure P rather than T, because P is readily quantitated in the animal and human uteri, while T is not. However, as long as the techniques used in measuring P are accurate, the fetal membranes are intact, preventing leakage of amniotic fluid, and the radius (R) of the uterine cavity is known, wT can be estimated (Coren and Csapo, 1963; A. B. M. Anderson *et al.*, 1967; Csapo, 1970) (in dynes/cm) from the theorem of Laplace: $P = 2wT/R$, thus $wT = PR/2$.

Since R is increasing considerably with the advance of pregnancy, during the first and second trimesters very significantly higher IUP values than those measured at term correspond with the wT of the parturient uterus. Of course, wT is controlled by a number of regulatory factors in addition to the physical parameter R. For example, stretch, progesterone, estrogen, and oxytocics profoundly modify myometrial excitability and thus wT in a complex manner. These effects are exerted through a control over the contractile capacity of the organ; the topographical site of the pacemaker area; the character, extent, and velocity of conduction of the train discharges; the ratio active/inactive regions; etc.

It is now well established (Kumar and Barnes, 1965; Csapo, 1969b; Zander *et al.*, 1969) that the pregnant uterus is endocrinologically and functionally asymmetric and that uterine stimulants may promote electrical activity in the nonplacental regions without affecting the placental region. It is probable that in early pregnant patients PG only affects restricted (low threshold) nonplacental uterine areas, and therefore, the wT of the organ as a whole fails in reaching parturient values. If so, the control over wT under massive pharmacological stimulation can be sustained by the inactive (high threshold) region, which remains refractory to the stimulant and by passive stretch restricts the full development of wT. The following is an examination of these premises.

Clinical trials revealed that before the intravenous infusion of PG F2α the pregnant human uterus displays all the characteristic symptoms of a blocked organ (Fig. 34) (Csapo *et al.*, 1971a). The IUP is limited to irregular, low-level oscillations; an expression of local, nonpropagating electrical activity of discrete uterine regions. PG F2α (infused at a massive rate of 25–50 µg/min) does not increase the cyclic active pressure (AP) when this stimulant reaches the uterus in a few seconds after the start of infusion. Only the resting pressure (RP) increases steeply, reaching maximum (44.5 ± 2.2) values within the initial hour of infusion and manifesting, therefore, in a sustained contracture. The mechanism by which this contracture develops is not yet resolved. It can be due to a primary F2α stimulation of discrete (low threshold) myometrial regions, or the smooth muscles of uterine blood vessels, or both.

Whatever is the primary target of F2α infusion, this (unphysiological) sustained contracture is present in all instances and is probably the primary cause of the PG-induced evolution of IUP. The region participating in sustained activity stretches the inactive uterine region, and by lowering its threshold, promotes its activity (as does volume increase). However, this stretch effect alone cannot explain the gradual evolution of IUP, which invariably occurs with the progress of time during infusion (Fig. 34). Therefore, additional effects should be considered in an attempt to explain the gradual increase in IUP.

The maximum IUP, reached during PG F2α infusion (or reinfusion) in a group of 14.7-weeks-pregnant patients (Table II) (Csapo *et al.*, 1971a) was found to be 113 mm Hg. This value is nearly twice as high as that (68.2 ± 1.1 mm Hg) (Csapo and Sauvage, 1971) recorded at term in one thousand patients who were induced to deliver by oxy-

TABLE II

INTRAUTERINE PRESSURE AND WALL TENSION OF TERM PREGNANT, OXYTOCIN-
STIMULATED PARTURIENT, AND PG F2α-STIMULATED 12–17 WEEKS PREGNANT
HUMAN UTERI

Type of case	Number of cases	Maximum IUP (mm Hg)	Wall tension (dynes/cm) ($\times 10^5$)	Delivery or abortion (%)
Term Pregnant	1000	30	1.43	0
Parturient	1000	68	3.24	98
12–17 Weeks	10	113	1.59	33

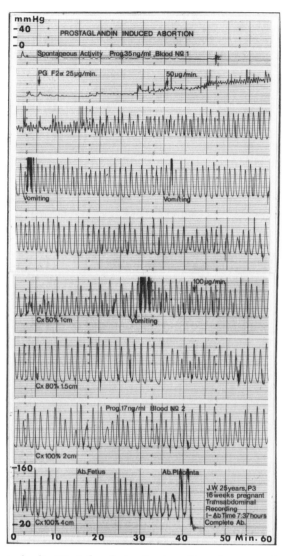

Fig. 34. Prostaglandin F2α-induced abortion in a 16-weeks-pregnant patient. Trans-abdominal recording. PG F2α, intravenous infusion at 25, 50, and 100 μg/min rates. Note that before PG F2α infusion the AP is low, the F is high, as characteristic of a blocked uterus. The plasma progesterone concentration is 35 ng/ml. PG F2α infusion (25–50 μg/min) during the first hour, only increases the RP (over 40 mm Hg) without significantly increasing the AP. During subsequent hours, the higher RP (contracture) is sustained, but on it are superimposed AP cycles of increasing magnitude. At 6–7 hrs after the start of infusion, the plasma progesterone is only 17 ng/ml, the IUP is over 80 mm Hg and clinical progress in abortion becomes apparent. After 7 hrs the fetal membranes rupture spontaneously and

tocin. However, as pointed out above, the functional significance of this relatively high IUP should be cautiously interpreted, since clinical progress in cervical dilation and in the emptying of the uterus is determined by wall tension (wT) rather than by IUP.

The analysis of one thousand cases of monitored induced labor (MIL) revealed (Csapo and Sauvage, 1971) that the IUP of term patients not in clinical labor is 30.0 ± 1.5 mm Hg (mean $\pm$ SE). Furthermore, that oxytocin stimulation (at an average infusion rate of 7.6 mU/min) increases this value to an average maximum of 68.2 ± 1.1 mm Hg and in so doing guarantees delivery in 6.3 hr. in 97.8% of the cases.

The calculated wT of these term patients when not in clinical labor is 1.43×10^5 dynes/cm, whereas when they progress in active labor, it is 3.23×10^5 dynes/cm (Table 2). The point of importance is that while the average maximum IUP induced by PG stimulation in the 14.7-weeks-pregnant patients is 113 mm Hg, their average wT is only 1.59×10^5 dynes/cm, higher than the value of the term pregnant patients, but significantly lower than the parturient patients.

Evidently, massive PG F2α stimulation of the 14.7-weeks-pregnant human uterus does not bring out the maximum working potential of the organ as a rule. This fact alone explains the 73% failure of PG F2α-induced abortions (Table I). What does remain unexplained, therefore, is the 27% success in complete abortion.

In attempting this explanation, the individual cases have to be examined in detail. In the series (of 10 patients) under consideration (Csapo *et al.*, 1971a), 3 patients had complete abortion and 3 incomplete abortion, e.g. they not only expelled the fetus, but also part of the placenta, or at least separated the placenta (group 1 "success"). The remaining 4 either failed totally, or only aborted the fetus (group 2 "failure"). The patients in group 1 displayed 80–180 mm Hg IUP during the initial 7–12 hrs of infusion. Apparently, the limited tensile strength of the fetal membranes has been overstepped at these high IUP values, the membranes ruptured, and therefore, in these leaky uteri the wT could no longer be calculated. However, it is reasonable to assume that wT continued to increase in these women, because they progressed in cervical dilatation and expelled not only the fetus, but also the placenta, completely or partly. Indeed, in those instances, when the detached placenta temporarily plugged the internal oss during its downward passage and in so doing closed the leaky uterine cavity, the temporary

the fetus is aborted. At this point, the leaky uterus loses pressure. However, when the placenta separates and closes the internal oss (during downward passage), over 160 mm Hg pressure is recorded transiently, until the placenta is expelled, completing abortion. (Csapo *et al.*, 1971a.)

rise of IUP to 200 mm Hg or higher documented that wT continued to increase after the rupture of the membranes. In contrast, patients in group 2 at best ruptured the membranes and through the partly open cervix discharged the fetus. However, they failed in continued cervical dilatation and placental separation. It is reasonable to assume, therefore, that the wT of these women did not increase after the rupturing of the fetal membranes.

Since PG F2α therapy was the same in both groups and they were also similar in gestational age and parity, their different response to F2α infusion remained to be explained. The fact, that the prolonged and massive PG F2α therapy compromised the fetoplacental unit of these patients was revealed by the observation (Csapo *et al.*, 1971a) that their preinfusion estradiol-17β level of 3.4 ± 0.8 ng/ml decreased at the peak of F2α infusion (200 μg/min) to 1.0 ± 0.4 ng/ml ($P < 0.025$). However, from the point of view of the mechanism of PG F2α action, additional observation (Csapo *et al.*, 1971a) appears to be even more relevant. In the patients of group 1, the preinfusion progesterone level of 38.0 ± 4.4 ng/ml decreased at the peak of F2α infusion to 11.8 ± 1.3 ng/ml ($P < 0.005$); while in group 2, the preinfusion progesterone level of 44.1 ± 7.5 ng/ml remained unchanged (44.4 ± 9.8 ng/ml).

These studies suggest that the abortifacient action of PG F2α may result from a suppression of placental endocrine function, brought about by the sustained high resting pressure of the uterus. This primary change is followed by a decrease in placental progesteronegenesis, a weakening of the progesterone block, an increase in wT, progress in cervical dilation, placental separation, and abortion. However, the causal relationship between these sequences of events is speculative, for the small number of observations stated above cannot settle this important issue.

The assumption that the prompt and sustained contracture response of the pregnant uterus to F2α is a significant initial step in the mechanism of abortifacient action and that the respective effects of progesterone and PG are related is supported by the following additional observations (Csapo *et al.*, 1971a). At 1 hr after F2α infusion, when the cyclic AP is only 13.2 ± 1.7 mm Hg, the sustained RP is 44.5 ± 2.2 mm Hg. This marked difference between the magnitudes of AP and RP is evidence that as late as 1 hr after the start of infusion, only a small fraction of the billions of myometrial cells are activated cyclically by F2α. Apparently, even at massive infusions (50 μg/min) F2α can only break through the progesterone block of low-threshold uterine regions, and the block has to be weakened with the progress of time before the full working potential of the organ (reflected by over 200 mm Hg IUP) can be released.

The unchanged plasma progesterone levels in group 2 present a puzzle, for the cyclic IUP of these patients increased considerably. An explanation may be found in myometrial progesterone levels, or in the effect of sustained stretch on cyclic activity, or both. Investigators are well advised to remember (Section X) that stretch is the most powerful stimulant of uterine function and that the progesterone requirement of the stretched uterus for an effective block is higher than that of the nonstretched uterus. Thus, in case of stretch (such as that induced by sustained F2α infusions), unchanged progesterone levels may well mean that the block has weakened.

The quantitative information about the F2α-induced changes in AP, RP, wT, estradiol, progesterone, cervical dilation, placental separation, etc., in cases of first trimester abortion is even more inadequate than in cases of second trimester abortion. The question of a direct luteolytic effect of PG F2α, has not been meaningfully examined as yet in patients. These studies demand special considerations in their design, for placental progesteronegenesis is already significant at the sixth week of pregnancy (Nagele) (Csapo *et al.*, 1971b), and it is difficult to separate the luteal and placental contributions to the progesterone block. However, model experiments in animals may promote the design of meaningful future clinical trials.

Once uterine activity has increased, a vicious circle can develop in that a further reduction in placental function would promote the evolution of IUP still further. Clinical success and failure may depend on the extent to which this vicious circle is put into operation, for it seems that placental separation and expulsion demand higher uterine wT than the rupturing of the membranes and the abortion of the fetus. In placental, rather than luteal) controlled species, the extensive block of the placental bed (sustained by a local effect of placental progesterone) offers extra protection against placental separation and complete abortion. This interpretation only has merits if it is validated by model experiments and if the evidence subsequently obtained aids investigators in designing successful clinical trials. The following is a summary of those model experiments that were focused upon the mechanism of PG action, in the framework of the above considerations.

If the PG-induced sustained contracture is the primary event in abortion, then the prolonged PG stimulation of the uterus may not be necessary for clinical success. The possibility has been considered, therefore, that abortion may be induced by a transient breakthrough of the progesterone block of a limited uterine area. Such an effect can create contracture, suppress placental endocrine function (and in so doing promote luteolysis if the corpus luteum is still functional), weaken the progesterone block, and thus allow the intrinsic activity of the uterus

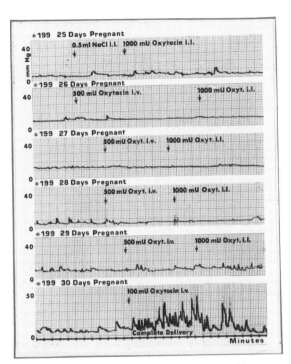

Fig. 35. The oxytocin response of the normal pregnant rabbit uterus. Extraovular microballoon technique. Note that the normal pregnant rabbit uterus is refractory to massive (intravenous and even intraluminal) oxytocin stimulation, until about 24 hrs before term, when the test dose of only 100 mU, intravenous, precipitates parturition. (Csapo, 1971c.)

to evolve. Once activity evolved, it should be possible to amplify it by low-level PG stimulation.

Thus, pregnant rabbits were equipped with extraovular microballoons, for the recording of IUP and with catheters for the intraluminal administration of either PG F2α (experimental), or its solvent, isotonic saline (control). In normal pregnant rabbits (Fig. 35) (Csapo, 1971c) oxytocin in massive intravenous doses (500–1000 mU) is ineffective until 24 hrs before term, even if the treatment is intraluminal rather than intravenous. This characteristic behavior of the normal pregnant rabbit uterus has been profoundly changed by PG F2α pretreatment. A single intraluminal dose of 500 μg F2α did break through the progesterone block at 24–27 days of gestation and induced (as it did in patients) a transient contracture on which were superimposed pressure cycles of limited magnitude (Fig. 36) (Csapo, 1971c). This effect increased with the progress in gestation, and at term even intravenous administration of F2α was effective in inducing delivery.

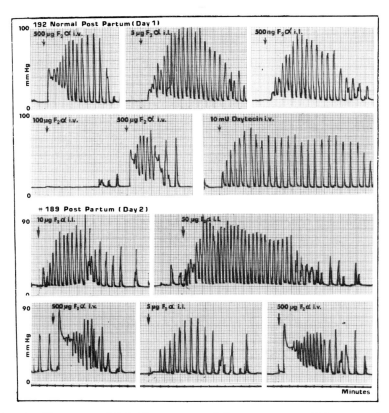

Fig. 39. The increased PG F2α response of the postpartum rabbit uterus. Intra-uterine microballoon technique. Note that the prepartum changes in the regulatory conditions of the myometrium continue during the postpartum period. While the difference in the effectiveness between intravenous and intraluminal PG F2α remains, maximum uterine responses are triggered by doses which are ineffective during pregnancy. (Csapo, 1971c).

gram per milliliter concentrations. It has been known for over a decade (Fig. 17) (Csapo, 1961b) that treatment of the excised uterus with calcium-free Krebs solution results in a rapid suspension of the generation and conduction of electric train discharges, a marked increase in threshold, and a suspension of membrane excitability and pharmacological reactivity. However, this effect was apparently limited to excitation and excitation–contraction coupling, for the calcium-deficient uterus fully retained its contractility and displayed maximum tension when its contractile system was activated directly in a strong electric field (Fig. 17) (Csapo, 1961b).

These effects were explained by a rapid loss of membrane calcium, considered already at this time (on the basis of the experiments illustrated by Fig. 17) as a key ion in normal excitation, coupling, and pharmacological response. The restoration of a small fraction of the normal calcium restored excitability and pharmacological reactivity in the calcium-deficient uterus to a limited degree. However, when working at a low calcium level, the isometric tension of the calcium-deficient uterus became a function of the oxytocin concentration of the Krebs solution, indicating that in the presence of oxytocics, the calcium-deficient uterus economizes with the activator, membrane calcium.

To these findings was added the recent demonstration (Fig. 40) (Csapo, 1971c) that PG F2α affects the excitability of the calcium-deficient uterus at picogram and fentogram per milliliter concentrations. When the field strength (of electric stimulation) was decreased in graded steps, the calcium-deficient uterus lost tension more steeply than the normal uterus, documenting reduced excitability. PG F2α not only restored in the calcium-deficient uterus normal electric excitability, but it decreased threshold below the level of the normal uterus. This finding (Fig. 40) that PG F2α controls threshold at picogram and fentogram per milliliter levels is significant, for uterine activity is primarily determined by changes in threshold, and presently there is no known oxytocic agent other than PG F2α that has a documented effect at this low concentration. Oxytocin, for example, is only effective in nanogram per milliliter concentrations (Fig. 40).

The bioassay method (Fig. 40) has been used in a pilot experiment (Csapo, 1971c) for the quantitation of PG F2α in the circulating blood of a 14-weeks-pregnant woman whose abortion has been induced successfully with F2α. Using 0.01 ml plasma samples (for sequential measurements) revealed that the F2α equivalents before infusion were 200 pg/ml, at 50 μg/min infusion rate 1000 pg/ml, at 100 μg/min rate 2 ng/ml, and at 200 μg/min rate 5 μg/ml. These results are preliminary, but they indicate that, at about 100 μg/min F2α infusion rate, the input of F2α is in balance with metabolic destruction. At lower than 100 μg/min infusion rates, considerably lower concentrations are present than those infused, while at higher rates F2α accumulates in the circulating blood. It is also apparent that normal pregnant patients have a concentration of circulating F2α, which may become biologically significant if the regulatory condition of the uterus changes.

The bioassay method (Fig. 40) does not identify F2α chemically; it only quantitates the active compound of the test sample in F2α equivalents. However, the exceedingly low concentrations in which the active agent is found in the plasma and at which it affects the threshold of the uterus suggest that it is PG F2α, that is being measured by the

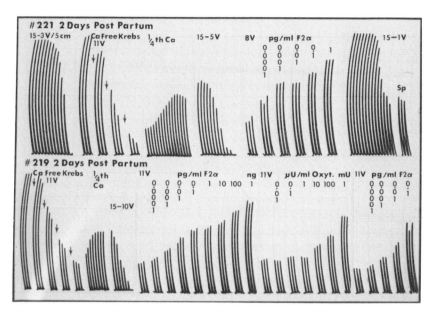

Fig. 40. A uterine bioassay method for the measurement of femtogram and pico-gram quantities of PG F2α and nanogram quantities of oxytocin. This *in vitro* bioassay method is based on the marked threshold changes to electric stimulation of the calcium-deficient uterus (Fig. 17). In the normal postpartum uterus, isometric tension changes as a function of the strength (15–3 V/5 cm) of electric stimulation (upper tracings). Repeated washing (↓) of the uterine strip with calcium-free Krebs solution reduces the effectiveness of 11 V/5 cm stimuli. The restoration of one-fourth of the normal [Ca] increases tension, but the threshold is high, and 5 V/5 cm stimulus fails in exciting the muscle. However, when the uterus is stimu-lated once a minute for 4 seconds at 8 V/5 cm field strength over 10 gm tension is recorded, which increases gradually as the PG F2α concentration is increased from 0.1 fg/ml to 1 pg/ml. In the presence of 1 pg/ml F2α, the calcium-deficient (one-fourth calcium) uterus not only responds to 4 V/5 cm stimulus, but even to 1 V/5 cm stimulus. Its threshold is lowered by F2α to such an extent that it contracts spontaneously, in spite of the calcium deficiency. In another prepara-tion (lower tracings) the same is shown as above, but it is also demonstrated that microunits (nanograms) per milliliter (rather than fentogram or picogram) quantities of oxytocin are needed for a distinct response. (Csapo, 1971c).

assay. It seems to be of considerable interest, therefore, to determine the plasma levels of PG F2α during the entire course of pregnancy and parturition and to establish also the relationship between therapeuti-cally increased plasma levels and the success in inducing abortion.

These *in vitro* studies served two functions. They provided a sensitive bioassay method for the measurement of PG in the plasma (body fluids and tissues) of untreated and PG-treated patients. In addition, they revealed that the myometrial effect of PG F2α is mediated by a control over excitability. Subsequent experiments (Csapo and Talo, 1971) with a multiple suction electrode assembly substantiated this latter evidence by showing that PG F2α promotes the generation and conduction of the electric train discharges that trigger the contractile response of the uterus.

It has been a rather unexpected turn that the intrinsically active, progesterone-blocked uterine model was again brought into focus, this time by the abortifacient action of a uterine stimulant, PG F2α. Apparently, what basic considerations failed to accomplish, repeated attempts at practical resolution did achieve. It is too early to say how this recognition will affect future developments in uterine physiology and clinical obstetrics. However, the evidence is there for those who wish to see it and having done so reconsider untenable concepts. This reassessment is mandatory, for no myometrial regulatory effects (including that of PG) can be explained unless the blocking action of progesterone is considered in the mechanism.

The extraordinary view that "in the human, progesterone has no inhibitory effect on myometrial activity or on its response to oxytocin" (Caldeyro-Barcia, 1964) delayed developments for too long already. The seconding of this motion lead to conceptional uncertainties and an unrewarding model, the intrinsically inert uterus, which demands external stimuli for emptying itself. In contrast, the intrinsically active, progesterone-blocked uterus empties itself without external stimuli, when its block weakens to such a degree that the activity-promoting action of its volume (amplified by intrinsic stimulants) is no longer balanced.

The studies with PG are of considerable theoretical significance. They revealed that the intrinsically inert uterus is an untenable model, for no matter how high the infusion rate is, the pregnant uterus could not release its maximum working capacity until it has undergone a profound regulatory change—the weakening of its progesterone block. These studies also uncovered a mystery, the uterine "sensitivity" to stimulants. This undefined term has been used in the protection of the oxytocin and prostaglandin theories of the initiation of labor. If the key to the efficacy of these stimulants (in term patients) is the "changing sensitivity" of the pregnant uterus, then this change in sensitivity is the dominant variable, a fact which should be considered in naming theories. Those attempting to promote the oxytocin or prostaglandin theories would be well advised to remember that 2000–12,000 mU/min oxytocin and 200 μg/min PG F2α do not promptly release the maximum working capacity of the pregnant human uterus, and in the majority of the cases, even prolonged infusions fail in inducing complete abortions. Thus, in the promotion of these theories, investigators must show that when labor is initiated quantities of oxytocics greater than those stated above are released, are present, and are effective without side effects. Short of this demonstration they have to acknowlege the crude reality that it is the regulatory condition of the uterus itself which is changing profoundly during pregnancy, amplifying the stimulatory effects of oxytocics present in minute quantities. Thus, the labeling of this well-documented process as a change in sensitivity hinders rather than promotes conceptual developments, because it obscures rather than exposes to further scrutiny the basic regulatory mechanism underlying the initiation of labor.

XV. A Model of Uterine Function

The results of the *in vitro* and *in vivo* experiments described above blend into a model of uterine function and regulation. The community of billions of smooth muscle cells creates a contractile coat around the uterine cavity that performs two opposite functions. It provides a protective shell for the developing embryo during pregnancy and, when term is reached, it delivers the newborn by powerful activity. This dual function is accomplished with the aid of an intricate regulatory mechanism.

Estrogen stimulation sustains the excitatory and contractile potentials of the nonpregnant uterus. These potentials are enhanced during pregnancy, by the combined stimuli of estrogen and the growing uterine volume. This promotion of activity must be balanced by opposing forces if the uterus is to perform its protective function, for without them these combined stimuli would force the myometrium to release its full working potential in the form of advanced IUP. Indeed, that is what happens every time the progesterone block weakens; pregnancy terminates.

In order to keep the contractile potential of the uterus as a potential (yet in complete readiness), the progesterone block is imposed upon the myometrium. The progesterone block does not affect the contractile potential of the uterus. It only restricts the generation and conduction of the activating train discharges to such a degree that the IUP is reduced to low level oscillations. Thus, the uterus is active, but functionally inefficient. During the critical period of nidation and placentation, the block is systemic and complete (in all the studied species), providing full protection against advanced uterine activity.

With the progress of pregnancy, the growing uterine contents increasingly promote activity (through a stretch effect). The uterine volume being increased, the balance between activity and block (demanded for the continuation of pregnancy) can only be sustained by increased progesteronegenesis. In some species, this demand is met by increased luteal activity. In others, placental progesterone-genesis supplements the need. In the former group, the block remains systemic and is withdrawn near term. In the latter, the luteoplacental shift (occurring at different times and to a different extent in different species) replaces the systemic control with a local progesterone effect. This change creates an endocrine and functional asymmetry in the uterus, and thus the nonplacental region can be active when the placental region is fully blocked.

In both instances the progesterone effect itself remains the same. Only the character of the block changes, according to the various combinations of systemic and local progesterone effects. The evolution of IUP (and OR) demands a partial withdrawal of the systemic block, but not of the local block, which is already partial. Therefore, in the luteally controlled species, the evolution of IUP (and OR) is delayed until shortly before delivery, when it occurs abruptly. In the placental controlled species, the local block permits an early evolution of IUP, due to the precocius activity of the nonplacental uterine region. Since the growing uterine contents increase the nonplacental area steadily, activity can only be limited by marked increase in placental progesteronegenesis. If progesterone support does not increase proportionately with the uterine volume, the acceleration of the evolution of IUP (and OR) signals placental compensatory failure and the advanced IUP eventually initiates labor.

For the entire duration of pregnancy, uterine function is sustained in complete functional readiness by the steady increase of the uterine volume. This maximum working potential can be sustained without the aid of external stimulants. However, the release of this potential is disallowed by the progesterone block. When, after having reached intrauterine maturity, the fetus no longer requires protection, in fact it demands delivery, the working capacity of the uterus is gradually released by a weakening of the progesterone block. In the fulfillment of its terminal activity, the uterus is probably aided by the effects of estrogen, oxytocin, and prostaglandins, for a rapid transformation from asynchronic to synchronic activity shortens labor, in the best interests of mother and newborn.

This model (Csapo, 1961a, 1969a, 1970, 1971a,b; Csapo and Wood, 1968) implies that the protection of prenatal development is accomplished by a balance between the respective effects of uterine volume (V) and progesterone concentration (P). When the ratio V/P increases, the evolution of IUP (and OR) progresses gradually. The terminal acceleration of this evolution process, culminating in the initiation of labor, results from a distinct imbalance between V and P due to a compensatory failure (in progesteronegenesis) and to the eventual activity promoting effects of terminal stimulants. The fetus also contributes to the timing of its own delivery, but it is uncertain whether its influence is mediated through the five direct regulatory factors of uterine function or some other mechanism. After delivery, the uterus involutes. However, if its volume is sustained, parturient activity becomes a chronic performance (Csapo and Takeda, 1965). The usefulness of this model has been examined through those clinical trials which were designed and conducted with help of its conceptual content.

Models are usually named so as to identify them. This uterine model may be called briefly the brake model, to emphasize the contention that the stimulatory effects of uterine volume, estrogen, and oxytocics, are balanced by the braking action of the progesterone block. Having developed the brake model, it is obligatory to ask, what is it good for? Obviously the model is crude, as models usually are. However, it brings out the concept of a delicate balance between the direct regulatory factors, which upgrade and suppress uterine activity. Of course, additional controlling agents are yet to be discovered. Therefore, the model should not be considered final or fully correct, but merely useful in designing new experiments, in predicting their outcome, and in absorbing new findings.

Personally, I intend to follow a sound advice. When I presented the embryonic form of this model to Albert Szent-Györgyi two decades ago, he remarked: "I am willing to listen; but please smile and don't take your model and yourself seriously."

XVI. Clinical Considerations

The brake model predicted that after menopause, the estrogen and volume stimulation of the uterus ceases, and only a low-level oscillation in IUP remains, which is not improved significantly by oxytocin. Clinical trials revealed (Csapo, 1970) that this is indeed the case. The model also predicted that if such a uterus is recycled by sequential estrogen and progesterone therapy (mimicking but not restoring the regulatory conditions of the normal menstrual cycle), on the withdrawal of hormone treatment, the patients may not only menstruate, but also display advanced IUP and OR. The clinical trials verified (Csapo, 1970) this expectation.

The model predicted that during normal menstrual cycles the stimulating effects on IUP (and OR) of estrogen and increased volume are balanced by progesterone, the hormone being provided for the preovulatory (interstitial) and postovulatory (luteal) steroidogenetic activity of the ovaries. It could be expected, therefore, that uterine activity would evolve at the end of the cycle when progesterone is withdrawn, the full evolution of IUP (and OR) being accomplished during the menstrual flow, when the hormone withdrawal is completed. Clinical trials verified this expectation (Fig. 41) (Csapo, 1970).

The model also predicted that if luteal progesteronegenesis is impaired, as is the case during anovulatory cycles, the IUP (and OR) is not suppressed to the same degree as it is during normal cycles. Conversely that the IUP (and OR) remains suppressed, if at the end of normal cycles (when endogenous progesterone is withdrawn) the patient is exposed to progesterone therapy. Clinical trials verified this expectation (Csapo, 1970).

However, the regulatory background of the myometrial cycle has not been fully explained by these demonstrations. The puzzle remained that after the second day of the menstrual flow, when the IUP reaches peak values, uterine activity decreases (Fig. 41) in spite of the low progesterone levels. It was the model which resolved this problem by predicting that this decrease of IUP is due to the shedding and discharge of the endometrial mass, e.g. the loss of uterine volume and the consequent cessation of the volume stimulation of uterine activity. The model predicted (Csapo, 1970) that the sustenance of this stimulation by an imposed volume (a balloon of 5 ml capacity) would maintain the high IUP until the imposed volume is withdrawn. It also predicted (Csapo, 1970) that this high IUP can be reduced by progesterone therapy. These experiments have been performed (Fig. 42) (Macedo-Costa and Darze, 1971) and the results are fully consistent with the predictions of the model. The decrease in IUP after the menstrual flow has been prevented by the imposition of a dummy volume (Fig. 42A), and the sustained IUP has been reduced by progesterone therapy (Fig. 42B). Evidently, the IUP of the nonpregnant human uterus (as that of the pregnant uterus) is powerfully controlled by the ratio V/P.

The model also predicted that if the patients become pregnant and, therefore, continued luteal function sustains the progesterone block, the IUP (and OR) remains suppressed, far beyond the normal length of the cycle. However, when during the latter half of the first trimester the luteoplacental shift is completed (and the systemic effect of luteal progesterone is withdrawn), then the local effect of placental progesterone permits a partial OR (because the block is no longer complete). Clinical trials verified this expectation (Csapo, 1969b).

The repeated success of these clinical trials increased confidence in

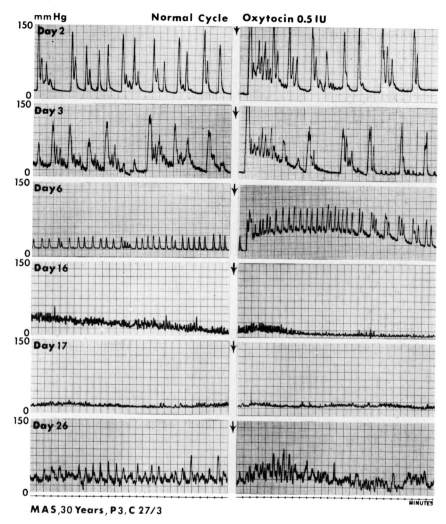

MAS, 30 Years, P 3, C 27/3

Fig. 41. The cyclic changes in IUP and OR during the normal menstrual cycle in patients. Immobilized microballoon technique. Note the decrease in AP and increase in F after the second day of the menstrual flow and the marked decrease of activity already at day 6 of the cycle. Note also the sustained suppression of IUP until 24 hrs before the onset of the menstrual flow. These cyclic changes in IUP constitute the myometrial cycle. (Csapo, 1970.)

their conceptual framework formulated in the model. More sophisticated trials were therefore in order, testing the full potential of the model in predicting their outcome. The acid test of the brake model is the demonstration that the pregnant uterus, being intrinsically active (due

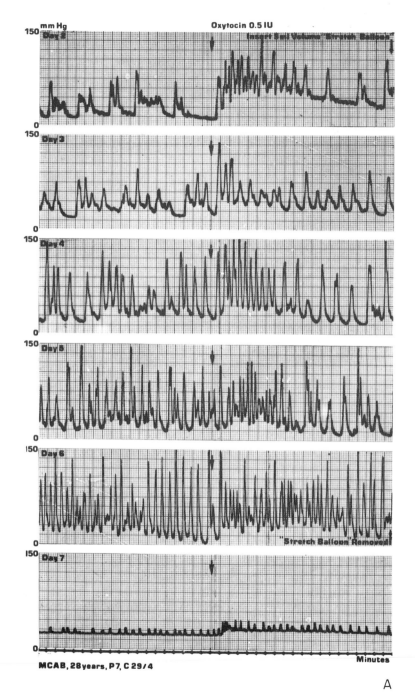

Fig. 42. (A) The effect of the replacement of the discharged uterine volume on the IUP during the early phase of the cycle. Immobilized microballoon technique. In contrast to the normal decrease (Fig. 41), the IUP (and OR) is sustained after the menstrual flow by the insertion into the uterine cavity of a 5 ml stretch balloon. When this dummy volume is removed, the IUP drops to its normal value. (Macedo-Costa and Darze, 1971.)

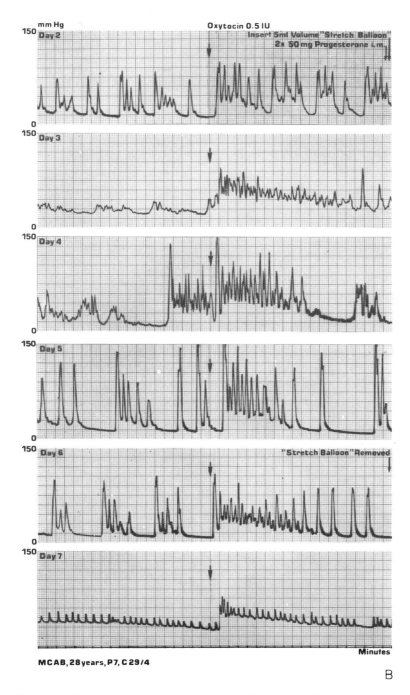

mm Hg — Oxytocin 0.5 IU

Day 2 — Insert 5ml Volume "Stretch Balloon" 2x 50 mg Progesterone i.m.

Day 3

Day 4

Day 5

Day 6 — "Stretch Balloon" Removed

Day 7

Minutes

MCAB, 28 years, P7, C 29/4

B

(B) The effect of progesterone replacement therapy on the volume sustained IUP during the early phase of the cycle. Technique and procedure as in Fig. 42A, except that 100 mg progesterone, intramuscular in oil, is administered to the patient at day 2 of the cycle. Note the suppression of IUP (and OR) at days 3 and 4. (Macedo-Costa and Darze, 1971.)

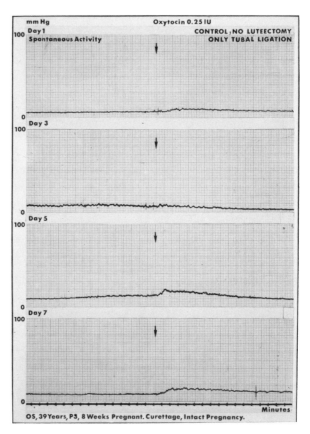

Fig. 43. The IUP after tubal ligation in early first trimester pregnant patients. Extraovular microballoon technique. Note that tubal ligation (and laparotomy) does not affect the IUP (and OR) recorded in control patients during 7 days. (Csapo *et al.*, 1971b.)

to sustained volume and estrogen stimulation, which may be amplified by prostaglandins), requires no extrinsic stimulants for delivering its contents. All that is needed for the activation of the uterus is the removal (or partial removal) of the braking action (Csapo, 1955, 1961a, 1962).

This prediction of the model has been examined recently (Csapo *et al.*, 1971b) in first trimester pregnant patients, for whom legal abortion and tubal ligation have been granted by the appropriate boards. It was expected that abdominal surgery by itself (demanded for tubal ligation) has no effect on progesteronegenesis, IUP (OR), and pregnancy (controls, Fig. 43). Only luteectomy does, provided it is performed at or before the seventh week of gestation (Nagele), when the corpus luteum

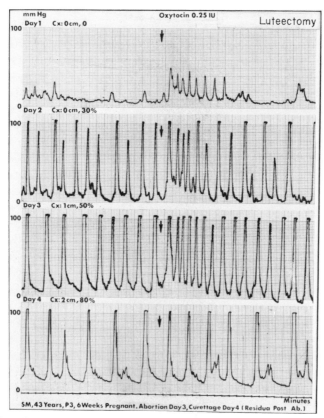

Fig. 44. The IUP after luteectomy in early first trimester pregnant patients. Techniques and procedures as in Fig. 43. Note that luteectomy precipitates the evolution of IUP (and OR) in 6-weeks-pregnant patients (Nagele) and in so doing induces clinical progress and abortion. (Csapo *et al.*, 1971b.)

is the dominant site of progesteronegenesis. The clinical trials fully realized these expectations in documenting that if the major source of progesterone is removed by luteectomy, the progesterone levels decrease and the IUP (and OR) evolves, triggering clinical abortion (Fig. 44).

However, the model also predicted that if luteectomy is performed at or after the eighth week of pregnancy, when the placenta is already the dominant source of progesterone, a transient drop in progesterone levels is followed by an increase, the IUP (and OR) does not evolve, and pregnancy continues undisturbed. The clinical trials verified this expectation (Fig. 45) and brought into focus the practical significance of the luteoplacental shift (Csapo, 1969b).

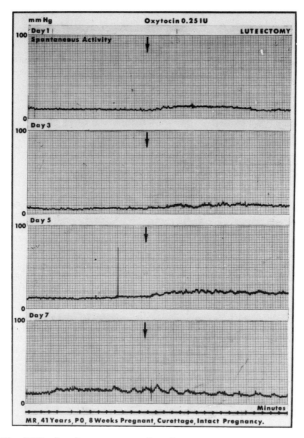

Fig. 45. The IUP after luteectomy in late first trimester pregnant patients. Techniques and procedures as in Figs. 43 and 44. Note that luteectomy fails (as a rule) in precipitating the evolution of IUP (and OR) if it is delayed until the eighth week of gestation (Nagele), and pregnancy is maintained. (Csapo *et al.,* 1971b.)

However, since it can be argued that luteectomy provokes regulatory changes other than that of the suspension of luteal progesteronegenesis, it was necessary to document that progesterone replacement therapy restores the status quo. This could be expected, since it has been predicted (Csapo, 1961a, 1969a, 1970) that as long as progesterone originates from an extrauterine source (corpus luteum) and is distributed by the circulating blood, progesterone substitution therapy is effective. The clinical trials documented that progesterone treatment sustains high plasma levels, suppresses the evolution of IUP (and OR), and prevents abortion (Fig. 46). Apparently, as long as the major source of progesterone is the corpus luteum, there are no marked species differences between rabbits and human patients in the regulation and function of the pregnant uterus (Csapo, 1963).

While the corpus luteum is indispensable during the first 7 weeks of gestation (i.e., during the first 5 weeks after ovulation), placental endocrine function retains full control over pregnancy maintenance, because luteal progesteronegenesis is dependent on placental support. Thus, the suppression or suspension of placental function leads to progesterone withdrawal during the entire duration of pregnancy. If proges-

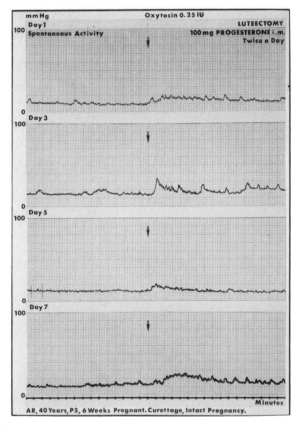

Fig. 46. The IUP after luteectomy and progesterone substitution therapy in early first trimester pregnant patients. Techniques and procedures as in Figs. 43, 44, and 45. Note that luteectomy fails in precipitating the evolution of IUP (and OR), and pregnancy is maintained if the progesterone block is sustained by replacement therapy. (Csapo *et al.,* 1971b.)

terone were the only regulator of uterine function, then the removal
of the uterine contents (by curettage) should precipitate the evolution
of IUP and OR. However, the model predicted that the removal of
the uterine volume and the consequent loss of its stimulatory effect
under these conditions disallows the evolution process, while the restitution (by a dummy) of the uterine volume precipitates it. Clinical trials
verified this expectation (Csapo, 1969a, 1970).

The model also predicted that if after the luteoplacental shift, placental progesteronegenesis is suppressed to such a degree as to result
in a significant increase in the V/P ratio, and specifically if V is increased
concomitantly, the IUP (and OR) evolves and pregnancy terminates
in abortion. Clinical trials examining the effects of intraamniotic hypertonic saline treatment fully verified this expectation. In so doing they
demonstrated the regulatory significance of the V/P ratio (Csapo, 1961a,
1969, 1970; Csapo and Wood, 1968; Csapo *et al.*, 1969, 1970) (Fig.
47). Indeed, the hypertonic saline technique became a broadly accepted
routine procedure for the termination of second trimester pregnancies
(Csapo, 1966). This demonstration was strengthened by the evidence,
that in case of endogenous failure in placental progesteronegenesis (for
example, fetal death *in utero*), a mere increase in V (by isotonic saline)
triggers the evolution of IUP and abortion (Csapo, 1966).

The results of these clinical trials were supplemented by a variety
of steroid chemical data, each of them predicted by the model. Carefully
conducted, sequential analysis (Yoshimi *et al.*, 1969) of plasma progesterone revealed a nadir at about the seventh week of pregnancy and
a continued fall in 17-hydroxyprogesterone. Thus, the luteoplacental shift
has been timed steroid chemically. Reliable measurements of myometrial
progesterone revealed the progesterone asymmetry of the rat (Csapo
and Wiest, 1969) and human (Kumar and Barnes, 1965; Zander *et al.*,
1969) uteri. The clinical significance of these demonstrations is considerable, as will be discussed presently.

To the reproductive biologist, the experiences of these clinical trials
were reassuring, for the outcome of these trials has been predicted by
a model, developed through laboratory experiments in animals. The obvious implication of this apparent exchangeability of concepts is that
in fundamental biological principles, there are no species differences.
This position (Csapo, 1963) has been repeatedly challenged during the
last 20 years by those investigators who technically restricted their in-
quiries to a single discipline or to the use of a single method. For example, some steroid chemists, who could not find a marked progesterone
withdrawal at times when uterine activity increased, dismissed the progesterone theory and ignored the fact that a slight decrease in plasma

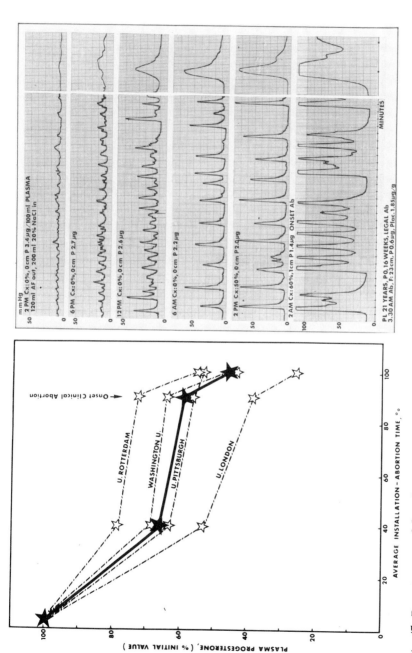

Fig. 47. Progesterone withdrawal and the accelerated evolution of uterine activity during hypertonic saline-induced legal abortions. Transabdominal recording technique. Note decrease of over 50% in peripheral plasma progesterone before the placenta is delivered and the accelerated evolution of IUP, triggering clinical abortion. (Csapo *et al.*, 1969.)

progesterone, indeed even steady rather than increasing levels, can be biologically significant if the uterine volume (and thus the V/P ratio) increases steadily. Others introduced catheters into the amniotic sac or the myometrial wall of human patients and on the basis of information obtained by this valuable but nevertheless limited approach developed sweeping conclusions about uterine function and its regulation. Thus, obscure phenomena occupied the foreground and controversies arose that hindered developments in reproduction and will remain obstructive for years to come.

For example, investigators using these catheters claimed to have located the pacemaker activity at the tubouterine junction. In reaching this conclusion, they overlooked the fact that pressure sensors (even if they work to perfection) cannot distinguish between the activity and passive stretch of the surrounding myometrial cells; only electrodes can. They also ignored the extensive electrophysiological work, conducted with a variety of techniques in animals, documenting that all uterine regions can act as pacemakers and that the low-threshold uterine region assumes this important function. This position was thought to be justified by species differences, and it was sustained in spite of the demonstration (Csapo and Takeda, 1963) that in the human, as in laboratory animals, an intimate relationship exists between the electric train discharges and the pressure cycles of the uterus, and that clinical progress depends on a change from asynchronic to synchronic electrical activity. What transpires from this phase of the study of uterine function, during which the emphasis was placed on the study of patients, rather than on the accuracy of techniques and on multidisciplinary approaches, is that model experiments and clinical trials are inseparable and that organized teamwork is the only hope for obtaining meaningful results.

This view emerges from the experience that the outcome of clinical trials in both nonpregnant and pregnant patients can be closely predicted with the help of a model, developed through a variety of laboratory experiments in animals. However, the full potential of this model was only displayed in the design and outcome of those clinical trials which examined the possibility that Nature's mechanism for the initiation and promotion of labor may be improved. These studies led to a form of preventive obstetrics.

The model predicted that the early completion of the luteoplacental shift in patients creates an endocrine and functional asymmetry in the human uterus, as early as the second trimester of pregnancy. Thus, as the uterine volume increases, with the advance of gestation, not only is the myometrium stretched increasingly, but the nonplacental region also increases. Therefore, an increasing number of cells can participate in spontaneous activity and pharmacological response. However, as long as the progesterone support of the placental sustains the block over the placental and nearby regions, this inactive uterine portion lowers (by passive stretch) the wall tension developed by the active region. These regulatory changes were expected to manifest in a gradual evolution of IUP and OR. Clinical trials revealed that both the AP and the AP/Tr (Tr = time of pressure rise)—the two parameters which reflect the balance between active and inactive cells the extent of propagating activity and synchrony—increase gradually in the unstimulated as well as in the oxytocin-stimulated uteri (Fig. 48) (Csapo and Sauvage, 1968).

The practical significance of these clinical trials (Csapo and Sauvage, 1968) is the demonstration that before the thirty-eighth week of normal pregnancy, oxytocin infusion rates which are permissible (because they do not increase the frequency of the pressure cycles and the resting

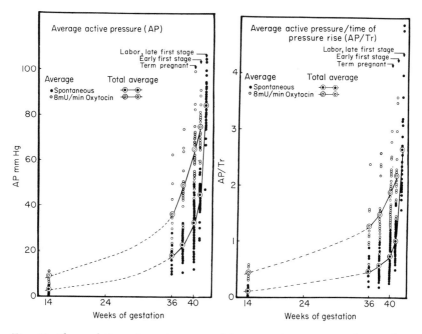

Fig. 48. The evolution of spontaneous activity, oxytocin response and rate of rise in pressure at the end of human pregnancy. Extraovular microballoon technique. Note the gradual evolution of IUP, reaching over 40 mm Hg values at term and increasing further during spontaneous labor. Note the similar evolution in the rate of rise in pressure AP/Tr. Note also than at about the thirty-eighth week, an oxytocin infusion of 8 mU/min rapidly elevates the IUP and brings it into the labor range. (Csapo and Sauvage, 1968.)

pressure beyond physiological limits) *do not* elevate the AP and AP/Tr into the spontaneous labor range. Evidently, normal pregnant patients do not become readily inducible by oxytocin before the thirty-eighth week of pregnancy.

The model predicted that the underlying mechanism of the accelerated evolution of IUP and OR in near-term patients is a failure of the placenta in compensating, by increased progesteronegenesis, for the activity promoting effects of the steadily increasing uterine volume, estrogen, and oxytocics. It is of considerable interest, therefore, that the sensitive and specific protein-binding assay method revealed (Csapo *et al.*, 1971c) increasing circulating progesterone levels only until 2–3 weeks before the spontaneous onset of clinical labor, and that the subsequent *plateau was followed by a slight decline* (Fig. 49). Apparently, labor is initiated long before its clinical symptoms manifest by the gradually evolving wall tension of the uterus due to the increase in the V/P ratio and the increasing effectiveness of terminal stimulants.

The transformation at the end of gestation of the uterus from a protective to an expulsive organ, which delivers a healthy newborn, presented Nature with a unique regulatory problem, for the cyclic uterine activity generated for the promotion of labor affects uterine blood flow and placental function, and in so doing exposes the fetus to considerable potential hazards. Nature resolved this problem by a compromise between rapidly evolving high activity and slowly evolving moderate activity. However, the experience of the centuries revealed that this compromise has not been resolved to perfection, since even in leading medical centers prolonged labor, dysfunctional labor, intrapartum fetal death, postpartum hemorrhage, the necessity of surgical intervention, etc., are frequent (Eastman, 1962).

The idea that Nature may be favorably assisted in finding the ideal compromise during the initiation and promotion of labor became a tangible reality (Page, 1946) when synthetic *oxytocin* was made available for therapeutic use. However, the controversies and uncertainties about the *basic mechanism* of the regulation of uterine function delayed the definition and the documentation of the proper therapeutic use of oxytocin, and the induction and promotion of labor remained in the realm of obstetric art, rather than of obstetric science (Theobald, 1968). The continuous monitoring of IUP throughout induced labor did provide a good start in definition and documentation, but the transabdominal recording technique (Caldeyro-Barcia, 1964) was unsuitable for routine use and the meaningful interpretation of the IUP tracings demanded more penetrating basic studies than those which were performed in patients.

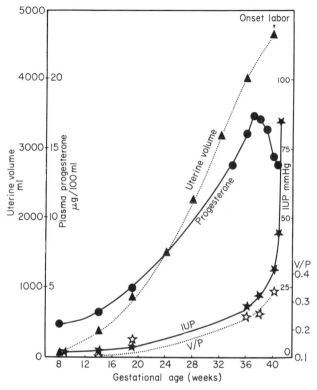

Fig. 49. The changes during normal human pregnancy in uterine volume (V), plasma progesterone concentration (P), the V/P ratio, and IUP. The changes in uterine volume are reconstructed from the relevant literature. Progesterone during early pregnancy is measured by a variety of methods (Csapo et al., 1969), while near term by the protein binding method (Csapo et al., 1971c). The IUP is measured by the extraovular microballoon technique (Csapo and Sauvage, 1968). The V/P ratio is calculated from the known values of V and P. Note that after midterm V increases more rapidly than P and that at 2–3 weeks before term, P fails to increase altogether, in fact slightly decreases. Note also that this disparity between the changes in V and P increases the ratio V/P and that until term (but not during labor) the changes in IUP parallel those of V/P. (Csapo and Sauvage, 1968; Csapo et al., 1971c.)

The development of the extraovular microsensor technique (Fig. 50) has been a forward step. It was found to be accurate, to sustain the hydrodynamic integrity of the amniotic sac, and to be suitable for routine use. Since the membranes are left intact, the theorem of Laplace remains applicable, and wT can be calculated from the IUP tracings. Furthermore, the accurately displayed RP, AP and AP/Tr offers reliable guidelines in

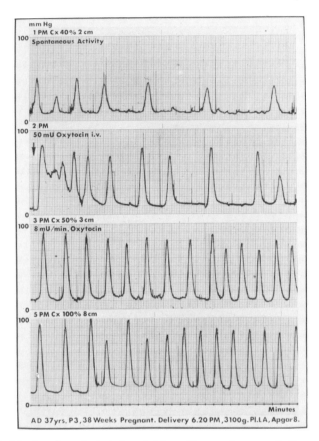

Fig. 50. Monitored oxytocin-induced labor. Extraovular microballoon technique. Note the spontaneous activity and the oxytocin test of this clinically inducible, obstetrically normal patient. Note also that an oxytocin infusion of 8 mU/min brings the IUP into the labor range and in so doing triggers the onset of labor, promotes normal and rapid clinical progress, and secures the safe delivery of the newborn. (Csapo and Sauvage, 1971.)

labor management, specifically in setting the oxytocin infusion rate. Therefore, this technical development combined with conceptual advances permitted extensive clinical trials of monitored induced labor (MIL).

It has been expected that near-term patients whose regulatory conditions are already unbalanced (as indicated by the increase of the V/P ratio, Fig. 49), are readily and predictably induced by low-level oxytocin infusions not exceeding 12 mU/min. Since the magnitude and character of IUP are readily displayed, the oxytocin infusion rate could be so adjusted as to generate an activity similar to that of advanced first stage labor. Thus, mother and fetus are spared from the unnecessary strains of the lengthy prelabor and prolonged (or dysfunctional) active labor. A study (Csapo and Sauvage, 1971) of one thousand consecutive patients delivered by MIL justified this expectation.

The patients were 27 years old and were induced at 3.5 days before term, on the average (Nagele). In spite of 26% nulliparity and 18.1% high risk patients, 97.8% of the study group were delivered per vaginam, reducing nonelective intrapartum cesarean section rate to 2.2%. Only 5% of the patients were induced twice, in spite of 13.6% medically indicated inductions conducted 8 days before term, on the average, and the limitation of the average oxytocin induction rate to 7.6 mU/min. The newborns had an average birth weight of 3289 gm and an Apgar score of 9.0 when delivered after 6.3 hr. of uncomplicated labor.

Oxytocin elevated the average active pressure from 30.0 to 68.2 mm Hg, the rate of rise in pressure from 0.9 to 2.2 mm Hg/sec, and the frequency from 10.8 to 13.1 pressure cycles/30 minutes. These oxytocin-induced IUP values are similar to those of advanced spontaneous labor.

In comparison with the performance of the routine service (or that of the obstetrical statistical cooperative), incidences of postpartum hemorrhage, midforceps, and intrapartum cesarean section were reduced. Sections indicated by dysfunctional labor were markedly reduced, and none was performed for fetal distress or second stage arrest. There was no intrapartum fetal death, in contrast to the 0.5% rate of the routine service, or the 0.7% rate, experienced by the same study patients during previous deliveries elsewhere. Thus, it appears that properly timed and conducted MIL, in the hands of trained personnel, is a desirable choice of labor management and a form of preventive obstetrics.

The clinical trials described suggest that the model developed through basic experiments in laboratory animals, has been useful. While much remains to be learned about the function and regulation of muscle, what is already known substantially advanced academic and clinical developments.

REFERENCES

Anderson, A. B. M., Turnbull, A. C., and Murray, A. M. (1967). *Amer. J. Obstet. Gynecol.* 97, 992.

Anderson, G. G., Speroff, L., and Hobbins, J. (1971). *Abstr, 18th Ann Meet., Soc. Gynecol. Invest.* p. 15.

Barr, L. (1963). *J. Theor. Biol.* 4, 73.

Bergstrom, S., Carlson, L. A., and Weeks, J. R. (1968). *Pharmacol. Rev.* 20, 1892.

Bergstrom, S., Bygdeman, M., Samuelson, B., and Wiqvist, N. (1971). *Hosp. Pract.* 6, 51.

Structure and Function of Muscle (1960), Vol. I, II & III, Ed. G. H. Bourne, Academic Press, New York & London.

Bozler, E. (1962). *Physiol. Rev.* **42**, Suppl. 5, Part II, 179.

Brotanek, V., Hendricks, C. E., Brenner, W., and Ekbladh, L. (1971). *18th Ann. Meet., Soc. Gynecol. Invest.*, p. 15.

Bulbring, E. (1962). *Physiol. Rev.* **42**, Suppl. 5, 160.

Bulbring, E. (Ed.). (1964). "Pharmacology of Smooth Muscle". Pergamon, Oxford.

Bulbring, E., Brading, A. F., Jones, A. W., and Tomita, T., eds. (1970). "Smooth Muscle." Williams & Wilkins, Baltimore, Maryland.

Burnstock, G., Holman, M. E., and Prosser, C. L. (1963). *Physiol. Rev.* **43**, 482.

Bygdeman, M., and Wiqvist, N. (1970). "Abstracts for Conference on Prostaglandins." New York Academy of Sciences, New York.

Caldeyro-Barcia, R. (1964). *In* "Muscle" (W. M. Paul *et al.*, eds.), p. 317. Pergamon, Oxford.

Coren, R., and Csapo, A. I. (1963). *Amer. J. Obstet. Gynecol.* **85**, 470.

Corner, G. W., and Allen, W. M. (1930). *Proc. Soc. Exp. Biol. Med.* **27**, 403.

Csapo, A. I. (1955). *In* "Modern Trends in Obstetrics and Gynecology" (K. Bowes, ed.), p. 20. Butterworth, London.

Csapo, A. I. (1959). *In* "Cell, Organism and Milieu" (D. Rudnick, ed.), p. 107. Ronald Press, New York.

Csapo, A. I. (1960). *In* "Physiology of Prematurity" (M. Kowlessar, ed.), p. 139. Josiah Macy, Jr. Found., New York.

Csapo, A. I. (1961a). *Ciba Found. Study Group* [Pop.] **9**, 3.

Csapo, A. I. (1961b). *In* "Oxytocin: An International Symposium" (R. Caldeyro-Barcia and H. Heller, eds.), p. 100. Pergamon, Oxford.

Csapo, A. I. (1962). *Physiol. Rev.* **42**, 7.

Csapo, A. I. (1963). *Amer. J. Obstet. Gynecol.* **85**, 359.

Csapo, A. I. (1966). *In* "Year Book of Obstetrics & Gynecology" (J. P. Greenhill, ed.), p. 126. Yearbook Publ., Chicago, Illinois.

Csapo, A. I. (1969b). *Postgrad. Med. J.* **45**, 57.

Csapo, A. I. (1969a). *Ciba Found. Study Group* [Pop.] **34**, 13.

Csapo, A. I. (1970). *Obstet. Gynecol. Surv.* **25**, No. 5, 403; **25**, No. 6, 515.

Csapo, A. I. (1971a). *In* "The International Encyclopedia of Pharmacology and Therapeutics" (M. Tausk, ed.), Sect. 48, Chapter 5, 123. Pergamon, Oxford.

Csapo, A. I. (1971b). *In* "Contractile Proteins and Muscle" (K. Laki, ed.), Chapter 14, p. 413. Dekker, New York.

Csapo, A. I. (1973). [Wiley Series on Problems in Reproduction. Vol. I. (J. B. Josimovitch, ed.)], p. 233. Wiley and Sons, New York.

Csapo, A. I., and Sauvage, J. P. (1968). *Acta Obstet. Gynecol. Scand.* **47**, 181.

Csapo, A. I., and Sauvage, J. P. (1971). *Advan. Obstet. Gynecol.*, **2** (in press).

Csapo, A. I., and Takeda, H. (1963). *Nature (London)* **200**, 680.

Csapo, A. I., and Takeda, H. (1965). *Amer. J. Obstet. Gynecol.* **91**, 221.

Csapo, A. I., and Talo, A. (1971). In preparation.

Csapo, A. I., and Wiest, W. G. (1969). *Endocrinology* **85**, 735.

Csapo, A. I., and Wood, C. (1968). *In* "Recent Advances in Endocrinology" (V. H. T. James, ed.), Chapter 7, p. 207. Churchill, London.

Csapo, A. I., Takeda, H., and Wood, C. (1963). *Amer. J. Obstet. Gynecol.* **85**, 813.

Csapo, A. I., Knobil, E., Pulkkinen, M. O., Van der Molen, H. J., Sommerville, I. F., and Wiest, W. G. (1969). *Amer. J. Obstet. Gynecol.* **105**, 1132.

Csapo, A. I., Sauvage, J. P., and Wiest, W. G. (1970). *Amer. J. Obstet. Gynecol.* **108**, 950.

Csapo, A. I., Sauvage, J. P., and Wiest, W. G. (1971a). *Amer. J. Obstet. Gynecol.* **111**, 1059.

Csapo, A. I., Pulkkinen, M. O., Ruttner, B., Sauvage, J., and Wiest, W. G. (1972). *Amer. J. Obst. Gynecol.* **112**, 1061.

Csapo, A. I., Knobil, E., Van der Molen, H. J., and Wiest, W. G. (1971c). *Amer. J. Obstet. Gynecol.* **110**, 630.

DeMattos, C. E. R., Kempson, R. L., Erdos, T., and Csapo,, A. I. (1967). *Fertil. Steril.* **18**, 545.

Eastman, N. J. (1962). *In* "Williams Obstetrics," 13th ed., p. 1126. Appleton, New York.

Ebashi, S., Ebashi, F., and Kodama, A. (1967). *J. Biochem. (Tokyo)* **62**, 137.

Embrey, M. P. (1966). Cited by Pickles *et al.* (1966).

Garamvolgyi, N., Vizi, E. S., and Knoll, J. (1971). *J. Ultrastruct. Res.* **34** ,135.

Goto, M., and Csapo, A. I. (1959). *J. Gen. Physiol.* **43**, 455.

Goto, M., and Tamai, T. (1960) "Modern Aspects of the Electrophysiology of Involuntary Muscles." Kinpodo Publ. Co. Ltd., Tokyo and Kyoto.

Goto, M., Kuriyama, H., and Abe, Y. (1959). *Jap. J. Physiol.* **21**, 880.

Hill, A. V. (1953). *Proc. Roy. Soc., Ser. B* **141**, 104.

Huxley, A. (1956). *Brit. Med. Bull.* **12**, 167.

Ichikawa, S., and Bortoff, A. (1970). *Amer. J. Physiol.* **219**, 1763.

Jennerick, H. P., and Gerard, R. W. (1953). *J. Cell. Comp. Physiol.* **42**, 79.

Jewelewicz, R., Cantor, B., Dyrenfurth, I., Warren, M., Pattner, A., Murray, T., Bowe, T., and Vande Wiele, R. L. (1971). *Abst., 18th Ann. Meet., Soc. Gynecol. Invest.*, p. 16.

Johansson, E. D. B. (1970). *In* "Steroid Assay by Protein Binding" (E. Diczfalusy, ed.), p. 188. Karolinska Inst., Stockholm.

Jung, H. (1964). *In* "Pharmocology of Smooth Muscle" (E. Bulbring, ed.), p. 113. Pergamon, Oxford.

Kao, C. Y. (1967). *In* "Cellular Biology of the Uterus" (R. M. Wynn, ed.), Chapter 11, p. 386. Appleton, New York.

Karim, S. M. M., and Filshie, G. M. (1970). *Lancet* **1**, 157.

Kumar, D., and Barnes, A. C. (1965). *Amer. J. Obstet. Gynecol.* **92**, 717.

Kuriyama, H. (1961). *Ciba Found. Study Group [Pop.]* **9**, 51.

Kuriyama, H., and Csapo, A. I. (1961). *Endocrinology* **68**, 1010.

Lancet (1970). **1**, 223.

Lepage, F., and Sureau, C. (1964). *Encycl. Med. Chir.* **B10**, 1-20, 5017.

Macedo-Costa, L. F., and Darze, E. (1971). University of Bahia, 1971.

Marshall, J. M. (1962). *Physiol. Rev.* **42**, Suppl. 5, 213.

Marshall, J. M. (1964). *In* "Pharmacology of Smooth Muscle," p. 143. Pergamon, Oxford. (E. Bulbring, ed.)

Needham, D. M., and Schoenberg, C. F. (1967). *In* "Cellular Biology of the Uterus" (R. M. Wynn, ed.), p. 291. Appleton, New York.

New York Academy of Sciences. (1970). "Abstracts, Conference on Prostaglandins," N.Y. Acad. Sci., New York.

Page, E. (1946). *Amer. J. Obstet. Gynecol.* **52**, 1014.

Pickles, V. R., Hall, W. J., Clegg, P. C., and Sullivan, T. J. (1966). *Mem. Soc. Endocr.* **14**, 89.

Porter, D. G. (1969). *Ciba. Found. Study Group [Pop.]* **34**, 79.

Ramwell, P. W., and Shaw, J. E. (1970). *Recent Prog. Horm. Res.* **26**, 139.
Reynolds, S. R. M. (1949). "Physiology of the Uterus," 2nd ed., Harper (Hoeber), New York.
Rushton, W. A. H. (1930). *J. Physiol.* (*London*) **70**, 317.
Salgo, M. (1971). [quoted by Csapo 1971b.]
Schofield, B. M. (1963). *In* "Recent Advances in Physiology" (R. Crease, ed.) Vol. 7, p. 222. Churchill, London.
Short, R. V. (1961). *In* "Hormones in Blood" (C. H. Gray and L. Bacharach, eds.), p. 379. Academic Press, New York.
Sureau, C., Chavinie, J., and Cannon, M. (1965). *In* "Electrophysiologie Uterine," 21st Congres de la federation des sociétés de gynécologie et d'obstetrique de langue française (Rapports, Discussions & Communications). Masson, Paris.
Suzuki, T. (1966). "The Basic and Clinical Aspects of the Electrophysiology of Smooth Muscle." Publ. Kanehara, Tokyo.
Talo, A., and Csapo, A. I. (1970). *Physiol. Chem. Phys.* **2**, 489.
Theobald, G. W. (1968). *Obstet. Gynecol. Surv.* **23**, 109.
Wiqvist, N., and Bygdeman, M. (1970). *Lancet* **2**, 716.
Yoshimi, T., Strott, C. A., Marshall, J. R., and Lipsett, M. B. (1969). *J. Clin. Endocrinol. Metab.* **29**, 225.
Zander, J., Holzmann, K., von Munstermann, A. M., Runnebaum, B., and Sieber, W. (1969). *In* "The Foeto-Placental Unit" (A. Pecile and C. Finzi, eds.), Int. Congr. Ser. No. 183, p. 162. Excerpta Med. Found., Amsterdam.

Addendum

Since the preparation of this chapter, developments in the topics discussed, specifically in that of prostaglandin, have been so great that the updating of the text did not seem feasible. It is recommended that the reader consult the following articles:

Csapo, A. I. (1972). "On the mechanism of the abortifacient action of prostaglandin F2α," Brook Lodge Symposium on Prostaglandins. *J. Reprod. Med.* **9**, 400.
Csapo, A. I. (1973). "Prospects of prostaglandins in postconceptional therapy." *Prostaglandins,* March 1973.

2

CAPACITIVE, RESISTIVE, AND SYNCYTIAL PROPERTIES OF HEART MUSCLE—ULTRASTRUCTURAL AND PHYSIOLOGICAL CONSIDERATIONS

ERNEST PAGE and HARRY A. FOZZARD

I. Introduction

Physiologists who study heart muscle seek correlations between the quantities they measure with physiological techniques and the structures to which these quantities correspond. The physiologist measures resting potentials, action potentials, and the cable properties of the membrane; he observes that the action potential is propagated from cell to cell.

The electron microscopist observes a plasma membrane lining the surface of cardiac cells and finds that this membrane may be modified in characteristic ways when two cells are in apposition.

It is now widely recognized that the plasma membrane and its junctional modifications seen in electron micrographs of heart muscle cells are, respectively, the ultrastructures concerned with the membrane capacitance and potential and with cell to cell propagation as studied by cell physiologists. Nevertheless, the correlation of ultrastructure with physiological measurements of membrane properties is perhaps less complete for heart muscle than for other types of striated muscle. Both the physiological phenomena and the structural organization of heart muscle are more complex than in skeletal muscle. Moreover, the techniques of cell physiology and of research on ultrastructure differ, as do the questions these two approaches have sought to answer. The cell physiology of heart muscle and the study of its ultrastructure have progressed in different directions because the techniques available to each allowed certain questions to be answered by appropriate experiments; questions experimentally inaccessible by one or the other technique have sometimes remained unasked.

These difficulties notwithstanding, it seems appropriate to ask ultrastructural questions about cardiac cellular physiology and physiological questions about cardiac ultrastructure. This review will illustrate this approach with respect to a feature of heart muscle for which this dual set of questions seems particularly useful: the ultrastructure and physiology of the cardiac cell boundaries. In Sections II and III, we shall summarize separately the ultrastructural and electrophysiological information about the cell boundary. In Section IV, the conclusions of the two techniques will be reconsidered together, and the experimental problems that emerge from this dual approach will be formulated.

To facilitate access to the cardiac ultrastructure literature, which forms a background for much of the discussion that follows, some of the more recent electron microscopic observations on vertebrate and invertebrate hearts have been compiled in an appendix to this article (Section V).

II. Ultrastructure of the Boundaries of Heart Muscle Cells

A. *Limitations of Electron Microscopic Techniques for the Study of Physiological Problems in Heart Muscle*

Most of the available information on cardiac ultrastructure has been obtained by conventional transmission electron microscopy on ultrathin sections of material fixed with aldehydes, osmium tetroxide, or both, and stained with salts of uranium and lead. More recently, this information has been confirmed and extended with the method of freeze-cleaving. Newer electron microscopic techniques (e.g., optical diffraction of electron micrographs, scanning electron microscopy, or the use of phase contrast or interference electron optics) have not yet been applied to heart muscle. The discussion in this chapter will therefore be based primarily on electron micrographs of thin sections prepared from fixed material.

Although modern transmission electron microscopes can resolve structures as small as 3 Å (Ruska, 1965), at least four factors limit the usefulness of these instruments for structural correlations with physiological phenomena in heart muscle (Page, 1969):

1. The most commonly used methods of preparing samples of heart muscle for electron microscopic examination (fixation, dehydration, embedding, utlramicrotomy, staining), as well as electron microscopy itself, preserve fine structure incompletely and distort the structures that are preserved.

2. Electron microscopy necessarily yields an image of the tissue at a particular moment; fine structures may, however, change in shape, volume, or location within the cell. Unless the tissue can be fixed and sampled at time intervals that correspond to these changes, the changes may not be recognized by this method.

3. Conventional transmission electron microscopy deals with two-dimensional sections of structures that are normally three-dimensional. To obtain information about the extent and shape of fine structures, it is necessary to have recourse to the tedious techniques of reconstruction from ultrathin serial sections or to quantitative morphometric cytology.

4. Because electron microscopic sections are necessarily thin, it is difficult to obtain a representative sample. Only an infinitesimal fraction of a heart can be sampled and examined; it is therefore often impossible to rule out that the fraction selected may be nonrepresentative.

These factors suggest that the ultrastructure of any single type of heart muscle is at best incompletely defined. In addition, generalizations about cardiac ultrastructure based on examination of a single type of heart muscle are probably unwarranted. Within any one species, the ventricular and atrial myocardium and the various portions of the systems that generate and conduct the cardiac impulse differ in ultrastructure. Moreover, the hearts of the various invertebrate and vertebrate phyla of the animal kingdom (see Section V) are as diverse in ultrastructure as in physiological characteristics; among mammals, about whose hearts the most information is available, significant differences are commonly found between different genera within the same family.

B. Classification of the Boundaries of the Heart Muscle Cell

We have chosen to consider in detail the ultrastructural correlates of the membrane capacity and of the syncytial behavior of heart muscle. To discuss this subject from the point of view of the relation between ultrastructure and physiology, it is essential to define the boundaries of a heart muscle cell. In this section, we propose a somewhat unconventional scheme for classifying the various structural specializations that make up this boundary. An advantage of this scheme (Table I) is that it stresses those aspects of boundary structure that seem to us most important for the physiological questions under discussion.

The scheme shown in Table I divides the cell boundary according to the presumed ionic composition of the solutions or gels bathing the two faces of the boundary: Class I boundaries, bathed on the external face with extracellular fluid and on the internal face with cytoplasmic

TABLE I
CLASSIFICATION OF STRUCTURAL SPECIALIZATIONS
OF THE CARDIAC CELL BOUNDARY

A. *Class I boundary:* Sarcolemmal membrane bathed on one face with a solution having the ionic composition of extracellular fluid and on the other face with a medium having the ionic composition of the cytoplasmic "solution."
    1. External sarcolemma
    2. Internal sarcolemma
       a. Sarcolemmal lining of the T system or of T system-like invaginations
       b. Sarcolemmal lining of extracellular fluid-filled spaces between cells at the transverse cell boundary, except where adjacent sarcolemmas are apposed to form a Class II boundary (nexus).
B. *Class II boundary:* Modified membrane (nexus) at the area where the sarcolemmas of two cells are in maximally close apposition. This membraneous structure separates the cytoplasms of two cells and is bathed on both faces with identical solutions having the ionic composition of cytoplasm.
    1. Located at the transverse cell boundary in continuity with the Class I boundaries defined in A2b above
       a. Predominant orientation transverse
       b. Predominant orientation longitudinal
    2. Located at areas of apposition of two external sarcolemmas.

"solution," and Class II boundaries, or nexuses, bathed on both faces with identical cytoplasmic solutions. It is known that Class I boundaries have a relatively low passive permeability to ions and a relatively high electrical resistance; and there is substantial experimental evidence that Class II boundaries are highly permeable to ions and have a low electrical resistance. A corollary of this assumption is that the syncytial behavior of heart muscle is a consequence of the properties of the Class II boundary; that is, that the properties of the nexus are responsible for the electrical continuity between the identical cytoplasmic solutions (or gels) bathing the two faces of the boundary. Before examining the physiological evidence for this assumption, it is useful to review the ultrastructural characteristics of the two types of boundary in heart muscle.

C. Ultrastructure of Class I Boundaries

1. General Features

Since the ultrastructure of the sarcolemma in fixed sections of heart muscle differs in no essential respect from that of the plasma membrane of many other animal cells, the reader is referred to recent reviews for a detailed discussion (Korn, 1969; Stoeckenius and Engelman, 1969). We shall confine ourselves to a description of structural features relevant to the physiological discussion to follow.

The word sarcolemma is used here in the restricted sense recommended by Fawcett and McNutt (1969). It refers to the plasma membrane and does not include the protein–polysaccharide coating that makes up the basal lamina or the associated reticulum of fine collagen fibrils. The sarcolemmal membrane is about 90 Å thick. When stained with heavy metals, it has the classic three-layered appearance in osmium- or aldehyde-fixed thin sections, the dimensions varying somewhat with the fixation technique (Sjöstrand, 1963, 1968). In freeze-cleaved material (fixed or unfixed, with and without etching) the plasma membrane of heart muscle, like that of other tissues, is less than 100 Å in thickness and appears to be covered with particles (Rayns *et al.*, 1968; Leak, 1970; McNutt, 1970; McNutt and Weinstein, 1970; Sommer and Johnson, 1970). The particles are 70–100 Å in diameter; Rayns *et al.* (1968) found that their density is significantly greater (400–$700/\mu^2$) at the fracture surface facing the extracellular space than at the fracture surface facing the cytoplasm (80–$120/\mu^2$) (see Figs. 1 and 2). It is at present an open question whether the particles observed in freeze-cleaved plasma membranes extend to the external and cytoplasmic surfaces of the membrane or whether they lie entirely within the membrane; in the latter case, the two fracture planes visualized with this technique would both pass inside and roughly parallel to the external and cytoplasmic surfaces [see discussion by Chalcroft and Bullivant (1970)].

2. CLASSIFICATION OF CLASS I BOUNDARIES

High resolution electron microscopic techniques applied to Class I membranes have so far yielded little information about the special physiological properties these membranes manifest in heart muscle. Useful information has instead come from a consideration of the distribution of Class I membranes in the heart muscle cell as revealed by electron micrographs of tissue sections or of freeze-cleaved tissue examined at relatively low magnification. Such electron micrographs show that in the most highly developed kinds of heart muscle (e.g., the ventricular myocardium of mammals) the sarcolemma can be divided arbitrarily into two parts. We have designated these parts, respectively, as the external and internal sarcolemmas (Table I).

A. THE EXTERNAL SARCOLEMMA. The external sarcolemma is that part of the sarcolemma which faces the tissue interspaces. In mammalian ventricular muscle these interspaces consist predominantly of intercellular clefts 0.15–0.3 μ in diameter. The range of diameters for these clefts *in vivo* is as yet undetermined; it is entirely possible that the fixation techniques used to date have significantly distorted the dimensions of the clefts by enlarging them. The clefts are continuous with much wider spaces, roughly polygonal in shape, and of the order of 5–12 μ in their largest transverse dimension. Within these clefts course the terminal arterioles, venules, and capillaries of the coronary vascular and lymphatic systems, as well as nerves and connective tissue. In most forms of heart muscle, the external sarcolemma is coextensive with the cell envelope (exclusive of the intercalated disk) seen by light microscopists, except for two differences—it does not include the basal lamina and reticulum, and it excludes areas of the longitudinally oriented cell surface where the apposed sarcolemmas of two cells lying parallel to each other have formed Class II boundaries (nexuses), whose occurrence is not detectable by light microscopy.

In certain forms of heart muscle, the external sarcolemma may manifest specializations that greatly increase its area. Examples of such specializations are found in the Purkinje fibers of sheep false tendon

(E. Page *et al.*, 1969; Mobley and E. Page, 1972). In this tissue, the external sarcolemma pouches outward into the interstitial space, and its surface presents multiple projections. Evaginations of the external sarcolemma have also been reported in the heart of the cuttlefish, where they are described as resembling microvilli about 1 μ in length (Kawaguti, 1963b).

In addition to these somewhat unusual specializations, the external sarcolemma of mammalian ventricular muscle commonly shows two types of small invaginations (Forssmann and Girardier, 1966; Fawcett and McNutt, 1969)—vesicular structures ~800 Å in diameter, possibly functioning for vesicular transport, and alveolate or spiny vesicles about 800–1000 Å in diameter, possibly implicated in the selective uptake of protein.

B. THE INTERNAL SARCOLEMMA. The internal sarcolemma is particularly highly developed in mammalian hearts and in the hearts of some invertebrates. It can be arbitrarily divided into two parts: (1) tubules originating as infoldings of the external sarcolemma; as will be clear from the discussion to follow, this portion includes both the transversely and longitudinally oriented portions of the T system in mammalian ventricular muscle, and analogous structures in lower vertebrates and in invertebrates; (2) the sarcolemmal lining of the spaces between cells at the transverse cell boundary, except where adjacent sarcolemmae are apposed to form Class II boundaries (nexuses).

3. COMPARATIVE ULTRASTRUCTURE OF THE INTERNAL SARCOLEMMA—
THE T SYSTEM OF MAMMALS

A. MAMMALIAN VENTRICULAR MYOCARDIUM. To the extent that it has been studied, the internal sarcolemma of heart muscle appears similar to other plasma membranes both in thin sections of fixed material and in freeze-cleaved ventricular muscle (Figs. 1 and 2) (Rayns *et al.*, 1968). In avian and mammalian ventricular cells, the tubular invaginations of the external sarcolemma that make up the T system fold inward from the sarcolemmal surface. The infolding occurs at points that are in register with the Z bands of the myofibrils that run under and over the T tubules. The sarcolemmal infoldings extend inward in a direction roughly at right angles to the external sarcolemma; their lumina are continuous with those of otherwise similar channels, which course longitudinally (Fig. 2), that is parallel to the long axis of the cell (Nelson and Benson, 1963; Forssmann and Girardier, 1966, 1970; E. Page, 1967b; Rayns *et al.*, 1968; Meddoff and Page, 1968; Fawcett and McNutt, 1969).

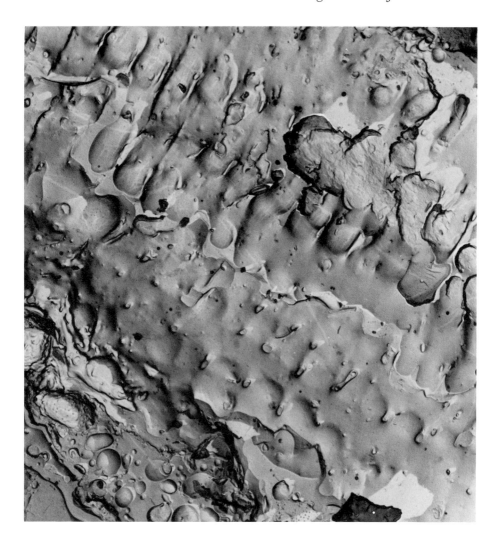

Fig. 1. Electron micrograph of guinea pig ventricle prepared by the freeze-etch technique. The myofibrillar axis lies from upper left to lower right. The upper half of the micrograph presents the external surface of the sarcolemma of one cell, the lower half of the internal surface of the sarcolemma of a second cell. A narrow extracellular space runs between the two cells. The apertures of the T tubules appear as depressions of the surface when viewed from the outside of the cell as in the upper cell, and as short stumps when viewed from the inside of the cell, as in the lower cell. Structure in lower left corner is a blood capillary. (×6900.) Reproduced from Rayns *et al.* (1968) through the courtesy of the authors.

The term "transverse tubular system," which was applied to heart muscle by analogy with the T system of skeletal muscle (Andersson-Cedergren, 1959) is thus inappropriate for mammalian heart muscle. We shall therefore refer to all of the membrane-limited volume enclosed by the internal sarcolemma (except that between two transverse cell boundaries) as the T system (E. Page, 1967b; Sommer and Johnson, 1968; Simpson and Rayns, 1968; Fawcett and McNutt, 1969).

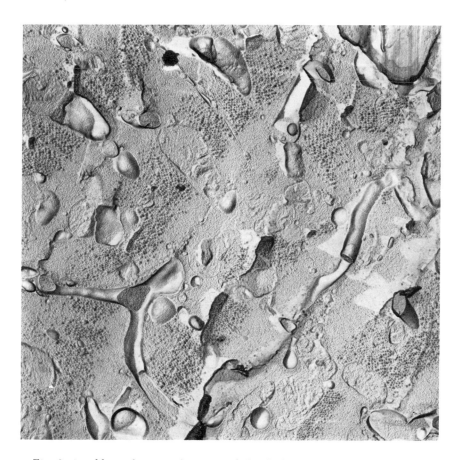

Fig. 2. An oblique fracture of a myocardial cell showing in particular the course of the T tubules in relation to the A band. The T tubule at lower left exhibits branching. The rough surface at the point of branching represents the tubular contents; to the left of this rough surface is the cytoplasmic face of the tubular membrane; below and to the right of the rough surface may be seen the depressions which remain after the tubule contents have been removed by the fracturing process. (×15,400.) Reproduced from Rayns *et al.* (1968) through the courtesy of the authors.

It is not known how deeply the transversely oriented parts of the
T system penetrate the cells of mammalian ventricular muscle. In particular, one would like to know whether and how often the two ends
of a T tubule open out, respectively, on the two opposite surfaces of
the external sarcolemma. It remains to be shown whether any T-tubules
run as straight cylinders between two external sarcolemmal surfaces.
However, even if, as is probable, their path is tortuous (i.e., if it includes
longitudinally oriented segments), the existence of "through-and-
through" tubules would represent a physiologically significant extracellular pathway for both electrical current and solutes, a pathway lined
throughout by a Class I membrane and containing within its lumen
a solution whose ionic composition is presumably that of extracellular
fluid.

The most detailed study of the cardiac T system to date has recently
been carried out in rat heart by Forssmann and Girardier (1970). These
investigators opacified the extracellular space and the T system with
which that space is in continuity by means of the horseradish peroxidase
method of Karnovsky (1967). Forssmann and Girardier found that the
diameter, shape, and distribution of the ventricular T system was much
less uniform than that of skeletal muscle. At the level of the myofibrillar
Z band, the system had multiple branches, usually confined to the third
of the sarcomere nearest the Z band. It was possible to distinguish three
types of longitudinally oriented specializations of the T system: (1)
tubules >1000 Å in diameter, located where the Z bands of two adjacent
myofibrils (as seen in longitudinal section) are out of register; (2)
tubules about 1000 Å in diameter, and circular in cross section, located
next to the A band; and (3) tubules 500–850 Å in diameter characterized
by a tortuous course.

The as yet incomplete observations on mammalian ventricular myocardium thus suggest that this tissue is honeycombed with extracellular
channels. Published ultrastructural studies do not yet permit a quantitative description of this honeycombing. Such a description would include estimates of the relative contributions of internal and external
sarcolemmas to the total area of Class I membrane; of the fraction
of the external surface occupied by the openings of the T tubules; of
the relative areas of internal sarcolemma oriented, respectively, in the
transverse and longitudinal directions, and of their relative cross-sectional areas and volumes; and of the relative volumes of the T system
and the cellular contents enclosed by the sarcolemma. When such estimates become available, it will be of interest to investigate whether
and how the diameters and volumes of the various components of the

TABLE II

ULTRASTRUCTURAL DIMENSIONS OF RAT LEFT VENTRICLES[a]

| | Fraction of cell volume | | | Membrane area/unit cell volume (μ^2/μ^3) | | |
Mito-chondria	Myofibrils	T system	Sarco-tubules	Sarco-lemma	Sarco-lemma + T system	Sarco-tubules[b]
0.34	0.481	0.012	0.035	0.27	0.34	1.3
±0.01	±0.009	±0.002	±0.002	±0.01	±0.01	±0.1

[a] 200 gm female Sprague Dawley rats. Total of twenty-five electron microscopic fields at final magnifications of either ×8600 or ×33,000.
[b] The dyads (or terminal cisternae) make up 0.08 ± 0.02 of total sarcotubular volume and 0.12 ± 0.02 of total sarcotubular membrane area. The fraction of external sarcolemmal arrea involved in contacts with subsarcolemmal dyads (cisternae) is 0.14 ± 0.01. Page, McCallister and Power (1971).

T system are affected by contraction, relaxation, pressure, stretch, and other factors.

Page, McCallister and Power (1971) have recently estimated the volume and membrane area of the T system in rat left ventricles by stereological measurements on electron micrographs using the equations of Sitte (1967). Table II shows the results of this technique, which are in agreement with independent measurements by Pager (1971).

E. Page and McCallister (1973) and McCallister and Page (1971, 1973) have recently studied by stereological methods the relationship of the external sarcolemma and T system in the rat left ventricle under conditions in which cells are stimulated to grow. They examined the relative contributions of the external sarcolemma and T system to the total cell surface (a) during hypertrophy of myocardial cells produced by constricting the ascending aorta (Page *et al.*, 1972) and (b) during growth of left ventricular myocardial cells in response to injection of approximately physiological doses of thyroxine into immature, hypothyroid rats whose growth had previously been arrested by extirpation of the thyroid gland. These studies have revealed a remarkable mechanism by which the myocardial cell keeps constant the total Class I membrane area per unit of cell volume. In response to both aortic constriction and thyroxine, myocardial cells increased in diameter; at the same time the area of external sarcolemmal membrane per unit cell volume fell significantly. Such a reduction is to be expected for a roughly cylindrical body in which the ratio of the curved cylindrical surface to the volume of the cylinder is given by $2/r$, where r = the radius of a cross section of the cylinder. By contrast, the ratio (external sarcolemmal area + T system area)/(cell volume) remained constant in the face of a 1.4-fold increase in the volume of a one sarcomere-long segment of the cell. It was possible to conclude that additional T system membrane had been synthesized to such an extent as to keep the total plasma membrane area per unit of cell volume from falling. It may be conjectured that the constancy of this ratio is essential for the nutritional requirements of the cell and for the preservation of the excitatory and conductive functions of the cell.

The discussion to this point has assumed that both the transverse and longitudinal components of the T system are roughly cylindrical, hollow structures filled only with an electrolyte solution having the ionic composition of extracellular fluid. This assumption may be an oversimplification for two reasons. First, the ionic composition of the T-tubular fluid has yet to be directly determined. Second, it has recently been found that the lumina of the T tubules in rat heart may contain membrane-limited structures (Fig. 3) whose origin will be discussed on page 111 (Meddoff and Page, 1968). At this writing, it seems possible that the fraction of T-tubular volume occupied by these structures must be considered variable.

B. MAMMALIAN ATRIUM. Several groups of investigators have reported that T tubules are absent or rare in mammalian atrium (Hibbs and Ferrans, 1969; McNutt and Fawcett, 1969). By contrast, Forssmann and Girardier (1970), using the peroxidase technique to opacify tubular extensions of the extracellular compartment in the rat atrium, found that there were two populations of cells, one with and one without a T system. In those cells containing a T system, the T tubules had a uniform diameter of about 800 Å and were oriented longitudinally, that is, parallel to the myofibrils and the long axis of the cell.

C. CELLS OF THE MAMMALIAN SINOATRIAL NODE, ATRIOVENTRICULAR NODE, BUNDLE OF HIS, AND INTRAVENTRICULAR CONDUCTION SYSTEM. The T system appears to be absent in the sinoatrial and atrioventricular nodes. It is also absent in cells of the Bundle of His and its major branches. These cells characteristically have a poorly developed contractile apparatus, as evidenced by the relative paucity and primitive state of organization of the myofibrils (Caesar et al., 1958; Rhodin et al., 1961; Sommer and Johnson, 1968; E. Page et al., 1969; Martinez-Palomo et al., 1970). However, in the peripheral Purkinje cells of cat right ventricle, there appears to be a correlation between the thickness of the myofibrillar bundles and the presence or absence of transversely oriented components of a T system. Cells in which many parallel myofibrillar bundles intervene between parallel sarcolemmal surfaces show the presence of transversely oriented T tubules; cells in which the myofibrillar mass intervening between sarcolemmas is narrow lack these transversely oriented components (E. Page, 1967b; see also Girardier, 1965). The question of defining the minimal thickness of myofibrillar bundles necessary for the presence of transversely oriented components of the T system is complicated by the organization of cardiac myofibrils into the well known *Felderstruktur* (Fawcett and McNutt, 1969).

Sommer and Johnson (1968) proposed that in mammalian ventricle all portions of the Purkinje system can be distinguished from ventricular tissue by the absence of transversely oriented T tubules in the Purkinje fibers. As a rigorous criterion for the differentiation of peripheral Purkinje fibers from ventricular myocardium, this proposal has not yet found wide acceptance.

4. Comparative Ultrastructure of the Internal Sarcolemma— Occurrence of a Cardiac T System in the Animal Kingdom

The T system has been found in the ventricular myocardium of all mammals so far examined. It is well developed in the working ventricular myocardium of some, but not all birds (Gossrau, 1969; Sommer and Johnson, 1968). It is absent in the ventricles of cold-blooded vertebrates, including the goldfish (Yamamoto, 1967), frog (Staley and Benson, 1968, Sommer and Johnson, 1968), toad (Grimley and Edwards, 1960; Nayler and Merrillees, 1964), boa constrictor (Leak, 1967), and turtle (Hirakow, 1970); the ventricles of some of these animals (e.g., the boa constrictor) do, however, have sarcolemmal invaginations which may penetrate inward as far as the first myofibrillar bundle. In all of the lower vertebrates, the diameter of ventricular myocardial cells is significantly smaller than that of mammalian ventricular myocardial cells. The observations on ventricles of lower vertebrates are thus consistent with the interpretation that the thickness of the myofibrillar mass intervening between two external sarcolemmas may determine the presence or absence of a T system. Reservations about this generalization have, however, been expressed by Sommer and Johnson (1968).

Tubular sarcolemmal infoldings that can be considered a T system have been described in several invertebrate hearts (see Appendix); a partial list includes the hearts of the moth (Sanger and McCann, 1968), mantis shrimp (Irisawa and Hama, 1965), cockroach (Edwards and

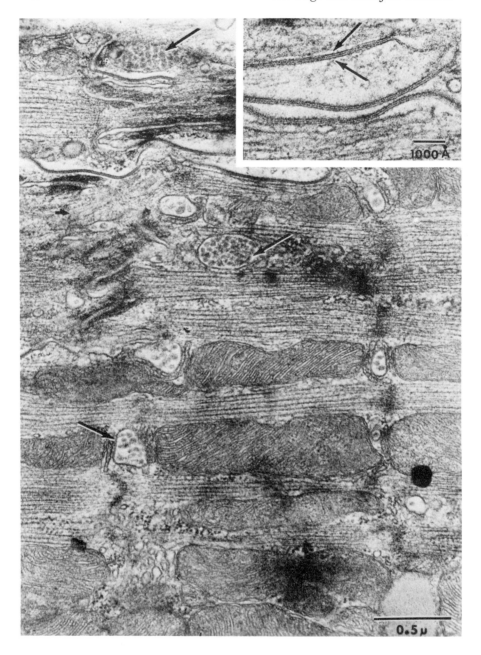

Challice, 1960), snail (North, 1963; Nisbet and Plummer, 1969), water flea (R. S. Stein *et al.*, 1966), squid (Schipp and Schäfer, 1969a,b), and grasshopper (Baccetti and Bigliardi, 1969b).

D. Ultrastructure of Class II Boundaries (Nexuses)

In many tissues, the plasma membranes of adjacent cells are in contact over a significant proportion of their surface areas. Close apposition of two plasma membranes may be associated with various electron microscopic appearances of the membranes involved in the apposition. The interpretation of these appearances is at present a subject of particularly active investigation. The problem is not yet solved. At the time this review is being written, there is no entirely satisfactory correlation between the ultrastructure and the physiological properties of Class II boundaries in heart muscle. The subject has recently been reviewed for vertebrate hearts by M. Dewey (1969) and McNutt (1970).

1. Historical Background

The decisive electron microscopic observation of Sjöstrand and Andersson (1954) and of Sjöstrand and Andersson-Cedergren (1960) showed conclusively that the intercalated disks described by light microscopists are true cell boundaries. Heart muscle is therefore composed of discrete cells and is not, as had been previously supposed, a syncytium—at least by structural criteria. Electron microscopic proof that heart muscle is cellular did not answer the structural questions of interest to physiologists, but merely changed the nature of these questions. If heart muscle cells are structurally single units, it becomes necessary to find an ultrastructural basis for the physiological observation that many heart muscle cells can behave as a unit with respect to their electrical properties. It became

---

Fig. 3. Electron micrograph of rat left ventricular myocardium fixed by perfusion with electrolyte solution containing osmium tetroxide as described by Meddoff and Page (1968) and stained with uranyl acetate and lead. Wavy boundary at left (short arrow) is classic intercalated disk. Lower long arrow points to cross section of T tubule containing numerous membrane-limited inclusions, which may also be seen within T tubular cross section one sarcomere to the right. Upper two long arrows point to multiple membrane-limited inclusions characteristically found near the transverse cell boundary. (×42,300.) *Insert:* Five-layered appearance of a gap junction from the same heart. The central dark layer can be shown to be permeable to extracellular lanthanum. (×84,600.) Electron micrograph prepared by Debra Meddoff Sherman.

evident early in the course of electron microscopic studies that the critical ultrastructural correlations required an investigation of the areas where heart muscle cells are in contact. What was the ultrastructural meaning of "contact" and what special ultrastructural features of the area of contact were responsible for the low resistance of the boundary to the flow of electric current (Barr *et al.*, 1965) and of ions (Weidmann, 1966)?

The electron microscopic investigation of these questions in heart muscle proceeded stepwise and in several directions, the approaches becoming more diversified as more and better electron microscopic techniques became available. The initial direction came from physiological and ultrastructural studies of the lateral cell junctions in epithelial cells. The laboratory of Loewenstein discovered that the junctional membrane between the lateral surfaces of epithelial cells has a low resistance to electric current, is highly permeable to ions, and permits the diffusion of fluorescent tracers from the cytoplasm of one cell to that of its neighbor (Loewenstein and Kanno, 1964). The high permeability of the junction is depressed under various physiologically interesting conditions, including cell injury and intracellular injection of calcium ion, as well as by exposure to anisotonic media (Loewenstein *et al.*, 1967), to calcium- and magnesium-free media (Rose and Loewenstein, 1969), to low temperature (Politoff *et al.*, 1969), and to metabolic inhibitors (Politoff *et al.*, 1969). The depression of permeability can, under certain conditions, be restored by passing a repolarizing current through the nonjunctional (Class I) membrane of the cell (Rose, 1970).

At about the same time as these unusual permeability properties were being discovered by physiologists, Farquhar and Palade (1963) described the ultrastructural appearance of the junctional complexes of epithelial cells in fixed tissue that has been sectioned at right angles to the planes of the apposed membranes. They recognized three types of structural specialization at the junctional surface—a zonula occludens or tight junction, a zonula adherens or intermediate junction, and a macula adherens or desmosome. Physiological interest centered on the zonula occludens, which was widely held to be the locus of the special permeability properties of the junctional complex. In Farquhar and Palade's electron micrographs of osmium-fixed material, it appeared as a five-layered zone about 90 Å wide—the outer layers were two osmiophilic (dark) lines corresponding to the cytoplasmic leaflets of the plasma membranes of the adjacent cells; a central dark line, interpreted as resulting from the fusion of the external (noncytoplasmic) leaflets of the two plasma membranes; and two light lines between the central fusion line and the two osmiophilic cytoplasmic leaflets. The zonula occludens formed a continuous beltlike attachment around the cell. Farquhar and Palade were able to show that this belt effectively sealed the space between cells; extracellular protein molecules diffused up to but not through the belt.

Until 1967, electron microscopists working with heart muscle believed that the zonula occludens of epithelial cells was an appropriate ultrastructural model for the junctional area that constitutes the low resistance pathway between cardiac cells. This belief became untenable (at least without significant modifications) after Revel and Karnovsky (1967) showed that the apparent fusion between the outer (noncytoplasmic) leaflets of the contiguous plasma membranes (Fig. 3, insert) was an artifact of the technique of fixation and staining. Instead, they were able to demonstrate that a 20 Å gap exists between the two membranes. They also found a characteristic subunit structure in the junctional area. Because of the 20 Å gap, Revel and Karnovsky suggested the term "gap junction" for junctional areas of this type. A similar subunit structure had previously been found by Robertson (1963) in the Mauthner cell synapses of the goldfish brain.

Subunit structures of the type characteristic of the Class II junctions between mammalian heart muscle cells have now been found in many tissues. Particularly extensive studies of such junctions have been made on plasma membranes isolated from liver (Benedetti and Emmelot, 1968; Goodenough and Revel, 1970; Chalcroft and Bullivant, 1970) and from vertebrate central nervous system (Brightman and Reese, 1969). The subunit structure has been demonstrated by several additional techniques, including the use of lanthanum to delineate the subunits, negative staining with phosphotungstic acid, and freeze-cleaving. These studies lead to the conclusion that the gap junction and the zonula occludens (tight junction) are morphologically distinct (Goodenough and Revel, 1970; Chalcroft and Bullivant, 1970); although both may occur in same tissues (e.g., in mouse liver), only gap junctions have to date been demonstrated in mammalian heart muscle.

2. TERMINOLOGY

We have defined the Class II junction or nexus operationally on the basis of the ionic composition of the electrolyte solutions between which the junction intervenes. This definition requires no assumptions about the ultrastructure or physiological properties of the junction. In the evolution of current ideas about junctional membranes in heart muscle, electron microscopists have used various descriptive terms to identify what they believed to be the junctional membrane corresponding to the low-resistance pathway between cells. The descriptive terms were based on the techniques available; it is therefore not surprising that the terminology has been nonuniform, as pointed out by McNutt (1970). The terms pentalaminar junction, tight junction, and zonula occludens were originally applied by analogy to the zonula occludens of epithelial cells. The expressions close junction and gap junction took cognizance of the 20 Å gap discernible between cells with certain electron microscopic techniques and emphasized that the Class II junctions between cardiac cells are distinct from tight junctions. McNutt has objected to the term "gap junction" on the grounds that this expression does not do justice to the subunit structure of the junction when it is visualized *en face* and that it is inconsistent with a direct contact between plasma membranes.

In view of the fact that the ultrastructures of both tight and gap junctions are as yet incompletely worked out, it seems premature to attribute the physiological properties of the nexus to a unique structure. In this regard, it is especially important to consider the possibility that Class II (nexal) junctions may be made and unmade under physiological conditions. For example, there may exist conformations of the membrane intermediate between what are now thought to be typical nexal and nonnexal conformations. In the following discussion, we shall retain the term "gap junction" for the typical structure delineated by lanthanum, without prejudice as to the uniqueness of this structure as a low-resistance pathway.

3. Distribution of Gap Junctions in Heart Muscle

A. Distribution in the Animal Kingdom. Junctions having the morphological characteristics of the gap junctions described by Revel and Karnovsky in the mouse heart have been found in the atrial and ventricular myocardium and throughout the Purkinje system of all mammalian hearts examined to date. The available electron micrographs of mammalian sinoatrial node do not show gap junctions, but the evidence for this tissue is not conclusive and requires reexamination. Nexal junctions were described in the hearts of adult birds by Hama and Kanaseki (1967) and of fetal chick heart cells in culture by Fischman and Zak (1970), but are reported to be absent in the ventricle of adult chickens by Sommer and Johnson (1969). On the basis of the available electron micrographs of fixed, thin-sectioned material, nexuses are either absent in the hearts of vertebrates other than mammals (and of at least some birds), or they are not demonstrable by conventional electron microscopic techniques. By contrast, nexal junctions have been found in electron micrographs of some invertebrate hearts, including those of the mantis shrimp (Irisawa and Hama, 1965) and tunicates (Kriebel, 1968). To date, none of the nexal junctions in the hearts of birds and of invertebrates have been examined with lanthanum, negative staining, or freeze-cleave techniques; the existence of a gap or of a subunit structure in these hearts is therefore not yet established.

The apparent absence of nexal junctions in many types of heart muscle that exhibit syncytial behavior with respect to their electrical properties poses a fundamental problem in the correlation of physiology with ultrastructure. On the one hand, nexal junctions of the kind readily demonstrable in electron micrographs of mammalian heart muscle may not be necessary for electrical continuity between heart muscle cells. Alternatively, they are necessary, but the electron micrographs fail to show their presence because of some technical failure in the preparation or sampling of the tissue. For example, the cells that appear to lack nexal junctions might form such junctions transiently, or the area of junctional membrane might be so small relative to that of nonjunctional membrane as to escape detection at the usual thickness of randomly sectioned material. While it is too early to rule out technical problems in preparation or sampling, it should be pointed out that direct contact between cardiac cells appears to be unnecessary for their synchronous contraction. In monolayers of tissue-cultured embryonic mouse heart, synchronization of beating occurs when two cells previously beating asynchronously are mutually in contact with the same fibroblast-like cell (derived from human amnion) but do not directly touch each other (Goshima and Tonomura, 1969).

B. DISTRIBUTION IN A PARTICULAR TYPE OF MAMMALIAN CARDIAC CELL. It is of interest to consider the distribution of nexal junctions in electron micrographs of mammalian heart muscle. The following discussion deals primarily with mammalian ventricular muscle and sheep Purkinje fibers, chiefly because the information available about these tissues is most extensive. Mammalian atrial tissue will also be briefly considered.

Nexal contacts between the plasma membranes of ventricular myocardial cells were formerly thought to be limited to the transverse boundary between the cells, that is, to the intercalated disks of the classical morphologists. M. M. Dewey and Barr (1964) first showed the existence of lateral nexal contacts between guinea pig ventricular cells. Since then, it has become evident that the classic transverse boundary or intercalated disk does not extend in a straight line across the entire width of the cell.

Instead, after running transversely for a certain distance, the boundary "turns a corner," runs longitudinally for a variable distance, and then turns again to continue in its original transverse directions. The result is that the transverse boundary has a steplike course. As pointed out by Fawcett and McNutt (1969), the effect of the alternately transverse and longitudinal courses resembles the treads and risers of a staircase. One type of lateral nexal contact between cells occurs where the transverse boundary turns in a longtitudinal direction.

Since this chapter was submitted for publication, E. Page and Mc-Callister (*J. Ultrastruct. Res.*, 1973, in press) have made measurements on electron micrographs of the intercalated disks of rat left ventricular myocardial cells by a combination of morphometric and serial section techniques (Sjöstrand, 1958; Weibel *et al.*, 1966; Sitte, 1967; Mobley and Page, 1972). These measurements show that gap junctions at the transverse cell boundary completely transect the cells and are arranged in two mutually perpendicular planes. These planes lie above and beside the myofibrillar insertions and parallel to the direction of myofibrillar pull; this arrangement minimizes distortion of the gap junctional surface during contraction. Because of extensive folding, the membrane area at the disk is about ninefold greater than that of a flat surface in hearts from rats weighing >200 gm. Of this surface, 0.10–0.13 consists of gap junction. There is less folding in small (38 gm) rats, in which gap junction makes up 0.17 of the disk area. For a 100 μ long cell with a radius of 6.66 μ, the fractions of total plasma membrane made up of external sarcolemma, two transverse boundaries, and T system are, respectively, 0.52, 0.31, and 0.17; the surface/volume ratio of extracellular spaces lined by nongap junctional membrane at the disk is 183 μ^2/μ^3, a value much greater than that for the T system.

M. Dewey (1969) quotes Spira (1967) as having observed that lateral nexal junctions are more common between canine Purkinje fibers (from the false tendon) than between atrial cells; he states that in dog hearts lateral nexal contacts are in turn more extensive in atrial cells than in ventricular cells.

Mobley and Page (1972) have recently measured the nexal fraction of the total junctional area in sheep cardiac Purkinje fibers in which the extracellular space was infiltrated with lanthanum. As shown in Table III, the nexus occupies 17% of the junctional area between the tightly packed Purkinje cells which make up the cell bundle or "fiber." Preliminary measurements on myocardial cells from guinea pig atria suggest that the corresponding gap junctional fraction for this tissue is comparable to that in Purkinje fibers (L. P. McCallister and E. Page, unpublished observations).

TABLE III

STEREOLOGICAL MEASUREMENTS ON SHEEP CARDIAC PURKINJE FIBERS

Fraction of cell volume	
Myofibrils	0.234 ± 0.009
Mitochondria	0.103 ± 0.006
Nuclei	0.009 ± 0.002
Fraction of Purkinje cell bundle volume	
Interstitial spaces between adjacent Purkinje cells	0.0023 ± 0.0002
Fraction of intercellular surface area	
Nexus (gap junction)	0.17 ± 0.03
Desmosome	0.023 ± 0.006
Fascia adherens	0.014 ± 0.005
Class I membrane area per unit Purkinje cell volume for a cell	
bundle 100μ in diameter	$0.39 \ \mu^2/\mu^3$
Area of Class I membrane lining spaces between Purkinje cells	
per unit of interspace volume for a fiber bundle 100μ in	
diameter	$170 \ \mu^2/\mu^3$
Folding factor[a] of Class I membrane	
Membrane facing connective tissue sheath	1.82 ± 0.09
Membrane facing adjacent cells	3.6 ± 0.3

[a] Folding factor defined as the ratio of the area actually measured on the electron micrograph to the area of a smooth (fold-free) surface. Data compiled from Mobley and E. Page (1972).

In mammalian ventricular muscle, nexal contacts may not be confined to the two sites discussed so far—the transverse cell boundary at the intercalated disk between cells geometrically in series with each other and the areas of lateral apposition between parallel cells. A third area of nexal contacts has been described in the rat ventricle (Meddoff and Page, 1968) (Fig. 3). This tissue was fixed by perfusion of the coronary vessels with buffered osmium tetroxide—a fixative that stabilizes nexal junctions (Goodenough and Revel, 1970). Under these conditions, it was possible to demonstrate that fingerlike processes of plasma membrane extend from the transverse boundary at the intercalated disk into the T system of the adjacent cell. The extensions take place through extracellular channels that are oriented parallel to the long axis of the adjacent cell and connect the T system of that cell with the tissue interspaces at the intercalated disk. The extensions have an irregular surface, with projections varying in height from knobs to longer structures with stalks. At intervals, the membranes lining these projections form nexal junctions with the membranes lining the T system of the adjacent cell, as well as with other projections. The extensions and the nexal junctions they form occur most frequently in the T tubules nearest the intercalated disk (Fig. 3), but can also be seen in parts of the T system apparently remote from the disk. These observations indicate that nexal contacts between cardiac cells may be distributed over a wider area of the internal sarcolemma than previously suspected.

4. FINE STRUCTURE OF GAP JUNCTIONS

A. RELATIONSHIP TO MACULA ADHERENS AND FASCIA ADHERENS. Electron micrographs of tissue sections or of freeze-cleaved heart muscle show that only part of the transverse boundary between adjacent myocardial cells consists of nexal junctions. Much of the remainder of the transverse boundary is made up of two other junctional specializations, which, following the suggestion of Fawcett (1966), we shall refer to, respectively, as maculae adherentes, or desmosomes and fasciae adherentes. At both of these specialized regions, the plasma membranes of adjacent cells run parallel to each other; the appearance of these membranes in thin sections does not differ from that of typical nonjunctional sarcolemma; it may, however, have a different structure, since freeze-cleaved fascia adherens is reported to have a lower density of particles on its external surface than nonjunctional sarcolemma (McNutt, 1970). In thin sections, the membranes of the macula adherens and fascia adherens differ from each other and from the nexus with respect to (1) the closeness of approach between adjacent sarcolemmas, (2) the characteristics of the intermembrane space, and (3) the structure of the cytoplasm immediately adjacent to the cytoplasmic face of each membrane.

The fascia adherens accounts for part of the nonnexal area of the transverse cell boundary. The space between the outer leaflets of the two adjacent sarcolemmas is 250–350 Å wide and contains a finely fibrillar material. The cytoplasmic faces of the adjacent membranes are occupied by a thick layer of interwoven filaments which McNutt (1970) has called the filamentous mat. The mat serves as the insertion of the thin filaments from the terminal myofibrils of the cell. The structure of the macula adherens, or desmosome, in heart muscle is usually distinct from that of the fascia adherens. In some species, including the pigeon heart (McNutt, 1970), the two junctional areas are similar. Characteristically, the desmosomes lie between the insertions of the terminal myofibrils into the fasciae adherentes; at the macula adherens the intercellular space of about 300 Å is bisected by a densely staining layer, presumably made up of a fibrillar protein. The filamentous material that occupies the sarcoplasm immediately adjacent to the inner leaflet of the plasma membrane is usually more condensed than the filamentous mat of the fascia adherens. The fascia and the macula adherens do not form a continuous belt around the junction, and both are readily penetrated by extracellular markers like lanthanum or horseradish peroxidase.

Table III shows that the fractions of junctional area made up of desmosomes and fasciae adherentes are, respectively, 0.023 and 0.014 for the cells of sheep cardiac Purkinje fibers. About 79% of the junctional region consists of tissue interspaces lined by Class I membranes which are neither gap junctions, desmosomes, nor fasciae adherentes. The very small diameter of these interspaces accounts for their large surface to volume ratio of 170 μ^2/μ^3. It is apparent from Table III that the interspaces make up a minute fraction of the volume of the fiber; this fraction is about 1/50 that occupied by the T system in a ventricular myocardial cell (Table II).

B. SUBUNIT STRUCTURE OF THE NEXUS. The subunit structure of the nexus in heart muscle emerged from the examination of *en face* thin sections and freeze-fractured specimens, i.e., from tangential thin sections or from fractures parallel to the planes of the two membranes in the area of confrontation between them. Revel and Karnovsky (1967) introduced colloidal lanthanum, an electron opaque medium, into the interstitial space, whence it diffuses into the extracellular space at the maculae and fasciae adherentes. These investigators found that the extracellular lanthanum does not stop at the nexus; instead, it penetrates between the apposed plasma membranes at the nexus. In thin sections at right angles to the plane of confrontation between the membranes (i.e., the plane passing perpendicularly through the nexal complex from the cytoplasm of one cell to the cytoplasm of its neighbor), the lanthanum defines an approximately 20 Å wide gap between the external leaflets of the two adjacent membranes. In sections thus oriented, the gap outlined by lanthanum appears to extend through the entire length of the nexus. In this respect, the nexus differs from a tight junction, which represents a true seal; i.e., the tight junction prevents the penetration of lanthanum (and other extracellular tracers) between cells (Goodenough and Revel, 1970), while the gap junction allows such penetration.

When thin sections of the nexus are oriented *en face*, the lanthanum-containing gap is seen to have a more complex structure. The lanthanum delineates packed polygonal subunits. The spacing of the polygonal subunits is about 90–100 Å and their diameter is 70–75 Å. The subunits are themselves impermeable to lanthanum from the lanthanum-filled space that surrounds them. This space is 20–30 Å wide in the plane of the nexal membrane complex and extends 55–60 Å in the plane perpendicular to that membrane complex. In the center of each subunit, Revel and Karnovsky found an electron-opaque core with a diameter of 10 Å or less; the opacity to electrons presumably signifies that lanthanum has penetrated the core. From the data on lanthanum-treated mouse hearts, Revel and Karnovsky therefore concluded that the subunit is a hollow prism about 50 Å tall (the 50 Å is the sum of the thicknesses of two external leaflets, each about 15 Å wide and a gap 20 Å wide). The wall of the prism is 30–35 Å thick. The core is of particular interest, since it could conceivably represent the low-resistance channel between the cytoplasms of the two cells. The fact that the core is filled with lanthanum is, however, paradoxical; if lanthanum is, as it would seem to be, confined to the extracellular space and its extensions, one would not expect this tracer to be present in a region that should be inaccessible to extracellular tracers.

The technique of freeze-cleaving confirms at least part of the conclusions derived from lanthanum-treated thin sections (Goodenough and Revel, 1970; Chalcroft and Bullivant, 1970). Fracture faces through the nexus itself manifest two characteristic appearances. If the fracture is coplanar with the extracellular face of the membrane, the appearance is that of a lattice studded with polygonal particles; if the fracture is coplanar with the intracellular membrane face, the appearance is that of a lattice studded with polygonal pits or depressions. Goodenough and Revel have concluded that the pits correspond to the spaces between the particles but are not merely depressions molded by the adjacent particles. They suggested that on the basis of the available evidence, the subunits do not appear to be openings of channels passing through the junctional membranes, but that the evidence does not suffice to rule out this possibility.

Studies on preparations of nexuses from liver tissue negatively stained with sodium phosphotungstate confirm the results of lanthanum treatment and of freeze-cleaving in showing a polygonal lattice (Benedetti and Emmelot, 1968). The results of all three of these techniques make up the ultrastructural information about which current speculation on the low-resistance pathway between cardiac cells revolves. It should be emphasized that the ultrastructure of this pathway is still conjectural and that decisive ultrastructural evidence is as yet lacking. At least three alternative views are tenable on the basis of the data so far published on the subunit; (1) the low resistance pathway is the 10 Å core of the subunit demonstrated with lanthanum; (2) the pathway passes through the substance of the polygonal subunit; (3) the pathway passes through channels corresponding to the depressions seen in freeze-cleaved specimens. A pathway through the core would be an attractive explanation; a decision about the core presents the difficulty that such a channel should not be permeable to lanthanum from the extracellular space and yet it is found to be filled with lanthanum. A pathway through the substance of the subunit would entail assumptions about the nature of the subunit that have yet to be specified; and a pathway through channels corresponding to the depressions in freeze-cleaved tissue raises questions about the relationship of these depressions to the lanthanum-permeable 20 Å channels of the gap.

C. Extracellular Gap at the Nexus. At the nexus, the gap delineated by lanthanum in thin sections cut at right angles to the nexus is thus seen to be made up of the 20–30 Å spaces surrounding the subunits. It is entirely clear that the nexus does not constitute a beltlike seal to the diffusion of extracellular solutes comparable to the beltlike seal at the zonula occludens of epithelial cells. The nexus is not a seal for two reasons: (1) nexuses occur as plaques and not as continuous belts around the cell; (2) the gap appears to be permeable to extracellular solutes, at least as far as can be judged from the diffusion of lanthanum. The existence of nexuses as plaques and not as continuous belts means that extracellular solutes that are diffusing between the cells can utilize pathways outside and in parallel with the nexus, pathways which presumably have a lower resistance to electric current and to the diffusion of solutes than do the intercellular gaps at the nexus. At present there exists no information about the impedance properties or the diffusion coefficients of solutes within the gap. The fact that lanthanum diffuses into the gap suggests but does not prove conclusively that this gap is filled *in vivo* with electrolyte solution.

Certain of the properties of the gap have been examined in heart muscle and liver tissue. If the hearts of adult rats are perfused for 5–20 min with a solution in which the calcium concentration has been reduced to 0.005 mM, the membranes of adjacent cells separate at both the maculae and fasciae adherentes; however, the membranes of adjacent cells continue to adhere to each other at the nexuses (Muir, 1967). By contrast, M. Dewey (1969) reports that after exposure of guinea pig ventricular muscle to solutions whose osmolality is increased to three times isotonic with sucrose, rupture occurs at the nexuses but not at the maculae adherentes. Treatment of mouse liver with 60% acetone, 70% methanol, or 95% ethanol produces a characteristic change in the morphology of gap junctions (Goodenough and Revel, 1970). In thin sections, the gap disappears, and in freeze-cleaved or negatively stained tissue specimens, the polygonal subunit structure is no longer demonstrable. These effects on the morphology of the gap junction do not occur if the junction is first fixed with osmium tetroxide, which apparently stabilizes the gap structure. The organic reagents were found to extract several phospholipids, some neutral lipids, and some protein. Twenty millimolar EDTA and 6 M urea produced no detectable change in the structure of the gap.

In a description of apposed membranes studied in thin sections of fixed vertebrate brain, Brightman and Reese (1969) made a distinction between gap junctions and junctions which they call "labile appositions." Unlike gap junctions, labile appositions lacked any associated cytoplasmic fuzz. They appeared to consist of five layers with an exceptionally thick median layer. Brightman and Reese present evidence that such labile appositions may not be present *in vivo* and may therefore be artifacts of fixation. They found that labile appositions were very rare in brains fixed initially with osmium tetroxide and postfixed *en bloc* with uranyl acetate. The labile appositions described by Brightman and Reese resemble the pentalaminar junctions found by Meddoff and Page (1968) in the T system of rat hearts fixed with osmium tetroxide and postfixed *en bloc* with uranyl acetate. Since these junctions in the rat heart are demonstrable under conditions which do not favor the formation of the (presumptively artifactual) junctions in vertebrate brain, it seems more probable that the appositions in rat heart are in fact physiologically significant.

The interpretation of junctions of the type designated labile appositions by Brightman and Reese is also important in evaluating the significance of such junctions in mammalian fetal hearts. For example, small regions of focal membrane contact, sometimes without associated fasciae adherentes, are demonstrable in human fetal hearts at 9 weeks of gestation (McNutt, 1970) and in other fetal tissues (Hay, 1968).

III. Capacitive and Conductive Properties of Cardiac Cell Boundaries

A. Introduction

The cell boundary is characterized electrically by its conductance and capacitance. These are general terms to describe the ability of the cell boundary to store electric charge or to allow passage of charge. These properties influence any electrical process of the heart muscle cell, and to a large measure characterize the electrical process. Of particular importance to the present discussion is the role of these two properties as membrane markers; i.e., as markers that provide structural information about the cell and its functions. Section III,B will discuss some concepts related to capacitance and conductance as applied to biological systems as a basis for understanding the experimental work to be presented subsequently. A discussion of experimental studies on cardiac Purkinje fibers, ventricular muscle, atrial muscle, and tissue cultured cells will follow (Section III,C). Emphasis in the discussion of electrical properties will be placed on the Purkinje fiber (false tendon) only because more is known about this tissue. A table is included (Table IV) that tabulates the membrane measurements made in the various tissues studied.

B. Basic Concepts

1. CAPACITANCE

The capacitance in an electrostatic system is defined by the relation of the potential difference between two conductors and the amount of charge that can be held by the conductors.

$$Q = CV \tag{1}$$

where Q is charge and V is potential difference. C is the proportionality factor and is called capacitance. Capacitance may be expressed in coulombs per volt, but it has been given its own unit, the farad. Since in practice a farad is a very large amount of capacitance, measurements are usually in microfarads.

Equation (1) shows that the greater the amount of charge that can be stored up for each unit of voltage, the larger is the capacitance. Any device that stores charge is acting as a capacitor. One example of a capacitor is two parallel conductive plates separated by a dielectric. A dielectric is an insulator, or a material that does not allow free movement of electrons. The charges in a dielectric are tightly bound, and their movement is usually restricted to a rotation of the charges as a dipole. The ease of their rotation is described by the dielectric constant.

For the parallel plate capacitor, the relationship between its ability to store charge, the dielectric between its plates and its geometry is given by

$$C = \frac{k_0 k A}{d} \tag{2}$$

where k_0 is the permittivity of free space (this is simply a constant), k is the dielectric constant, A is the area of either plate in square meters,

TABLE IV

Passive Electrical Properties of Cardiac Muscle[a]

Tissue	Animal	R_i (Ω cm)	R_m (Ω cm²)	C_{total} (μF/cm²)	C_m (μF/cm²)	C_{foot} (μF/cm²)	C_s (μF/cm²)	R_s (Ω cm²)	R_{nl} (Ω cm²)	C_{nl} (μF/cm²)	Source
Purkinje	Kid	105	1,940	12.4	—	—	—	—	—	—	Weidmann (1952)
	Sheep	118	1,714	12.8	2.4	2.4	7.0	300	—	—	Fozzard (1966)
	Sheep	—	—	12.0	4	—	8	150	—	—	Dudel et al. (1966)
	Sheep	—	2,039	8.5	2.5	—	6.0	336	0.18	360	Freygang and Trautwein (1970)
	Sheep	181	—	—	—	—	—	—	—	—	Weidmann (1970)
	Sheep	—	—	—	3.0	—	—	—	—	—	Schoenberg and Fozzard (1971)
Ventricle	Dog	—	3,500	1.0	—	0.76	—	—	—	—	Sakamoto (1969)
	Sheep and calf	470	9,100	0.81	—	0.59	—	—	—	—	Weidmann (1970)
	Calf	—	14,000	—	—	—	—	—	3.0	—	Weidmann (1966)
	Dog	—	—	1-2	—	—	—	—	—	—	Beeler and Reuter (1970)
Atrium	Frog	2000	120	25	—	—	—	—	36	—	Woodbury and Gordon (1965)
	Rabbit	—	—	—	—	1.3	—	—	—	—	Paes de Carvalho et al. (1969)
Tissue culture	Chick	—	480	20	—	—	—	—	—	—	Sperelakis and Lehmkuhl (1964)

[a] R_i = specific resistance of the myoplasm; R_m = membrane dc resistance; C_m = total specific membrane capacitance; C_{total} = total specific membrane capacitance of the membrane in parallel with membrane resistance; C_{foot} = capacitance as calculated from the foot of the action potential; C_s = part of the total capacitance that has a resistance R_s in series with it (using the model in Fig. 4A); R_{nl} = total dc resistance of two apposing transverse cell boundaries referenced to 1 cm² of estimated transverse membrane; C_{nl} = capacitance associated with the two apposing transverse cell boundaries. The table does not include all measurements that have been reported, but simply lists those used in this chapter. These values depend heavily on the assumptions used in their calculation, and they should be interpreted in that context.

and d is the distance between the plates in meters. From Eq. (2) it is evident that the capacitance (charge storage) is larger, (1) the greater the dielectric constant, (2) the greater the area of the plates, (3) the smaller the distance separating the charges. For more detailed discussion, the reader is referred to a physics text (e.g., L. Page and Adams, 1958).

Using this simple relationship, Fricke (1925) made the first calculation of the thickness of cell membranes. He measured the charge storage property of red blood cell membranes and obtained a value for capacitance of 0.8 μF/cm^2. Then he assumed a dielectric constant of 3 (a representative value for the lipids that were thought to make up the membrane) and calculated that the charge separation was 22 Å. This value for charge separation is about the length of a molecule of 20–30 carbon atoms and is of the same order of magnitude as one would expect from electron microscopy of membranes. Similar values for capacitance (around 1 μF/cm^2) have been obtained on a wide variety of biological membranes, a remarkable result considering the variation in the permeability properties of membranes and, presumably, in their detailed structure.

Membrane capacitance has been measured by determining either the time constant or the phase angle. The time constant is a measure of how rapidly a linear system containing a capacitance C and a resistance R in parallel comes to a new steady state after a step change. Usually the step change employed is a step in current. If so, the new equilibrium voltage is not established instantaneously, but with a lag resulting from the time necessary to fill or to discharge the capacitance. This time constant has the form of the product $R \times C$, but its experimental meaning is related to the geometry of the tissue and the nature of its equivalent circuit. The test step involved may also be one in voltage, as discussed later. If the perturbation employed to study the capacitance of a system is alternating current, the factor measured is a phase angle. This represents the amount by which the current supplied leads the voltage produced across the capacitor. If there is no capacitance, the phase angle would be zero, and if there is a purely capacitative impedence (i.e., a capacitance without associated resistance) the angle would be 90°.

2. CONDUCTANCE

To facilitate development of the concept of conductance, its electrical counterpart, resistance, will be described first. The resistance in an electrostatic system is defined by the relation between the potential difference (the electrical driving force) and the current (the movement of charge). This is, of course, Ohm's law. The unit of resistance is volts per ampere and is called an ohm. Ordinarily, the resistive property or resistivity of a material, such as a homogeneous ionic solution, is measured by the current that flows between two parallel plates at different voltages, the plates being 1 cm² in area, and separated by 1 cm, so that the current passes through 1 cm³ of the material being studied. The unit of resistivity is the ohm-centimeter. In the case of membranes, where the resistive barrier is a thin sheet, such a measurement is not feasible, so resistivity is related to 1 cm² of membrane, or Ω cm². In using the resistive feature of membranes as a marker, it is important to bear in mind that 2 cm² of membrane have half the effective resistance of 1 cm².

Conductance G is often a more convenient way to think about the ability of charge to move in an electric field. It represents simply the inverse of resistance.

$$G = 1/R \tag{3}$$

where R is measured in ohms, G is measured in mhos. Conductivity is also the inverse of resistivity, and for membranes this is given in mhos per square centimeter. Conceptually, conductance indicates the ease with which charges move in an electric field; resistance indicates the difficulty with which they move. A convenience of conductance is that it is additive: 2 cm² of membrane have twice the conductance of 1 cm².

In biological systems, current is usually carried by translocation of ions. A general discussion of bioelectricity is beyond the limits of this chapter; the reader is referred to a recent review (Woodbury *et al.,* 1970). How ions move through the cell membrane is determined by its molecular structure, the details of which are only speculative at this time. The major concepts being considered as models of membrane ion transport include movement of ions through specialized regions of relatively high permeability (pores or channels), binding to fixed charges in the membrane (ion exchange), or binding to mobile organic carriers (neutral or charged). At present, the experimental results do not lend themselves unequivocably to a single model of the molecular events (Cole, 1965; Eisenman, 1968; Tosteson *et al.,* 1968).

The simplest indicator of cell boundaries and their function is the linear resistive or conductive property of the membrane. The use of this indicator presumes that the movement of ions across the membrane in question is a linear function of the driving force operating on these ions and avoids the problem of the molecular mechanism involved.

C. Membrane Capacitance and Conductance

1. PURKINJE FIBERS

A. MEASUREMENTS OF MEMBRANE CAPACITANCE. Although Purkinje fibers cannot be considered typical cardiac cells, their role in synchronizing activation of excitation in the ventricles makes their electrical behavior a matter of considerable importance. From the electrophysiological viewpoint, the analysis of Purkinje fibers is not as complex as that of ventricular cells, at least to a first approximation. We shall see later that the apparent simplicity of Purkinje fibers is deceptive. Because of their apparent simplicity, most of the information on capacitance and conductance in heart muscle has been derived from Purkinje fibers.

The first quantitative experiments were those of Weidmann (1952). Weidmann performed a cable analysis on Purkinje fibers using the method previously developed by Hodgkin and Rushton (1946) for nerve, but with use of intracellular electrodes. This technique assumed that the Purkinje fiber can be treated as a right circular cylinder in which the surface of the cylinder is the only limiting membrane for current flow. It also required that the cylinder be infinitely long relative to its diameter. By demonstrating that a cut end of the fibers would "seal" and develop a high electrical resistance (for more details, see Section III,C,1,d), Weidmann was able to modify this requirement and to apply the Hodgkin–Rushton technique to the case of a semiinfinite cylinder. Using square current steps, he determined the apparent membrane capacitance to be 12 $\mu F/cm^2$, a value about one order of magnitude greater than that in nerve and most biological tissues.

This large value for membrane capacitance posed several questions. First, since it is assumed that the charge separation of a membrane is similar to that of a parallel plate capacitor, does it follow that the dielectric constant of cardiac membrane is twelve times that of nerve? This would make the membrane dielectric constant approach that of water. Or is the charge separation only one-twelfth of that in nerve? This would be in the range of 5 Å, probably an unreasonable value. It is also possible that calculation of the area of the membrane is not accurate. Weidmann pointed out that the surface of the fiber appeared to be folded when it was viewed in the light microscope. He estimated that this folding might double the actual surface membrane area, but this would still leave the specific membrane capacitance far too large.

A charge storage capacity of 12 μF/cm^2 in the Purkinje fiber membrane would have an important influence on the excitatory process that generates the upstroke of the action potential. The action potential is generated by entry of positively charged ions from outside the membrane barrier to the inside of the barrier. The amount of positively charged ions required would therefore be greatly increased by the large capacitance. The minimal amount required can be approximated by use of Eq. (1). Assuming the magnitude of the action potential to be 120 mV and the capacitance to be 12 μF/cm^2, the charge required would be

$$(1.2 \times 10^{-1} \text{ V})(1.2 \times 10^{-5} \text{ F}) = 1.44 \times 10^{-6} \text{ coulombs/cm}^2$$

A mole of ions with single positive charges contains 96,500 coulombs (the Faraday). The required amount of univalent cation is thus

$$(1.44 \times 10^{-6})/(9.65 \times 10^4) = 1.4 \times 10^{-11} \text{ moles/cm}^2$$

or 14 picomoles/cm^2 of membrane. This is twelve times the charge and twelve times the amount of univalent cation needed for squid axon membrane. Since the rates of depolarization during the upstroke of the action potential in the Purkinje fiber and the squid axon are approximately the same, the intensity of the inward current must be correspondingly greater in the Purkinje fiber. While it has been difficult to obtain quantitative information on the fast sodium entry system in Purkinje fibers, the available results do not extrapolate to such high values (Dudel *et al.*, 1966), nor have computations of the action potential using reasonable values yielded a sufficiently high rate of depolarization (Noble, 1962a).

An additional consequence of the large apparent membrane capacitance is that the conduction velocity of the calculated Purkinje action potential has been too low (Noble, 1962b). Assuming that the Purkinje fiber is reasonably approximated by a cable, it is possible to measure how much capacitance is being charged during the onset of depolarization of the action potential (Hodgkin and Huxley, 1952; Tasaki and Hagiwara, 1957). This method is dependent on the fact that the source of the action potential can be considered constant if the action potential is conducted at a constant rate. The action potential is then preceded by a passive depolarization that extends ahead of it and is called a foot. The foot can be described by the exponential relation

$$V(t) = A e^{Kt} \tag{4}$$

where $V(t)$ is the voltage time course of the foot, A is an arbitrary constant, t is time, and K is defined (within a small error) as

$$K = \frac{2R_i C_m \theta^2}{a} \tag{5}$$

where R_i is resistivity of the myoplasm in ohm-centimeters, C_m is specific capacitance of the membrane in microfarads per square centimeter, θ is constant propagation velocity of the action potential in centimeters per second, and a is the radius of the fiber in centimeters. K, which can be measured as the inverse of the time constant of the action potential foot, was found to be about 1×10^4 sec^{-1} for a Purkinje fiber (Fozzard, 1966). Associated measurements of the internal resistance, conduction velocity, and radius, of the fiber permitted the calculation that the capacitance apparently filled by the advancing front of the Purkinje action potential is 2–3 μF/cm^2 (Fozzard, 1969). The discrepancy between this measure of membrane capacitance and the 12 μF/cm^2 found using a square wave technique will be discussed later.

B. Measurements of Membrane Conductance. At the same time that he measured the membrane capacitance, Weidmann (1952) measured the resistance of the membrane. As we noted previously, this technique depends on the assumption that the Purkinje fiber can be treated as a right circular cylinder that is long relative to its diameter, and whose limiting membrane (for inside–outside currents) is given by its cylindrical surface only. Weidmann partially justified this assumption by examining the decline in steady-state voltage with distance from a microelectrode that passed a small, steady, hyperpolarizing current. He found that the decline in potential was exponential, as predicted by the one-dimensional cable equation (Hodgkin and Rushton, 1946). Measurement of the space constant (the distance required for the steady state voltage to fall to $1/e$ of its value at the input electrode), of the diameter of the fiber and of its input resistance made it possible for Weidmann to calculate a specific membrane resistance for the fiber. The value he obtained was 1940 Ω cm^2 (51×10^{-5} mho/cm^2). These figures have been confirmed by other investigators (Fozzard, 1966; Reuter, 1967). It must be emphasized that this calculation is based on the assumption that the obstacle to current flow from within the cell column is localized to the surface of the column, and that this surface is not folded or invaginated. If the actual surface area is larger than assumed in this calculation, the membrane specific resistivity would be correspondingly higher, and the specific conductivity would be correspondingly lower. The recent determination of Mobley and Page

(1972) that the surface membrane is more than 10 times that predicted by the right circular cylinder model supports this suggestion, and results in values for specific membrane resistance of 20,000 Ωcm^2.

C. INTERPRETATION OF MEMBRANE PROPERTIES BY CIRCUIT ANALOGS. The large discrepancy between the membrane capacitance of 2–3 μF/cm^2 derived from the foot of the action potential and that of 12 μF/cm^2 found by Weidmann with a method using a square current step raised the possibility that much of the capacitance measured with Weidmann's technique might be associated with membranes that are not localized to the surface of a right circular cylinder—perhaps with membranes within the cylinder. A typical fiber is a multicellular bundle, with each cell about 100 μ in length and perhaps 20–30 μ in cross section. Little resistance could be attributed to the transverse boundaries of the cells at the intercalated disks (Weidmann, 1952). However, it was possible that the longitudinal cell boundaries face intercellular clefts that are connected to the outside (Caesar *et al.*, 1958; E. Page, 1967b); these longitudinal membranes might therefore have a transmembrane potential difference across them. A second possibility for additional membrane responsible for the large apparent capacitance is the T system (Section II,C,2), which represents the invaginations of the external sarcolemma. Although a T system has been clearly shown in skeletal and cardiac muscle (Huxley, 1964; Simpson and Oertelis, 1962), it is probably sparse or absent in the kind of Purkinje fibers used for the determination of membrane capacitance. These two possibilities represent the two types of internal sarcolemma defined in Section II. The internal membranes might not participate in the upstroke of the propagated action potential, because the narrow extracellular spaces connecting them to the outside could constitute sufficient resistance to produce a time lag before they were depolarized. On the basis of such considerations, the membrane capacitance may be separated into two parts (Fig. 4A) denoted C_m and C_s. C_m is in parallel with the membrane resistance R_m. This two time constant model was used by Falk and Fatt (1964) to separate membrane components by examining the response to varying frequencies of alternating current. At low frequencies, the apparent capacitance would be $(C_m + C_s)$, and at high frequencies it would approach C_m. Careful study with extracellular and with intracellular electrodes revealed that for frog skeletal muscle the equivalent circuit in Fig. 4A agreed reasonably well with experimental observations. The actual values found for frog skeletal muscle were $C_m = 2.6$ μF/cm^2, $C_s = 4.1$ μF/cm^2, and $R_s = 330$ Ω cm^2 (all expressed with reference to the apparent surface area of the fiber calculated by assuming a right circular cylinder). C_m could be

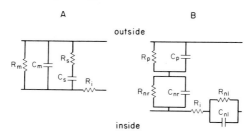

Fig. 4. Two possible equivalent circuits used for analysis of the membrane properties of cardiac Purkinje fibers. The symbols are the same as in Table IV.

related to the surface membrane (external sarcolemma), and C_s to the transverse tubular membrane. R_s appeared to be a lumped resistance; the lumped resistance was not well explained by the fluid in the tubular spaces, but was probably inside the cell. The structural equivalent of R_s in skeletal muscle is not yet completely settled. Several studies of passive properties suggest that it may be entirely explained by the resistance that would occur in the lumina of a network of transverse tubules (Schneider, 1970; Adrian, Chandler and Hodgkin, 1969; Hodgkin and Nakajima, 1972). However, the tubular network does not account for all properties of the R_s, and there may also be a lumped "access" resistance, perhaps at the mouth of the tubules (Eisenberg *et al.*, 1972).

Another way of evaluating the equivalent circuit is by a voltage step. A technique is available for achievement of reasonably uniform voltage control along a 1 mm segment of Purkinje fiber (voltage clamp). Application of this technique to record the capacitive transient in the Purkinje fiber (Fozzard, 1966) showed that two phases of capacitive current flowed in response to the voltage step. The first phase was equivalent to the charging of a capacitor of about 2 $\mu F/cm^2$ and had to be completed before the transmembrane voltage change was achieved. An additional capacitive current flowed with a time constant of about 2 msec. Analysis of these two capacitive currents according to the circuit of Falk and Fatt (Fig. 4A) indicated that C_m was 2.4 $\mu F/cm^2$, C_s was 7 $\mu F/cm^2$, and R_s was 300 Ω cm^2. Substitution of sucrose for sodium chloride in the external solution resulted in a marked increase in R_s, suggesting that R_s was at least partly localized in the intercellular space. At the time it was first proposed, the most convenient explanation of this equivalent circuit was one similar to that proposed by Falk and Fatt for skeletal muscle; i.e., to assume that the C_s in heart muscle was the capacitance of the transverse tubules. Subsequent ultrastructural observations (E. Page *et al.*, 1969; Mobley and Page, 1972) favor placing this capacitance in clefts between cells, with the possibility of an additional contribution from specialized areas of the longitudinal cell surface. This equivalent circuit of two time constants also would allow the rapid upstroke of the action potential to occur with less entry of positive charge and permit conduction at the measured rates without unusual requirements for inward current intensity. Additionally, this circuit would favor the rapid repolarization associated with the initial upstroke of the action potential in the Purkinje fiber, and it might even contribute to the notch often seen during this part of the action potential (McAllister, 1968). Although the circuit favors conduction in the Purkinje fiber, it also makes voltage clamping in that preparation more difficult. The voltage clamp is important because it has been the most useful tool so far found for studying the ionic events underlying the action potential. The potential difference across the part of the membrane in series with the resistance R_s is not well controlled by the voltage clamp. At best, it may be several milliseconds before this part of the membrane is completely depolarized, and at worst, its voltage may never be uniform (McAllister, 1969).

Subsequently, this complex capacitance has been seen by others (Dudel *et al.*, 1966). It has been carefully restudied by Freygang and Trautwein (1970) using alternating current and intracellular microelectrodes. These latter investigators addressed themselves to two questions: (1) In light of the paucity or absence of transverse tubules in Purkinje fibers, can the radial complex capacitance be explained by current passing serially through two cell membranes? (2) Is there any capacitive component to the flow of current longitudinally along the fiber axis?

Another possible equivalent circuit for radial current is shown in Fig. 4B. Here R_p and C_p represent the surface membrane, and R_{nr} and C_{nr} represent the second cell membrane equivalent. The second equivalent would have different values even if the characteristics of the membranes involved were the same, because their area (and therefore the effective resistance and capacitance) would be different from that of the equivalent circuit shown in Fig. 4A. Under resting conditions, there would presumably be no voltage difference across the circuit (R_{nr} and C_{nr}), but radial currents would have to flow through it to reach the surface membrane and the intercellular space. The two models shown in Fig. 4 are electrically identical; i.e., electrical measurements across them (inside–outside) can never distinguish between them. Furthermore, each model represents the minimal number of circuit elements that can be used to represent the total circuit, and therefore each model represents lumped resistances, capacitances, and series resistance–capacitance units. For this reason, an exact relation between measurements of these lumped values and morphologically identifiable membranes is not to be expected, although useful approximations may be found. For example, a more accurate circuit for that shown in Fig. 4A might be that shown in Fig. 5A, but this circuit would reduce to that in Fig. 4A. The measurements of Freygang and Trautwein confirmed the previous description of the behavior of radial currents. If analyzed according to the circuit in Fig. 4A, values obtained by Freygang and Trautwein were the same as those found by Fozzard (1966). If analyzed according to the circuit in Fig. 4B (and with the assumption that R_p is 1700 Ω cm^2), $R_{nr} = 240$ Ω cm^2, $C_p = 9$ μF/cm^2, and $C_{nr} = 0.3$ μF/cm^2; again, all values are related to surface area of the fiber calculated from overall length and diameter and assuming a right circular cylinder. To distinguish between circuit models in Fig. 4, it would be necessary to make a known alteration in an element of the circuit. Freygang and Trautwein attempted to accomplish this by altering the external potassium ion concentration (to change the membrane resistance), by treatment with cocaine (to reduce the membrane resistance), and by substitution of sucrose for sodium chloride (to increase the resistance of any extracellular current path). Unfortunately, the results are not subject to a unique interpretation; they fit both circuit models in Fig. 4 about equally well.

The membrane models in Fig. 4 are also useful for the understanding of excitation. For a current to excite the muscle membrane, it must alter the membrane voltage sufficiently to bring the membrane into a voltage region where the conductances are changed regeneratively. Less current is needed to achieve the excitation if the resting membrane conductance is lower ($V = I/G$) and if the charge storage of the membrane (i.e., its capacitance) is smaller. The classic way of describing

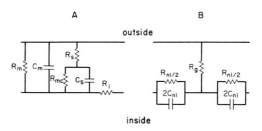

Fig. 5. (A) Modification of the equivalent circuit shown in Fig 4A to resemble more closely one possible anatomical arrangement. In this model, R_{mc} is the resistance of the membrane indicated by the capacitance C_s. R_m, C_m, R_s, and C_s are altered by this change from the values they would have in the circuit in Fig. 4A, but a simple relationship exists between them. (B) An equivalent circuit used in analysis of current flow between two cardiac cells. R_g is the resistance of the leakage pathway to the outside between transverse boundaries of two cells at the intercalated disk.

the response to a stimulating current is by a strength–duration curve. The strength–duration curve is determined by passing currents of various durations and strengths and plotting for each duration the strength of the current just sufficient to excite an action potential. If the membrane is represented by only a resistance and a capacitance (R_m and C_m), as appears to be true for nerve, then the amount of charge required to bring the membrane to a threshold voltage for excitation (current times its duration) will approach a constant value with short stimuli. This expectation is approximately true for nerve, and the curve corresponds to the equation

$$I = I_{Rh}/(1 - e^{-t/\tau}) \qquad (6)$$

where I is the current required for excitation for any pulse of duration t, I_{Rh} is rheobasic current, and τ is defined as

$$\tau = \lim_{t \to 0} \frac{It}{I_{Rh}} \qquad (7)$$

If, on the other hand, the surface membrane (C_m or C_p) can be brought rapidly to a threshold voltage value with discharge of less capacitance, as in either Fig. 4A or 4B, then the charge requirement for excitation ought not to approach a constant value at short times. On the contrary, it should fall to a value for charge that is equivalent to the discharge of C_m (if the circuit model in Fig. 4A is used).

Experiments on very short Purkinje fibers, to reduce the complications of cable geometry as much as possible, show that the charge required is fairly uniform until very short times. Further, the amount of charge is consistent with depolarization of all of the membrane capacitance. (Schoenberg and Fozzard, 1971). Why the fibers behaved as predicted from the simple R-C model is not entirely clear. One possible explanation is that the covert membrane is not simply a passive membrane that would absorb charge to be depolarized, but is capable of contributing to the action potential by conductance change in the same fashion as the surface membrane.

From the recent work of Mobley and Page (1972), it now appears possible to interrpret the electrical measurements in terms of structure (see Table III). The total surface membrane is approximately 12 times the value that would be calculated by the right circular cylinder model. Approximately one-fourth of it faces the outside of the fiber bundle, and corresponds to the overt membrane which is not in series with any resistive path other than the bulk extracellular space. The other three-fourths of the surface membrane of the cells faces narrow longitudinal clefts between cells within the bundle, and represents the covert membrane. This membrane shows no special features that would distinguish it from any other part of the cell membrane, and it might be expected to participate in the action potential. This covert membrane is connected to the bulk extracellular space through the narrow clefts that appear to be filled with extracellular solution. The cleft would constitute a resistance in series with this membrane, and the resistance would be distributed in much the same fashion as has been suggested for skeletal muscle.

D. PROPERTIES OF TRANSVERSE CELL BOUNDARIES. Physiologists have long believed that heart muscle is an electrical syncytium. Their reason was simply that if one stimulates any part of a piece of heart muscle, the electrical event is rapidly conducted throughout the rest of the muscle. Shortly after micropipettes that could be introduced into a cell without damaging it became available, Weidmann (1952) measured the resistance to the passage of current along a Purkinje column. By cable analysis, he could determine that the specific internal resistivity was about 100–150 Ω cm. For the purpose of this measurement, the resistances of the myoplasm and the transverse cell boundaries are lumped together. The specific resistivity thus obtained was not much different from that of the Tyrode solution used to bathe the tissue (51 Ω cm). This value was also lower than that measured by Katz (1948) for skeletal muscle, a cell type that is thought to lack transverse membrane boundaries. The measurements of internal resistance in Purkinje fibers were confirmed by Fozzard (1966), and by Weidmann himself (1970) using a different technique. The low resistivity of the myoplasm and the transverse cell boundaries suggests that these transverse cell boundaries must have a much higher conductance than surface membranes. But if internal membranes make up a part of the radial equivalent circuit of the Purkinje fiber (see preceding section), the membranes at the transverse cell boundaries ought to represent similar resistance–capacitance (R–C) elements. Quantitative considerations (Fozzard, 1966) appear to rule out the possibility that transverse cell membranes are the only basis of the complex circuit (e.g., of R_s and C_s in the Fig. 4A model). However, in Fozzard's study it was not possible to exclude any additional capacitive element in longitudinal current circuit. Freygang and Trautwein (1970) used external electrodes to perform an alternating current analysis of longitudinal currents in Purkinje fibers. They found a capacitance as well as a resistance, represented by a time constant of 60 μsec, that might be ascribed to the transverse membranes. If the transverse cell boundary is assumed to be a flat plate that crosses the fiber at right angles to the surface about every 100 μ (an approximation comparable to that of the right circular cylinder), the resistance of this boundary would correspond to 0.18 Ω cm² and the capacitance to 360 μF/cm². If the membrane is folded, so that the effective area is really ten times this crude estimate, then the results would be 18 Ω cm² and 36 μF/cm², respectively.

Whatever the actual values, it seems reasonable to conclude that there are resistive and capacitive couplings between Purkinje cells. Such couplings almost certainly correspond to the transverse boundaries at the intercalated disks. The more important question is how these couplings relate to longitudinal conduction of the action potential. In the case of the Fig. 4A model, the capacitance that must be discharged quickly by local circuit currents is predominantly C_m with very little charge requirement by C_s. The reason is that the upstroke of the action potential is a rapid, or high frequency, event. The change in voltage across C_s or the discharge of C_s, would lag behind, and it would be only half finished by the time of the peak of the action potential. The value for C_m and the capacitance filled by the foot (page 123) are about equal. For conduction in the Fig. 4B model, the elements R_{nr} and C_{nr} are not involved, because longitudinal current flow is parallel to such internal membranes. The circuit would contain R_p and C_p and R_{nl} and C_{nl}. Intuitive analysis of this circuit is not so obvious as in Fig. 4A, but one can see that at high frequencies, R_{nl} would be shunted by C_{nl}; the impedance to longitudinal current flow would therefore approach R_i. This reduction in longitudinal impedance would tend to speed conduction in spite of the higher value of surface membrane capacitance C_p. Other factors being unchanged, the effect would be limited by how much internal impedance changed, and this change would be from $(R_i + R_{nl})$ at d.c. to R_i at high frequencies.

The experiments of Délèze (1970) further illustrate that the cells in the Purkinje fiber are electrically continuous across the transverse cell boundaries. His experiments are concerned with the electrical response of the fiber to membrane injury. If a hole is made in the surface membrane of the fiber, a low resistance path is established. As a result, current flows into the cell interior (which is initially at a negative potential), thereby depolarizing the cell. Délèze injured the membrane by cutting or by coagulation with a laser. He observed that under these conditions the membrane potential at distances of more than 1 mm fell to less negative values. If calcium ion was present in the bathing solution, the potential returned to its previous level within several minutes. Measurement of input resistance showed that injury did indeed produce a low resistance shunt across the cell membrane, and that recovery was associated with a restoration or increase of the input resistance.

These results are consistent with the interpretation that the transverse boundaries of Purkinje fibers normally have a high conductance. Initially the shunt affects cells distant from the site of injury. Subsequently the low-resistance shunt to the extracellular space disappears, perhaps by a resealing of the damaged membrane or by formation of a new high-resistance transverse boundary near the site of injury. A new high-resistance transverse boundary might arise by the conversion of a previously high-conductance membrane into one characterized by a low conductance. A possible mechanism for this conversion is the exposure of the membrane to the high (extracellular) concentration of ionized calcium. However, there is as yet no direct evidence to establish whether the new electrical boundary is at the old transverse cell boundary or whether it is at a newly formed membrane (see Baldwin, 1970).

2. Ventricular Muscle

A. The Geometrical Problem. Quantitative measurements of membrane capacitance and conductance have proved to be even more difficult in ventricular muscle than in Purkinje fibers. First, the anatomy of ventricular muscle is complex, and a simple assumption to calculate its surface area, comparable to the right circular cylinder assumption for Purkinje fibers, is not available (but see Table II). The result is that measurements of specific membrane capacitance and conductance are even less reliable than in Purkinje fibers and are difficult to compare with other measurements. Second, introduction of a well defined electrical perturbation (e.g., a current step) has been difficult because of two- or three-dimensional spread of the current. It was shown by Noble (1962c) that if a two-dimensional model is used to evaluate the effect of a point source of current (corresponding to two parallel plates of membrane extending in all directions for distances large relative to the separation of the plates), the influence of membrane properties on voltage distribution is quite small.

Noble's model might be applicable to atrial muscle (Section III,C,3), but it is probable that in ventricular muscle there is three-dimensional spread of current from a point source. The problem of three-dimensional spread is discussed thoroughly by Eisenberg and Johnson (1970). By an ingenious technique, Tanaka and Sasaki (1966) demonstrated the three-dimensional spread of current from an intracellular microelectrode in ventricular muscle of the mouse. These investigators introduced a current-carrying micropipette into the cells; through this pipette they then inserted a smaller micropipette for recording of potentials. They could thus sample the potential at various radial distances from the current source. In this way, they showed that the potential falls as one would expect from three-dimensional spread of current.

There are ways to approximate the one-dimensional cable case for ventricular muscle. Some portions of the ventricle (trabeculae and, to a lesser degree, papillary muscles) are composed of cells that lie approximately parallel to each other. If each end of such a preparation is placed in a separate chamber, so that the muscle is the only connection between the two chambers, current can be made to flow fairly uniformly through the muscle. This arrangement is known as a gap. Usually the muscle passes between the two chambers through a hole in the partition between the chambers. To prevent or reduce flow of current between the chambers through the extracellular space around the muscle, the gap may simply be made tight by fitting and insulating with petrolatum, or it may be a separate chamber perfused with a solution of an isosmotic nonelectrolyte (e.g., sucrose). Depending on the magnitude of the extracellular leak of current, the measurements will be more or less quantitative.

The gap technique would at least in theory produce a uniform intracellular voltage within a cross section of the muscle in the plane of its exit from the gap. If this were so, and if the muscle were composed of uniform cell columns running parallel to one another at right angles to the gap, no currents would flow radially; moreover, the temporal and spatial distribution of voltage within the muscle would be as described by the one-dimensional cable theory used for analysis of the properties of Purkinje fibers. Furthermore, the surface area of the cells could be approximated by assuming an average radius for the cell columns, an average cell area of cross section, and cell membranes without folds. In reality the, cell area and radius of ventricular cells are not uniform. In addition, the cell membranes are not smooth, but folded and invaginated, and the geometry is complicated by the existence of a T system (Section II,C,2). Nevertheless, the predictions based on these assumed approximations can be tested experimentally and the errors introduced by them can be examined.

B. LONGITUDINAL MEMBRANE CONDUCTANCE. Two approaches have been used to examine longitudinal and transverse membrane conductances. The first uses extracellular electrodes to pass current into parallel cell columns, as described above. The second is based on the longitudinal diffusion of radioactive potassium ion along a fiber divided by a gap.

Sakamoto (1969) used a petrolatum gap to isolate two chambers and measured the membrane potential along the ventricular fibers at various distances from the gap. The decline in potential with distance during a constant current pulse was approximately exponential, as predicted by the one-dimensional cable theory, with a length constant of 1.2 mm. Assuming that the cells had an average radius of 8 μ, and that the myoplasmic resistance was 100 Ω cm (Weidmann, 1952), Sakamoto calculated that specific membrane resistance was 3500 Ω cm^2. The calculation also included the assumption that the cell membrane was not folded or invaginated.

Using an extracellular electrode technique with extracellular resistance made large, Weidmann (1970) measured the spatial decrement of potential in trabecular muscle and found the length constant to be 880 μ. Employing the same assumptions about cell uniformity and membrane area as Sakamoto, he calculated that membrane resistance was 9100 Ω cm^2. The larger value for membrane resistance in Weidmann's experiments was attributable partly to the fact that the measurements gave a calculated myoplasmic resistance of 470 Ω cm.

Weidmann (1966) approached the measurement of membrane properties by an independent method using radioactive potassium ion. He found that intracellular ^{42}K diffuses farther in the longitudinal direction of the cells than in the radial direction, as if the potassium were able to move intracellularly and across the transverse cell boundary more easily than through the surface membranes. A close analogy can be made between the radioactive potassium distribution and potential distribution after onset of a constant current pulse. From this analogy one can determine a length constant of 1.55 mm. Assuming the same cell properties as mentioned above, and also that the resting membrane conductance is entirely due to potassium ions, Weidmann calculated the membrane resistance to be 14,000 Ω cm^2. Again, any folding or invagination of the surface membrane would result in an increase in this calculated value.

C. TRANSVERSE CELL BOUNDARY CONDUCTANCE. The spread of electronic potential in ventricular muscle (e.g., 1.2 mm) under conditions approximating a one-dimensional cable is similar to that in Purkinje fibers. To explain this similarity, it is necessary that current in ventricular muscle flow more readily across transverse than across longitudinal cell boundaries, since the observed spread greatly exceeds the average cell length (assumed to be approximately 100 μ). Weidmann (1970) did find a much higher myoplasmic resistivity in ventricular muscle (470 Ω cm) than in Purkinje fibers (about 150 Ω cm). The difference could have been the result of a technical problem in the measurement. However, a technical problem seems improbable in view of the fact that in the same apparatus Weidmann found a value of 181 Ω cm for the myoplasmic resistivity of Purkinje fibers. An additional technical problem is that the cells in the trabecular muscle used by Weidmann varied in size; their nonuniform diameter would have the effect of making myoplasmic resistance appear higher. However, if Weidmann's value is accurate, it may indicate that the transverse cell boundaries in ventricular muscle have a higher resistance, or that the myoplasm itself is less conductive. A lowered myoplasmic conductivity might result from reduction in the cross-sectional area for current flow due to closely packed myofibrils and mitochondria (Table II), although an analogy to skeletal muscle (Katz, 1948) suggests that this explanation is not probable.

The diffusion of radioactive potassium ion (Weidmann, 1966) is also consistent with high conductance of the transverse cell boundaries. Weidmann calculated that if conductance of the transverse cell boundary were due solely to potassium ion, and if complete mixing of the isotope occurred between cell boundaries, then the conductance of the disk would be 0.33 mho/cm^2 (a resistance of 3 Ω cm^2). The membrane area of the transverse boundary was assumed to be that of the (flat) base of a right circular cylinder. Even if the membrane area were ten times the value calculated, the value so obtained would still indicate that the conductance of the transverse cell membrane is much higher than that of the longitudinal cell membrane. If intracellular mixing was incomplete, or if the transverse cell membrane was permeable to ions other than potassium, the conductance would be even higher.

Weidmann (1970) tried to evaluate the possibility that transverse membranes store charge by comparing the voltage response between two intracellular micropipettes during current flow. He was not able to find a significant lag of potential change between the two electrodes, but his recording equipment would not allow for a determination of a time constant shorter than about 100 μsec. He was therefore unable either to confirm or to deny the existence in ventricular muscle of the 60 μsec time constant found by Freygang and Trautwein (1970) for longitudinal currents in Purkinje fibers.

Not all investigators have agreed that the transverse cell boundary has a high conductance in ventricular muscle. Sperelakis and colleagues (e.g., Tarr and Sperelakis, 1964) by a variety of techniques have failed to find distant spread of electrotonic potential. It is not clear whether this discrepancy is due to the special properties and unusual geometry of the preparations they used or to difficulties in interpreting the results because of complex recording methods.

D. MEMBRANE CAPACITANCE. Sakamoto (1969) found a time constant for ventricular muscle of about 3 msec using the 84% point of the electronic potential measured near the gap. As a second method he plotted the time-to-half maximal potential against the distance at which the potential was measured; this second method gave a time constant of 4.2 msec. Weidmann had also used the time course of the electrotonic potential and had found a time constant of 4.4 msec. In order to calculate membrane capacitance from these values, the membrane resistance needs to be known, since

$$C_m = \tau_m / R_m \qquad (8)$$

Both Sakamoto and Weidmann thus calculated a specific membrane capacitance of about 1 $\mu F/cm^2$. It should be remembered that the measurement of specific membrane resistance is dependent on several assumptions, including that of a smooth, unfolded membrane; the values obtained must therefore be considered uncertain. Nevertheless, this figure for C_m is strikingly different from the values determined for the Purkinje fiber (Weidmann, 1952; Fozzard, 1966).

For the Purkinje fiber, a different way of calculating the membrane capacitance from the time constant of the foot of the propagated action potential yielded a much smaller value than cable analysis (Section III,C,1a). This technique requires a value for the fiber diameter and the measurement of a constant conduction velocity. Myoplasmic resistivity must also be either measured or assumed. For his studies on ventricular muscle, Sakamoto (1969) assumed a myoplasmic resistivity of 105 Ω cm and calculated a membrane capacitance of 0.76 $\mu F/cm^2$. Weidmann (1970) used his measured value of 470 Ω cm for myoplasmic resistivity in ventricular muscle and calculated a capacitance by this method of 0.59 $\mu F/cm^2$. Both values are similar to the ones obtained by square current pulse analysis in ventricular muscle, but are quite small.

If the membranes involved in storing charge are folded or invaginated, as appears to be the case, the specific capacitance for an actual unit area of membrane is even smaller than the measured value of less than 1 $\mu F/cm^2$. This value would be then much lower than membrane capacitance measured in other tissues (e.g., nerve) and would require that the membrane be thicker or that the dielectric constant be smaller in ventricular muscle than in other tissues. There is no evidence that the sarcolemma of heart muscle is thinner or thicker than that of other muscles. The most attractive explanation for the wide variation in specific membrane capacitance is therefore that the calculation of area is incorrect. This suggestion requires that the actual area in the ventricular muscle preparations be less than that calculated by assuming a set of parallel right circular cylinders, if the capacitance is to be in the range of 1 $\mu F/cm^2$.

In ventricular muscle, a voltage clamp has been achieved by Beeler and Reuter (1970), who used a single sucrose gap method. From their capacitive current they calculated a crude value for membrane capacitance of 1–2 $\mu F/cm^2$, but no details of how they estimated the involved surface area of the fiber were given. The capacitive charging was slow in their voltage clamps. They concluded that this slowness resulted from charging of the R–C of the entire membrane through a series resistance. The resistance could be interpreted in terms of the model in Fig. 4A, but they felt that a more probable site of origin of the series resistance was the extracellular space between cells within the trabecular muscle used. In addition to making it difficult to study the membrane capacitance, the resistance demonstrated by Beeler and Reuter will also provide a handicap for investigators studying fast currents with the sucrose gap technique. See also Tarr and Trank (1971).

3. ATRIAL MUSCLE

Only a few quantitative measurements have been made on atrial muscle, which shares the geometrical problems already discussed for ventricular muscle. The first quantitative measurement was by Trautwein *et al.* (1956) on a trabeculum of frog atrial muscle. These investigators found the space constant to be about 0.4 mm and the time constant about 6 msec. Similar values were found by Woodbury and Gordon (1965) using rabbit atria. When the results of Woodbury and Gordon were analyzed according to the one-dimensional cable model the membrane resistance was 120 Ω cm^2 and the membrane capacitance was 25 $\mu F/cm^2$. The authors emphasize that these values should not be taken as quantitative, since calculations of area, as well as other assumptions, are in doubt.

The method of measuring the time constant of the foot of the propagating action potential was used in rabbit atrial muscle by Paes de Carvalho *et al.* (1969), who calculated a value of 1.3 $\mu F/cm^2$. Again, the usual assumptions were required for this computation. Rougier *et al.* (1968) studied frog atrial muscle with a double sucrose gap technique and found that the capacitive transient of current at the onset of a voltage clamp declined with a time constant of 1–5 msec. No quantitative measure of capacitance could be obtained from these determinations because of uncertainty regarding the membrane area involved, but the authors did succeed in demonstrating that the decay of capacitive current was smooth. This single-phased decline of capacitive current differs from that seen by Fozzard (1966) for Purkinje fibers. However, the figure illustrating this effect (Fig. 3 of Rougier *et al.*, 1968) shows that membrane clamp voltage was achieved before the end of the capacitive current. If this is not an artifact of the recording system, then some resistance in series with membrane capacitance must be present. To resolve these questions will require a more complete analysis of the properties of atrial fibers, perhaps employing techniques that were successful for ventricular muscle.

Transverse cell boundaries also exist in atrial muscle; their low resistance to passage of action currents was shown by Barr *et al.* (1965). Their observation provided direct confirmation that conduction in heart muscle is by local circuit currents. The anatomical meaning of this observation is, unfortunately, not yet clear. Baldwin (1970) has recently reported an attempt to correlate morphological changes with injury current in damaged frog atrial muscle—see discussion of the work of Délèze (1970) in the section on Purkinje fibers. While she could prove that cells isolated themselves electrically after injury and that signs of injury were apparent in the electron microscope, she was not able to see any changes in intercellular junctions.

A problem not so far considered is that the junctions between cardiac cells do not appear to represent myoplasmic continuity. As discussed in Section II, the junctions in heart muscle are called gap junctions because a distance of 20–30 Å separates the membranes of the two apposed cells. Since this interposed space appears to be continuous with the extracellular space, it represents an alternative path for current flowing from one cell to another. This shunt is represented in Fig. 5B, where R_g is the resistance of the gap between cells (shunt resistance), and $\frac{1}{2}R_{n1}$ is the resistance of one of the two cell membranes at the junction. R_g can be calculated if one assumes that the gap has a conductivity like extracellular fluid and chooses an average cell diameter and distance between the cells. A solution for the transmission of an action potential using Noble's (1962a) modification of the Hodgkin–Huxley equations was obtained by Heppner and Plonsey (1970) in terms of a ratio of R_g and $\frac{1}{2}R_{n1}$.

If R_g is calculated on the basis of cells of 8 μ radius and a gap between cells of 80 Å, the highest junctional membrane resistivity that will permit reliable conduction is about 4 Ω cm². Woodbury and Crill (1970) have developed an approximate closed solution of the problem posed by Heppner and Plonsey, which allows one to understand better the calculated values of R_g and $\frac{1}{2}R_{n1}$. While these calculations are certainly dependent on many assumptions, they point the way to how we may understand the significance of measurements of transverse membrane resistance and to the geometry of the cellular junction.

4. OTHER STUDIES

Two additional topics, which are only indirectly related to those already reviewed, merit discussion. The first topic is the membrane properties of cells cultured from heart tissue. Only a few observations have been made using cultured cells, partly because such cells are difficult to use for electrophysiological study. However, the possibility of achieving a simple geometry makes tissue culture a promising area for future investigation. A second topic is the problem of the junction between cardiac tissues with different membrane properties. This problem is becoming apparent at the junction between Purkinje fibers and ventricular muscle. Although little is presently known about these two subjects, they deserve mention because they represent important areas of future study.

Using an a.c. bridge, Sperelakis and Lehmkuhl (1964) measured resistance and capacitance of cultured myocardial cells. With a reasonable guess for surface area, they calculated membrane resistance as 480 Ω cm² and membrane capacitance as 20 μF/cm². Electrical interconnection between tissue cultured cells is used by De Haan and Gottlieb (1968) to explain why cell groupings maintain a normal resting potential better after impalement than do single cells. These tissue culture studies are complicated by a variety of technical problems, but further effort may make exact measurements possible (Eisenberg and Johnson, 1970).

An interesting geometrical problem becomes apparent if one visualizes the Purkinje fiber as a one-dimensional cable making connection with ventricular muscle, which is more like a three-dimensional structure. Under these conditions it would be far easier for electrical events in the ventricular muscle (e.g., action currents) to influence the Purkinje fiber than the reverse. For this geometrical reason, one would expect that conduction in the normal direction from Purkinje to muscle is more easily blocked then from muscle to Purkinje. Mendez *et al.* (1970) have demonstrated that this prediction is correct for dog Purkinje–myocardial junctions. They have begun a careful study of the nature of the connection between these cell types.

IV. Directions for Future Experiments

The foregoing discussion suggests that future electrophysiological work on the cardiac cell boundary will require a quantitative description of that boundary and its geometrical relations to intracellular and extracellular structures. At present, the most powerful method available for this purpose is a combination of point counting with line integration like that used to obtain the data in Tables II and III. The extension of this method to the transverse cell boundary is feasible, although technically difficult. The method is potentially applicable to electron micrographs obtained with the newer electron microscopic techniques, including electron microscopy of freeze-cleaved sections and scanning electron microscopy. The ultimate goal is a three-dimensional model based on experimental observations which will presumably include reconstructions from ultrathin serial sections.

Future experimental work will also be directed at the molecular basis of the low-resistance junctions between cardiac cells. Biochemical methods, high resolution electron microscopy, X-ray diffraction, and other physical methods like those that have already been applied to Class I membranes may be expected to yield a clearer insight into the problem of the low-resistance pathway in what is currently called a gap junction. This problem is part of the larger question of the nature of contacts between heart muscle cells. In addition to its biochemical aspects, this question includes the development of contacts between embryonic heart muscle cells, as well as the investigation of the possibility that the area and type of contacts between cells of mature heart muscle are variable under physiological or abnormal conditions.

Few quantitative electrical measurements are available in heart muscle, and more are needed. Because the heart is a heterogeneous structure, each part has posed technical difficulties to quantitative electrical study. These technical difficulties have led to a necessary but unfortunate nonuniformity in the electrophysiological techniques that have been used. Some of the apparent differences between the membrane properties of atrial muscle, Purkinje fibers, ventricular muscle, and other special areas may be the result of systematic errors related to the difference in techniques. A more careful comparison of several tissues by the same method of measurement will be helpful. In addition, many of the apparent differences in membrane electrical properties may be the result of the different geometries of the tissues rather than intrinsic differences in the membrane structure itself.

Sometimes the complexities of the cells and of their associated electrical measurements are such that accurate assessment of the electrical role of a particular membrane may be impossible, no matter how careful the measurements. One possible way around this problem is to locate cells with simpler membrane systems to study. Numerous laboratories have begun the time-consuming task of investigating the comparative morphology and electrophysiology of cardiac cells. It seems desirable to concert the efforts of the laboratories pursuing these different methods of study into a cooperative effort. The goal of such an effort would be to determine the morphology of tissues that are electrophysiologically unique, and to measure the membrane electrical properties of tissues that are structurally unique. As tissue culture techniques are perfected, this approach may also provide simpler geometrical arrangements of cells for combined ultrastructural and electrophysiological study.

Development of electrical models and their analysis will continue to be fruitful; indeed these models are essential for analysis of electrical measurements. As more quantitative structural and electrical data are obtained, mathematical biologists will have the opportunity to contribute significantly to an integration of cardiac membrane structure and function.

One aspect of the transverse cell boundary between cardiac cells that deserves special attention is the investigation of physiological conditions that may alter the resistance of the junctions between cardiac cells. Models for such studies are available in published work from the laboratory of Loewenstein on junctions between epithelial cells as described on page 106. The possible effects of the intracellular concentration of divalent cations and of repolarizing membrane currents seem particularly interesting as modifiers of the syncytial properties of cardiac muscle.

V. Appendix: Compilation of Articles on the Ultrastructure of Vertebrate and Invertebrate Heart Muscle

Tables IV and V are a compilation of articles on the ultrastructure of vertebrate and invertebrate heart muscle. The emphasis in selection of references for the tables was on publications which have appeared since the first edition of this treatise in 1960; a few classic articles from the decade 1951–1960 are included.

TABLE IV

COMPILATION OF RECENT ARTICLES ON ULTRASTRUCTURE OF VERTEBRATE HEARTS

Species	Chamber	ID	SR	TT	MITOS	MF	Other	References
					Mammals			
Bat	Atrium						Specific granules	Jamieson and Palade (1964)
	Ventr.					+		Hama and Kanaseki (1967)
Beef	Atrium						Specific granules	Jamieson and Palade (1964)
	Ventr.		+	+				Simpson (1965)
	Ventr.				+	+	Micropinocytosis	Girardier et al. (1967)
	Ventr.					+		Pape, Kübler, and Smekal (1969)
	SAN	+				+		Rhodin et al. (1961)
	SAN	+	+		+	+	Nucleus	Hayashi (1962)
	AVN					+		Rhodin et al. (1961)
	AVN		+		+	+		Hayashi (1962)
	AVN	+				+		Maekawa et al. (1967)
	Purk.	+				+		Rhodin et al. (1961)
	Purk.	+	+			+	Nucleus	Hayashi (1962)
	Purk.				+	+		Pape, Kübler, and Smekal (1969)
Cat	Atrium						Specific granules	Jamieson and Palade (1964)
	Atrium		+		+	+	Nucleus, specific granules	McNutt and Fawcett (1969)
	Atrium	+						McNutt (1970)
	Atrium					+		Sperelakis et al. (1970)
	Ventr.	+				+		Dewey and Barr (1964)
	Ventr.					+		Spotnitz et al. (1966)
	Ventr.					+		Fawcett (1968)
	Ventr.	+	+	+		+	Nucleus, glycogen	Sommer and Johnson (1968)
	Ventr.	+	+	+	+	+		Fawcett and McNutt (1969)
	Ventr.	+				+		McNutt (1970)
	Ventr.	+			+	+		Sperelakis et al. (1970)
	Purk.	+				+		E. Page (1967a)

TABLE IV (*Continued*)

Species	Chamber	ID	SR	TT	MITOS	MF	Other	References
Cat	Purk.	++	+	+				E. Page (1967b)
	Purk.	++	+	+	+			E. Page (1968)
	Purk.	+	+			+		Sommer and Johnson (1968)
Dog	Atrium						Specific granules	Jamieson and Palade (1964)
	Ventr.	+++	++	++	++	+		Lindner (1957)
	Ventr.	++		+		+		Moore and Ruska (1957)
	Ventr.					+		Kawamura (1961)
	Ventr.		++		+++	+	Granules, glycogen	Sommer and Spach (1964)
	Ventr.				++	+	Glycogen	Tubbs, et al. (1964)
	Ventr.	+	+		+	+		R. B. Jennings et al. (1965)
	Ventr.							Martin and Hackel (1966)
	Ventr.	+	+		+	+	Nucleus	Spotnitz et al. (1966)
	Ventr.	+			+	+		Legato et al. (1968)
	Ventr.				+++	+		Sommer and Johnson (1968)
	Ventr.	+			++		Nucleus, granules	Bishop and Cole (1969)
	Ventr.					+		Herdson et al. (1969)
	Ventr.		+	++	+++	+		R. B. Jennings et al. (1969)
	Ventr.				++	+	Glycogen	Legato and Langer (1969)
	Ventr.	+			++	+	Nucleus, nerve	Sommer and Johnson (1970)
	SAN		+++++		+++++	+++		Kawamura (1961)
	SAN				++++	+		James et al. (1966)
	SAN	+				+	Lysosome	James and Sherf (1968a,b)
	SAN				+	+		James (1970)
	AVN	+				++	Nerve, nucleus	Kawamura, (1961)
	AVN					++	Nucleus	Maekawa et al. (1964)
	AVN						Glycogen	Hashiba (1966)
	AVN	+				++	Nerve	Maekawa et al. (1967)
	AVN					++		Challice (1969)
	Purk.	+	+			++	Nucleus	Kawamura (1961)

Animal	Tissue	Markers	Structure	Reference
	Purk.	+		Sommer and Johnson (1968)
	Purk.	+	Intercalated disks	Martinez-Palomo et al. (1970)
	Purk.	+		Sommer and Johnson (1970)
Dormouse	Ventr.	++		Poche (1959)
	Ventr.	++		Poche and Mönkemeier (1962)
Ferret	Ventr.	+ +		Simpson and Rayns (1968)
Goat	Ventr.	+		Hama and Kanaseki (1967)
	Purk.	+		Sommer and Johnson (1968)
Guinea pig	Atrium	+ ++	Specific granules	Jamieson and Palade (1964)
	Ventr.	+ ++		Lindner (1957)
	Ventr.	+		Sjöstrand and Andersson-Cedergren (1960)
	Ventr.	+++ ++		M. M. Dewey and Barr (1964)
	Ventr.	+++ ++		Barr et al. (1965)
	Ventr.	+++ ++		Denoit and Coraboeuf (1965)
	Ventr.	+++		Simpson 1965
	Ventr.	+++ ++ +		Rayns et al. (1968)
	Ventr.	+++ +		Sommer and Johnson (1968)
	Ventr.	+++ +		M. Dewey (1969)
Hamster	Atrium	+ + + +	Specific granules	Jamieson and Palade (1964)
Man	Atrium	+ +	Specific granules	Fawcett (1961)
	Atrium	+ +		Jamieson and Palade (1964)
	Atrium	+		Challice (1969)
	Atrium			Burch and Sohal (1969)
	Ventr.	++ + +		Nelson and Benson (1963)
	Ventr.	+++ + +		Richter and Kellner (1963)
	Ventr.	+++ +++ +++	Glycogen	Kawamura et al. (1964)
	SAN	+++ +++ +++		James et al. (1966)
	SAN	+++ +++ ++		James and Sherf (1968b)
	Embryonic	+	Granules	Leak and Burke (1964)

TABLE IV (*Continued*)

Species	Chamber	ID	SR	TT	MITOS	MF	Other	References
Man	SAN					+	Nucleus, lysosomes	James (1970)
	AVN				+	+	Nucleus	James and Sherf (1968a)
	AVN		+			+	Nucleus, lysosomes	Challice (1969)
	AVN				+	+		James (1970)
	Purk.	+				+	Nucleus	Maekawa et al. (1964)
	Purk.	+				+		Kawamura and Hayashi (1966)
	Purk.	+				+		Maekawa et al. (1967)
	Purk.	+	+					Challice (1969)
	Emb. Ventr.	+						McNutt (1970)
Monkey	Ventr.	+				+		Burch and Sohal (1969)
	SAN							Kawamura and Hayashi (1966)
	SAN							Maekawa et al. (1967)
	AVN							Maekawa et al. (1967)
	AVN						Leptofibrils	Viragh (1968)
	Purk.		+		+			Sommer and Johnson (1970)
Mouse	Atrium						Specific granules	Jamieson and Palade (1964)
	Atrium				+			Mizuno et al. (1965)
	Atrium						Capillaries	M. A. Jennings and Florey (1967)
	Atrium						Capillaries	Karnovsky (1967, 1968)
	Atrium		+		+	+	Capillaries	Lockwood et al. (1969)
	Ventr.	+	+		+	+		Moore and Ruska (1957)
	Ventr.	+		+				Sjöstrand and Andersson-Cedergren (1960)
	Ventr.		+			+		Challice and Edwards (1961)
	Ventr.				+		Granules	Moor et al. (1964)
	Ventr.				+			Mizuno et al. (1965)
	Ventr.							Ruska (1965)
	Ventr.					+	Capillaries	Karnovsky (1967)

Species	Tissue	1	2	3	4	5	Component	Reference
	Ventr.		+	++		+	Nuclei	Revel and Karnovsky (1967)
	Ventr.		+	++		+	Nuclei	Blailock et al. (1968)
	Ventr.	+						Burch and Sohal (1969)
	Ventr.							Schulze and Wollenberger (1969)
	Ventr.	++					Nerve	Shiina et al. (1969)
	SAN	++	+			++	Nerve	Maekawa et al. (1967)
	AVN					+	Nerve–muscle junction	Thaemert (1970)
Rabbit	Atrium							Trautwein and Uchizono (1963)
	Atrium							Jamieson and Palade (1964)
	Ventr.	+	++	+		+	Specific granules	Sjöstrand and Andersson-Cedergren (1960)
	Ventr.	+	++	+	++	++	Nucleus	Nelson and Benson (1963)
	Ventr.	++++	++	+		++	Nerve	Bozner et al. (1965)
	Ventr.	+++	++			+++		Hadek and Talso (1967)
	Ventr.			++		++++		Shimamoto and Hiramoto (1967)
	Ventr.					++		Sommer and Johnson (1968)
	SAN		+	+			Nerve	Torii (1962)
	SAN					+	Nucleus	Trautwein and Uchizono (1963)
	SAN	+					Nucleus, nerve	Maekawa et al. (1964)
	SAN					+	Nucleus	Challice (1965)
	SAN						Nucleus, nerve	Kawamura and Hayashi (1966)
	SAN							Challice (1969)
	AVN		+		+	+	Nerve	Torii (1962)
	AVN	+			+	+	Nerve	Maekawa et al. (1967)
	Purk.				+			Torii (1962)
	Purk.						Nerve	Maekawa et al. (1967)
	Purk.					+		Sommer and Johnson (1968)
Rat	Atrium						Capillaries	Palade (1961)
	Atrium	+		+	+	+	Capillaries	M. A. Jennings et al. (1962)
	Atrium							Tice and Barnett (1962)

TABLE IV (*Continued*)

Species	Chamber	ID	SR	TT	MITOS	MF	Other	References
Rat	Atrium		+				Specific granules	Jamieson and Palade (1964)
	Atrium						Capillaries	M. A. Jennings and Florey (1967)
	Atrium					+		Burch and Sohal (1969)
	Atrium					+		David et al. (1969)
	Atrium						Lysosomes	Hendy et al. (1969)
	Atrium		+		+	+	Specific granules	Hibbs and Ferrans (1969)
	Atrium		+			+	Specific granules	Ferrans et al. (1969)
	Atrium		+++	+++	+++	+++	Specific granules	Forssmann and Girardier (1970)
	Ventr.	+++	++	++	+++			Moore and Ruska (1957)
	Ventr.	++			++	+		Porter and Palade (1957)
	Ventr.					+		Stenger and Spiro (1961)
	Ventr.				++	+	Nerves	Viragh and Porte (1961a)
	Ventr.				+			Poche and Moenkemeier (1962)
	Ventr.							Hibbs and Black (1963)
	Ventr.				++	++		Karnovsky and Hug (1963)
	Ventr.		+		++	++		Niehaus et al. (1963)
	Ventr.			+				Girardier et al. (1964)
	Ventr.				+++			Karnovsky (1964)
	Ventr.			+	+++	+++		Denoit and Coraboeuf (1965)
	Ventr.							Mizuno et al. (1965)
	Ventr.			+++				Poche and Hausamen (1965)
	Ventr.	+	+++					Rostgaard and Behnke (1965)
	Ventr.	+++	+++	+	+	+	Nucleus	Simpson (1965)
	Ventr.	+++	+++			+		Bózner et al. (1966)
	Ventr.		+++					Forssmann and Girardier (1966)
	Ventr.					++		Hadek and Talso (1966)
	Ventr.		+		+	+		Trillo et al. (1966)
	Ventr.		+++					Wegner and Moelbert (1966)

Region							Remark	Reference
Ventr.					+		Nerves	Hadek and Talso (1967)
Ventr.				+	+		Capillaries	M. A. Jennings and Florey (1967)
Ventr.		+ +		+	+			Muir (1967)
Ventr.		+ +	+	+	+		Endocardium	Melax and Leeson (1967)
Ventr.		+ +	+ +		+		Capillary	Bruns and Palade (1968)
Ventr.					+			Denoit and Coraboeuf (1968)
Ventr.						+		Meddoff and Page (1968)
Ventr.		+		+	+ +		Capillary	Palade and Bruns (1968)
Ventr.					+ + +			Poche et al. (1968)
Ventr.		+ +	+ +	+		+		Sohal et al. (1968)
Ventr.		+ + +		+		+ +	Lipid synthesis	O. Stein and Stein (1968)
Ventr.		+ + +		+		+ + +		David et al. (1969)
Ventr.			+	+				Emberson and Muir (1969a,b)
Ventr.		+	+	+	+	+	Nucleus	Lee (1969)
Ventr.		+ + +						Novi (1968a,b)
Ventr.		+	+	+		+		Novi (1969)
Ventr.		+ +	+ +					Poche et al. (1968)
Ventr.		+				+		Berry et al. (1970)
Ventr.					+ +		Nucleus	S. Bloom (1970)
Ventr.		+ +	+ +	+ + + +	+ +	+ +		Forssmann and Girardier (1970)
Ventr.		+	+	+	+	+		King and Gollnick (1970)
Ventr.		+						Walker et al. (1970)
SAN		+	+	+	+			Viragh and Porte (1961a)
SAN							Nerve	Maekawa et al. (1967)
AVN		+	+		+		Nerve	Viragh and Porte (1961b)
AVN								Maekawa et al. (1967)
Purk.		+	+	+				Bompiani et al. (1959)
Purk.							Nerve	Maekawa et al. (1967)
Embryonic		+ +	+					Melax and Leeson (1969)
Sheep Ventr.		+ +	+	+ +	+ +	+ +	Lysosome	Simpson and Oertelis (1962)
Ventr.					+ +			Simpson (1965)

TABLE IV (*Continued*)

Species	Chamber	ID	SR	TT	MITOS	MF	Other	References
Sheep	SAN	+					Nerve	Maekawa et al. (1967)
	AVN						Nerve	Maekawa et al. (1967)
	Purk.		+			+		Caesar et al. (1958)
	Purk.					+		Ruska (1965)
	Purk.	++	+			+	Nerve	Maekawa et al. (1967)
	Purk.	++				+	Leptofibrils	E. Page et al. (1969)
	Purk.				+			Mobley and E. Page (1972)
							Birds	
Budgerigar	Ventr.	+	+	+	+	+	Capillaries, nerve	Gossrau (1968)
Canary	Ventr.				+	+		Slautterback (1965)
	Ventr.					+		Hama and Kanaseki (1967)
Chicken	Ventr.		+		+	+		Hama and Kanaseki (1967)
	Ventr.		+			+		Mizuhira et al. (1967)
	Ventr.				+	+		Sommer and Johnson (1969)
	Ventr.				+			Hirakow (1970)
	Ventr.					+		Sommer and Johnson (1970)
	Purk.	+						Sommer and Johnson (1970)
Finch	Ventr.				+	+		Slautterback (1965)
	Ventr.					+		Mizuhira et al. (1967)
	Ventr.					+		Hirakow (1970)
Goose	Ventr.				+			Slautterback (1965)
Humming-bird	Ventr.		+		+	+	Nucleus	DiDio (1967)
Lovebird	Ventr.					+		Hama and Kanaseki (1967)

Animal	Region					Notes	Reference
Pigeon	Atrium			++			Gossrau (1969)
	Ventr.			++			Gossrau (1969)
	Ventr.		+				McNutt (1970)
	Purk.		+	+		Leptofibrils	Gossrau (1969)
Quail	Ventr.			+			Hama and Kanaseki (1967)
	Ventr.						Mizuhira *et al.* (1967)
Reptiles							
Boa constrictor	Ventr.		+	+		Nerve, nucleus, glycogen	Leak (1967)
Turtle	Atrium	+		++			Fawcett and Selby (1958)
	Ventr.	+		++			Trillo *et al.* (1966)
	Ventr.	+		++		Nucleus	Hirakow (1970)
Amphibians							
Frog	Atrium	+++		++			Barr *et al.* (1965)
	Atrium			++			M. Dewey (1969)
	Ventr.	+	+	++			Lindner (1957)
	Ventr.		+				Sjöstrand and Andersson-Cedergren (1960)
	Ventr.		+	+		Nerve	Ruska (1965)
	Ventr.	+	+			Nerve	Trillo and Bencosme (1965)
	Ventr.	+++	+++	++		Nucleus	Thaemert (1966)
	Ventr.	+++	++	++		Dense bodies, lysosomes	Staley and Benson (1968)
	Embryonic						Sommer and Johnson (1969)
							Huang (1967)
Toad	Ventr.	++	++	++			Grimley and Edwards (1960)
	Ventr.		+				Nayler and Merrillees (1964)

TABLE IV (Continued)

Species	Chamber	ID	SR	TT	MITOS	MF	Other	References
Toad	Ventr.	+	+			+		Trillo et al. (1966)
	Ventr.	+		+	+			De Mello et al. (1969)
Fish								
Goldfish	Ventr.	+	+			+		Yamamoto (1967)
Hagfish	Atrium	+	+			+		G. D. Bloom (1962)
	Ventr.				+	+		G. D. Bloom (1962)
	Ventr.	+			+	+	Granules, Golgi, crystalline inclusion bodies (CIB)	Leak (1969)
Lamprey	Atrium	+	+			+		G. D. Bloom (1962)
	Ventr.	+			+	+		G. D. Bloom (1962)
	Ventr.				+	+		Trillo et al. (1966)
Chordates								
Tunicate								
Ciona								Kriebel (1968)

Key to abbreviations: SAN = Sinoatrial node, AVN = atrioventricular node, Purk = Purkinje fibers or cells, ID = intercalated disk, SR = sarcoplasmic reticulum, TT = transverse tubules or T system, MITOS = mitochondria, MF = myofibrils.

TABLE V

COMPILATION OF RECENT ARTICLES ON ULTRASTRUCTURE OF INVERTEBRATE HEARTS

Species	Common name	Reference
Insects		
Hyalophora cecropia	Moth	Sanger and McCann (1968); McCann and Sanger (1969)
Aiolopus strepens	Grasshopper	Baccetti and Bigliardi (1969b)
Locusta migratoria	Grasshopper	Hoffmann and Levi (1965)
Drosophila repleta	Fly	Burch *et al.* (1970)
Blatella germanica	Cockroach	Edwards and Challice (1960)
Crustaceans		
Procambarus clarkii	Crayfish	Kawaguti (1963c) Komuro (1969) Howse *et al.* (1970)
Homarus americanus	Lobster	Smith (1963)
Palinurus vulgaris	Lobster	Baccetti and Bigliardi (1969a)
Tachypleus tridentatus Leach	Horseshoe crab	Kawaguti (1963d)
Limulus polyphemus	Horseshoe crab	Leyton and Sonnenblick (1971); Sperelakis (1971)
Squilla	Mantis shrimp	Irisawa and Hama (1965)
Daphnia pulex	Sand flea	R. S. Stein, *et al.* (1966)
Mollusks		
Sepia esculenta	Cuttlefish	Kawaguti (1963b)
Sepia officinalis	Cuttlefish	Schipp and Schaefer (1969a,b)
Euhadra hickonis	Snail	Kawaguti (1963a)
Helix aspersa	Snail	R. J. North, (1963)
Archachatina marginata	Snail	Nisbet and Plummer (1966)
Achatina fulica hamillei	Snail	Nisbet and Plummer (1966)
Achatina panthera lamarkiana	Snail	Nisbet and Plummer (1966)
Achatina achatina monochromatica	Snail	Nisbet and Plummer (1966)
Venus mercenaria	Clam	Hayes and Kelly (1969) Kelly and Hayes (1969)

ACKNOWLEDGMENTS

Portions of this work were supported by U.S.P.H.S. Grants HE 10503 and HE 11665 and by Myocardial Research Unit Contract PH43681334. Dr. Page was a recipient of a U.S.P.H.S. Career Development Award.

REFERENCES

Adrian, R. H., Chandler, W. K., and Hodgkin, A. L. (1969). *J. Physiol. (London)* **204**, 207.
Andersson-Cedergren, E. (1959). *J. Ultrastruct. Res. Suppl.* 1.

Baccetti, B., and Bigliardi, E. (1969a). *Z. Zellforsch. Mikrosk. Anat.* **99**, 25.

Baccetti, B., and Bigliardi, E. (1969b). *Z. Zellforsch. Mikrosk. Anat.* **99**, 13.

Baldwin, K. M. (1970). *J. Cell Biol.* **46**, 455.

Barr, L. M., Dewey, M. M., and Berger, W. (1965). *J. Gen. Physiol.* **48**, 797.

Beeler, G. W., and Reuter, H. (1970). *J. Physiol. (London)* **207**, 165.

Benedetti, E. L., and Emmelot, P. (1968). *In* "Ultrastructure in Biological Systems" (A. S. Dalton and F. Haguenau, eds.), Vol. 4, pp. 33–123. Academic Press, New York.

Berry, M. N., Friend, D. S., and Scheuer, J. (1970). *Circ. Res.* **26**, 679.

Bishop, S. P., and Cole, C. R. (1969). *Lab. Invest.* **20**, 219.

Blailock, Z. R., Robin, E. R., and Melnick, J. L. (1968). *Exp. Mol. Pathol.* **9**, 84.

Bloom, G. D. (1962). *Z. Zellforsch. Mikrosk. Anat.* **57**, 213.

Bloom, S. (1970). *J. Cell Biol.* **44**, 218.

Bompiani, G. D., Rouiller, C., and Hatt, P. Y. (1959). *Arch. Mal. Coeur Vaiss.* **52**, 1257.

Bózner, A., Barta, E., Pavlovikova, H., and Mrena, E. (1965). *Cor Vasa* **7**, 227.

Bózner, A., Inczinger, F., and Mrena, E. (1966). *Folia Morphol. (Prague)* **14**, 400.

Bózner, A., Knieriem, H.-J., Meesen, H., and Rainauer, H. (1969). *Virchows Arch., B* **2**, 125.

Brightman, M. W., and Reese, T. S. (1969). *J. Cell Biol.* **40**, 648.

Bruns, R. E., and Palade, G. E. (1968). *J. Cell Biol.* **37**, 244.

Burch, G. E., and Sohal, R. S. (1969). *Amer. Heart J.* **78**, 358.

Burch, G. E., Sohal, R., and Fairbanks, L. D. (1970). *Arch. Pathol.* **89**, 128.

Caesar, R., Edwards, G. E., and Ruska, H. (1958). *Z. Zellforsch. Mikrosk. Anat.* **48**, 698.

Carney, J. A., and Brown, A. L. (1964). *Amer. J. Pathol.* **44**, 521.

Chalcroft, J. P., and Bullivant, S. (1970). *J. Cell Biol.* **47**, 49.

Challice, C. E. (1965). *J. Roy. Microsc. Soc.* [3] **85**, 1.

Challice, C. E. (1969). *Ann. N.Y. Acad. Sci.* **156**, 14.

Challice, C. E., and Edwards, G. A. (1961). *In* "Specialized Tissues of the Heart" (A. P. De Carvalho, W. C. de Mello, and B. F. Hoffman, eds.), pp. 44–76. American Elsevier, New York.

Cole, K. S. (1965). *Physiol. Rev.* **45**, 340.

David, H., Kleinau, H., Ruff, D., and Manowski, B. (1969). *Acta Biol. Med. Ger.* **23**, 907.

De Haan, R. L., and Gottlieb, S. H. (1968). *J. Gen. Physiol.* **52**, 643.

Délèze, J. (1970). *J. Physiol. (London)* **208**, 547.

De Mello, W. C., Motta, G. E., and Chapeau, M. (1969). *Circ. Res.* **24**, 475.

Denoit, F., and Coraboeuf, E. (1965). *C. R. Soc. Biol.* **159**, 2118.

Denoit, F., and Coraboeuf, E. (1968). *J. Microsc. (Paris)* **7**, 245.

Dewey, M. M. (1969). *In* "Comparative Physiology of the Heart: Current Trends" (F. V. McCann, ed.), pp. 10–28. Birkhäeuser, Basel.

Dewey, M. M., and Barr, L. (1964). *J. Cell Biol.* **23**, 553.

DiDio, L. A. J. (1967). *Anat. Rec.* **159**, 335.

Dudel, J., Peper, K., Rüdel, R., and Trautwein W. (1966). *Pfluegers Arch. Gesamte Physiol. Menschen Tiere* **292**, 255.

Edwards, G. A., and Challice, C. E. (1960). *Ann. Entomol. Soc. Amer.* **53**, 369.

Eisenberg, R. S., and Johnson, E. A. (1970). *Progr. Biophys. Mol. Biol.* **20**, 1.

Eisenberg, R. S., Vaughan, P. C., and Howell, J. N. (1972). *J. Gen. Physiol.* **59**, 360.

Eisenman, G. (1968). *Fed. Proc., Fed. Amer. Soc. Exp. Biol.* **27**, 1249.

Emberson, J. W., and Muir, A. R. (1969a). *J. Anat.* **104**, 411

Emberson, J. W., and Muir, A. R. (1969b). *Quart. J. Exp. Phsyiol.* **54**, 36.

Falk, G., and Fatt, P. (1964). *Proc. Roy. Soc., Ser. B* **160**, 69.

Farquhar, M. G., and Palade, G. E. (1963). *J. Cell Biol.* **17**, 375.

Fawcett, D. W. (1961). *Circulation* **24**, 336.

Fawcett, D. W. (1966). "An Atlas of Fine Structure." Saunders, Philadelphia, Pennsylvania.

Fawcett, D. W. (1968). *J. Cell Biol.* **36**, 266.

Fawcett, D. W., and McNutt, N. S. (1969). *J. Cell Biol.* **42**, 1.

Fawcett, D. W., and Selby, C. C. (1958). *J. Biophys. Biochem. Cytol.* **4**, 63.

Ferrans, V. J., Hibbs, R. G., and Buja, L. M. (1969). *Amer. J. Anat.* **125**, 47.

Fischman, D. A., and Zak, R. (1970). *Abstr. 3rd Annu. Meet. Int. Study Group Rese. Cardiac Metab., 1970*, Stowe, Vermont, p. 81.

Forssmann, W. G., and Girardier, L. (1966). *Z. Zellforsch. Mikrosk. Anat.* **72**, 249.

Forssmann, W. G., and Girardier, L. (1970). *J. Cell Biol.* **44**, 1.

Forssmann, W. G., Siegrist, G., Orci, L., Girardier, L., Pictet, R., and Rouiller, C. (1967). *J. Microsc. (Paris)* **6**, 279.

Fozzard, H. A. (1966). *J. Physiol. (London)* **182**, 255.

Freygang, W. H., and Trautwein, W. (1970). *J. Gen. Physiol.* **55**, 524.

Fricke, H. (1925). *J. Gen. Physiol.* **9**, 137.

Furshpan, E. J., and Potter, D. D. (1959). *J. Physiol. (London)* **145**, 289.

Girardier, L. (1965). *In* "Electrophysiology of the Heart" (B. Taccardi and G. Marchetti, eds.), pp. 53–70. Pergamon, Oxford.

Girardier, L., Dreifuss, J. J., Haenni, B., and Petrovici, A. (1964). *Pathol. Microbiol.* **27**, 16.

Girardier, L., Derifuss, J. J., and Forssmann, W. G. (1967). *Acta Anat.* **68**, 251.

Goodenough, D. A., and Revel, J. P. (1970). *J. Cell Biol.* **45**, 272.

Goshima, K., and Tonomura, Y. (1969). *Exp. Cell Res.* **56**, 387.

Gossrau, R. (1968). *Histochemie* **13**, 111.

Gossrau, R. (1969). *Z. Anat. Entwicklungs-Gesch.* **128**, 163.

Grimley, P. M., and Edwards, G. A. (1960). *J. Biophys. Biochem. Cytol.* **8**, 305.

Hadek, R., and Talso, P. J. (1966). *Nature (London)* **212**, 944.

Hadek, R., and Talso, P. J. (1967). *J. Ultrastruct. Res.* **17**, 257.

Hama, K., and Kanaseki, T. (1967). *In* "Electrophysiology and Ultrastructure of the Heart" (T. Sano, V. Mizuhira, and K. Matsuda, eds.), pp. 27–40. Grune & Stratton, New York.

Hashiba, K. (1966). *Jap. Circ. J.* **30**, 151.

Hay, E. D. (1968). *In* "Epithelial-Mesenchymal Interactions" (R. Fleischmayer and R. Billingham, eds.), pp. 31–55. Williams & Wilkins, Baltimore, Maryland.

Hayashi, K. (1962). *Jap. Circ. J.* **26**, 765.

Hayes, R. L., and Kelly, R. E. (1969). *J. Morphol.* **127**, 151.

Hendy, F. J., Abraham, R., and Grasso, P. (1969). *J. Ultrastruct. Res.* **29**, 485.

Heppner, D. B., and Plonsey, R. (1970). *Biophys. J.* **10**, 1057.

Herdson, P. B., Kaltenbach, J. P., and Jennings, R. B. (1969). *Amer. J. Pathol.* **57**, 539.

Hibbs, R. G., and Black, W. C. (1963). *Anat. Rec.* **147**, 261.

Hibbs, R. G., and Ferrans, V. J. (1969). *Amer. J. Anat.* **124**, 251.

Hirakow, R. (1970). *Amer. J. Cardiol.* **25**, 195.

Hodgkin, A. L., and Huxley, A. F. (1952). *J. Physiol. (London)* **117**, 500.

Hodgkin, A. L., and Nakajima, S. (1972). *J. Physiol. (London)* **221**, 121.

Hodgkin, A. L., and Rushton, W. A. H. (1946). *Proc. Roy. Soc., Ser. B* **133**, 444.

Hoffman, J. A., and Levi, C. (1965). *C. R. Acad. Sci. Ser. D* **260**, 6988.

Howse, H. D., Ferrans, V. J., and Hibbs, R. G. (1970). *J. Morphol.* **131**, 237.

Huang, C. Y. (1967). *J. Ultrastruct. Res.* **20**, 211.

Huxley, H. E. (1964). *Nature (London)* **202**, 1067.

Irisawa, A., and Hama, K. (1965). *Z. Zellforsch. Mikrosk. Anat.* **68**, 674.

James, T. N. (1970). *Amer. J. Cardiol.* **25**, 213.

James, T. N., and Sherf, L. (1968a). *Circulation* **37**, 1049.

James, T. N., and Sherf, L. (1968b). *Amer. J. Cardiol.* **22**, 389.

James, T. N., Sherf, L., Fine, G., and Morales, A. R. (1966). *Circulation* **34**, 139.

Jamieson, J. D., and Palade, G. (1964). *J. Cell Biol.* **23**, 151.

Jennings, M. A., and Florey, H. (1967). *Proc. Roy. Soc., Ser. B* **167**, 39.

Jennings, M. A., Marchesi, V. T., and Florey, H. (1962). *Proc. Roy. Soc., Ser. B* **156**, 14.

Jennings, R. B., Baum, J. H., and Herdson, P. B. (1965). *Arch. Pathol.* **79**, 135.

Jennings, R. B., Herdson, P. B., and Sommers, H. M. (1969). *Lab. Invest.* **20**, 548.

Johnson, E. A., and Sommer, J. R. (1967). *J. Cell Biol.* **33**, 103.

Karnovsky, M. J. (1964). *J. Cell Biol.* **23**, 217.

Karnovsky, M. J. (1967). *J. Cell Biol.* **35**, 213.

Karnovsky, M. J. (1968). *J. Gen. Physiol.* **52**, 645.

Karnovsky, M. J., and Hug, K. (1963). *J. Cell Biol.* **19**, 255.

Katz, B. (1948). *Proc. Roy. Soc., Ser B* **135**, 506.

Kawaguti, S. (1963a). *Biol. J. Okayama Univ.* **9**, 140.

Kawaguti, S. (1963b). *Biol. J. Okayama Univ.* **9**, 27.

Kawaguti, S. (1963c). *Biol. J. Okayama Univ.* **9**, 1.

Kawaguti, S. (1963d). *Biol. J. Okayama Univ.* **9**, 11.

Kawamura, K. (1961). *Jap. Circ. J.* **25**, 594.

Kawamura, K., and Hayashi, K. (1966). *Jap. Circ. J.* **30**, 149.

Kawamura, K., Hayashi, K., and Maekawa, M. (1964). *Proce. Asian Pac. Congr Cardiol., 3rd, 1964* pp. 136–140.

Kelly, R. E., and Hayes, R. L. (1969). *J. Morphol.* **127**, 163.

King, D. W., and Gollnick, P. D. (1970). *Amer. J. Physiol.* **218**, 1150.

Komuro, T. (1969). *J. Electronmicrosc.* **18**, 291.

Korn, E. D. (1969). *Anun. Rev. Biochem.* **38**, 263.

Kriebel, M. E. (1968). *J. Gen. Physiol.* **52**, 46.

Leak, L. V. (1967). *Amer. J. Anat.* **120**, 553.

Leak, L. V. (1969). *J. Morphol.* **128**, 131.

Leak, L. V. (1970). *J. Ultrastruct. Res.* **31**, 76.

Leak, L. V., and Burke, J. F. (1964). *Anat. Rec.* **149**, 623.

Lee, S. H. (1969). *Histochemie* **19**, 99.

Legato, M., and Langer, G. A. (1969). *J. Cell Biol.* **41**, 401.

Legato, M., Spiro, D., and Langer, G. A. (1968). *J. Cell Biol.* **37**, 1.

Leyton, R. A., and Sonnenblick, E. H. (1971). *J. Cell Biol.* **48**, 101.

Lindner, E. (1957). Z. *Zellforsch. Mikrosk. Anat.* **45**, 702.

Lockwood, W. R., Clower, B. R., and Hetherington, F. (1969). *Amer. J. Anat.* **126**, 185.

Loewenstein, W. R., and Kanno, Y. (1964). *J. Cell Biol.* **22**, 565.

Loewenstein, W. R., Nakas, M., and Socolar, S. J. (1967). *J. Gen. Physiol.* **50**, 1865.

McAllister, R. E. (1968). *Biophys. J.* **8**, 951.

McAllister, R. E. (1969). *Biophys. J.* **9**, 571.

McCallister, B. D., and Brown, A. L. (1965). *Lab. Invest.* **14**, 692.

McCallister, L. P., and Page, E. (1971). *Amer. Soc. Cell Biol. Abstr.* p. 183.

McCallister, L. P., and Page, E. (1973). *J. Ultrastruct. Res.* **42**, 136.

McCann, F. V., and Sanger, J. W. (1969). In "Comparative Physiology of the Heart: Current Trends" (F. V. McCann, ed.), pp. 29–46. Birkhuenser, Basel.

McNutt, N. S. (1970). *Amer. J. Cardiol.* **25**, 169.

McNutt, N. S., and Fawcett, D. W. (1969). *J. Cell Biol.* **41**, 46.

McNutt, N. S., and Weinstein, R. S. (1970). *J. Cell Biol.* **47**, 666.

Maekawa, M., Kawamura, K., Torii, H., and Hayashi, K. (1964). *Proc. Asian Pac. Congr. Cardiol., 3rd, 1964* pp. 1703–1707.

Maekawa, M., Nohara, Y., Kawamura, K., and Hayashi, K. (1967). In "Electrophysiology and Ultrastructure of the Heart" (T. Sano, V. Mizuhira, and K. Matsuda, eds.), pp. 41–54. Grune & Stratton, New York.

Martin, A. M., and Hackel, D. B. (1966). *Lab. Invest.* **15**, 243.

Martinez-Palomo, A., Alanis, J., and Benitez, D. (1970). *J. Cell Biol.* **47**, 1.

Meddoff, D. A., and Page, E. (1968). *J. Ultrastruct. Res.* **24**, 508.

Melax, H., and Leeson, T. S. (1967). *Cardiovasc. Res.* **1**, 349.

Melax, H., and Leeson, T. S. (1969). *Cardiovasc. Res.* **3**, 261.

Mendez, C., Mueller, W. J., and Urguiaga, X. (1970). *Circ. Res.* **26**, 135.

Mizuhira, V., Hirakow, R., and Ozawa, H. (1967). In "Electrophysiology and Ultrastructure of the Heart" (T. Sano, V., Mizuhira, and K. Matsuda, eds.), pp. 15–26. Grune & Stratton, New York.

Mizuno, S., Araya, K., and Takahashi, A. (1965). *Jap. Circ. J.* **29**, 1261.

Mobley, B. A., and Page, E. (1972). *J. Physiol. (London)* **220**, 547.

Moor, H., Ruska, C., and Ruska, H. (1964). *Z. Zellforsch. Mikrosk. Anat.* **62**, 581.

Moore, D. H., and Ruska, H. (1957). *J. Biophys. Biochem. Cytol.* **3**, 261.

Muir, A. R. (1967). *J. Anat.* **101**, 239.

Nayler, W. G., and Merrillees, N. C. R. (1964). *J. Cell Biol.* **22**, 533.

Nelson, D. A., and Benson, E. S. (1963). *J. Cell Biol.* **16**, 297.

Niehaus, H., Poche, R., and Reimold, E. (1963). *Virchows Arch. Pathol. Anat. Physiol.* **337**, 245.

Nisbet, R. H., and Plummer, J. M. (1966). *Proc. Malacol. Soc. London* **37**, 199.

Nisbet, R. H., and Plummer, J. M. (1969). In "Comparative Physiology of the Heart: Current Trends" (F. V. McCann, ed.), pp. 47–68. Birkhaenser, Basel.

Noble, D. (1962a). *J. Physiol. (London)* **160**, 317.

Noble, D. (1962b). *Excerpta Med., Sect. 1* **II**, 177.

Noble, D. (1962c). *Biophys. J.* **2**, 381.

North, R. J. (1963). *J. Ultrastruct. Res.* **8**, 206.

Novi, A. M. (1968a). *Experientia* **24**, 1231.

Novi, A. M. (1968b). *Anat. Rec.* **160**, 123.

Novi, A. M. (1969). *Virchows Arch., B* **2**, 24.

Paes de Carvalho, A., Hoffman, B. F., and de Paula Carvalho, M. (1969). *J. Gen. Physiol.* **54**, 607.

Page, E. (1967a). *J. Ultrastruct. Res.* **17**, 63.

Page, E. (1967b). *J. Ultrastruct. Res.* **17**, 72.

Page, E. (1968). *J. Gen. Physiol.* **51**, Part 2, 221s.

Page, E. (1969). *Ala. Med. Sci.* **6**, 85.

Page, E., Power, B., Fozzard, H. A., and Meddoff, D. A. (1969). *J. Ultrastruct. Res.* **28**, 288.

Page, E., McCallister, L. P., and Power, B. (1971). *Proc. Nat. Acad. Sci. U.S.* **68**, 1465.

Page, E. and McCallister L. P. (1973). *Amer. J. Cardiol.* **31**, 172.

Page, E., Polimeni, P., Zak, R., Earley, J., and Johnson, M. (1972). *Circ. Res.* **30**,

Page, L., and Adams, N. I., Jr. (1958). *In* "Principles of Electricity," 3rd ed., Univ. Phys. Ser. Van Nostrand-Reinhold, Princeton, New Jersey.

Pager, J. (1971). *J. Cell Biol.* **50**, 233.

Palade, G. E. (1961). *Circulation* **24**, 368.

Palade, G. E., and Bruns, R. R. (1968). *J. Cell Biol.* **37**, 633.

Pape, C., Kübler, W., and von Smekal, P. (1969). *Beitr Pathol. Anat. Allg. Pathol.* **140**, 23.

Poche, R. (1959). *Z. Zellforsch. Mikrosk. Anat.* **50**, 332.

Poche, R., and Hausamen, T. U. (1965). *J. Ultrastruct. Res.* **12**, 579.

Poche, R., and Mönkemeier, D. (1962). *Virchows Arch. Pathol. Anat. Physiol.* **335**, 271.

Poche, R., Maltos, C. M. D. M., Rembarz, H. W., and Stoepel, K. (1968). *Virchows Arch., A* **344**, 100.

Politoff, A. L., Socolar, S. J., and Loewenstein, W. R. (1967). *Biochim. Biophys. Acta* **135**, 791.

Politoff, A. L., Socolar, S. J., and Loewenstein, W. R. (1969). *J. Gen. Physiol.* **53**, 498.

Porter, K. R., and Palade, G. E. (1957). *J. Biophys. Biochem. Cytol.* **3**, 269.

Rayns, D. G., Simpson, F. O., and Bertaud, W. S. (1968). *J. Cell Sci.* **3**, 467.

Reuter, H. (1967). *J. Physiol. (London)* **192**, 479.

Revel, J. P., and Karnovsky, M. J. (1967). *J. Cell Biol.* **33**, C7.

Rhodin, J. A. G., del Missier, P., and Reid, L. C. (1961). *Circulation* **24**, 349.

Richter, G. W., and Kellner, A. (1963). *J. Cell Biol.* **18**, 195.

Robertson, J. D. (1963). *J. Cell Biol.* **19**, 201.

Rose, B. (1970). *Science* **169**, 607.

Rose, B., and Loewenstein, W. R. (1969). *Biochim. Biophys. Acta* **173**, 146.

Rostgaard, J., and Behnke, O. (1965). *J. Ultrastruct. Res.* **12**, 579.

Rougier, O., Vassort, G., and Stämpfli, R. (1968). *Pflüegers Arch. Gesamte Physiol. Menschen Tiere* **301**, 91.

Ruska, H. (1965). *In* "Electrophysiology of the Heart" (B. Taccardi and G. Marchetti, eds.), pp. 1–19. Pergamon, Oxford.

Sakamoto, Y. (1969). *J. Gen. Physiol.* **54**, 765.

Sanger, J. W., and McCann, F. V. (1968). *J. Insect Physiol.* **14**, 1105.

Schipp, R., and Schäfer, A. (1969a). *Z. Zellforsch. Mikrosk. Anat.* **98**, 576.

Schipp, R., and Schäfer, A. (1969b). *Z. Zellforsch. Mikrosk. Anat.* **101**, 367.

Schneider, M. F. (1970). *J. Gen. Physiol.* **56**, 640.

Schoenberg, M., and Fozzard, H. A. (1971). Unpublished observations.

Schulze, W., and Wollenberger, A. (1969). *Histochemie* **19**, 302.

Shiina, S. I., Mizuhira, V., Uchida, K., and Amakawa, T. (1969). *Jap. Circ. J.* **33**, 601.

Shimamoto, T., and Hiramoto, Y. (1967). In "Electrophysiology and Ultrastructure of the Heart" (T. Sano, V. Mizuhira, and K. Matsuda, eds.), pp. 55–64. Grune & Stratton, New York.

Simpson, F. O. (1965). *Amer. J. Anat.* **117**, 1.

Simpson, F. O., and Oertelis, S. J. (1962). *J. Cell Biol.* **12**, 91.

Simpson, F. O., and Rayns, D. G. (1968). *Amer. J. Anat.* **122**, 193.

Sitte, H. (1967). In "Quantitative Methoden in der Morphologie" (E. R. Weibel and H. Elias, eds.), pp. 167–198. Springer-Verlag, Berlin and New York.

Sjöstrand, F. S. (1958). *J. Ultrastruct. Res.* **2**, 122.

Sjöstrand, F. S. (1963). *J. Ultrastruct. Res.* **9**, 561.

Sjöstrand, F. S. (1968). In "Ultrastructure in Biological Systems" (A. J. Dalton and F. Haguenau, eds.), Vol. 4, pp. 151–210. Academic Press, New York.

Sjöstrand, F. S., and Andersson, E. (1954). *Experientia* **10**, 369.

Sjöstrand, F. S., and Andersson-Cedergren, E. (1960). In "The Structure and Function of Muscle" (G. H. Bourne, ed.), Vol. 1, pp. 421–446. Academic Press, New York.

Slautterback, D. B. (1965). *J. Cell Biol.* **24**, 1.

Smith, J. R. (1963). *Anat. Rec.* **145**, 391.

Sohal, R. S., Sun, S. C., Colcolough, H. L., and Burch, G. E. (1968). *Lab. Invest.* **18**, 49.

Sommer, J. R., and Johnson, E. A. (1968). *J. Cell Biol.* **36**, 497.

Sommer, J. R., and Johnson, E. A. (1969). *Z. Zellforsch. Mikrosk. Anat.* **98**, 437.

Sommer, J. R., and Johnson, E. A. (1970). *Amer. J. Cardiol.* **25**, 184.

Sommer, J. R., and Spach, M. S. (1964). *Amer. J. Pathol.* **44**, 491.

Sperelakis, N. (1971). *Z. Zellforsch. Mikrosk. Anat.* **116**, 443.

Sperelakis, N., and Lehmkuhl, D. (1964). *J. Gen. Physiol.* **47**, 895.

Sperelakis, N., Rubio, R., and Redick, J. (1970). *J. Ultrastruct. Res.* **30**, 503.

Spira, A. (1967). Ph. D. Thesis, University of Michigan.

Spotnitz, H. M., Sonnenblick, E. H., and Spiro, D. (1966). *Circ. Res.* **18**, 49.

Staley, N. A., and Benson, E. B. (1968). *J. Cell Biol.* **38**, 99.

Stein, O., and Stein, Y. (1968). *J. Cell Biol.* **36**, 63.

Stein, R. S., Richter, R. A., Zussman, R. A., and Brynjolfsson, G. (1966). *J. Cell Biol.* **29**, 168.

Stenger, R. J., and Spiro, D. (1961). *J. Biophys. Biochem. Cytol.* **9**, 325.

Stoeckenius, W., and Engelman, D. M. (1969). *J. Cell Biol.* **42**, 613.

Tanaka, I., and Sasaki, Y. (1966). *J. Gen. Physiol.* **49**, 1089.

Tarr, M., and Trank, J. (1971). *J. Gen. Physiol.* **58**, 511.

Tarr, M., and Sperelakis, N. (1964). *Amer. J. Physiol.* **207**, 691.

Tasaki, I., and Hagiwara, S. (1957). *Amer. J. Physiol.* **188**, 423.

Thaemert, J. C. (1966). *J. Cell Biol.* **29**, 156.

Thaemert, J. C. (1970). *Amer. J. Anat.* **128**, 239.

Tice, L. W., and Barnett, R. J. (1962). *J. Cell Biol.* **15**, 401.

Torii, H. (1962). *Jap. Circ. J.* **26**, 39.

Tosteson, D. C., Andreoli, T. C., Treffenberg, M., and Cook, P. (1968). *J. Gen. Physiol.* **51**, 3735.

Trautwein, W., and Uchizono, K. (1963). *Z. Zellforsch. Mikrosk. Anat.* **61**, 96.

Trautwein, W., Kuffler, S. W., and Edwards, C. (1956). *J. Gen. Physiol.* **40**, 135.

Trillo, A., and Bencosme, S. A. (1965). *Arch. Inst. Cardiol. Mex.* **35**, 803.

Trillo, A., Martinez-Palomo, A., and Bencosme, S. A. (1966). *Arch. Inst. Cardiol. Mex.* **36**, 54.

Tubbs, F. E., Crevasse, L., and Wheat, M. W., (1964). *Circ. Res.* **14**, 236.

Viragh, S. (1968). *In* "Symposium on Muscle" (E. Ernst and F. B. Straub, eds.), pp. 67–68. Akadémiai Kiadó, Budapest.

Viragh, S., and Porte, A. (1961a). *Z. Zellforsch. Mikrosk. Anat.* **55**, 263.

Viragh, S., and Porte, A. (1961b). *Z. Zellforsch. Mikrosk. Anat.* **55**, 282.

Walker, S. M., Schrodt, G. R., and Edge, M. B. (1970). *Amer. J. Anat.* **128**, 33.

Wegner, G., and Mölbert, E. (1966). *Virchows Arch. Pathol. Anat. Physiol.* **341**, 54.

Weibel, E. R., Kistler, G. S., and Scherle, W. F. (1966). *J. Cell Biol.* **30**, 23.

Weidmann, S. (1952). *J. Physiol. (London)* **118**, 348.

Weidmann, S. (1966). *J. Physiol. (London)* **187**, 323.

Weidmann, S. (1970). *J. Physiol. (London)* **210**, 1041.

Woodbury, J. W., and Crill, W. E. (1970). *Biophys. J.* **10**, 1076.

Woodbury, J. W., and Gordon, A. M. (1965), *J. Cell. Comp. Physiol.* **66**, 35.

Woodbury, J. W., White, S. H., Mackey, M. C., Hardy, W. L., and Chang, D. B. (1970). "Physical Chemistry," Vol. 9B, Chapter 11. Academic Press, New York.

Yamamoto, T. (1967). *In* "Electrophysiology and Ultrastructure of the Heart" (T. Sano, V. Mizuhira, and K. Matsuda, eds.), pp. 1–14. Grune & Stratton, New York.

3

CONTRACTILE STRUCTURES IN SOME PROTOZOA (CILIATES AND GREGARINES)

E. VIVIER

In Metazoa the function of contractility belongs to certain specialized cells organized into contractile tissue and forming muscles: striated muscles and smooth muscles.

In Protozoa the single cell has to carry out all the functions of relation to environment, nutrition and reproduction. Many Protozoa in every group react to various stimuli by contraction. This contractility, however, is never the basic function of the cell; it is merely one of many. Consequently the investigation and study of contractile structures in Protozoa runs into numerous difficulties. While in the specialized cell of Metazoa the contractile structures can easily be located, since they represent the sole function, in the unicellular organism these structures are closely intermixed with many others engaged in other functions. For this reason possible contractile elements, may be found beside sensory or conducting skeletal or elastic organites or the ones of nutrition or defense; thus only in a few rare favorable cases is it possible to assume with any chance of success that some structures are contractile elements, but never with absolute certainty.

The contractile elements have been sought among the fibrillar structures, both the microtubular and the microfilamentous type, but these structures are extremely abundant and varied in Protozoa and have proved to be connected with numerous functions (see Pitelka, 1968).

In flagellates few species have revealed the existence of contractile structures with a well established function; the best observations in this area are those of Grassé (1956), followed by those of Grimstone and Cleveland (1965) on the axostyle of certain zooflagellates. *In vivo* observations have shown that the axostyle performs undulating movements inside the cell; electron microscopy has revealed an organized system of groups of microtubules linked by bridges; the authors believe that this is a genuine contractile organite.

In Actinopoda the microtubular systems found in the axopodia have also been thought to be contractile, but in Rhizopoda and more particularly in Amoebae, as well as in slime molds, microfilaments have been found and these are considered to be responsible for movement (Wohlfarth-Botterman, 1964; Schaffer-Danneel, 1967; Pollard and Ito, 1970).

Contractile structures have best been characterized in ciliates and Sporozoa (gregarines), but numerous problems still arise, first of all as to their identification as such, and next, as to their mode of action. Only these two groups will be dealt with in the following account.

I. Contractile Structures in Ciliates

Ciliated Protozoa move by means of their cilia. The problem of ciliary movement is the same for them as for the ciliated cells of Metazoa and will not be discussed here. Many of these Infusoria, however, have demonstrated contractile properties either of their entire cell, or of specialized parts of their organism. In some cases these properties are truly spectacular and the ciliates that possess them furnish favorable material for the study of contractile structures.

A. *The Contractility of Ciliates*

Only ciliates revealing sizable contractile properties will be considered here. These are encountered chiefly in the following two groups: peritrichs and heterotrich spirotrichs.

Fig. 1. Photograph of a colony of *Epistylis anastatica,* showing extended and contracted animals. Scanning electron microscopy; ×600. (E. Small, Univ. Illinois, unpublished photograph.)

Peritrichs are ciliates attached to the substratum by a stalk, sometimes grouped in colonies (Fig. 1). Contraction is sometimes manifested on the body of the Infusorium and in certain species on the peduncle. Upon contraction of the body, the disk supporting the ciliated peristomial fringe is retracted and the collar tightens over it, while the cell assumes an ovoid shape (Fig. 1a and b); this contraction is produced by the simultaneous action of 3 endoplasmic myoid systems the general arrangement of which (Fig. 2) has long been known (see Fauré-Fremiet *et al.,* 1956):

1. a group of longitudinal myonemes tighten between the upper portion of the scopula and the inner surface of the epiplasm;

2. a kind of annular sphincter which comes behind the inner edge of the collar;

3. the retractor of the disk, which is a bundle going from the dorsal epiplasm to the deep portion of the peristomal ciliary fringe.

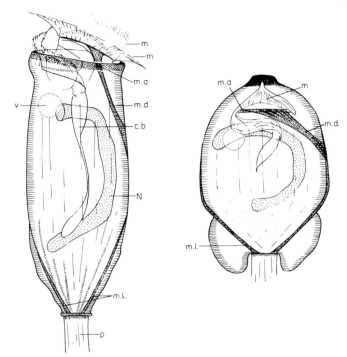

Fig. 2. Schemes showing an extended animal of *Epistylis* (left) and a contracted one (right); m = membranelle, m.a. = annular myoneme, m.d. = retractor myoneme of the disk, m.l. = longitudinal myoneme, N = nucleus, c.b. = buccal cavity, v = pulsatile vacuole, p = peduncle (stalk). (After Faure-Fremiet *et al.,* 1962.)

In the species with a contractile stalk (*Vorticella, Carchesium, Zoothamnium,* etc . . .), a voluminous myoneme, also known as a spasmoneme, produces retraction. These peduncles contract by performing an extremely quick spiral movement. After contraction the body, like the peduncle, progressively resumes its characteristic form; the elongation appears to be produced by some elastic structure, a myonemic antagonist.

Stentor and above all *Spirostomum* are among heterotrich spirotrichs, the most intensively studied species, in which contraction is the greatest. While *Stentor*, when expanded, is of a varyingly elongated conical form, the contracted animals becomes spherical and reduces its volume. This, however, has not been precisely evaluated, but the reduction appears to be by about $\frac{1}{3}$ (according to measurements furnished by Randall and Jackson, 1958).

The contraction of *Spirostomum* is even more spectacular, for it is accompanied by spiralization of the body (Fig. 3): there is a 75% reduction in length and a roughly 40% reduction in volume (according to Vivier *et al.*, 1969); it appears that the reduction in volume is accounted for by an evacuation of water through the pulsating system.

The mechanism of contraction in heterotrich ciliates, and more particularly in *Spirostomum*, has so for not been the subject of much study, but it appears to be extremely interesting from several points of view. Studies by Jones *et al.* (1966) and Sleigh (1969, 1970) have shown that no action potential is propagated in *Spirostomum* and they have established that excitation is from the anode. Legrand (1972) has demonstrated that "tetanization" does not appear to be possible and that experimental chemical excitation will only produce contraction-relaxa-

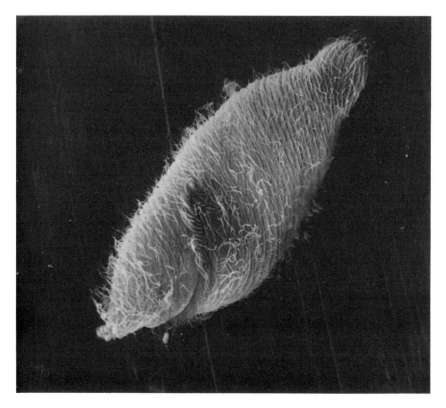

Fig. 3. Photograph of contracted *Spirostomum*. Note the spiralization of the body. The extended animal is about four times longer. Scanning electron microscopy; ×850. (Vinckier and Legrand, France, unpublished photograph.)

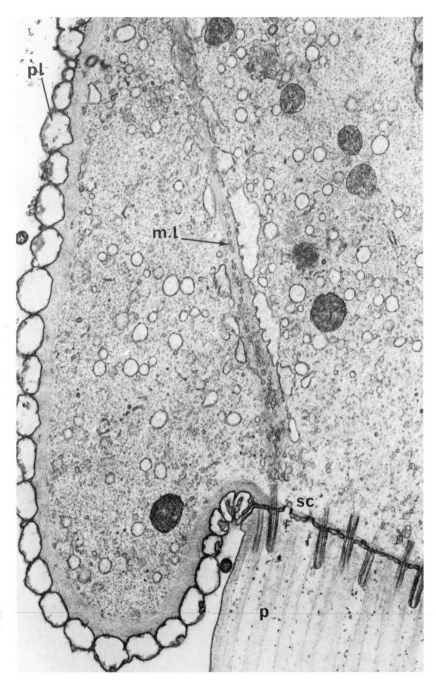

Fig. 4.

tions. This author, using a saline solution containing ATP, was able to obtain up to 40 contraction-relaxations in 3 minutes, a figure which would seem to be maximal.

Determination of the structures responsible for this contraction has been the subject of many controversies.

B. Contractile Structures

The fine structure of myonemes has only been observed in the state of contraction produced by fixatives.

In peritrichs, various works (Fauré-Fremiet *et al.*, 1956, 1962; Randall, 1956; Sotello and Trujillo-Cernez, 1959; Randall and Hopkins, 1962) have established that the different myonemes are made up of bundles of microfibrils running along the axis of the fiber which in some cases are somewhat wavy (Figs. 4 and 5); the diameter of the microfibrils is 30 to 40 Å. In the peduncle (Fig. 6) these bundles of microfibrils are obliquely inclined in relation to the axis of the stalk; the degree of obliqueness varies according to the species and in a given species also varies according to the area of the spasmoneme and the degree of its contraction.

On their endoplasmic surface the bundles of microfibrils have a border of large vesicles, so-called perimyal vesicles, belonging to the cellular endoplasmic reticulum; these vesicles have ribonucleoprotein granules on the cytoplasmic side, but are smooth on the microfibrillar side. In the peduncle (Favard and Carasso, 1965) a group of tubes runs through the fibrillar mass (Figs. 6 and 7), arranged parallel to the fibrils and regularly spaced; these tubes belonging to the endoplasmic reticulum have a diameter of around 500 Å and are about 800 Å distant from one another; they are sometimes ramified and run roughly alongside the periphery of the spasmoneme.

Endomyal vesicles (Fig. 5) have also been described (Fauré-Fremiet *et al.*, 1962) inside the fibrillar bundles, but their significance remains obscure.

In Heterotrichs several types of myonemes have been described with the aid of light microscopy (see Randall and Hopkins, 1962); electron

---

Fig. 4. Longitudinal section in the lower part of *Epistylis;* p = peduncle (stalk), m.l. = longitudinal myonemes, sc = scopula, pl = pellicle. Electron microscopy; ×11,000. (After Faure-Fremiet *et al.*, 1962.)

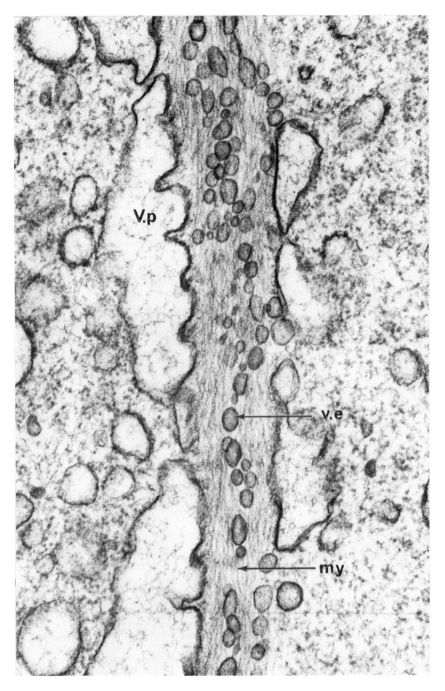

Fig. 5.

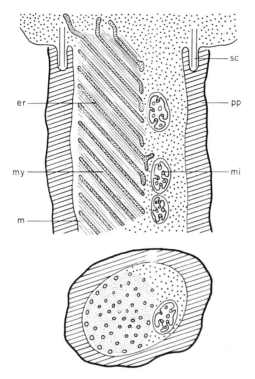

Fig. 6. Scheme showing the structural organization of the contractile stalk of *Carchesium;* longitudinal (top) and transverse sections (bottom); sc = scopula, er = tubular endoplasmic reticulum, m = plasmatic membrane, my = myofibrils of the spasmoneme, mi = mitochondria, pp = wall of the peduncle. (After Favard *et al.*, 1965.)

microscopy (Fig. 8) has shown the existence of several fibrillar systems in the subpellicular and ectoplasmic regions (Fauré-Fremiet *et al.*, 1956; Fauré-Fremiet and Roullier, 1958; Randall and Jackson, 1958; Finley, 1951; Randall, 1956; Yagin and Shigenaka, 1963; Finley *et al.*, 1964; Grain, 1966, 1968): (1) microtubular fibers (km fibers according to Randall and Jackson; ectoplasmic myonemes according to Fauré-Fremiet and Rouiller; ectomyonemes according to Finley *et al.*), grouped according to different systems and particularly abundant in the superficial

Fig. 5. View, at high magnification, of the longitudinal myoneme of *Epistylis.* Note the myofibrils (my), the perimyal vesicles of the endoplasmic reticulum (vp), and the endomyal vesicles (ve). Electron microscopy; ×50,000. (After Faure-Fremiet *et al.*, 1962.)

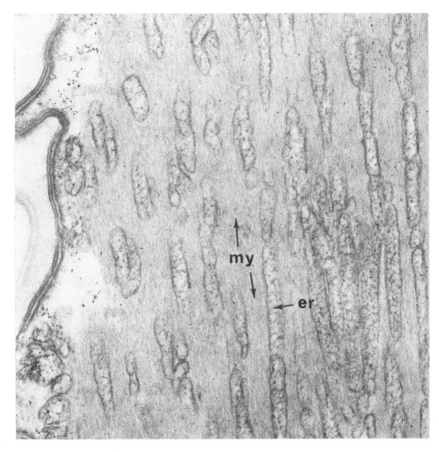

Fig. 7. Section of the peduncle of *Carchesium* showing the myofibrils (my) and the tubular system of the endoplasmic reticulum. Electron microscopy; ×70,000. (After Favard *et al.*, 1966.)

ectoplasm, more or less related to the kinetosoma (Grain, 1968); (2) microfibrils grouped in bundles (M bands according to Randall and Jackson; endoplasmic myonemes according to Fauré-Fremiet and Rouiller; endomyonemes according to Finley *et al.*) at the ectoendoplasmic border.

These two types of fibers are considered by some authors to be equally contractile (Fauré-Fremiet and Rouiller, 1958; Randall and Jackson, 1958); the latter authors in particular attribute an important role to superficial microtubular fibers in minor contractions and extension, with the deep microfibrils forming an additional system capable of reinforcing the contractile and extensible properties. Recent observations, however

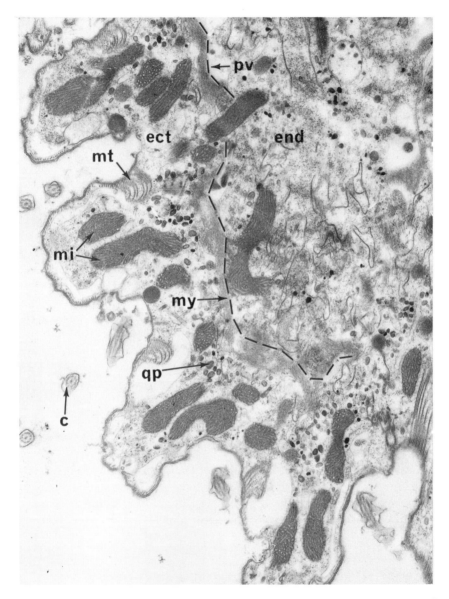

Fig. 8. Transverse section of a superficial region of *Spirostomum*. The layer of myonemes is shown by a broken line; c = cilia, mi = mitochondria, my = myoneme, gp = paraglycogen granules, mt = ectoplasmic and pellicular microtubules, end = endoplasm, ect = ectoplasm. Electron microscopy; ×14,000. (Vivier, unpublished photograph.)

(Grain, 1968; Vivier *et al.*, 1969; Legrand, 1972) tend to attribute the contractile function solely to the microfibrillar bundles at the ectoplasmic-endoplasmic border; in particular the experiments and studies of Legrand (1970) have shown that after glycerin extraction in *Spirostomum* only these structures are still intact and capable of contracting.

The studies of Bannister and Tatchell (1968) on *Stentor*, of Vivier *et al.* (1969), Legrand (1970) on *Spirostomum*, have established that the microfibrillar bundles are surrounded by vesicles identical to the perimyal vesicles of Peritricha; as to the constituting fibrils, these, according to Bannister and Tatchell, reach a diameter of 88 to 100 Å (in other words greater than that of the fibrils of Peritricha) and are tubular (the wall being 30 Å thick).

It would thus seem that the contractile elements in ciliates are made up of microfibrillar bundles fairly analogous to those existing in the smooth muscle cells of Metazoa. The perimyal vesicles correspond to the sarcoplasmic reticulum; the presence of calcium has been demonstrated in the spasmonemes of Peritricha by Carasso and Favard (1966). In *Spirostomum*, which has extraordinary contractile properties, there is even a group of structures (Fig. 9) corresponding to a remarkably functional system; Vivier *et al.* (1969) have shown that the myonemes are closely adjacent to the mitochondrial trabeculae that rise from the perimyal vesicles up to the peak of the ectoplasmic crests and that these mitochondria, extremely rich in tubules, are situated in an area abundantly furnished with reserve polysaccharides (paraglycogen).

II. Contractile Structures in Sporozoa (Gregarines)

Gregarines are the only Protozoa that show active movements in the vegetative stage. All Sporozoa, however, have a mobile stage in common, that which corresponds to the dissemination form, called sporozoite, schizozoite, merozoite or endozoite according to species (see Porchet-Hennere and Vivier, 1971).

A. *Contractility and Movement in Sporozoa*

A distinction must be made between the dissemination forms of Sporozoa and the gregarines, whose movements are more varied and complex.

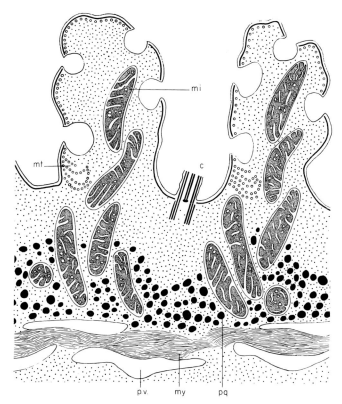

Fig. 9. Scheme of the superficial region of *Spirostomum* showing, in particular, the association of myofibrils, perimyal vesicles, mitochondria, polysaccharides granules (same symbols as in Fig. 8). (After Vivier *et al.*, 1969.)

1. DISSEMINATION FORMS

The dissemination forms of Sporozoa are very small organisms of around 10μ; they ensure propagation of the parasitic infection, either inside the same host or in different hosts.

These mobile stages are animated by torsion movements, flexion or slight undulation and they are able to displace themselves, although the mechanism of this movement is as yet inadequately known.

2. GREGARINES

In the vegetative stage (trophozoite) gregarines may reveal several types of movements. Some involve no appreciable displacement of the

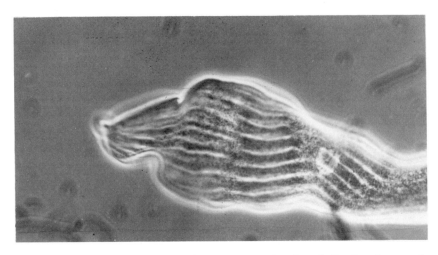

Fig. 10. Photograph *in vivo* of the gregarine *Selenidium hollandei,* showing the wavelike movements and contractions. Phase contrast microscopy; ×1,400. (Hildebrand, Lille, France, unpublished photograph.)

animal and take place on the spot; these are as follows (Fig. 10): deformation movements, torsion or flexion, in the form of localized irregular contractions; peristaltic movements, in the form of circular contractions, moving from one end to the other in alternating directions; pendulum or rolling as well as sigmoid movements, in the form of alternating and rhythmic flexing movements.

The other movements are connected with displacement and thus have the task of locomotion, which, for example in the case of *Lecudina* is 4 to 5 μ per second (Vivier, 1968). They are known as sliding or translation movements; they take place without appreciable deformation of the body and apparently without movement of the superficial structures on the standard scale of investigation *in vivo*. The mechanism of this movement remained obscure until recent years, when by means of electron microscopy on sections or scanning electron microscopy new hypotheses (Vivier, 1968) suggested a logical solution. Following the examination of transverse and tangential sections of the gregarines *Lecudina,* fixed while moving, Vivier (1968) has shown that the deformation of the pellicular folds (epicyte) corresponds to the requirements for locomotion by creeping or swimming; scanning microscopy (Vavra and Small, 1969; personal observations) has confirmed (Fig. 11) the existence of undulating movements of the epicytic folds. It would thus now seem that displacement by translation must be due to move-

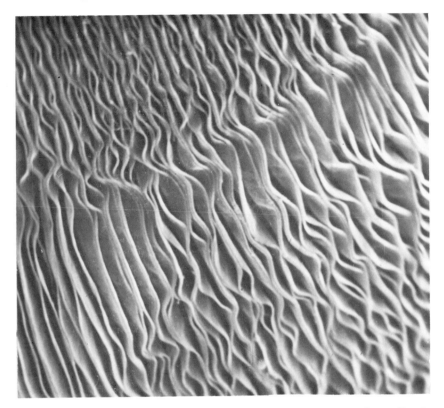

Fig. 11. Superficial view of the pellicular folds of the gregarine *Lecudina pellucida.* Note the undulations that explain the locomotion of the animal. Scanning electron microscopy; ×2900. (Vinckier, Lille, France, unpublished photograph.)

ment of the folds, which naturally may vary in detail from one species to the other. This concept, however, has so far only been supported by examination of static pictures, due to the small size of the structures involved.

In some gregarines there may exist contractile elements related to a function different of locomotion, this is the case, for instance, with the suckers of the genus *Lecudina* (Schrevel and Vivier, 1966).

B. Contractile Structures

The structures responsible for contraction have only been very tentatively identified in Sporozoa. They have obviously been sought among

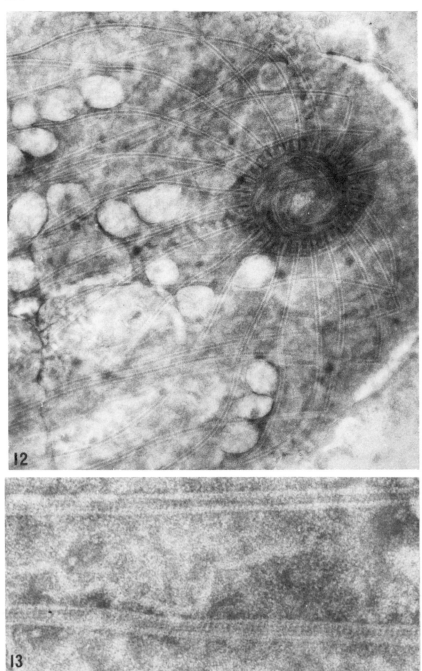

Figs. 12 and 13.

the fibrillar elements, but in the majority of cases no direct proof could so far be furnished as to their possible contractile properties.

1. CONTRACTILE STRUCTURES OF INFECTIOUS GERMS

The ultrastructure of various infectious germs has been examined in many Sporozoa by numerous authors (see Porchet-Hennere and Vivier, 1971). The only fibrillar structures existing in the cell which could give rise to movements are made up of longitudinal microtubules on the periphery, underneath the pellicle, apparently joined by a ring on the anterior part of the organism (Fig. 12).

These microtubules, which often number about 20, are not accompanied by any other kind of fibril. Their diameter is about 250 Å. They appear to be made up (unpublished work still in progress) of a wall composed of longitudinal elements (protofibrils analogous to those demonstrated in many types of microtubules) and by a central region with a periodic structure suggested a pile of disks (Fig. 13).

While most authors are in agreement that these are contractile elements, no proof of their properties has been furnished and it is possible that the movements are due to contraction of cytoplasmic proteins, in which case the microtubules would play the role of elastic or skeletal structures.

2. CONTRACTILE STRUCTURES OF *Selenidium*

Selenidium are gregarines of a primitive type (*Archigregarina*), parasites of polychaet annelids worms, which do not displace themselves, but reveal pendulum or rolling movements. Their ultrastructure has been the subject of various works (Vivier and Schrevel, 1964; Schrevel, 1971); it reveals fibrillar longitudinal bundles situated immediately beneath the pellicle in the form of superficial bulges called myonemes (Fig. 14).

These bundles are made up of one, two, sometimes three layers of microtubules. In transverse sections (Fig. 15) the microtubules, whose

---

Fig. 12. View, in negative staining, of the anterior region of merozoite of the Coccidia *Eimeria necatrix*. Note the under-pellicular fibers radiating from the apical ring. Electron microscopy; ×71,000. (Dubremetz, Lille, France, unpublished photograph.)

Fig. 13. View, at top magnification, of the fibers underneath the pellicle in *Eimeria necatrix*. Note the longitudinal and transverse complex structure. Electron microscopy; ×240,000. (Dubremetz, Lille, France, unpublished photograph.)

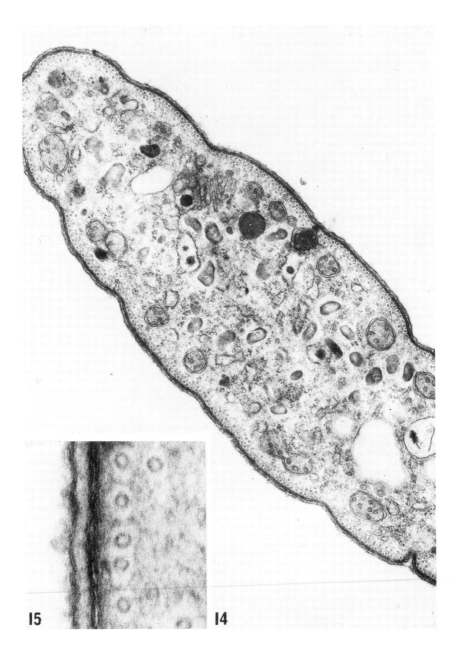

Figs. 14 and 15.

diameter is about 300 Å, appear to be surrounded by a light zone out-lined by a dark hexagonal edge. The microtubules reveal the usual fine structure. Examination of their wall suggests the presence of structural longitudinal protofibrils.

The dark edge which outlines the hexagons, at the center of which are the microtubules, may correspond to the transverse section of the microfibrils which seem to appear in longitudinal view (Fig. 16); how-ever, there has so far been no confirmation of this.

Thus in *Selenidium* there is a well defined organization of joined fibrillar elements. No direct proof has been furnished to confirm the contractile properties of these "myonemes", but various arguments may be taken into account; among them mention should be made of the observations drawn from the comparative study of the various species of *Selenidium* (Schrevel, 1971). It appears that the number of tubular fibers and the number of fiber layers is directly related to the locomotive activity of Gregarines. In *S. hollandei* and *S. pendula,* which are active species, the number of fibers is considerable, while in the Selenidiidae that are parasites of polychaetes Cirratulidae, which are slow-moving species, the number is smaller.

Some other observations (Figs. 15 and 16) that show the occasional presence of straight longitudinal fibers underneath a folded pellicle also suggest the possibility of retraction of the "myoneme" as being the cause of the folding. Finally, it is particularly interesting to note that in the active species, these "myonemes" are in the immediate vicinity of a dense mitochondrial layer (mitochondria combined to form a reticulum) and capable of providing the necessary energy, since this mitochondrial layer (Fig. 18) is in contact with the abundant reserves of polysaccharides of the endoplasm (Vivier and Schrevel, 1966).

In any event it is interesting to find (Fig. 19) an association in this arrangement of microtubules on the one hand, and structures that are undoubtedly microfibrillar at the hexagonal border on the other; this could be the equivalent of the thick and fine filaments of the striated muscle of Metazoa.

Fig. 14. Transverse section of the gregarine *Selenidium hollandei,* showing in particular the layer of myonemes which covers the inside of the pellicle. Electron microscopy; ×18,000. (After Vivier *et al.,* 1964.)

Fig. 15. View, at top magnification, of the peripheral region of the gregarine, showing especially the organization of the microtubular fibers (their wall has a substructure) in the center of the hexagonal area. Electron microscopy; ×120,000. (After Vivier *et al.,* 1964.)

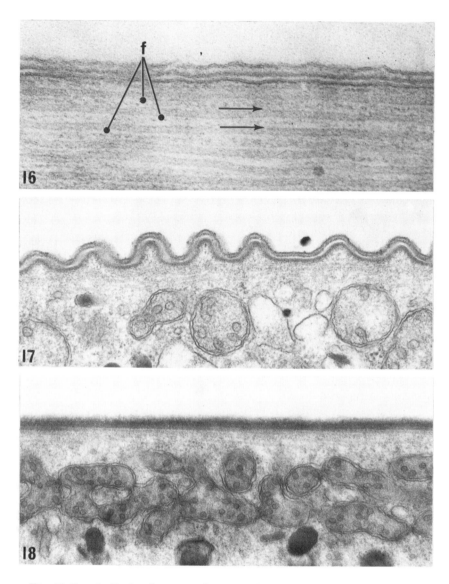

Fig. 16. Longitudinal and tangential section of the gregarine *Selenidium filiformis*, showing the fibrils (f), parallel to the surface, separated from one another by a clear area; at the middle of this area, there is a thin dense line (arrows). Electron microscopy; ×80,000. (After Schrevel, 1970.)

Fig. 17. Longitudinal and oblique section of the gregarine *Selenidium pendula* showing some roughly straight fibers underneath of a folding pellicle. Electron microscopy; ×36,900. (After Schrevel, 1970.)

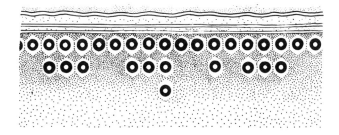

Fig. 19. Scheme showing in transverse section the superficial region of a gregarine *Selenidium* with the three membranes of the pellicle and, underneath, the layers of microtubules. (After Vivier *et al.*, 1964.)

3. Contractile Structures of Adhering Organs

In some cases, in particular in the genus *Lecudina*, it has been demonstrated that the anterior mucron corresponds to a sucker (Schrevel and Vivier, 1966).

The inside of the mucron is occupied by a multitude of small fibrils (Figs. 20 and 21) with a diameter of 60–70 Å; they have a wavy course and appear to be arranged in slightly oblique and crossed bundles. They seem to be attached directly to the lower surface of the anterior cell wall and stretch out towards the rear over a distance of about 2 μ.

4. Contractile Structures of Eugregarines

The majority of eugregarines have the property of displacing themselves by moving the epicytic folds (see above); they also frequently reveal peristaltic movements, flexion, or localized contractions. Thus here we have complex movements probably corresponding to different contractile structures.

The most recent examinations of the pellicular and ectoplasmic ultrastructure (Reger, 1967; Vivier, 1968; Vinckier, 1969; Warner, 1968; Desportes, 1969) have shown the existence of several fibrillar systems (Fig. 22) capable of acting in these different contractions:

Fig. 18. Superficial and oblique section of *Selenidium hollandei* showing the reticulated mitochondria underneath the myonemes. Electron microscopy; ×64,000. (After Schrevel, 1970.)

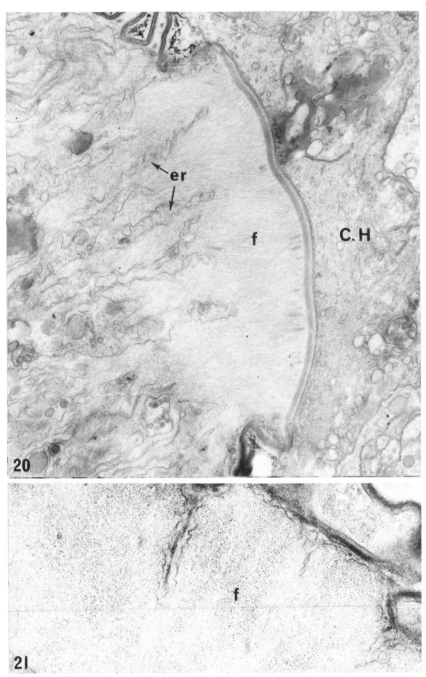

Figs. 20 and 21.

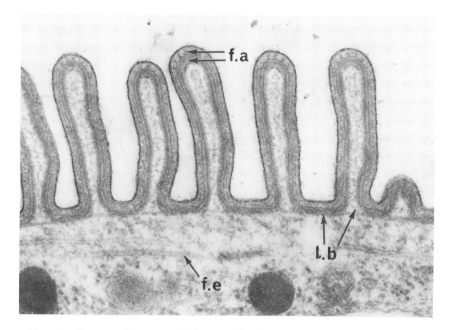

Fig. 22. Transversal section of the superficial region of the gregarine *Lecudina pellucida*. Electron microscopy; ×45,000. f.e. = ectoplasmic fibers, l.b. = basal lamina or epicytic layer, f.a. = apical fibrils. (Vivier, unpublished photograph.)

a. Ectoplasmic microtubular fibers. These are generally isolated and varyingly abundant in different species; they are situated some distance beneath the pellicle and are circular in gregarines. They have the usual fine morphology of microtubules; they may be in the form of circular contractile systems and thus could play a part in, for instance, the peristaltic contractions, but there has been nothing so far to establish their possible contractile properties. In any event they never reveal any strict overall organization and cannot therefore be considered as myonemes.

b. Ectoplasmic periodic structures. These have only been encountered in some *Monocystidea* (Vinckier, 1969). These are more or less fusiform

Fig. 20. Longitudinal section in the sucker of the gregarine *Lecudina tuzetae*. The picture shows the bundles of microfibrils starting from the anterior pellicle; C.H. = host-cell, er = endoplasmic reticulum. Electron microscopy; ×24,000. (Porchet-Hennere, Lille, France, unpublished photograph.)

Fig. 21. Transverse section in the sucker of the gregarine *Lecudina pellucida*, showing the section of the fibrils (f). Electron microscopy; ×30,000. (Vivier, unpublished photograph.)

formations, roughly 1 μ in length, with a thickness of 80 to 200 mμ, made up of alternatingly light and dark bands roughly 200 Å wide, a size reminiscent of the appearance of striated muscle in light microscopy. So far no role could be attributed to these structures.

c. *Dense subepicytic layer.* This layer, often designated as the basal layer, is immediately beneath the innermost membrane of the pellicle in many species and often surrounds the ectoplasm as a circular band that does not penetrate into the folds. Even though the fine structure could never be perfectly determined, this layer could be made up of fine microfibrils.

As in the preceding cases, however, no function could be attributed to this dense subepicytic layer, and it is possible that it plays either a skeletal or an elastic or a contractile role.

d. *Fibrils of the apex of the folds.* On transverse sections dense dots are observed at the apex of the folds and sections tangential to the apex reveal longitudinal fibrillar structures. These structures appear to be situated between and beneath the membranes of the pellicle in a well defined arrangement ("apical arch" and "apical densification" according to Vivier, 1968).

These may be contractile structures responsible for the wavy movements of the folds, but as yet few arguments can be advanced to prove their contractile properties. It appears, however, (unpublished work) that glycerinated extraction keeps these structures intact at the apex of the folds, which would be in accord with the properties of contractile proteins already recognized elsewhere.

III. Conclusion

Protozoa totipotential cells can reveal contractile properties and thus may possess contractile structures. In no case do we find a well differentiated cellular system in them identical to those known in cells of Metazoa. Nevertheless, in these unicellular organisms there are elementary contractile systems which sometimes are sufficiently organized to justify the term myoneme often applied to them.

The best characterized myonemes have chiefly been identified in ciliates, in which their structure recalls that of smooth muscle fibers of Metazoa cells with the usual association of the endoplasmic reticulum, mitochondria and even sometimes reserve polysaccharide substances. The properties of these myonemes, recognized as such, are currently the subject of cytophysiological and cytochemical studies that reveal the analogies and differences between them and the systems of specialized cells of Metazoa.

In Sporozoa, and more particularly in gregarines, which reveal movements of different types, the structures having supposed contractile properties are essentially of two kinds; one, of the microfibrillar type (suckers) is comparable in appearance to the myonemes of ciliates and thus the myofibrils of cells of smooth muscles of Metazoa; the other is microtubular. In the latter case it is sometimes the only visible organized structure capable of having this property, but it is still possible that contraction is carried out by unorganized cytoplasmic protein not visualized by the cytological techniques used. If such microtubules are the originators of movement, its mechanism is not known and the problem seems to be the same as the one of movements of cilia; thus we are probably dealing with a mechanism different from that which is currently accepted for the myofibrils of striated muscle. The case of Gregarinae of the genus *Selendium* may, however, prove to be interesting, for it has a fairly complex organization of "myonemes", if it is really proved to be contractile, could be compared with the organization of the cell of striated muscle.

Numerous problems remain, however, among which mention must be made of the associated structures, particularly the reticulum, which, in the case of gregarines, has not generally been found to be in contact with supposedly contractile structures; there is also the problem of the points of attachment.

In summary, while in certain cases an analogy could be found between the contractile structures of Protozoa and those of smooth muscle fibers, no true analogy could be established with the myofibrils of striated muscle. In elementary organisms these are elementary contractile structures, in other words not yet specialized and highly differentiated, but they are no less interesting for all that and this lightens the difficulty of their study.

REFERENCES

Bannister, L. H., and Tatchell, C. E. (1968). *J. Cell Sci.* 3, 295–308.
Carasso, N., and Favard, P. (1966). *J. Microsc. (Paris)* 5, 759–770.
Desportes, I. (1969). *Ann. Sci. Nat. Zool. Biol. Anim.* [12] 11, 31–96.
Fauré-Fremiet, E., and Rouiller, C. (1958). *Bull. Microsc. Appl.* 8, 117–119.
Fauré-Fremiet, E., Rouiller, C., and Gauchery, M. (1956). *Arch. Anat. Microsc. Morphol. Exp.* 45, 139–161.
Fauré-Fremiet, E., Favard, P., and Carasso, N. (1962). *J. Microsc. (Paris)* 1, 287–312.
Favard, P., and Carasso, N. (1965). *J. Microsc. (Paris)* 4, 567–572.
Finley, H. E. (1951). *Ann. N.Y. Acad. Sci.* 62, 229–246.
Finley, H. E., Brown, C. A., and Daniels, W. A. (1964). *J. Protozool.* 11, 264–280.
Grain, J. (1966). *Ann. Stat. Biol. Besse-en-Chandesse* 1, 71–76.

Grain, J. (1968). *Protistologica* 4, 27–35.

Grassé, P.-P. (1956). *Arch. Biol.* 67, 595–611.

Grimstone, A. V., and Cleveland, L. R. (1965). *J. Cell Biol.* 24, 387–400.

Inaba, R. (1961). *Bull. Biol. Soc. Horoshima Univ.* 10, 35–43.

Jones, A. R., Jahn, T. L., and Fonseca, J. R. (1966). *J. Cell. Physiol.* 68, 127–134.

Legrand, B. (1970). *Protistologica* 6, 283–300.

Pitelka, D. R. (1968). *In* "Research in Protozoology" (T. T. Chen, ed.), Vol. 3, pp. 279–388. Pergamon, Oxford.

Pollard, T. D., and Ito, S. (1970). *J. Cell Biol.* 46, 267–289.

Porchet-Hennere, E., and Vivier, E. (1972). *Annee Biol.* 10, 77–113.

Randall, J. T. (1956). *Symp. Soc. Exp. Biol.* 10, 185–196; *Nature* (*London*) 178, 9–14.

Randall, J. T., and Hopkins, J. M. (1962). *Phil. Trans. Roy. Soc. London Ser. B* 245, 59–79.

Randall, J. T., and Jackson, S. F. (1958). *J. Biophys. Biochem. Cytol.* 4, 807–830.

Reger, J. F. (1967). *J. Protozool.* 14, 488–497.

Schaffer-Danneel, S. (1967). *Z. Zellforsch. Mikrosk. Anat.* 78, 441–462.

Schrevel, J. (1971). *Protistologica* 7, 101–130.

Schrevel, J., and Vivier, E. (1966). *Protistologica* 2, 17–28.

Sleigh, M. A. (1969). *Proc. Intern. Congr. Protozool. 3rd 1969*, pp. 165–166.

Sleigh, M. A. (1970). *Acta Protozool.* 7, 335–352.

Sotello, J. R., and Trujillo-Cernez, O. (1959). *J. Biophys. Biochem. Cytol.* 6, 126–128.

Vavra, J., and Small, E. B. (1969). *J. Protozool.* 16, 745–757.

Vinckier, D. (1969). *Protistologica* 5, 505–517.

Vivier, E. (1968). *J. Protozool.* 15, 230–246.

Vivier, E., and Schrevel, J. (1964). *J. Microsc.* (*Paris*) 3, 651–670.

Vivier, E., and Schrevel, J. (1966). *J. Microsc.* (*Paris*) 5, 213–228.

Vivier, E., Legrand, B., and Petitprez, A. (1969). *Protistologica* 5, 145–159.

Warner, F. D. (1968). *J. Protzool.* 15, 59–73.

Wohlfarth-Bottermann, K. E. (1964). *In* "Primitive Motile Systems in Cell Biology" (R. D. Allen and N. Kamiya, eds.), pp. 79–109 Academic Press, New York.

Yagin, R., and Shigenaka, Y. (1963). *J. Protozool.* 10, 364–368.

THE MICROANATOMY OF MUSCLE

R. P. GOULD

I. Introduction

A universal characteristic of cells is their ability to respond to various kinds of stimuli. One form of their responsiveness is cell movement. Either a part of the cell, e.g., a cilium or mitochondrion, or the whole cell itself reacts by moving. Thus, cytoplasmic movement appears to be a fundamental feature of all cells. In multicellular animals some cells become highly specialized with respect to cytoplasmic movement and are called muscle cells or fibers. In man, a multicellular organism, where there are about 10^{15} cells but only a 100 or so different cell types (Sonneborn, 1964), there are three such specialized kinds of muscle cell—smooth muscle and two varieties of striated muscle, skeletal and cardiac.

One interesting general question that can be asked, but which as yet has no certain answer, is whether the mechanisms involved in muscle fiber movement or contraction are different in kind from the mechanisms encountered in other forms of cell movement, e.g., ameboid movement, cyclosis, or whether the mechanisms of muscle contraction are merely a specialized form of a more general process employed by all cells during cytoplasmic movement? Since smooth muscle appears to be the least specialized of the three kinds of muscle, it will be considered first. Although the emphasis in this chapter will be on muscle microanatomy, correlations both with the gross anatomy of muscle and its fine structure will be made where useful.

II. Smooth Muscle

Structural studies of smooth muscle may be divided into two main periods. First, the initial descriptions of smooth muscle structure made in the midnineteenth century (Henle, 1841; Schwann, 1847; Kölliker, 1849) were followed by a long period of light microscope study, with particular emphasis on the innervation of smooth muscle using silver and methylene blue techniques. This vast literature on smooth muscle innervation has been summarized by Hillarp (1959, 1960), Jabonero (1959, 1965), Richardson (1960), Taxi (1965), Botar (1966), and Burnstock (1970), and these works should be consulted for further details.

Second, with the application of the electron microscope to biological tissues in the early 1950s, studies on the fine structure of smooth muscle began to be published in the latter part of the decade and have continued up to the present time. Excellent summaries of this literature may be found in the reviews of Dewey and Barr (1968), Barr (1969), Burnstock (1970), and in this treatise (Csapo).

A. Distribution

Smooth muscle is known also as unstriped, unstriated, involuntary, plain, or visceral muscle. It has a widespread distribution in the mammalian body, since its presence in blood vessels ensures its presence in most organs. Because of this, it has an essential role to play in the functions of all organs and organ systems and may be said to subserve the body's vegetative processes. While somatic or striped muscle adjusts the organism to its environment, plain muscle may be said to fulfill an important role in maintaining the internal environment or physiological balance of the body.

Besides forming the contracting walls of blood vessels and large lymphatic vessels (vascular smooth muscle), it is also found in the walls of most hollow viscera (visceral smooth muscle), where it is usually arranged as an outer longitudinal coat and an inner circular coat of densely packed fibers. In some hollow organs, this arrangement may be reversed, e.g., possum bladder (Burnstock and Campbell, 1963), or a third, inner coat may be present of oblique (e.g., human stomach) or longitudinal fibers (e.g., human ureter).

In the human alimentary tract, smooth muscle is present from about the middle third of the esophagus to the anal canal, and by its activity ensures that the ingested food is carried along its length and is thoroughly mixed with the digestive juices. Finally, it participates in the excretion of the residue. In the respiratory system, it occurs in the airways from the trachea above to the alveolar ducts below.

The complement of smooth muscle in the walls of the urinary tract is concerned in the excretion of urine, while its presence in the nipple, areola of the nipple, stroma of the ovary, Fallopian tubes, uterus, vagina, erectile tissue of the clitoris and penis, parts of the male and female urethra, deep fascia of the penis, stroma of the prostate, seminal vesicles, ductus deferens, ductus epididymis, ductuli efferentes, seminiferous tubules, and dartos muscle ensures the survival of the race.

Other sites of smooth muscle occurrence are the dermis of the skin where it forms the arrectores pilorum (the little muscles which pull on the hair follicles), the subcutaneous tissue of the scrotum and penis (glans and prepuce), the perineum, the eyeball (in which it forms the musculature of the iris and ciliary body), the capsule and trabeculae

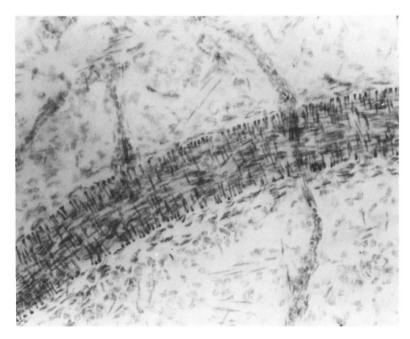

Fig. 1. Part of arteriole from frog bladder showing circularly arranged smooth muscle cells (×115).

of the spleen (in which the visceral muscle component is much less in man than in the pig, dog, or cat, and less than in the ox or sheep), and finally it may be mentioned that smooth muscle also occurs in the walls of the larger ducts of glands.

In blood vessels, the smooth muscle component is usually confined to the tunica media, although in the larger veins it is found in considerable amounts in the tunica adventitia. The muscle fibers in vessel walls are arranged in a circular or spiral fashion, with the circular orientation dominating (Strong, 1938; Pease and Paule, 1960; Keech, 1960); but as the blood vessels get smaller, the helically arranged fibers of the large vessels gradually assume a completely circular orientation (Fig. 1) in small arterioles (Rhodin, 1962, 1967).

B. Size, Shape, and Arrangement of Smooth Muscle Cells

While the study of fresh voluntary or skeletal muscle fibers teased apart from one another in, for example, Ringer's fluid has provided useful information, such a procedure using smooth muscle is not so successful, and indeed it is very difficult to view smooth muscle fibers satisfactorily. The main difficulty lies in the fact that the small fragile tips of the cells may be broken off during teasing of the fresh tissue or in the maceration of potassium hydroxide or nitric acid–glycerin-treated material. Because the cells may be broken, assessment of both cell shape and length becomes uncertain. Further, in the case of freshly teased fibers in physiological media, such isolated cells, because of their loss of tone, are longer than the resting cell in physiological conditions.

In fixed, stained sections of smooth muscle for light microscopy the changes introduced into the cells through the dehydration and embedding procedures make the sectioned tissue of only moderate value in the accurate assessment of both fiber shape and length. Determination

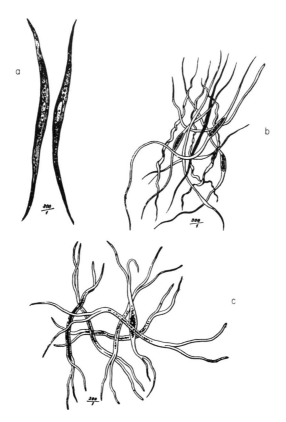

Fig. 2. Drawings of early studies of smooth muscle fibers: (a) muscle fibers treated with serum; (b) fibers from intestinal smooth muscle isolated by means of nitric acid; (c) dichotomously divided fibers isolated from a pleuritic membrane. Arnold (1870).

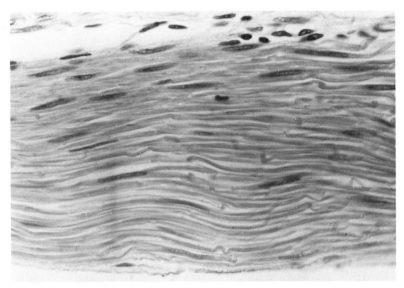

Fig. 3. Longitudinally sectioned intestinal smooth muscle fibers. Note their fusiform shape and elongated nuclei. (×1200.)

of the fiber tip in such sections is difficult since the fibers may be cut obliquely. However, serial sectioning of tissue fixed for electron microscopy helps, in part, to overcome this problem. The use of special stains such as those of Van Gieson, which colors collagen fibers red and smooth muscle yellow, or Mallory's triple stain, which yields blue and red, respectively, also helps not only in observing the structure of the muscle fibers, but also in avoiding the mistake of confusing smooth muscle with collagen.

The classic light microscope studies over the last 100 years or so have drawn for us a picture (Figs. 2–4) of the smooth muscle cell as an elongated, spindle-shaped element, usually with sharply pointed ends, although occasionally bifurcated, with an approximately centrally-placed nucleus. However, recent electron microscope examination of serially sectioned fibers (Taxi, 1965; Thaemert, 1966; Merrillees, 1968) have revealed several unusual features different from the previously accepted description. These differences have been summarized by Burnstock (1970) as follows:

1. The nucleus of smooth muscle fibers is not always centrally placed but varies up to 50 μ from the center point of the fiber.

2. Cells appear to vary in size, but this may be due to different degrees of contraction of different cells during fixation (Table I).

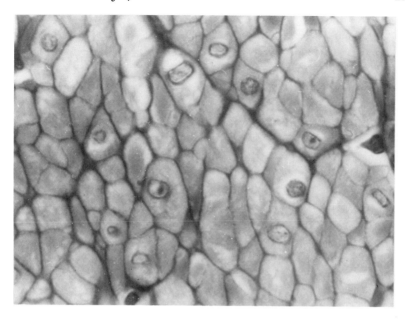

Fig. 4. Transversely sectioned visceral smooth muscle fibers. Note the approximately central nuclei and the different diameters of the fibers according to the level at which they were sectioned. (×1200.)

3. Cells are not really fusiform of double cones, but show extremely uneven contours along their length. For example, in the vas deferens, few fibers had a regular polyhedral profile in cross section, and the shapes changed considerably from level to level. Such transverse sections were polyhedral, triangular, or even extremely flattened ellipses, and many cells were ribbon- or rodlike in outline (Fig. 4).

The smallest fibers, such as those found in blood vessels, may be rather irregular in shape, short, broad, and usually flattened, with their ends deeply forked or even more extensively branched (Keech, 1960; Pease and Paule, 1960; Prosser *et al.*, 1960).

Other smooth muscle fibers, which do not conform to the conventional bipolar ellipsoidal shape, are those from the muscularis mucosae of the mouse intestine, which have many processes (Lane and Rhodin, 1964); the outer surface of smooth muscle cells from the rat small mesenteric veins (Devine and Simpson, 1967); and finally the iris dilator muscle, which consists of very narrow myoepithelial cells with unusually complicated outlines closely packed together with interlocking processes (Richardson, 1964).

TABLE I
Sizes of Typical Smooth Muscles

Tissue	Animal	Technique[a]	Length (μm)	Diameter (μm)
Intestine	Mouse	EM	400	—
	Cat	Mac	120	6
Longitudinal coat jejunum	Mouse	EM	150 (relaxed)	2–3.5
			70 (contracted)	2–3.5
Circular coat	Axolotl	EM	—	5
		Mac	1200	12
Taenia coli	Guinea pig	Mac	150	6
		EM	70–90	3–4
		EM	200 (relaxed)	2–4
Ciliary muscle	Cat	Mac	100	—
Nictitating membrane	Cat	EM	350–400	—
Retractor penis	Dog	Mac	90	6
Vas deferens	Guinea pig	EM	450 (slightly contracted)	—
Vascular smooth				
Small arteries	Mouse	EM	60	1.5–2.5
Arteriole	Rabbit	EM	30–40	5
	Rat	Living cell	—	2.08–2.78
Renal vein	Pig	Mac	110	3

[a] EM = electron microscope; Mac = nitric acid maceration. After Burnstock, (1970).

For the sake of completeness, it might be proper at this point to mention in more detail this type of cell which is also found in association with the secretory elements of the mammary glands, sweat glands, lacrimal glands, and the glands of the oral cavity. These are known as myoepithelial, basal, or basket cells, and they are able to act like smooth muscle cells and so aid movement of the secretion into the excretory ducts. They are situated between the secretory cells and the basement membrane and are either spindle-shaped or branched with cylindrical nuclei. They can be well seen if Masson staining is employed, when they then appear bright red. Unlike smooth muscle cells, myoepithelial cells are derived from ectoderm, not mesenchyme, and doubt has been expressed as to whether they are true unstriped muscle cells.

In the past, this difficulty has been debated with respect to the nature of the musculature of the iris (Vinnikow, 1938; Lowenstein and Lowenfield, 1962), which differentiates from the ectoderm of the optic cup. It has been suggested that the iridial muscles are really myoneural elements rather than true smooth muscle cells (Vinnikow, 1938), but

Richardson (1964) has now firmly established the myoepithelial character of these cells.

Smooth muscle cell dimensions show a considerable range, although not as great as that found in striped muscle. The largest fibers are found in the mammalian pregnant uterus (up to 12 μm $\times$ 600 μm) (Csapo, 1962) and urodele intestine (12 μm $\times$ 1200 μm) (Burnstock, 1970), and the smallest in arterioles (approximately 2 μm $\times$ 10–15 μm).

The arrangement of smooth muscle cells varies with the tissue studied. But in most visceral muscle coats (Figs. 5 and 6), the cells are arranged in branching bundles of irregular diameter and length. As neighboring bundles of cells anastomose with each other, the relative positions of individual muscle fibers within a single bundle change through the losses and gains of fibers within the bundle (Csapo, 1955, 1959, 1962; Prosser *et al.*, 1960; Merrillees *et al.*, 1963; Evans and Evans, 1964; M. R. Bennett and Rogers, 1967).

The fascicular arrangement of smooth muscle fibers correlates well with the physiological findings that muscle bundles rather than individual muscle cells are the effector units in smooth muscle systems. The minimum size for an effector bundle appears to be about 100 μm in

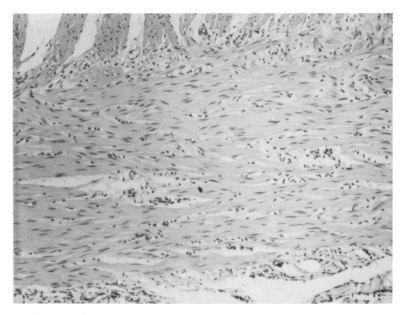

Fig. 5. Visceral smooth muscle. Densely packed longitudinally and transversely sectioned fibers may be seen. The longitudinal fibers are arranged in anastomosing fasciculi. ($\times$100.)

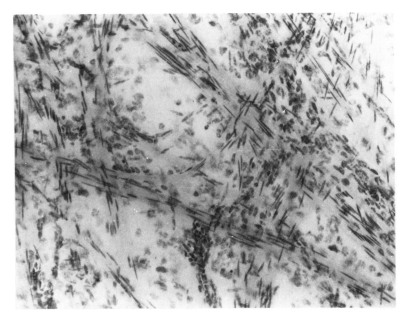

Fig. 6. Loosely arranged fasciculi of smooth muscle fibers in frog bladder (×115).

diameter (M. R. Bennett and Burnstock, 1968). The smooth muscle cells of arterial media have also been shown to have a fascicular arrangement (Boucek *et al.*, 1963).

C. Cytology of Smooth Muscle Cells

In the usual arrangement of smooth muscle in sheets, the fibers are so disposed that the thickest part of one fiber lies alongside the tapering ends of its neighbors (Fig. 4). In a study of the taenia coli, M. R. Bennett and Rogers (1967) have concluded that each muscle cell is surrounded by 10–12 others. Exceptions to this condition are the small muscles of the hairs, in which the nuclei lie approximately side by side, producing a localized thickening of muscles, and in many blood vessels where branched muscle cells appear to meet end to end (Rhodin, 1962). It follows, then, that in a cross section of smooth muscle, relatively few nuclei are seen (Fig. 4); some of the fibers are of course cut through their widest part and so show nuclei, but the remainder are cut elsewhere, and accordingly, fiber sections of differing shapes and sizes are present. The very smallest sections are those of the pointed ends of the fiber.

Each muscle fiber has a single elongated, ellipsoidal nucleus (Fig. 3) which occupies its thickest part. In a cross section of the fiber, the nucleus commonly is seen in an eccentric position. The length of the nucleus varies with that of the fiber, but not to the same extent, and its average dimensions are about 10–25 μm $\times$ 1–3 μm. The nucleus is elongated in the long axis of the fiber and has pointed or rounded ends, but owing to contraction of their containing muscle fibers, nuclei are frequently seen in stained preparations to be contorted to a greater or lesser degree. Spiral twisting is very common. Within the nucleus, there are two to five nucleoli and a chromatin network, which is mainly peripheral in its disposition. Adjacent to the nucleus, in a small depression in its membrane, a pair of centrioles have been observed.

At the ends of the nucleus, the sarcoplasm has a tendency to accumulate and small, spherical, lipid granules about 0.3 μm in diameter may be observed. Spherical to elongate mitochondria 0.5–0.8 μm are found throughout the cytoplasm, but also have a tendency to concentrate at the ends of the nucleus. The Golgi complex and lysosomes are also juxtanuclear in position. Glycogen granules, unless separately stained, will be lost from the cytoplasm, and their former sites will be indicated by empty spaces.

The most important components in the cytoplasm or sarcoplasm of smooth muscle cells are the contractile proteins. Under the light microscope, the cytoplasm looks fairly homogeneous in texture, but under certain conditions, fine longitudinal threads may be observed. These are called myofibrils. Less than 1 μm in diameter, they run the whole length of the cell, are doubly retractile under the polarizing microscope, and are quite lacking in the cross banding seen in striated muscle fibrils. From numerous electron microscope studies (Burnstock, 1970; Lowy and Small, 1970), it is clear that the myofibrils seen under the light microscope are due to aggregations of much smaller myofilaments of myosin and actin into the larger fibrils.

D. Intercellular Relationships in Smooth Muscle

Kölliker (1849) concluded from his studies on smooth muscle that the cells were discrete and separate. During the rest of the nineteenth century, despite repeated descriptions of protoplasmic strands connecting smooth muscle cells as intercellular bridges (McGill, 1909), Kölliker's views continued to prevail. With the work of McGill (1907, 1909) on the histogenesis of smooth muscle, the idea of protoplasmic continuity between the cells began to be generally accepted and was eventually fully developed by Hagquist (1931, 1956).

However, the real difficulty for these early workers in deciding the question of whether smooth muscle is a syncitium or not was a technical one. The limit of resolution of the light microscope is such that no decision on this problem could be made with certainty, and only with the application of the electron microscope could the problem be finally resolved. The studies of Mark (1956), Caesar *et al.* (1957), and Rhodin (1962) clearly established that smooth muscle is not a syncitium; and those of Dewey and Barr (1962, 1964) and Dewey (1965) established that, although there is no cytoplasmic continuity between cells, at discrete points a smooth muscle cell's plasma membrane appears to fuse with that of an adjacent cell, such a point of apparent fusion being called a nexus.

Although over much of their surface smooth muscle cells may be in close approximation to each other (500–800 nm) a variety of elements, including collagen, elastic fibers, mucopolysaccharides, blood vessels, nerves, Schwann cells, macrophages, and fibroblasts, may be seen in the extracellular space (Caesar *et al.*, 1957).

The connective tissue elements, mainly reticular (Fig. 7) and elastic fibers and PAS-positive mucopolysaccharides, form intimate sheaths around the muscle fibers supporting them and binding them together.

Fig. 7. Longitudinally sectioned visceral smooth muscle stained to show the fine reticular fibers forming delicate sheaths around the fibers (×1200).

These fine connective tissue fibers become continuous with the stouter strands of connective tissue (Fig. 5) which separate the layers or bundles of smooth muscle fibers and conduct the blood vessels and nerves through the tissue.

Generally, the blood supply of smooth muscle is thought to be sparser than that of skeletal muscle, but in a recent study of the blood supply of the human uterus (Farrer-Brown *et al.*, 1970) made by injecting the lower end of both uterine arteries with radiopaque medium and examining thick sections of tissue microradiographically, a profuse network of capillaries and venules closely related to groups of smooth muscle fibers was observed (Fig. 8).

E. Growth and Hypertrophy of Smooth Muscle

It is generally agreed that visceral smooth muscle cells are derived embryologically from mesenchyme cells (McGill, 1907; Yamamoto, 1961; Arey, 1965; Leeson and Leeson, 1965a,b; Yamauchi and Burnstock, 1969). The mesenchyme cells retract their several processes and gradually assume a fusiform shape. The cells then approximate to each other and form fasciculae. In Yamauchi and Burnstock's (1969) study of developing smooth muscle in the mouse vas deferens, the rapid increase in the length of the fiber during the first 20 days after birth was followed by a marked increase in diameter during the next 5 days. Fully developed muscle fibers were present in the mouse vas 1–3 months after birth. This is in contrast to rat ureter muscle cells, which are fully developed 1–5 days after birth (Leeson and Leeson, 1965a).

Increase in size and possibly number of smooth muscle cells in adult tissues may be induced either by hormones or, in blood vessels, by changes in blood pressure. The spectacular increase in size and weight of the human uterus during pregnancy from an organ some 8 cm or less in length and weighing about 30 gm to one of 32 cm length and 1000 gm weight represents the physiological response of its tissues, including the smooth muscle component to hormonal stimulation. But that smooth muscle may be stimulated to increase in bulk by less happy circumstances is witnessed all too frequently by the thickened and tortuous arteries of the hypertensive patient. With regard to the means whereby smooth muscle increases in bulk, there are clearly three possibilities. First, the individual fibers may increase in size, i.e., hypertrophy. That this does happen is undoubted, and according to Lange (1939), it is the enlargement of the individual fibers rather than an increase in their number that accounts for almost all the hypertrophy shown

Fig. 8. A microradiograph of human uterine myometrium showing the radiopaque filled capillaries forming a dense network round the smooth muscle cells (×270). By kind permission of G. Farrer-Brown.

by smooth muscle. Second, the muscle fibers may increase in number, i.e., proliferate by mitotic division. This also is known to happen (Muir, 1929), but how important smooth muscle hyperplasia is in an enlarging mass of smooth muscle is not certain. In a virgin rabbit, after injection of the female sex hormones, mitotic proliferation of the smooth muscle cells of the uterus is said to take place (Bloom and Fawcett, 1968). However, Barr (1969), asserts that there are no data that indicate a significant amount of cell division in smooth muscle. Finally, smooth muscle may increase in mass through new cells differentiating from undifferentiated mesenchyme cells, which are thought to persist in many parts of the body into adult life, e.g., the undifferentiated perivascular cells found alongside capillaries. However, absolute proof of such a transformation is difficult to obtain.

The capacity of smooth muscle to regenerate is small. Large defects in sheets of smooth muscle are usually filled with scar tissue, although at sites of injury, mitosis has been observed in the smooth muscle cells.

III. Voluntary, Somatic, or Skeletal Muscle

A. General Architecture

In considering the detailed structure of voluntary muscle, it is helpful to begin by examining a section through a whole muscle, such as the superior rectus muscle of the eyeball (Figs. 9 and 11). Surrounding the muscle, there is a connective tissue sheath called the epimysium, from the deep surface of which septa pass into the muscle at irregular intervals. These septa, which constitute the perimysium, invest bundles (fasciculi) of muscle fibers, the bundles being of somewhat angular outlines in section and of different sizes. Subdivision of the larger bundles into two or three successively smaller bundles is frequently observed. The name perimysium is retained for the connective tissue surrounding the lesser bundles. Finally, from the perimysium, very delicate extensions of fine connective tissue called the endomysium pass in to surround each muscle fiber (Fig. 10).

The size of the fasciculi composing a muscle determines its texture. Thus, in muscles capable of very finely judged movements (e.g., the ocular muscles), the texture is fine, whereas in muscles that perform movements demanding less precision but greater power (e.g., those of the buttock), the fasciculi are much larger and the texture is correspondingly coarse. It is of interest to note that this distinction between mus-

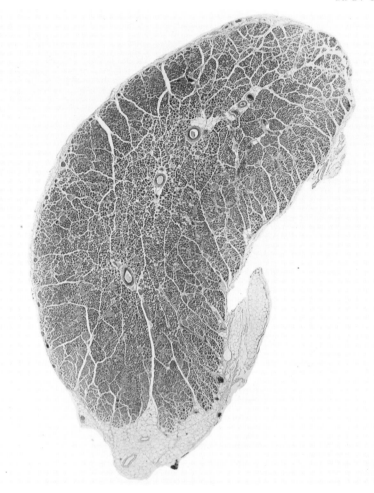

Fig. 9. Cross section of superior rectus muscle of 65-year-old human adult showing the general structure of a voluntary muscle (×19).

cles of fine and coarse texture is further instanced by differences in the sizes of their motor units.

In muscles that lie immediately under the deep fascia, the relation the epimysium bears to the tissue varies in different parts of the body. Thus, where the deep fascia is a well developed layer, as in the thigh, loose areolar tissue is present between the fascia and the epimysium of the subjacent muscles; in regions where the deep fascia is more delicate in texture, however, it is not clearly to be distinguished from the epimysium.

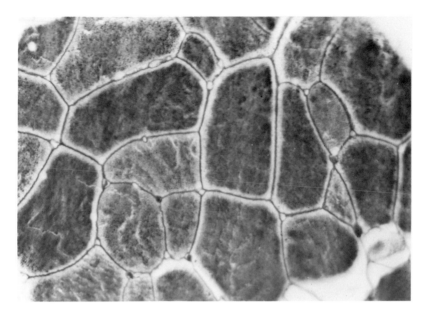

Fig. 10. Cross section of human voluntary muscle stained to show the thin sheath of reticular fibers round each muscle fiber forming the endomysium (×1000).

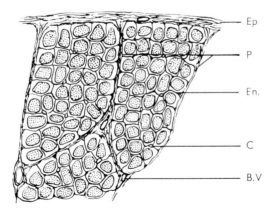

Fig. 11. Diagram of the connective tissue framework of a voluntary muscle based on part of Fig. 9. From the epimysium (Ep), which surrounds the whole muscle, septa pass inward to surround variously sized bundles of fibers. The connective tissue sheaths of the bundles constitute the perimysium (P), and from it delicate strands continue between the individual muscle fibers as the endomysium (En). Larger blood vessels (BV) are present within the perimysium, and capillaries (C) lie between the individual muscle fibers.

In sites where muscles take origin from the overlying deep fascia, it is not possible to demonstrate epimysium. Here, the process of cleaning a muscle, which elsewhere simply means the removal of epimysium whereby the run of the muscle fibers may be revealed, is tedious and at best unsatisfactory, involving as it does some cutting across the muscle fibers.

B. Connective Tissue Component of Voluntary Muscle

The amount of connective tissue relative to muscle fibers is much greater in some muscles than in others, a fact which to some extent explains why some cuts of meat are tougher than others. The lower the collagen content, it is supposed, the more tender is the cut, and vice versa. In lean meat, the proportion of the total protein formed by collagenous tissues ranges from 3% to 30%. However, the work of Locker (1960) followed by a number of other studies (Locker and Haggard, 1963; Herring *et al.*, 1965; Marsh and Leet, 1966; Cassens, 1970) has shown that the degree of muscle contraction is also directly related to meat tenderness.

The elements composing the connective tissue of muscle are collagen fibers, reticular fibers, elastic fibers, and several varieties of cells such as fibroblasts, histiocytes (macrophages, also called clasmatocytes or resting wandering cells), and fat cells. The amount of elastic tissue present varies with the type of action of the muscle, being greatest in those muscles attached to soft parts, e.g., the muscles of the tongue and of the face. It has, moreover, been shown by Bucciante and Luria (1934) that there is an increase in the elastic tissue component of the superior rectus muscle with aging. The connective tissue between the individual muscle fibers consists of reticular fibers and of fine collagen fibers; often very scanty in amount, it conveys the blood capillaries and the smallest branches of the nerves. The larger blood vessels and nerves lie within the perimysium between adjacent fasciculi, a situation in which muscle spindles are also commonly found. The manner of attachment of the connective tissue component to the tissue that serves as the origin of insertion of the muscle will be considered below in relation to the muscle–tendon junction. First, however, the mechanical role connective tissue plays within the substance of the muscle must be discussed. To some degree, the sheaths that it forms both for the individual muscle fibers and for the muscle fasciculi (interconnected as they are with one another) must regulate and control the extent of contraction. It is of interest that the proportion of connective tissue present is greater in muscles that are capable of finely graded movements (Fernand, 1949). Further, as Banus and Zetlin (1938) have shown, the initial tension that develops in a relaxed muscle when it is extended is due solely to its connective tissue content.

It is appropriate to mention here that during muscle regeneration, the endomysial sheaths of the individual fibers, in other words their investing connective tissue tubes, serve as guides for the new fibers, which sprout out from the stumps of the old.

C. Muscle Fibers of Voluntary Muscle

1. FIBER SHAPE

Within its connective tissue framework, the tissue of voluntary muscle is formed by independent, cylindrical fibers which are in fact elongated, multinucleated cells. This last character may allow each fiber to be regarded as a syncytium, but actually each is a structural unit with no other evidence of a composite cellular origin.

In mammals, it is rare to find the fibers anastomosing with their neighbors, but in invertebrates this is not uncommon. Generally speaking, too, voluntary muscle fibers do not branch, but in the tongue of the frog, the muscle fibers divide into numerous branches as they approach their insertion to the mucous membrane. This has also been observed in other animals, particularly those with freely mobile tongues, and to a lesser extent in man. In mammals, those fibers of the facial musculature that are attached to the skin show this same feature.

The shape of the fibers of striated muscle has been studied by many workers, who by dissecting individual fibers established that they were elongated elements, more or less tubulular, but of varying appearance. Bardeen (1903), investigated in this way the external oblique muscle of the abdomen in the dog, rabbit, and man, and classified the fibers into three categories: (1) fibers of nearly cylindrical shape, with conical ends, running from origin to insertion; (2) fibers of fusiform shape, with fine tapering extremities ending within the belly of the muscle; and (3) fibers of conical shape, the broad base being attached to a tendon and the other end terminating in a fine attenuated extremity. Bardeen's findings confirmed those of several earlier workers and later were to be confirmed in turn by Huber (1916) for the rabbit and Lindhard (1926) for the frog.

When muscle fibers are sectioned transversely, the shape of their cross sections is oval or spherical if cut when fresh, but irregularly polyhedral

if cut after fixation; the difference is due to shrinkage both of their
surrounding connective tissue and of the muscle fibers themselves during
fixation. It is of interest that Watzka (1939) found less shrinkage in
the fibers of the wild rabbit than in those of the tame one, and less
in the red fibers of the domestic animal than in the white fibers of
the same animal; and further, that there was more connective tissue
in the muscles of the wild variety than in the tame variety, and more
in the red muscles than in the white ones of the latter. In brief, the
fibers of muscles with more connective tissue shrank less. The diameters
of isolated, freely suspended muscle fibers from the semitendinosus muscle of the frog were measured by Buchthal and Knappeis (1946) in
two planes at right angles to each other. Even under these conditions,
when external pressure plays no part, it was found that the cross section
of the fiber is usually somewhat oval.

2. Diameter of Voluntary Muscle Fibers

There is a considerable range in the diameter of different voluntary
muscle fibers, 10–100 μ commonly being accepted. First, it may be said
that their size differs in the major animal classes, as shown by Bowman
(1840), Mayeda (1890), and Schwalbe and Mayeda (1891). Using
a standard method of fixation, Mayedon dissected the fibers of a large
variety of muscles, especially those he considered to be homologous,
and obtained measurements that showed that the fiber diameter descends
in this order: fish, toads, reptiles, mammals, and birds. Second, the fibers
in one muscle may be generally larger than those of another muscle
in the same animal. Thus, the difference in the mean fiber diameter
of muscle samples from different parts of the body or from different
muscles of the same limb have been demonstrated in man (Halban,
1893), sheep (Hammond and Appleton, 1932), and rabbit (Graf, 1947).
Third, all observers are agreed that in any cross section of a muscle,
the range of fiber diameters is very wide. It could be argued that the
thin fibers seen are in fact the tapering ends of thick fibers, but Schiefferdecker (1909) thought this played only a small and unimportant role
when comparison is made between the caliber of the fibers of two different muscles. He therefore concluded that muscle fibers of different cross
section were present in the same muscle, and this is now the majority
view. Huber (1916) considered dissection of separate fibers the method
most likely to solve the problem, and Lindhard (1926) did this in teased
preparations of frog muscle and showed thick and thin fibers lying side
by side.

The spatial distribution of fibers of different diameter seen in cross sections of muscle has received little attention. Garven (1925) noted that the smaller calibered fibers were situated at the periphery and Herrick (1902) made a similar observation in the extrinsic eye muscles of fishes. From a study of the many photographs and tracings of muscle cross sections in the literature, Fernand (1949) states that thick and thin fibers appear side by side in all areas of the muscle cross section, but that generally the largest fibers are situated in the depth of the muscle. In his own work on the rabbit, Fernand found three groups of muscles: (1) those with large fibers at the periphery of their cross section, (2) those with large fibers in the depth of the cross section, and (3) those with no characteristic spatial distribution.

Many attempts have been made to establish the factors that may be correlated with the varying diameters of muscle fibers. Thus, there is some support for the idea that longer fibers have a larger diameter (Tsuruyama, 1937), but there is no direct correlation. A quotation from Fernand (1949) puts the matter well: "Thus in the normally developed muscles of the rabbit, there appears to be no correlation between the length of the fibers in the different muscles and the diameter of the fibers. However, the observation that the fibers of the largest caliber are found in the muscles with the longest fibers, as for instance the biceps femoris and semimembranosus suggests that longer fibers have the potentiality for developing a greater girth than shorter fibers." In man, Kohashi (1937) found that the most powerful muscles have the thickest fibers, and Oshima (1938) reached the same conclusion for the cat. Fernand agrees that there is a correlation but points out that other factors, such as the number of muscle fibers contained in the muscle and the functional axis of the muscle also contribute to the final determination of fiber caliber.

Concerning the part muscle size may play, the evidence is conflicting. In the sheep, Hammond and Appleton (1932) found that the actual size (weight) of the muscle has "little or no effect on the size of the fiber," but in the muscles of man, Halban's work does suggest a link, although not a constant one, between muscle and fiber size. His illustration contrasting fiber size in the superior rectus and gluteus maximus muscles is reproduced in Fig. 12. Body size in farm animals cannot be correlated with fiber diameter, but nutritional state is apparently of importance; not only does inanition lead to a decrease in fiber caliber, but full fed steers have larger fibers than those rough fed (Robertson and Baker, 1933; Lister, 1970), and pigs that enjoy better feeding in their first 16 weeks achieve a bigger fiber diameter when adult than control animals (McMeekan, 1941). The facts that muscle fibers increase

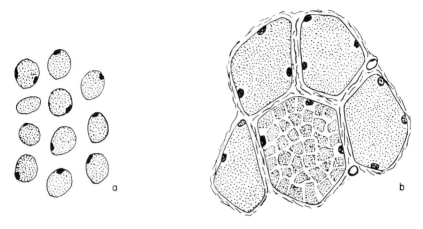

Fig. 12. Diagram of the difference in fiber size which may be met with in two muscles of the same subject; (a) ocular muscle, (b) gluteus maximus. (×340). After Halban. In one fiber group of b, Cohnheim's areas have been indicated. Compare with Fig. 13.

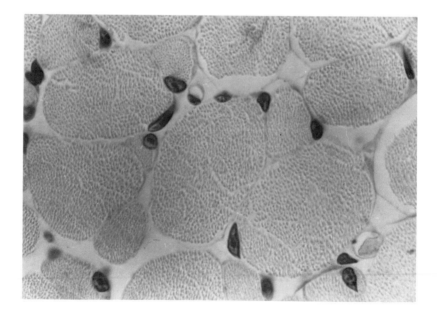

Fig. 13. Cross section of voluntary muscle fibers from dogfish embryo. Note the myofibrils filling the sarcoplasm, the peripherally located nuclei, and the nearby capillaries. (×1200.)

in diameter from birth to maturity and also in response to exercise are well known. Further, during the past decade, the importance of hormones and particularly the sex hormones as stimulants to muscle growth has become widely realised (Lister, 1970). Thus, androgens and estrogens are known to stimulate growth and nitrogen retention to an extent that varies from one species to another. On the other hand, castration decreases the rate of growth of animals (Prescott, 1963; Robinson, 1964; Lister, 1965). It is generally accepted that the fibers of the same muscles are larger in human males than females (Bowman, 1840; Schwalbe and Mayeda, 1891), and in the sheep, Hammond and Apple· ton (1932) found the ram to have larger fibers than the ewe. However, Elliot *et al.* (1943) could not detect any difference in fiber size between male and female rats kept under identical conditions. It may be that this matter should be further investigated.

3. LENGTH OF VOLUNTARY MUSCLE FIBERS

It is generally agreed that in short muscles fibers may run from the origin of the muscle to its insertion; that is, the fibers and muscle belly may be coextensive. Such is the case, for example, in the tensor tympani, where the fibers are only a few millimeters long. With regard to the length of the fibers in longer muscles, the consensus of opinion seems to be that the fibers do not run from end to end of the muscle. Nevertheless, there is evidence that in long muscles with parallel-running fibers, such as the sartorius, this may be the case—at any rate, in man. The classification of fibers by shape drawn up by Bardeen to which reference has been made shows that in the external oblique muscle some fibers do run from end to end, while others have one or both extremities terminating within the muscle. The findings made in the human sartorius will now be considered. In an adult human sartorius muscle 52 cm long, Lockhardt and Brandt (1938) were able to isolate fibers 34 cm in length, and even at that the ends were broken across. These authors also state that in a fetus with a sartorius 5 cm in length, the fibers when isolated by teasing could be seen to run the complete length of the muscle, and that this was also the case with the semitendinosus. They also found that in adult eye muscles the fibers run all the way. Of several earlier workers who dissected the human sartorius, Felix (1887) obtained the longest fibers, but his figure of 12.3 cm falls far short of that given by Lockhart and Brandt. The conflict in findings must be related to the technical difficulties involved in this type of investigation.

In a rabbit's thigh muscle Huber (1916) teased out a fasciculus 3.4

cm long, comprising 26 muscle fibers, and found only 1 fiber passing from end to end; of the remainder, 10 fibers reached one tendon end, 12 reached the other, and 3 were spindle-shaped reaching neither end. He also observed that the disposition of the fibers was such that their fine filamentous intrafascicular endings were usually applied to the thicker parts of neighboring fibers. Van Harreveld (1947) states that in the rabbit sartorius only about two-thirds of the total number of muscle fibers present in the muscle appear in any single cross section, and that in the length of the muscle there are three separate fibers arranged in series. It is quite clear that the longer muscle fasciculi tend to have longer fibers, even though the proportion of these that end intrafascicularly is not yet agreed upon. The great length of the fibers in a muscle like the sartorius and their arrangement in longitudinally disposed fasciculi is typical of muscles that act over freely movable joints, allowing as it does the maximum displacement of the parts to which they are attached. The amount of this displacement corresponds, of course, to the degree of shortening possible for a muscle fiber to undergo during contraction; for human muscle, Haines (1934) states that shortening to 57% of its length when fully stretched can occur.

4. Muscle Fiber Investments

In the past, considerable confusion has arisen over the nature of the membranes coating the striated muscle fiber (Barer, 1948). Covering the surface of each muscle fiber and closely and completely investing it is a thin delicate membrane. Just visible with the compound microscope, it is only 1.0 μ in thickness and forms a resistant sheath enclosing the soft protoplasmic contents. Forming as it does "a tubular membranaceous sheath of the most exquisite delicacy" (Bowman, 1840), it is too thin to be studied in ordinary fixed and stained preparations. It is especially well seen in fish and amphibia, in which it is thicker and stronger than in the mammals (Schafer, 1912). Electron microscopy apart, the best method of demonstrating this membrane is to examine a few roughly teased fresh striated muscle fibers in Ringer's fluid (Barer, 1948). Where fibers have been damaged, retraction clots form (for details of their formation, see Speidel, 1939) and at the side of each, an empty space is left within the intact membrane that can then be studied free of muscle substance. Several other ways of achieving the same object are also mentioned by Barer. Thus, anything that causes a violent contraction of the muscle fiber—e.g., treatment with caffeine, nicotine, or saponin —may cause the membrane to bulge away from the contracted muscle substance; colloidal swelling agents, such as dilute acids or alkalies, may produce swelling of the muscle substance with subsequent rupture of the membrane, while treatment with trypsin digests the muscle substance, leaving the membrane seemingly intact. In the past, this membrane has been named the sarcolemma (Bowman, 1840). Thus revealed, this investment appears as a homogeneous, apparently structureless membrane.

However, electron microscopic examination of this thin film (Bloom and Fawcett, 1968) has shown that it is not a single structure but is composed of two main elements. First, its innermost surface is made up of the plasmalemma, or cell membrane, of the muscle fiber, and in current usage, the term sarcolemma should be reserved for this structure alone. It appears to be similar in all respects to other cell membranes, and like them, because it is about 75 nm thick, it would be unresolvable in the light microscope. External to the sarcolemma but closely adherent to it is a thicker layer (up to 1000 nm) composed of a fuzzy, amorphous, mucopolysaccharide-like material (Pellegrino and Franzini-Armstrong, 1969). Modern terminology would designate this structure the fiber's basement membrane, or better, the basal lamina (Fawcett, 1966). Attached to this membrane are fine reticular fibers which run into and mingle with those of the surrounding endomysium.

Concerning the mechanical properties of this compound membrane, Ramsay and Street (1940) found that following the production of retraction clots in single muscle fibers, the empty membrane reached to stretch in the same way as the intact fiber; that is, the same increase in load produced the same percentage elongation, a result that suggests that the resting tension of the muscle fiber is governed by this membrane alone. The finding of Ramsay and Street brings to mind the observation made by Banus and Zetlin (1938) that the connective tissue sheath of a whole muscle, when dissected free, gave the same tension–length curve as the whole muscle. However, Sichel (1941) obtained a quite different result, finding that the elongation of empty lengths of membrane averaged more than twice that of intact regions. If correct, this would mean that the intact muscle fiber would resist extension by a tension fully twice that of the membrane alone.

According to Buchthal (1942), it is not possible to measure the true resting tension of this membrane in the region of a retraction clot, since it is already elongated. Barer (1948) would allow the force of this argument when the clot is large and hard, but points out that it is possible to produce empty lengths of fiber membrane without such clots and without any apparent lengthening.

How the membrane serves to transmit the contractile force of the muscle fibers to the connective tissue and muscle tendon will be considered later; that its elasticity may account for the appearance of "active" relaxation in single muscle fibers is discussed by Fenn (1945).

5. NUCLEI OF STRIATED MUSCLE FIBERS

Each striated muscle fiber possesses many nuclei (Fig. 14), and in a fiber some centimeters long, there may be several hundred. In the great majority of fibers of human and mammalian muscle, the nuclei lie at the surface of the fiber immediately under the sarcolemma surrounded by a zone of protoplasm. It follows that opposite the nucleus, the fibrils (shortly to be described) do not quite reach to the inner surface of the sarcolemma. The nuclei are of an ovoid form, elongated in the long axis of the fiber, and average 8–10 μ in length, although extremes of 5–17 μ are met with (Sobotta, 1930). In the fresh state, the nuclei are difficult to see, but in fixed and stained preparations, they are readily demonstrated. Along the fiber, their distribution is fairly regular, but toward the tendinous attachment, they become more numerous and are more irregularly distributed. The position of the nuclei beneath the sarcolemma, i.e., hypolemmal, usual in adult human muscle differs from that found in the embryo, in which the nuclei occupy a position in the middle of the fiber. In many lower vertebrates (fishes, amphibia, and reptiles), the nuclei lie within the fiber, either centrally or eccentrically, but seldom just under the sarcolemma. This is commonly said also to be the case for the so-called red muscles of mammals, but this point will be discussed later.

It should be emphasized that the usual position of the nuclei on the periphery of the fiber is best appreciated when transverse sections through muscle fibers are examined. In longitudinal section, many nuclei do appear in this position, but clearly, those nuclei that lie on the upper or lower surface of longitudinally disposed fibers will appear as if they were in the middle of the fibers. Each nucleus usually contains one or two nucleoli and its heterochromatin is usually found just under the nuclear membrane.

6. THE SARCOPLASM

It is a convenient simplification to regard striated muscle fibers as being composed of four main constituents—sarcolemma, nuclei, sarcoplasm, and myofibrils. Of these, the last two remain to be discussed. Sarcoplasm is commonly regarded as undifferentiated protoplasm of a semifluid consistency in which the myofibrils are embedded. The obser-

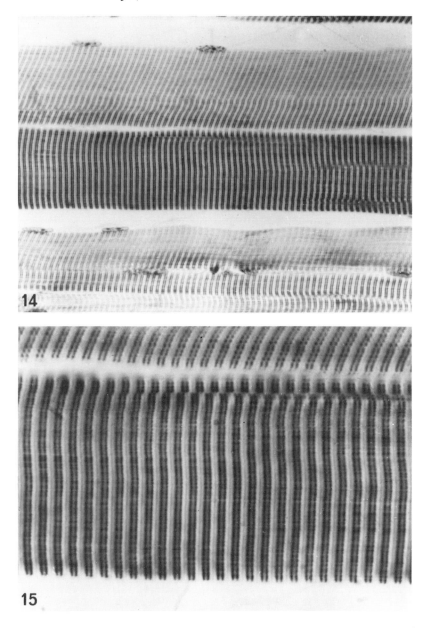

Fig. 14. Longitudinal section of several voluntary muscle fibers showing the cross banding and peripheral nuclei (×500).

Fig. 15. High power of voluntary muscle fiber showing longitudinally arranged myofibrils and the cross striations. Note the dark A bands in which H bands can be distinguished and the light I bands bisected by the thin Z line (×1200.)

vations of Speidel (1939) on living muscle suggest that in normal fibers the sarcoplasm and the myofibrils are in the gel state and that following injury this may change to that of a sol. In an ordinary histological preparation showing the muscle fiber cut in transverse section, it can be seen that within the fiber the space bounded by the sarcolemma is not completely occupied by the myofibrils; the residual space is presumably filled by sarcoplasm. Not all fibers show the same relative amount of sarcoplasm to myofibrils, and Knoll (1891), who investigated the matter very thoroughly, divided fibers into those rich and those poor in sarcoplasm. This point will be returned to when red and white muscle fibers are considered.

The advent of the electron microscope has resulted in a much more detailed understanding of sarcoplasmic organization (Pellegrino and Franzini-Armstrong, 1969), but it tends to be forgotten, and H. S. Bennett (1956) has done well to remind us, that Retzius and Cajal, among others, had observed with the light microscope extremely delicate longitudinal and transverse filaments within the interstitial substance between the myofibrils. This sarcoplasmic reticulum was also described and indeed figured by Heidenhain. The early suggestions that the sarcoplasmic reticulum functions as a means of exchange of substances between the myofibrils and the sarcoplasm during contraction and relaxation and that it serves as a system for the rapid conduction of the excitatory impulse to contraction both transversely across the fiber and longitudinally have been fully substantiated (Pellegrino and Franzini-Armstrong, 1969).

7. THE MYOFIBRILS

The name "striated" as applied to muscle refers to the cross banding which is such a striking feature of voluntary (and see later cardiac) muscle when its fibers are examined microscopically. However, another type of striation is also noticeable, i.e., longitudinal, which is due to the longitudinal disposition within the sarcoplasm of the fiber of its constituent myofibrils, also called sacrostyles. In a cross section of a muscle fiber, these elements can be seen as fine dots (Fig. 13), which may be uniformly distributed in the sarcoplasm or gathered into groups, the polygonal areas or fields of Cohnheim (1865), separated by greater amounts of clear sarcoplasm. It has long been debated whether Cohnheim's area represents the true distribution of the myofibrils or whether they are artifactual consequences of the histological techniques employed. The weight of evidence now favors its interpretation as an artifact and no functional significance is now attached to these groupings of myofibrils (Bloom and Fawcett, 1968). Teasing of a muscle fiber after treatment with, say, chromic acid, will split it up into bundles of myofibrils, and in favorable cases individual fibrils can be separated off. In Fig. 15, a muscle fiber cut in longitudinal section is seen and some of its constituent myofibrils appear clearly. That myofibrils exist in living muscle and are not caused to form by subsequent treatment with chemicals seems no longer in doubt (Jordan, 1933; Spiedel, 1939). Thus, in the tails of young frogs and salamander tadpoles, Jordan studied striated muscle under high magnification and was fully satisfied that myofibrils and cross striations exist in living muscle fibers. Tissue culture studies have given mixed results, but Hogue (1937) demonstrated cross-striated myofibrillae in living cultures of cardiac muscle cells. Microdissection of fresh muscle fibers as a means of demonstrating myofibrils is exceedingly difficult (Barer, 1948), the trouble being that in the living fiber the stickiness of the sarcoplasm tends to cause the myofibrils to join together in irregular bundles. By using microelectrodes, however, Barer (1948) was able to produce contraction of some but not all of the myofibrils within a single fiber, clear evidence of their existence as independent contractile elements. In other experiments, Barer was able to produce asynchronous contraction of myofibrils, an occurrence which naturally poses the question of what mechanism ensures the normal synchronous contraction of all the myofibrils in a fiber. As noted earlier, the sarcoplasmic reticulum may provide the answer.

Myofibrils are elongated threadlike structures which measure about 1-2 μ in thickness; small as that may seem, the electron microscope has revealed each to be packed solidly with still smaller elements called myofilaments.

Just as the muscle fiber shows alternate light and dark bands, so does each myofibril, and it appears that the appearance of cross striation presented by the whole fiber results from the fact that the light and dark bands of its component myofibrils are in register with one another; that is, they lie in approximately the same transverse plane (Fig. 16).

Since the myofibrils are the contractile units of muscle, their structure must be considered in detail, and since a wealth of terms has been employed in the past to describe their structure, some pains will be taken with nomenclature. For further details Jordan (1933) and Høncke (1947) should be consulted. In Fig. 17, the named portions of one myofibril are shown. As mentioned earlier, each myofibril when viewed with the light microscope consists of alternate dark and light portions (Figs. 14 and 15). When examined by polarized light, however, the dark portions are clearly birefringent and are therefore called anisotropic bands,

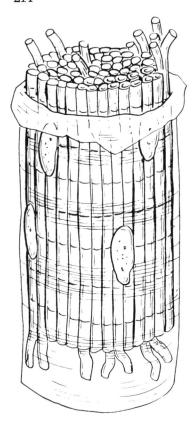

Fig. 16. Diagrammatic representation of the structure of one voluntary muscle fiber. Features to be noted are (1) the cross banding of the fiber, which results from the striations of its contained myofibrils being in register, (2) the close-packed myofibrils, some of which are spraying out at the ends, (3) the delicate sarcolemma, which has been turned back at one end of the fiber, and (4) the position of the nuclei immediately under the sarcolemma. Adapted, with kind permission, from a figure in Sir Wilfred Le Gros Clark's "Tissues of the Body."

or in short, the A bands. They are also frequently referred to as Q bands, from the German *querscheibe,* meaning transverse disk. The light portions, often said to be nonbirefringent, are in fact weakly birefringent, but the difference is sufficiently marked to merit their being called isotropic, or in short, the I bands; they are also called J disks (the German

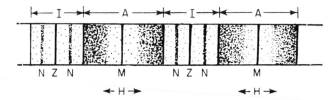

Fig. 17. Diagram of the main striations seen in a single myofibril (rabbit). This figure is adapted from Barer (1948), whose illustration was based on the electron microscope photographs of Hall *et al.* (1946). The lettering indicates the named bands. The extent of one sarcomere is from one Z band to the next.

equivalent); further alternative names are clear disks, light bands, and hyaline substance. Other than the main A and I bands, the only other striation which can be recognized with any certainty in the living fiber is the Z disk. This appears as a narrow dark line in the middle of the I band and was formerly thought to be evidence of a membrane called Krause's membrane.

The length of myofibril bounded by two Z bands is called a sarcomere, and it is regarded as a structural unit within which certain changes are postulated to occur during contraction. The designation Z band comes from *zwischenscheibe*, which means intervening disk; other names by which this feature of the myofibril has been known are telophragma, Dobie's line, end disk and, alas, several more.

Additional features of the myofibril that may be detected from an examination of fixed and stained material (Fig. 15) will now be described, and it is of interest to note that the detailed observations made by the pioneer histologists in this field have been confirmed by present day workers using the electron microscope. It was known long ago that the appearances presented by the myofibrils are variable (Bowman, 1840), depending on the degree of fiber contraction.

Within the A band there may be recognized a lighter region, which is known as Hensen's disk or the H band, *heller* meaning clearer or brighter. In stretched myofibrils, the H band is broadest, but in the contracted state, it is either very narrow or entirely absent. Bisecting the A band, the M band may be seen; this is also known as the *mittelscheibe*, i.e., intermediate disk, mesophragma, or median membrane.

About midway between the Z and A bands, there is less frequently observed a narrow dark striation known as the N band or disk (sometimes called the accessory disk of Engelmann), so that each I band will then show three striations, a central Z band flanked by an N band on each side. Since the Z bands are taken as limiting the sarcomeres, each sarcomere is sometimes described as possessing a terminal disk at each end between the N and Z bands. This may be called the E disk of Merkel. The name N derives from *nebenscheibe, neben* meaning next to or beside.

Electron microscope analysis of myofibril structure has enabled measurements of the different bands to be made with an accuracy previously denied to light microscopists. The A band, which appears to remain constant in length throughout the contraction process, is about 1.5 μ long. The I band, whose myofilaments stretch about 1 μ in either direction from the Z line, varies in width according to the degree of overlap between the I and A myofilaments. Thus, in the relaxed state, the I band is at its greatest width, but is minimal in the fully contracted condition. Fuller details of the fine structure of myofibrils and their biochemical and biophysical organization may be found in other chapters of this treatise.

8. Organelles and Inclusions in the Striped Muscle Fiber

Thus far it has been noted that the striated muscle fiber consists of a thin membrane, the sarcolemma, within which are nuclei, sarcoplasm, and myofibrils. The sarcoplasm is simply the cytoplasm of the muscle fiber within which, as in cells elsewhere, two kinds of formed elements can be recognized, organelles and inclusions. The name organelle is given to those bodies that are structurally specialized parts of the cytoplasm and could therefore be regarded as part of the living substance of the cell; the term inclusion embraces the various particles of nonliving material that may be present.

Of the organelles within striped muscle fibers, the myofibrils and sarcoplasmic reticulum have been discussed. In suitable preparations, a small Golgi apparatus can also be frequently seen at each pole of the nucleus. Mitochondria, sometimes called sarcosomes, are also present and were successfully demonstrated by the light microscope more than seventy years ago by Retzius. The use of the electron microscope in recent years has added greatly to our knowledge of these small but important structures. A fuller discussion of the numbers and distribution of mitochondria will be found in the section on red and white muscle in this chapter. Suffice to say that they are present in considerable number in muscle fibers and are usually located in close proximity to the myofibrils. Other sites where mitochondria are found are the paranuclear sarcoplasm just off the poles of the nuclei and in the subsarcolemmal sarcoplasm where dense masses may be present between the sarcolemma and the myofibrils. Lysosomes are usually juxtanuclear in position.

Inclusions that may be present in striped muscle fibers include glycogen granules of irregular size and shape situated in the sarcoplasm near the junctions of the A and I bands of the myofibrils, and fat granules, which will be further mentioned when red muscle is discussed.

D. Hypertrophy of Voluntary Muscle

The capacity of voluntary muscle to increase in size with exercise is common knowledge, and it would appear that such hypertrophy results not from an increase in the number of fibers present in the muscle but from enlargement of the diameter of the individual fibers (Morpurgo, 1897). In the enlarged fibers, the number of myofibrils increases (Morpurgo, 1897; Denny-Brown, 1961). As previously mentioned, androgen administration to the castrated rat causes a rapid muscle hypertrophy, a noticeable weight change occurring 2–3 days after treatment (Venable, 1966). Other striking cytoplasmic changes that occur during fiber hypertrophy are an increase in fiber basophilia (ribosomes), glycogen, and mitochondria. The highly specialized nature of striated muscle fibers when fully formed is in harmony with the belief that they probably do not undergo proliferation by cell division, and indeed there is evidence that their number does not increase in the human embryo after it has reached the age of 4 or 5 months, when it is 17 cm in length (MacCallum, 1898).

E. Blood Supply of Voluntary Muscle

Voluntary muscle has a rich blood supply derived from branches of neighboring arteries. The vessels enter with the nerves along a line that is frequently definite enough to receive the name of neurovascular hilum (Brash, 1955). From the epimysium, the arteries travel into the substance of the muscle along the strands of perimysium, dividing as they do so, the various branches of the entering vessels establishing free anastomoses with one another. Curiously, these anastomoses, which show up well in injected material, do not function very efficiently should one of the supply vessels be ligated in an experimental animal (Le Gros Clark and Blomfield, 1945). The finer branches of supply come to lie transversely to the long axes of the muscle fibers and from them arise the capillaries which course between the fibers, lying in the endomysium. It is noteworthy that the arteries and veins run together up to the point at which the terminal arterioles and venules are given off, but thereafter these small vessels arise alternately; this arrangement presumably allows the intervening capillary to run in a relatively direct course from arteriole to venule and so expedites rapid removal of metabolites (Le Gros Clark, 1952). For the most part, capillaries run longitudinally between the individual muscle fibers. With their frequent linkage by transverse vessels, which pass over or under the intervening fibers, a fine capillary network with narrow oblong meshes is formed (Figs. 18 and 19).

In the immediate neighborhood of motor end plates, the capillary anastomoses are especially well developed (Wilkinson, 1929), an arrangement in harmony with the conception of these sites being particularly active metabolically. In the red muscles of the rabbit, Ranvier (1874)

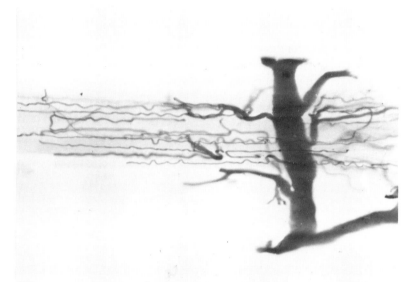

18

Fig. 18. Injected blood vessels in a teased portion of voluntary muscle. The larger arteriole can be seen giving rise to the longitudinally arranged and anastomosing capillaries. (×115.)

Fig. 19. Longitudinal section of fetal dog voluntary muscle showing the widely dilated capillaries in close proximity to the fibers (×500).

stated that the capillaries run a more tortuous course than in white muscle and also that the cross connecting vessels frequently show considerable dilatations. These enlargements have been considered to act as reservoirs for oxygen from which the muscle fibers can be supplied during sustained muscular contraction, at a time when the capillary circulation is impeded. The tortuosities of the capillaries may also permit their accommodation to changes in length of the muscle fiber, by straightening during elongation and contorting during contraction.

Lymphatic vessels are present in the epimysium and perimysium but seem to be absent from the endomysium between the individual fibers. If this is true, and Aagaard (1913) seems to be alone in describing a lymphatic capillary plexus around the fibers of voluntary muscle, then voluntary muscle might be said not to possess lymphatic vessels.

F. Red and White Muscle

It is common knowledge that muscle color is not uniform throughout the animal kingdom, or indeed in the muscles of the same animal, and the white breast muscle and pigmented leg muscle of the domestic hen are familiar to all.

Light microscopists have long been aware that muscles are made up of at least two kinds of fiber which differ structurally as well as tinctorially (Grützner, 1884; Knoll, 1891; Ranvier, 1873, 1874; Bell, 1911; Bullard, 1912–1913; Denny-Brown, 1929). The pigmented fibers, smaller in diameter and rich in sarcoplasm, mitochondria, and fat droplets, are called red fibers. The larger-diametered fibers, or white fibers, have less sarcoplasm and fewer mitochondria and lipid droplets. Before the last edition of this treatise, the brief statements on red and white muscle that appeared in standard textbooks were based mainly on the observations of Ranvier (1873, 1874), who investigated the rabbit adductor magnus (white) and semitendinosus (red) muscles. He believed that red muscle fibers are more distinctly striated longitudinally but that their transverse striation is less regular. Also, he observed that the nuclei are more numerous in red fibers than white and that nuclei of red fibers occur within the depth of the fibers as well as beneath the sarcolemma. The features of the capillary blood vessels of red muscle as described by Ranvier have already been mentioned earlier in this chapter.

Further, although before Ranvier's day it had been shown that the redness of muscle was due to its content of myoglobin and not to the richness of the blood supply, he was the first to provide evidence of a correlation between the contraction speed of a fiber and its histological appearance and pigment content; he showed that red fibers have a slower, more prolonged contraction.

However, because these fiber differences are not always readily demonstrable in routine histological preparations, their heterogeneity has received little attention for many years. But in the 1950s, new cytochemical methods, especially those for demonstrating enzyme activity, revealed striking differences among fiber types in many muscles and convincingly confirmed the earlier observations. The advent of these new cytochemical methods and more recently the application of the electron microscope to the study of muscle morphology led to a great burst of investigations which cannot be easily summarized here, but for full references, see Padykula and Gauthier (1967), Engel (1965, 1970), Dubowitz, (1970), and Gauthier (1970). One result of these numerous studies was not only to confirm the observations of the early workers, but also to show that the structural and cytochemical variety among muscle fibers is greater and more subtle than was originally envisaged. Thus, not only are there differences even within one class of animals, e.g., the Mammalia, but within a single mammalian species the appearance and distribution of fiber types differs significantly from muscle to muscle (Beatty *et al.*, 1966; Gauthier and Padykula, 1966). As Padykula and Gauthier (1970) assert, "failure to recognize the heterogeneity of skeletal muscles is misleading to investigators studying physiological or pathological conditions in individual fibers."

The morphological and cytochemical criteria used to distinguish fiber types have been listed by Padykula and Gauthier (1970):

Fiber diameter
Mitochondrial content and distribution
Size and shape of myofibrils
Ultrastructure of the sarcomere
Form of the sarcoplasmic reticulum
Mitochondrial enzymic activity
Phospharylase activity
Glycogen content
Triglyceride content

In their analysis of the albino rat diaphragm using the single criterion of mitochondrial differences (Padykula and Gauthier, 1963; Gauthier and Padykula, 1966) these authors distinguished three fiber types:

1. *Red fibers.* These form 60% of the muscle, have an average diameter of 27 μ, and are rich in mitochondria, which form a broad layer between the plasma membrane and the contractile elements. Large mitochondria also form chains between the myofibrils and have a characteristic form and arrangement at each I band.

2. *White fibers.* These form 20% of the fiber population and are the largest in diameter, 44 μ. Unlike the red fibers, the mitochondria do not form subsarcolemmal aggregations, and interfibrillar chains are inconspicuous or absent. Nearly all the mitochondria are associated with the I band. Thus, the mitochondrial content of white fibers relative to its myofibrillar mass is low.

3. *Intermediate fibers.* These are approximately 34 μ in diameter, and although at first appearance they seem similar to the red variety, subsarcolemmal accumulations of mitochondria and interfibrillar chains are less conspicuous because the mitochondria are smaller. Cytochemical observations on mitochondria enzymes activity also reveal the intermediate nature of this fiber.

Several classifications of fiber types are to be found in the literature, and it is not always possible to marry the different categories from the different systems. Padykula and Gauthier (1967) have abandoned their 1962 classification of rat muscle fibers into A (white), B (intermediate), and C (red) groups in favor of the older descriptive terminology of red, white, and intermediate, since they feel that the alphabetical terminology can be confused with similar letter designations for nerves. Dubowitz (1970) classifies fibers into Type I (red, or B and C) and Type II (white, or A) according to whether they are high in mitochondrial oxidative enzymes and low in phosphorylase activity (Type I) and vice versa for Type II fibers. Other classifications have used size and shape of myofibrils, and it seems likely that as morphological differences become more firmly established (Gauthier, 1970), they will become the principle criteria for distinguishing different kinds of fibers.

Although direct correlations have been made between the morphology and physiology of twitch and slow fibers of frog muscle (Peachey and Huxley, 1962; Page, 1965), it is much more difficult to make such correlations in mammalian muscle as the physiological data are derived primarily from studies of whole muscle and not individual fibers (Gauthier, 1970).

G. Tendons and Tendon Sheaths

The relation of tendons to muscle architecture and the advantages afforded by their presence having been discussed by R. D. Lockhart in Volume I, Chapter 1, their microscopic structure will be considered in this chapter. Tendons consist almost solely of white fibrous tissue, the collagen fibers of which form closely packed parallel bundles. The bundles do not remain separate throughout their extent, for fibers are given off to and received from neighboring bundles so that successive cross sections of a tendon show slightly differing appearances. The surface of a tendon has a characteristic white, shiny aspect; closer inspection, however, will often reveal light and dark streaks across its length, an appearance resembling that of watered silk (Schafer, 1912), which results from the somewhat wavy course taken by the collagen fibers when not under tension.

Surrounding the tendon, there is a sheath of dense areolar tissue from which septa carrying the blood vessels, nerves, and lymphatics pass between the secondary bundles into which the collagen fibers are grouped. The cellular component of tendon consists of fibroblasts, sometimes called tendon cells, arranged in single rows between the bundles of collagen fibers. In transverse section the cells appear stellate, since the middle of each cell containing the nucleus occupies the angular space between several bundles of fibers, while lamellar extensions proceed from the cells into the bundle interspaces. The processes of the tendon cells may be considered to mark off, indefinitely it is true, the primary collagen fiber bundles. For the most part, tendon cells are not connected with one another, but it is sometimes possible to follow a lateral extension of a cell in one row into continuity with a cell of another row. When seen in surface view the tendon cells may appear rectangular, rhomboidal, or irregular, but seen in profile as in a longitudinal section of tendon, they appear as rods owing to their thinness. Elastic fibers are present to some extent in the connective tissue septa, but otherwise tendon has very little elastic tissue.

In those parts of the body where tendons lie close to bone, friction is reduced by the presence of tendon sheaths. Essentially, each is a closed tubular sac into which the tendon is invaginated, so that it is closely invested by one layer of the sheath, which indeed follows it in its movements. The other layer of the sheath lines the thickened fibrous tissue which acts as a retaining band for the tendon. Within the tubular sheath, there is a little mucinous fluid, similar to the synovial fluid of joints, but means of which movements are facilitated. Not all tendons have well formed sheaths, and around those that lack them it will be found that there is loose connective tissue, the meshes of which contain the same kind of mucinous fluid.

Composed as it is predominantly of intercellular material, the metabolic activity of adult tendon is low and its blood supply correspondingly meager (Edwards, 1946). Blood vessels enter tendon both at its muscle end and at its periosteal end, and in the case of some tendons with sheaths, along reflections of the sheaths as well. Such reflections are called mesotendons. But, threefold though its blood supply may be, tendon is poorly vascularized. During development, tendon has quite a good blood supply, but with advancing age this becomes greatly reduced.

H. The Muscle–Tendon Junction

The precise manner in which the force of the contracting myofibrils is transmitted to the muscle tendon has been a matter of controversy for years. Two main theories have had their supporters: the first, or continuity theory, supposes that there is direct union of the myofibrils and the tendon fibrils, while according to the second, or sarcolemma theory, there is no such continuity and the sarcolemma forms a limiting membrane between the substance of the muscle fiber and the tendon elements. Long (1947) and Barer (1948) review the literature on this matter. Long, who studied the development of the muscle–tendon attachment in the rat, found no evidence in support of continuity. In both fetal and adult stages, the muscle fibers are enclosed by a limiting membrane both at their terminations and along their sides, and the myofibrillae end on the internal aspect of this membrane. In the attachment region, the reticular fibers that envelop the ends of the muscle fibers continue proximally as the sarcolemmal and endomysial reticulum, and distally they are continuous with the tendon fibers.

However, Barer does not feel that the protagonists of either view can prove their case by using conventional fixed and stained material, but rather that the observation of whole muscle fibers holds out the best means of a solution. In his studies of living muscle fibers, including microdissection, he obtained no evidence of any special connection between myofibrils and tendon fibrils. Thus, when retraction clots form in the junctional zone, a clear area of sarcolemma and basement membrane is left between the clot and the tendon fibrils, an occurrence which also follows spontaneous herniation of the muscle substance after treatment with dilute acids. From microdissection of such material, Barer believes that the sarcolemma (and basement membrane) is continuous over the end of the fiber and that the tendon fibrils are attached to the outside of the sheath.

The idea that there is any structural continuity between the muscle substance and the basement membrane is not supported by the facts. The statement has been made that the Z bands are really rigid membranes that stretch right across the fiber, linking the Z bands of neighboring myofibrils, and are circumferentially attached to the basement membrane. This would, if true, offer a means for the transmission of the contractile force to the sarcolemma and basement membrane, and beyond it to the endomysial connective tissue and so to the tendon. But the evidence against such Z band basement membrane continuity is overwhelming. As was noted in an earlier section, the basement membrane can be made to separate from the muscle fiber substance in a great many ways, and Bowman himself showed that the whole myofibrillar and sarcoplasmic content of a fiber could be expressed without difficulty through the basement membrane (in his terminology, sarcolemma). There is obviously no firm bond between basement membrane and the contractile elements. In any event, the Z bands are myofibril bands, and many features have been observed which dispute the notion of their being united uniformly across the fiber to form a continuous membrane. For example, columns of ice can advance between bundles of myofibrils without hindrance; nuclei have been seen to move between myofibrils without seemingly being hindered by cross membranes; oil droplets may be pushed along inside a muscle fiber with no difficulty; and finally, there is the time-honored observation of Kuhne (1863), who observed a tiny living parasite (a nematode) inside a fresh muscle fiber, moving freely up and down inside the fiber without permanently affecting the appearance of the cross striations. Actually, for reasons which he gives in full, Speidel (1939) thinks too much has been made of this. Finally, the electron microscope evidence (Pellegrino and Franzini-Armstrong, 1969) unequivocally demonstrates the discontinuity between one myofibril Z line and its neighbors.

Barer believes it possible to explain the transmission of contractile force from myofibrils to sarcolemma by a frictional or viscous relationship between these structures; a viscous gel when contracting would exert a pull on its surrounding membrane and thence to the endomysial connective tissue.

H. S. Bennett (1955) gives a brief account of such electron microscopic observations as have been made on the junctional zone between muscle and tendon. Simply put, the evidence points to a direct transmission of tension at the ends of the myofibrils. The myofibrils stop a short way from the sarcolemma, and tapering strands of fibers of unknown nature continue through the intervening sarcoplasm space, appearing as noncontractile continuations of the myofibrils. Inserting on the sarcoplasm opposite the attachments of these connecting fibrils are the tendon fibrils which therefore form an extension or extrapolation of them.

With regard to the Z bands, Bennett states that their dense material is confined to the myofibrils and does not extend across the sarcoplasm; such material as does extend from the Z band of one myofibril to that adjacent to it is weak and easily stretched, and moreover sarcoplasm

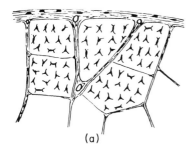
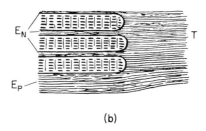

(a)   (b)

Fig. 20. Diagrams of (a) the structure of part of a tendon in cross section, and (b) part of a muscle–tendon junction. In the cross section, note the stellate appearance of the tendon cells whose extensions faintly demarcate areas occupied by the primary bundles of collagen fibers (not represented). The secondary bundles are surrounded by septa of fibrous tissue which penetrate the tendon from its investing connective tissue sheath. Blood vessels are present in the sheath and in the septa, but elsewhere are sparse. At the muscle–tendon junction, some of the connective tissue fibrils of the tendon (T) become attached to the sarcolemma of each muscle fiber; others become continuous with the connective tissue of the epimysium (Ep), perimysium (not represented), and endomysium (En).

of varying thickness intervenes between the sarcolemma and the myofibrils. Clearly, any view of contraction that invokes the attachment of complete Z membranes to the sarcolemma at the periphery of the fiber is untenable. Figure 20 is a diagrammatic representation of the muscle–tendon junction and of the structure of tendon.

IV. Cardiac Muscle

A. General Structure

Cardiac muscle tissue is found only in the heart and surrounding the mouths of the great veins which enter it. In some ways it resembles both voluntary and smooth muscle, yet in its rhythmic, unceasing activity from early embryonic life until death, it stands alone. Like voluntary muscle, its contractile elements show transverse striation; like visceral muscle, its fibers possess centrally placed nuclei, and further, it too has an autonomic innervation.

The composition of its fibers is essentially similar to that of voluntary muscle. It has the same basic constituents—sarcoplasm, myofibrils, nuclei, and sarcolemma—but in their form and caliber, the fibers show distinct differences. In transverse section (Figs. 21 and 22), they are

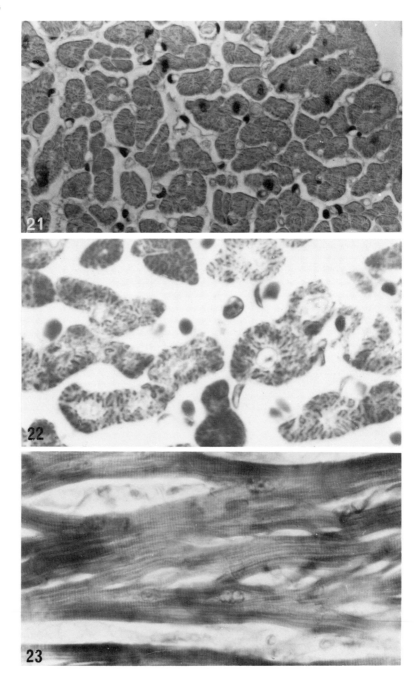

Figs. 21–23.

less regular in shape. Their most characteristic feature is that in longitudinal section they can be seen to give off branches at acute angles and so effect anastomoses (Fig. 23) with adjacent fibers. The branches are more slender than the parent fibers, which themselves seldom measure more than 15 μ in cross section; as will later be mentioned, this size may be greatly exceeded by those fibers of heart muscle that are believed to be specialized for the purpose of conducting the impulse for cardiac contraction. The branching nature of cardiac muscle fibers has in the past led to the belief that the whole heart musculature in fact formed a syncytium, but recent findings made with the electron microscope and shortly to be discussed have brought about a change in view.

The question of whether or not cardiac muscle fibers possess a sarcolemma has been much debated (Cohn, 1932), but electron microscope examination of cardiac muscle (Bloom and Fawcett, 1968) reveals that the surface structures of cardiac muscle are similar to voluntary muscle, except that the basement membrane is much thinner.

The sarcoplasm of cardiac muscle fibers is not only very abundant but also distinctly granular, and because of this, the transverse striation of the fibers is less distinct than that of voluntary muscle; the longitudinal striation, however, is very evident. Embedded in its sarcoplasm are myofibrils which are quite similar to those of voluntary muscle and show the same alternate sequence of light and dark bands that confers the cross-striated appearance upon the fibers as a whole. The A, I, and Z bands can all be recognized. As in voluntary muscle, the sarcoplasm contains mitochondria, fat droplets, and glycogen granules.

The nuclei, which are rather pale, occupy the middle of the fibers, in contrast to the subsarcolemmal position favored in voluntary muscle. In shape they are somewhat ellipsoidal, but it is not uncommon to see nuclei with square-cut ends. At each end of the nucleus there is a cone-shaped accumulation of sarcoplasm free from myofibrils, the polar cone, in which are present small yellowish brown pigment granules which increase in number with age. In the condition of brown atrophy, this pigment becomes very obvious within the shrunken cardiac muscle fibers.

Fig. 21. Transversely sectioned cardiac muscle showing the somewhat irregular outlines of the fibers, central nuclei, and numerous capillaries ($\times$250).

Fig. 22. Transversely sectioned cardiac muscle fibers showing their central nuclei and large myofibrils ($\times$600).

Fig. 23. Longitudinally sectioned cardiac muscle showing the branching and anastomosing fibers, their central nuclei, and cross striations ($\times$400).

B. Intercalated Disks

Separating the cardiac muscle fibers into segments some 50–120 μ in length, each of which usually contains one nucleus, seldom two, there are transverse bands less than a sarcomere in width known as intercalated disks. These frequently transverse the fiber in a series of short steps (Figs. 24–26). They increase in number with age and are in fact late in developing, for Cohn (1909) failed to find them in fetal hearts or in the hearts of children. In the hearts of other animals, however, they appear earlier and they have been found by Witte (1919) in fetal pigs of the 76 mm stage, and by Jordan and Banks (1917) in the beef heart at the second to the third month. Jordan and Steele (1912) observed them in the guinea pig during the last week of gestation, and also in the heart of a 4-year-old child.

Many views have been advanced with regard to the significance of the disks, and even yet the matter has not been settled. The idea that

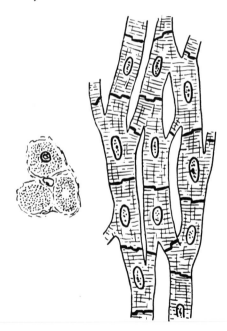

Fig. 24. Diagram of the structure of cardiac muscle in longitudinal and transverse section. Points to be noted are the branching of the fibers, which are striated both longitudinally and transversely, the centrally placed nuclei, and the intercalated disks. Of the three fibers shown sectioned transversely, one is cut through the nucleus, one through the polar cone of myofibril-free sarcoplasm, and one between this zone and a disk. Compare with Figs. 21–23.

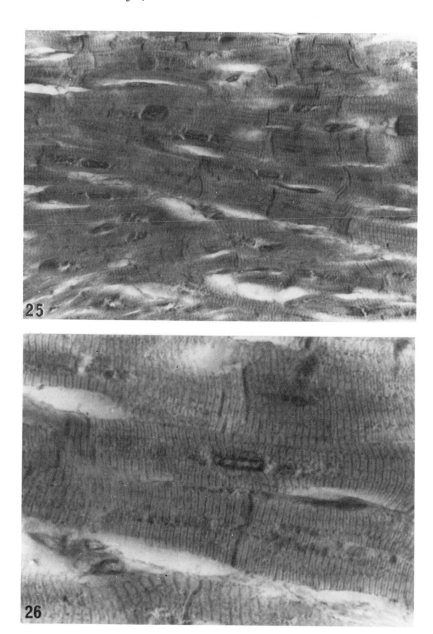

Fig. 25. Longitudinal section of cardiac muscle stained to show the numerous, thick intercalated disks traversing the fibers (×500).

Fig. 26. Longitudinally sectioned cardiac muscle showing the fiber's cross striations, central nuclei, and intercalated disks (×1200).

they represent cell membranes goes back almost a century, and indeed there are features of cardiac muscle that support this belief. Thus, they show up well after treatment with silver salts; the segments which they demarcate commonly contain one nucleus; fragmentation of the fibers, induced by maceration or as the result of pathological myocardial segmentation, occurs at the disks; they increase in number and complexity with growth and activity of the heart; there may be a different degree of contraction of the fiber on each side of a disk (Van Breeman, 1953); finally, the fact that the disks are attached to the sarcolemma would allow of the segments being regarded as individual cells bounded all around by a cell membrane.

Against the view that the disks are cell membranes are the facts of their late appearance in development and their numerical increase with age, unaccompanied, it is believed, by corresponding cell division. Moreover, on the basis of appearances as seen with the light microscope, it has long been believed that the myofibrils are continuous through the disks. However, electron microscopic studies have shown that this appearance is illusory (Van Breeman, 1953) and that they are in fact the specialized end walls of two adjoining cardiac muscle fibers, modified to form a desmosome-like structure (Sjöstrand *et al.*, 1958). They appear to be devices for maintaining firm cohension of adjoining cardiac muscle cells during the contraction process.

C. Blood Supply and Lymphatic Drainage

Cardiac muscle has an abundant blood supply, and as in voluntary muscle, the capillaries follow the general arrangement of the muscle fibers. Using an injection mass of radiopaque media and a microradiographic analysis of 300 μ to 5 mm thick sections, Farrer-Brown (1968a,b,c) has made a detailed study of the blood supply to human cardiac muscle. Branches from the main coronary artery on the epicardial surface come off at right angles and course directly through the myocardial wall toward the endocardium (Fig. 27). These branches, approximately 400–1500 μ in diameter, then divide into a cascading treelike pattern giving off many small branches with gradually diminishing calibre (Fig. 27). In addition to these branching arteries, a small number of straight ones with only a few branches pass through the myocardium and supply the anterior and posterior papillary muscles.

The final branching of small arteries shows two different patterns. In the first, the terminal small arteries arise in a brushlike manner from the main coronary branch. In the second, found only in the middle

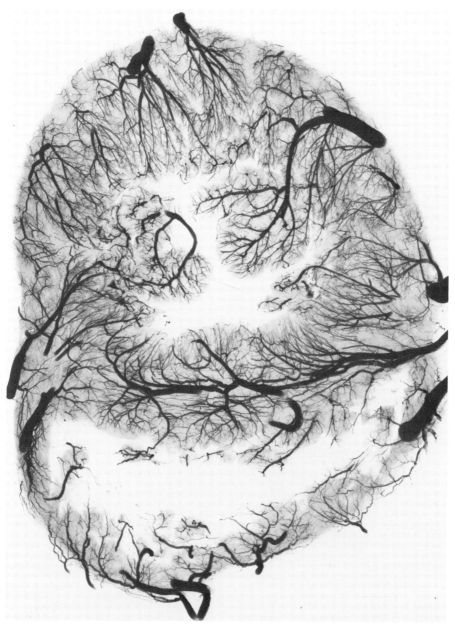

Fig. 27. A microradiograph of a midventricular heart of a 5-year-old boy showing the branches of the main coronary arteries penetrating the myocardium. (×3). By kind permission of G. Farrer-Brown.

of the myocardial wall, the small arteries turn at right angles (Fig. 28) to the line of direction of the main artery and then divide into a few branches which spread out sometimes over 180° to supply the muscle fibers in the area. The capillaries in all areas (Fig. 29) ran parallel with the muscle fibers and showed frequent anastomoses.

Unlike voluntary muscle, lymphatic vessels are plentiful and form a network through the intermuscular connective tissue.

D. Hypertrophy of Cardiac Muscle

Accompanying conditions such as hypertension and certain valvular diseases of the heart, when it is called upon to work against increased resistance, the cardiac muscle responds by an increase in size. This hypertrophic response is believed to be due solely to an enlargement of the existing cardiac muscle fibers and not to any increase in their number. The structure of heart muscle makes an investigation of a problem such as this very difficult, but the work of Karsner *et al.* (1925) seems to have established that no new fibers are formed. It may also be stated as collateral evidence that investigations into the regenerative capacity of heart muscle have shown that this is insignificant (Harrison, 1947; Walls, 1948).

E. Conducting System of the Vertebrate Heart

1. GENERAL OBSERVATIONS

In the hearts of mammals and birds, there is present a structurally specialized tissue that is generally considered to conduct the impulse for cardiac contraction. In support of the belief that this tissue is a neomorphic development that has undergone parallel evolution in response to functional requirement in these two vertebrate classes, and that it does in fact act as the conducting system, Davies and his co-workers have carried out extensive comparative and experimental studies that carry great conviction (Blair and Davies, 1935; Davies, 1930; Davies and Francis, 1946; Davies *et al.*, 1956). Forty years ago, it was held that the well developed conducting system of the mammal represented a remnant of more extensive tissue of similar structure in lower vertebrate hearts, but Davies and his colleagues have found themselves unable to support this view in any way. Prakash in 1954 and

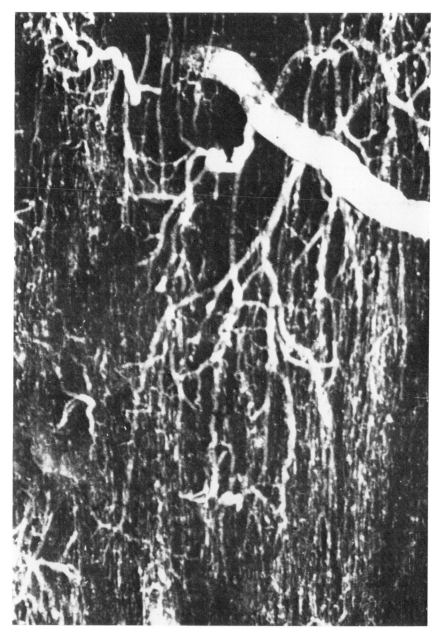

Fig. 28. A microradiograph of the middle of the human myocardium showing the final branches of small arteries leaving the main artery at right angles (×48). By kind permission of G. Farrer-Brown.

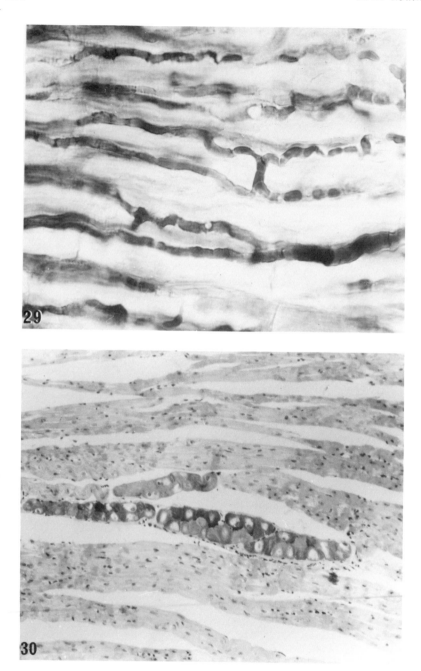

Figs. 29 and 30.

earlier papers and Bhatnagar (1957), however, dispute the neomorphism of the system in birds and mammals.

The specialized conducting tissue of the mammal, which is closely similar to that of the bird, will now be briefly described, but for fuller details the studies of Truex (1961; Truex and Smythe, 1964) should be consulted.

2. The Sinoatrial Node

First described by Keith and Flack (1907), this structure is now accepted as the site of initiation of the impulse for cardiac contraction and may be called the pacemaker. It is of horseshoe shape and embraces the ventral aspect of the termination of the superior vena cava. It is much more extensive than usually described, extending through the entire thickness of the arterial wall from epicardium to endocardium, and moreover, the anterior limb of the horseshoe extends fully halfway down the sulcus terminalis. It was originally throught to be composed of tissue intermediate in structure between nerve and muscle and accordingly was given the noncommittal name of nodal tissue. Its fibers, however, can be shown by silver impregnation to be specialized cardiac muscle fibers, completely cross striated, of fusiform shape, and branched. Intermingled with the nodal fibers is a considerable quantity of connective tissue. In diameter, the nodal fibers are smaller than ordinary neighboring atrial fibers. Data given by Blair and Davies for a 14-year-old human heart are nodal 2–7 μ and atrial 5–11 μ. Duckworth (1952) has shown that there is a considerable increase in the connective tissue content of the node just after birth. The nodal fibers are in direct continuity with the myocardial fibers of the right atrium, and these in turn with those of the left atrium. Although reports have been made from time to time of a direct connection between the sinoatrial and atrioventricular nodes, there seems to be no real evidence that such exists. The impulse therefore travels between the two nodes via ordinary heart muscle.

3. The Atrioventricular Node

First described by Tawara (1906), this structure, which is smaller than the sinoatrial node, lies in the lower part of the atrial septum.

Fig. 29. Capillaries of cardiac muscle outlined with peroxidase ($\times 500$).

Fig. 30. Periodic acid–Schiff-stained cardiac muscle showing a bundle of large (ts) positively stained Purkinje fibers lying between the ordinary cardiac fibers ($\times 115$).

Its specialized fibers are thus in continuity with the ordinary myocardial fibers of both atria. The nodal fibers, which show complete cross striation, differ from those of the sinuatrial node in being rather thicker, although more slender than the fibers of the ordinary myocardium, more cylindrical, and much more branched.

4. THE ATRIOVENTRICULAR BUNDLE

In direct continuity with the atrioventricular node is the atrioventricular bundle, a structure that can be dissected with ease in the ox, but with considerably more difficulty in man (Walls, 1948). The bundle runs forward under the septal cusp of the tricuspid valve to divide over the fleshy part of the interventricular septum into right and left limbs. The right limb forms a compact bundle on the right side of the septum from which it usually passes into the moderator band and so across the ventricular cavity to the base of the anterior papillary muscle. Arrived there, it breaks up into a subendocardial network of Purkinje fibers. The left limb is in the form of a broad, flattened band that descends for some distance on the left side of the septum before dividing into branches that pass across the ventricular cavity via the "false" tendons to reach the papillary muscles on the opposite wall. Once there, they break up into a plexus of Purkinje fibers beneath the endocardium.

5. PURKINJE FIBERS

In many mammals, the atrioventricular bundle and its limbs consist of large Purkinje fibers, first described by Purkinje in 1845. In the human heart, however, the fibers in the bundle and the proximal parts of its limbs are no larger than the ordinary fibers—indeed, they are slightly narrower—and it is about halfway down the septum before Purkinje fibers can be detected in the limbs of the bundle.

The special endocardial network observed by Purkinje in the ungulate heart was composed of large, pale fibers, often with two nuclei, in which striation was limited to the periphery of the fibers. This description still holds good, but it has recently been amplified considerably by Kugler and Parkin (1956), who reinvestigated the problem of whether or not there was direct continuity between ventricular Purkinje fibers and ordinary myocardial fibers in the hearts of dog, cow, and man. In all three hearts, it was found that three types of Purkinje fibers were present as follows:

Type 1 fibers have a large diameter, three to five times that of myocardial fibers, and have large spherical nuclei arranged in groups of two to four in the center of the fiber. Sarcoplasm is abundant, and in consequence, the myofibrils which are arranged mainly at the periphery appear relatively sparse. This type of Purkinje fiber is common under the endocardium but seldom penetrates the walls. In many cases, myofibrillar continuity of these fibers with those of type 2 was seen.

Type 2 fibers have a diameter about twice that of ordinary myocardial fibers (Fig. 30) and have less sarcoplasm than Purkinje fibers of type 1. Their myofibrils tend to be peripheral, although many are central, while their large and spherical nuclei are singly arranged. They can be observed to join with ordinary myocardial fibers or with type 3 fibers, and the transition is abrupt.

Type 3 fibers are only slightly wider than ordinary myocardial fibers and can be called a transition type. They have singly arranged, large, spherical nuclei, and the continuity they establish with ordinary ventricular fibers is abrupt.

REFERENCES

Aagaard, O. C. (1913). *Anat. Hefte, Abt. 2* **47**, 493.
Arey, L. B. (1965). "Developmental Anatomy," 7th ed., p. 245. Saunders, Philadelphia, Pennsylvania.
Arnold, J. (1870). *In* "Manual of Human and Comparative Histology" (S. Stricker, ed.), Vol. 1, p. 188. New Sydenham Society, London.
Banus, M. G., and Zetlin, A. M. (1938). *J. Cell. Comp. Physiol.* **12**, 403.
Bardeen, C. R. (1903). *Anat. Anz.* **23**, 241.
Barer, R. (1948). *Biol. Rev. Cambridge Phil. Soc.* **23**, 159.
Barr, L. (1969). *In* "The Biological Basis of Medicine" (E. E. Bittar and N. Bittar, eds.), Vol. 6, p. 95. Academic Press, New York.
Beatty, C. H., Basinger, G. M., Dully, C. C., and Bocek, R. M. (1966). *J. Histochem. Cytochem.* **14**, 590.
Bell, E. T. (1911). *Monatsschr. Anat. Physiol.* **28**, 297.
Bennett, H. S. (1955). *Amer. J. Phys. Med.* **34**, 46.
Bennett, H. S. (1956). *J. Biophys. Biochem. Cytol.* **2**, Suppl., Part 3, 171.
Bennett, M. R., and Burnstock, G. (1968). *In* "Handbook of Physiology" (Amer. Physiol. Soc., C. F. Code, ed.), Sect. 6, Vol. IV, p. 1709. Williams & Wilkins, Baltimore, Maryland.
Bennett, M. R., and Rogers, D. C. (1967). *J. Cell Biol.* **33**, 573.
Bhatnagar, S. P. (1957). *Indian J. Med. Sci.* **11**, 1.
Blair, D. M., and Davies, F. (1935). *J. Anat.* **69**, 303.
Bloom, W., and Fawcett, D. W. (1968). "A Textbook of Histology," p. 297. Saunders, Philadelphia, Pennsylvania.
Botar, J. (1966). "The Autonomic Nervous System." Akadémiai Kiadó, Budapest.
Boucek, R. J., Takashita, R., and Fojaco, R. (1963). *Anat. Rec.* **147**, 199.
Bowman, W. (1840). *Phil. Trans. Roy. Soc. London* **130**, 457.

Brash, J. C. (1955). "Neurovascular Hila of Limb Muscles." Livingstone, Edinburgh.

Bucciante, L., and Luria, S. (1934). Arch. Ital. Anat. Embriol. 33, 110.

Buchthal, F. (1942). Biol. Medd. (Copenhagen) 17, 1.

Buchthal, F., and Knappeis, G. C. (1946). Personal communication (cited by Høncke, 1947).

Bullard, H. H. (1912–1913). Amer. J. Anat. 14, 1.

Bullard, H. H. (1919). Johns Hopkins Hosp. Rep. 18, 323.

Burnstock, G. (1970). In "Smooth Muscle" (E. Bulbring, A. Brading, A. Jones, and T. Tomita, eds.), p. 1. Arnold, London.

Burnstock, G., and Campbell, G. (1963). J. Exp. Biol. 40, 421.

Caesar, R., Edwards, G. A., and Ruska, H. (1957). J. Biophys. Biochem. Cytol. 3, 867.

Cassens, R. G. (1970). In "The Physiology and Biochemistry of Muscle as a Food" (E. J. Briskey, R. G. Cassens, and B. B. Marsh, eds.), Vol. 2, p. 679. Univ. of Wisconsin Press, Madison.

Cohn, A. E. (1909). Verh. Deut. Pathol. Ges. 13, 182.

Cohn, A. E. (1932). In "Special Cytology" (E. V. Cowdry, ed.), p. 1159.

Cohnheim, J. F. (1865). Arch. Pathol. Anat. Physiol. Klin. Med. 34, 606.

Csapo, A. (1955). In "Modern Trends in Obstetrics and Gynaecology" (K. Bowes, ed.), 2nd ser., Chapter, 2, p. 20. Butterworth, London.

Csapo, A. (1959). Ann. N.Y. Acad. Sci. 75, 790.

Csapo, A. (1962). Physiol. Rev. 42, Suppl. 5, 7.

Davies, F. (1930). J. Anat. 64, 129.

Davies, F., and Francis, E. T. B. (1946). Biol. Rev. Cambridge Phil. Soc. 21, 173.

Davies, F., Francis, E. T. B., Wood, D. R., and Johnson, E. A. (1956). Trans. Roy. Soc. Edinburgh 63, 71.

Denny-Brown, D. (1929). Proc. Roy. Soc., Ser. B 104, 371.

Denny-Brown, D. (1961). Res. Publ., Ass. Res. Nerv. Ment. Dis. 38, 147.

Devine, C. E., and Simpson, F. O. (1967). Amer. J. Anat. 121, 153.

Dewey, M. M. (1965). Gastroenterology 49, 395.

Dewey, M. M., and Barr, L. (1962). Science 137, 670.

Dewey, M. M., and Barr, L. (1964). J. Cell Biol. 23, 553.

Dewey, M. M., and Barr, L. (1968). In "Handbook of Physiology" (Amer. Physiol. Soc., C. F. Code, ed.), Sect. 6, Vol. IV, p. 1629. Williams & Wilkins, Baltimore, Maryland.

Dubowitz, V. (1970). In "The Physiology and Biochemistry of Muscle as a Food, 2." (E. J. Briskey, R. G. Cassens and B. B. Marsh, eds.), p. 87. Univ. of Wisconsin Press, Madison.

Duckworth, J. W. A. (1952). M.D. Thesis, University of Edinburgh.

Edwards, D. A. (1946). J. Anat. 80, 147.

Elliot, T. S., Wigginton, R. C., and Corbin, K. B. (1943). Anat. Rec. 85, 307.

Engel, W. K. (1965). In "Neurohistochemistry" (C. W. M. Adams, ed.), p. 622. Elsevier, Amsterdam.

Engel, W. K. (1970). Arch. Neurol. 22, 97.

Evans, D. H. L., and Evans, E. M. (1964). J. Anat. 98, 37.

Farrer-Brown, G. (1968a). Cardiovasc. Res. 2, 179.

Farrer-Brown, G. (1968b). Brit. Heart J. 30, 527.

Farrer-Brown, G. (1968c). Brit. Heart J. 30, 679.

Farrer-Brown, G., Bielby, J. O. W., and Tarbit, M. H. (1970). *J. Obstet. Gynaecol. Brit. Commonw.* **77**, 673.

Fawcett, D. W. (1966). "The Cell: An Atlas of Fine Structure," Saunders, Philadelphia, Pennsylvania.

Felix, W. (1887). *Festschr. Albert von Kölliker,* p. 282. Englemann, Leipzig.

Fenn, W. O. (1945). *In* "Physical Chemistry of Cells and Tissues" (R. Hoeber, ed.), p. 445. Churchill, London.

Fernand, V. S. V. (1949). Ph.D. Thesis, University of London.

Garven, H. S. D. (1925). *Brain* **48**, 380.

Gauthier, G. F. (1970). *In* "The Physiology and Biochemistry of Muscle as a Food, 2." (E. J. Briskey, R. G. Cassens, and B. B. Marsh, eds.), p. 103. Univ. of Wisconsin Press, Wisconsin.

Gauthier, G. F., and Padykula, H. A. (1966). *J. Cell Biol.* **28**, 333.

Graf, W. (1947). *Acta Psychiat. Neurol.* **22**, 21.

Grützner, P. (1884). *Rec. Zool. Suisse* **1**, 665.

Haggquist, G. (1931). *In* "Handbuch der mikroskopischen Anatomie des Menschen" (W. von Möllendorff, ed.), Vol. 2, Part 3, p. 247. Springer-Verlag, Berlin and New York.

Haggquist, G. (1956). *In* "Handbuch der mikroskopischen Anatomie des Menschen" (W. von Möllendorff, ed.), Vol. 2, Part 4, p. 4. Springer-Verlag, Berlin and New York.

Haines, R. W. (1934). *J. Anat.* **69**, 20.

Halban, J. (1893). *Anat. Hefte, Abt. 2* **3**, 267.

Hammond, J., and Appleton, A. B. (1932). "Growth and Development of Mutton Qualities in Sheep." Oliver & Boyd, Edinburgh.

Harrison, R. G. (1947). *J. Anat.* **81**, 365.

Henle, J. (1841). "Allgemeine Anatomie, Lehre von den Mischungs und Vorbastandtheilen des Menschen Korpers," p. 573. Voss, Leipzig.

Herrick, C. J. (1902). *J. Comp. Neurol.* **12**, 329.

Herring, H. K., Cassens, R. G., and Briskey, E. J. (1965). *J. Sci. Food Agr.* **16**, 379.

Hillarp, N.-Å. (1959). *Acta Physiol. Scand.* **46**, Suppl. 175, 1.

Hillarp, N.-Å. (1960). *In* "Handbook of Physiology" (Amer. Physiol. Soc., J. Field, ed.), Sect. 1, Vol. II, p. 974. Williams & Wilkins, Baltimore, Maryland.

Hogue, M. J. (1937). *Anat. Rec.* **67**, 521.

Høncke, P. (1947). *Acta Physiol. Scand.* **15**, Suppl. 48, 12.

Huber, G. C. (1916). *Anat. Rec.* **11**, 149.

Jabonero, V. (1959). *Acta Neuroveg.* **19**, 276.

Jabonero, V. (1965). *Acta Neuroveg.* **27**, 101.

Jordan, H. E. (1933). *Physiol. Rev.* **13**, 301.

Jordan, H. E., and Banks, J. B. (1917). *Amer. J. Anat.* **22**, 285.

Jordan, H. E., and Steele, K. B. (1912). *Amer. J. Anat.* **13**, 151.

Karsner, H. T., Saphir, O., and Todd, T. W. (1925). *Amer. J. Pathol.* **1**, 351.

Keech, M. K. (1960). *J. Biophys. Biochem. Cytol.* **7**, 533.

Keith, A., and Flack, M. (1907). *J. Anat.* **41**, 172.

Knoll, P. (1891). *Denkschr. Akad. Wiss. Wien* **58**, 633.

Kohashi, Y. (1937). *Okajimas Folia Anat. Jap.* **15**, 175.

Köllicker, A. (1849). *Z. Wiss. Zool.* **1**, 48.

Kugler, J. H., and Parkins, J. B. (1956). *Anat. Rec.* **126**, 335.

Kuhne, W. (1863). *Arch. Pathol. Anat. Physiol. Klin. Med.* **26**, 222.

Lane, B. P., and Rhodin, J. A. G. (1964). *J. Ultrastruct. Res.* **10**, 489.

Lange, K. H. (1939). *Gegenbauers Jahrb.* **84**, 363.

Le Gros Clark, W. E. (1952). "The Tissues of the Body," p. 144. Oxford Univ. Press, London and New York.

Le Gros Clark, W. E., and Blomfield, L. B. (1945). *J. Anat.* **79**, 15.

Leeson, C. R., and Leeson, T. S. (1965a). *Anat. Rec.* **151**, 183.

Leeson, C. R., and Leeson, T. S. (1965b). *Acta Anat.* **62**, 60.

Lindhard, J. (1926). *In* "Physiological Papers" (R. Ege *et al.*, eds.), p. 188. Heinemann, London.

Lister, D. (1965). Ph.D. Thesis, University of Cambridge.

Lister, D. (1970). *In* "The Physiology and Biochemistry of Muscle as Food." (E. J. Briskey, R. G. Cassens, and B. B. Marsh, eds.), p. 705. Univ. of Wisconsin Press, Madison.

Locker, R. H. (1960). *Food Res.* **25**, 304.

Locker, R. H., and Haggard, C. J. (1963). *J. Sci. Food Agr.* **14**, 787.

Lockhart, R. D., and Brandt, W. (1938). *J. Anat.* **72**, 470.

Long, M. E. (1947). *Amer. J. Anat.* **81**, 159.

Lowenstein, O., and Lowenfeld, I. E. (1962). *Eye* 3, 231.

Lowy, J., and Small, J. V. (1970). *Nature (London)* **227**, 46.

MacCallum, C. P. (1898). *Bull. Johns Hopkins Hosp.* **9**, 208.

McGill, C. (1907). *Int. Monatsschr. Anat. Physiol.* **24**, 209.

McGill, C. (1909). *Amer. J. Anat.* **9**, 493.

McMeekan, C. P. (1941). *J. Agr. Sci.* **31**, 1.

Mark, J. S. T. (1956). *Anat. Rec.* **125**, 473.

Marsh, B. B., and Leet, N. G. (1966). *J. Food Sci.* **31**, 450.

Mayeda, R. (1890). *Z. Biol.* **27**, 119.

Merrillees, N. C. R. (1968). *J. Cell Biol.* **37**, 794.

Merrillees, N. C. R., Burnstock, G., and Holman, M. E. (1963). *J. Cell Biol.* **19**, 529.

Morpurgo, B. (1897). *Arch. Pathol. Anat. Physiol. Klin. Med.* **150**, 522.

Muir, R. (1929). "Textbook of Pathology," p. 158. Arnold, London.

Oshima, T. (1938). *Jap. J. Med. Sci.* I. **7**, 59.

Padykula, H. A., and Gauthier, G. F. (1963). *J. Cell Biol.* **18**, 87.

Padykula, H. A., and Gauthier, G. (1967). *In* "Exploratory Concepts in Muscular Dystrophy and Related Disorders" (A. J. Milhorat, ed.), p. 117. Excerpta Med. Found., Amsterdam.

Page, S. G. (1965). *J. Cell Biol.* **26**, 477.

Peachey, L. D., and Huxley, A. F. (1962). *J. Cell Biol.* **13**, 177.

Pease, D. C., and Paule, J..(1960). *J. Ultrastruct. Res.* **3**, 447.

Pellegrino, C., and Franzini-Armstrong, C. (1969). *Int. Rev. Exp. Pathol.* **7**, 139.

Prakash, R. (1954). *Proc. Zool. Soc. Bengal.* **7**, 27.

Prescott, J. H. D. (1963). Ph.D. Thesis, University of Nottingham.

Prosser, C. L., Burnstock, G., and Kahn, J. (1960). *Amer. J. Physiol.* **199**, 545.

Ramsay, R. W., and Street, S. (1940). *J. Cell. Comp. Physiol.* **15**, 11.

Ranvier, L. (1873). *C. R. Acad. Sci.* **77**, 1030.

Ranvier, L. (1874). *Arch. Physiol. Norm. Pathol.* [2] **1**, 1.

Rhodin, J. A. G. (1962). *Physiol. Rev.*, **42**, Suppl. 5, 48.

Rhodin, J. A. G. (1967). *J. Ultrastruct. Res.* **18**, 181.

Richardson, K. C. (1960). *J. Anat.* **94**, 457.

Richardson, K. C. (1964). *Amer. J. Anat.* **114**, 173.

Robertson, D. C., and Baker, D. D. (1933). *Mo., Agr. Exp. Sta., Res. Bull.* **200**, 1.

Robinson, D. W. (1964). *Anim. Prod.* **6**, 227.

Schafer, E. A. (1912). *In* "Quains Anatomy" (E. A. Schafer, J. Symington, and T. H. Bryce, eds.), 11th ed., Vol. 2, Part I, p. 111. Longmans, Green, New York.

Schiefferdecker, P. (1909). "Muskeln und Muskelkerne." Barth, Leipzig (quoted by Fernand, 1949).

Schwalbe, G., and Mayeda, R. (1891). *Z. Biol.* **27**, 482.

Schwann, T. H. (1847). "Microscopical Researches into the Accordance in Structure and Growth of Animals and Plants" (translated from German by Henry Smith). Sydenham Society, London.

Sichel, F. J. M. (1941). *Amer. J. Physiol.* **133**, 446.

Sjöstrand, F. S., Andersson-Cedergren, E., and Dewey, M. M. (1958). *J. Ultrastruct. Res.* **1**, 271.

Sobotta, J. (1930). "Textbook and Atlas of Human Histology and Microscopic Anatomy," p. 17. Stechert, New York.

Sonneborn, T. J. (1964). *Proc. Nat. Acad. Sci., U.S.* **51**, 915.

Speidel, C. C. (1939). *Amer. J. Anat.* **65**, 471.

Strong, K. C. (1938). *Anat. Rec.* **72**, 151.

Tawara, S. (1906). "Das Reizleitungssystem des Saugethierherzens." Fischer, Jena.

Taxi, J. (1965). *Ann. Sci. Natur. Zool. Biol. Anim.* [12] VII, 413.

Thaemert, J. C. (1966). *J. Cell Biol.* **28**, 37.

Truex, R. C. (1961). *In* "Specialized Tissues of the Heart" (A. P. de Carvalho, W. C. de Mello, and B. F. Hoffman, eds.), p. 22. American Elsevier, New York.

Truex, R. C., and Smythe, M. A. (1964). *In* "Electrophysiology of the Heart," p. 177. Pergamon, Oxford.

Tsurayama, K. (1937). *Jap. J. Med. Sci.* I. **6**, 249.

Van Breeman, V. C. (1953). *Anat. Rec.* **117**, 49.

Van Harreveld, A. (1947). *Amer. J. Physiol.* **151**, 96.

Venable, J. H. (1966). *Amer. J. Anat.* **119**, 271.

Vinnikow, J. A. (1938). *Dokl. Akad. Nanlc SSSR* **18**, 119.

Walls, E. W. (1948). *Brit. Heart J.* **10**, 188.

Watzka, M. (1939). *Z. Mikrosk-Anat. Forsch.* **45**, 688.

Wilkinson, H. J. (1929). *Med. J. Aust.* **2**, 768.

Witte, L. (1919). *Amer. J. Anat.* **25**, 333.

Yamamoto, T. (1961). *J. Electronmicrosc.* **10**, 145.

Yamauchi, A., and Burnstock, G. (1969). *J. Anat.* **104**, 1.

5

POSTMORTEM CHANGES IN MUSCLE

J. R. BENDALL

I. Introduction*

Stiffening and loss of extensibility of the musculature, that is, the onset of rigor mortis, are the most obvious postmortem changes in the physical characteristics of muscle. They are generally accompanied by the production of lactate and by marked acidification, although neither of these changes is a necessary concomitant of the process (Bernard, 1877; Hoet and Marks, 1926). The obligatory condition is the loss of adenosine triphosphate (ATP), which in the resting, recently excised muscle acts as a plasticiser to maintain it in a flexible and easily extensible state (Erdös, 1943; Szent-Györgyi, 1945; Bate-Smith and Bendall, 1947; Bendall, 1951; Lawrie, 1953). It is with this process and the underlying biochemical changes that we shall be mainly concerned here.

The simplest way in which to demonstrate the role of ATP as plasticiser and antirigor agent is to add it as magnesium salt to muscle fiber bundles which have previously been glycerolated (H. H. Weber, 1952; Szent-Györgyi, 1953) or frozen (Bendall, 1961) and then washed free of all substrates and enzymes, except those directly associated with the contractile apparatus. Providing that a calcium chelator, e.g. EGTA, is present, or alternatively the natural calcium pump of the sarcoplasmic reticulum is still mainly intact (Bendall, 1969), such muscle preparations, initially in full rigor, will relax on addition of an ATP solution. Figure 1 illustrates such an experiment with a muscle bundle, previously frozen at $-20°C$ and then washed in isotonic potassium chloride solution ($0.15\ M$). The bundle, loaded with a weight of about 500 gm/cm², is seen

---

* Abbreviations used in this chapter:

ΔH^*, ΔF^*, and ΔS^*	enthalpy, free energy, and entropy of activation (Eyring)
AMP	adenosine monophosphate
ADP	adenosine diphosphate
ATP	adenosine triphosphate
IMP, IDP, ITP	inosine mono-, di- and triphosphates
PC	phosphocreatine
P_i	inorganic phosphate
$\sim$P	energy-rich phosphate potential (see text)
NAD⁺, NADH	nicotine adenine dinucleotide—oxidized and reduced, respectively
EGTA	ethyleneglycolbis(aminoethyl ether)-N,N′-tetraacetic acid
PFK	phosphofructokinase
ETC	electron transport chain of mitochondria
LD	m. longissimus dorsi
Sterno	m. sternomandibularis (or sternocephalicus)

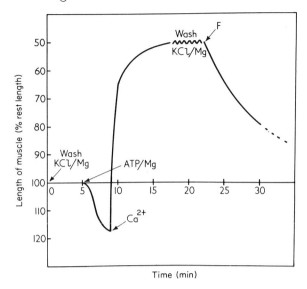

Fig. 1. Effect of ATP on a muscle bundle, frozen prerigor and subsequently allowed to thaw isometrically at rest length in isotonic potassium chloride solution. Cross-sectional area = 0.09 mm²; load = 500 gm/cm²; KCl/Mg = 150 mM KCl/4 mM MgCl₂; ATP/Mg = KCl/Mg + 5 mM ATP; Ca²⁺ = addition of CaCl₂ to 0.2 mM; F = preparation of sarcoplasmic reticular particles + the above ATP solution.

to lengthen immediately on addition of the ATP/Mg solution. On addition of Ca²⁺ to 0.2 mM, however, the contractile apparatus is activated, splitting of ATP sets in rapidly and the muscle shortens and lifts the load (see Bendall, 1969, p. 63). Washing out the ATP with potassium chloride solution now stops further shortening, but the loaded muscle remains fixed at its short length; i.e. it returns to the rigor state, as soon as ATP is withdrawn. Readdition of ATP plus a preparation of particles of sarcoplasmic reticulum (≡calcium pump) now resolves the rigor, and the muscle slowly lengthens back again to its original length.

Although the above experiment establishes that loss of ATP from the muscle is the event most closely linked to the onset of rigor mortis, it leaves many other questions open. For example, it implies that in the recently excised, resting muscle there must be active ATP hydrolases and other ATP-utilizing enzymes, the nature of which need to be determined if we are to be able to predict the time course of the rigor process. Moreover, we also need to know what resynthetic mechanisms will oppose this loss of ATP and thus postpone rigor onset. Of the latter, the chief are the creatine kinase reaction, by which ATP is resynthesized from ADP and phosphocreatine, and the combined reactions of anaerobic glycolysis which resynthesise $1\frac{1}{2}$ moles of ATP per mole lactate produced. The latter are also accompanied by the production of one proton (H^+) per lactate, so that during normal rigor in a well fed animal, where there is initially a large store of glycogen in the muscles, a considerable pH fall is the rule (from about pH 7.1 in the resting state to about pH 5.6 in full rigor).

The situation is, of course, quite different if oxygen is made freely available, because then the Krebs cycle replaces glycolysis and this, via the electron transport chain, resynthesises 18.5 moles of ATP per mole pyruvate oxidized. The result is that the glycogen store lasts about twelve times longer with a corresponding delay in the onset of rigor, but with little or no acidification nor fall in pH. Such a state of affairs is quite abnormal to rigor however, and does not occur to any appreciable extent in animal carcasses, except in a very thin superficial layer in the parts of the musculature exposed to air. The main bulk of muscle goes anaerobic within a few minutes of death, as can be easily calculated from the measured rates of oxygen uptake (Bendall and Taylor, 1972) and the diffusion coefficient of O_2 (Hill, 1965, p. 211).

The duration of the rigor process can vary enormously in practice, being dependent as it is on the magnitude of the initial phosphocreatine and glycogen stores (Bendall, 1951) and also, like any other chemical process, on the temperature. Thus, even at a constant temperature of 37°C, it can vary from less than $\frac{1}{2}$ hr in an exhausted animal to more than 4 hr in a well fed, rested animal. These times would be extended to about 1 and about 10 hr, respectively, at a temperature of 20°C. Such variability, which can now be completely accounted for in terms of the energy reserves in the muscle at death, was the source of much surprise in the early days of forensic medicine. For example, A. S. Taylor (1910) reports incredulously the case of a decapitated corpse which still reacted to electrical stimulation 20 hr after death and was therefore less than half way through the rigor process at this time (see Bendall, 1961, p. 229). This must indeed have been a very well-fed and rested criminal.

Similar variability occurs from species to species, and even between breeds of one species, the most notorious example being the pig (Ludvigsen, 1954; Briskey and Wismer-Pedersen, 1961; Bendall and Lawrie, 1964). In general, however, the chief muscles of the medium- and large-sized mammals behave very similarly to one another in pattern and duration, e.g. rabbit, ox, sheep, man, and the less highly selected breeds of pig. An exception is the whale which can show enormously long delays before rigor sets in in its musculature (Marsh, 1952b). Among other animals, there is a striking contrast between amphibia (frogs in particular), which show a long drawn out rigor process lasting more than 24 hr at 20°C, and some birds, e.g. the pectoral muscle of the pigeon, which goes into rigor in less than 4 hr at this temperature (Bendall, 1970).

We shall show in what follows, that apart from its relevance to forensic medicine, the variability in the duration of rigor plays an important and often dominant role in determining the quality of the meat we eat, one of the chief factors here being the temperature of the carcass during the commercial slaughtering and dressing process. It is appropriate, therefore, that we should begin this survey with a description of the physical changes, mainly in the extensibility of the muscle to an applied load, which occur during rigor, and the effect which temperature and the condition of the animal have upon their pattern and duration. For this purpose, we shall mainly cite the example of the psoas muscle of the rabbit concerning which extensive records exist in the archives of the British Meat Research Institute at Langford.

II. Physical Changes

A. Extensibility

Of the physical changes which take place in the muscle after death, the most easily measurable is the loss of extensibility to an applied load. This is so because only one parameter is involved, the stretch-deformation of the fibers. This method is to be preferred to other methods, such as penetrometer or sclerometer measurements (Mangold, 1927), which are far more complex and difficult to interpret. In what follows, therefore, extensibility changes will be taken as the main criteria of the onset of stiffening. It is more useful to express these changes as the reciprocal of the extensibility, which is an approximate measure of the elastic modulus and therefore more truly reflects the strength of the muscle.

1. METHODS OF MEASUREMENT

The simplest form of extensibility measurement consists in setting up a strip of muscle in a conventional kymograph and applying a load to it by hand, and this method was adopted in early studies by Bate-Smith (1939). The stretch deformation–time curve is long, drawn out, and continuous, but either pre- or post-rigor, can be analyzed in terms of three components, rapid, moderately rapid, and slow. These can be expressed in terms of three exponential curves, each with its own constant. Thus: deformation in time $t = A(1 - e^{-k_1 t}) + B(1 - e^{-k_2 t}) + C(1 - e^{k_3 t})$.

The effect of stiffening is then to reduce the overall deformation in a given time to one-twentieth to one-fortieth of the prerigor value. Analysis of the curves reveals that changes in components A and B are virtually completed 6 min after application of the load. By continuing the measurements for a further 2 min, the contribution of component C during the preceding time can be eliminated. It is also found that the resting prerigor muscle gives a recovery curve on release from the load nearly identical with the preceding stretch curve, so that the ratio $A/(A + B)$ is virtually the same for both curves. On the other hand, if active shortening occurs during the course of rigor, this ratio falls considerably, particularly for the recovery curve, but rises again to its prerigor level as stiffening is completed.

The method, as outlined, is unnecessarily cumbersome if a large number of measurements of extensibility are to be made during the course of rigor. For this reason, it was abandoned in later work in favor of a more convenient routine procedure, employing the mechanical loading device, described by Bate-Smith and Bendall (1949); see also, Briskey *et al.* (1962). This method consists essentially of applying and removing the load on the muscle by means of an electrically operated arm, so arranged that the muscle is alternately loaded and unloaded every 8 min until rigor is complete. The type of diagram obtained is illustrated in Fig. 2. It will be noted that the onset of stiffening is very clearly indicated in each case by the marked decrease of extensibility, of which the best measure is the deformation or recovery in the first 15 sec after application or removal of the load (Bate-Smith and Bendall, 1949). If longer times than this are taken, any shortening of the muscle during rigor will tend to be included in the measurement and will give an erroneously large value for the extension, as for example in diagrams b and d in Fig. 2, where considerable shortening occurred, as indicated by the upward displacement of the curves.

For the purpose of drawing smooth curves of the progress of stiffening, therefore, the 15 sec deformation is measured by extrapolation on each individual stretch and recovery curve during the course of rigor, its reciprocal is taken, and the results are then plotted as percentage gain in elastic modulus against time. In most experiments reported, the actual load on the muscle was set to give a prerigor extension corresponding to about 15% of the rest length. This load generally amounts to 50–90 gm/cm² of cross section.

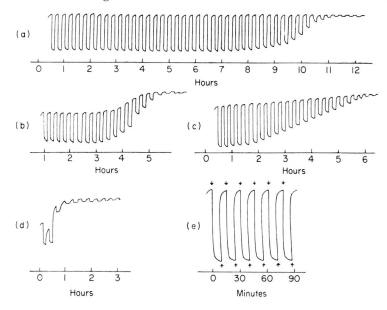

Fig. 2. Diagrams of extension changes during rigor in the psoas major muscle of the rabbit; curves recorded mechanically as described in the text. Muscles taken from: (a) Well fed, immobilized rabbit; temperature = 17°C; pH change, 7.00 to 5.70 during rigor. (b) As (a), but muscle held at 38°C. (c) As (a), but rabbit killed without prior immobilization; pH change, 6.52 to 5.70. (d) Rabbit exhausted by insulin convulsions; temperature = 17°C; pH 7.25 throughout. (e) Enlarged tracing of the early stage of rigor in muscle (a) above; ↓ means load on, ↑ means load off.

The relation between load and extension is complex, but can be empirically expressed, up to loads of 300 gm/cm², by:

$$\text{Extension} = k \log w - c \qquad (1)$$

where k and c are arbitrary constants, and w is the load.

When stiffening occurs, k diminishes twenty- to fortyfold, but c remains nearly constant. This equation is identical with that of Hill (1952), relating tension to the fractional change of length of resting muscle.

2. Increase of Modulus in Relation to Length of Muscle

From the experience of work on the rabbit psoas (Bate-Smith and Bendall, 1956) the extent of the modulus increase during rigor seems to be independent of any length change which may have occurred, at least at 17°C. Thus in normal acid rigor, occurring at 17°C without active shortening, the modulus increases to about 38× the prerigor value, whereas in cases of alkaline rigor, in which the muscle shortens to 0.81 of the rest length, it increases to 36×. These values do not differ significantly from one another, despite the very different pH values (5.7 and 7.3, respectively) and amounts of shortening.

On the other hand, Marsh (1953) has reported a series of experiments with beef LD muscle in which the modulus increase was found to be inversely related to the degree of active shortening which occurred. The muscles which showed the most change were, however, those which had been held at 37°C, where there is a tendency for lengthening to set in on the completion of rigor (Marsh, 1954). This lengthening can often begin while the modulus is changing most rapidly, and thus will tend to overshadow it. As we shall see in Section D, the phenomenon cannot be attributed to measurement artifacts, and probably indicates that series elastic components in the muscle rapidly become more fragile during and just after the completion of rigor at these higher temperatures, and yield more and more to stresses they would easily have been able to bear in the prerigor state.

3. Time Course of the Changes

The time course of the changes of extensibility are well illustrated by diagrams a, b, c, and d of Fig. 2, chosen from the large collection at the Meat Research Institute, Langford. The psoas muscle in diagram a was taken from a well fed rabbit, immobilized with Myanesin for 20 min before death, and was held thereafter at 17°C. It represents the maximum duration of rigor so far observed at this temperature. The diagram is seen to consist of three phases: (1) A phase during which the extensibility of the muscle remains constant and high, duration about 11 hr. This phase will be called the *delay period*. (2) In the next phase extensibility decreases rapidly, but with very little change of length of the unloaded muscle. This is the *rapid phase*. (3) In the last phase, extensibility has again become constant but at a lower level. This, the *postrigor phase*, lasts for many hours at 17°C and below without change of extensibility, until, in fact, putrefaction sets in.

In contrast to this protracted rigor at room temperature, diagram b illustrates the effect of holding a similar muscle at 37°C. There is still a characteristic delay period, with little or no change of the unloaded length, but it lasts for only about 4 hr against about 11 hr at 17°C, and is followed by a rapid phase during which the muscle shortens by about 15% of its rest length, as shown by the upward displacement of the curve. The extensibility falls to a minimum as the shortening is completed. Diagrams a and b are characteristic of well fed and immobilized animals, or of animals which have died quietly. On the other hand, if struggling occurs at death, as for example when a rabbit is killed by stunning and decapitation, the rigor diagrams of the muscles involved in the struggle, particularly the psoas, are more or less fore-shortened and resemble the latter stages of diagram a. An example of this type of rigor at 17°C is given in diagram c, where the extensibility decreases slowly at first and then ever more rapidly until rigor is complete. In such cases, it is often difficult to distinguish the delay period clearly from the rapid phase, most probably due to the unequal rates of stiffening among individual fibers of the muscle. As we shall see later, this type of rigor is characterized by a low initial pH, indicating the production of much lactate during the struggle at death, and also by a much reduced initial level of P-C and ATP, in contrast to quiescent muscles in which these three parameters are initially very high. The duration of rigor, however, can be reduced even further than this by exhausting the animals before death by means of insulin or strychnine convulsions or by subcutaneous injection of adrenalin (Bendall and Lawrie, 1962). Such cases are examples of the so-called alkaline rigor of Bernard (1877), where there is no production of lactate and no fall of pH, which remains high throughout at about 7.2. An extreme example is given in diagram d, where stiffening, accompanied by considerable shortening, occurred almost immediately after excision of the muscle. In this case, there was no P-C present initially, and the ATP level was already much reduced. Other cases of alkaline rigor have been described in which both these parameters were moderately high at death, with the result that there was a delay period of about 90 min before the rapid onset of shortening and stiffening (see Bate-Smith and Bendall, 1947, 1956). Alkaline rigor is encountered quite frequently in commercial practice, particularly in beef animals which have been under stress before slaughter. It is also characteristic of hunted animals, for example deer (venison). In such animals, where most of the musculature is of the red type, the alkalinity gives rise to a very dark red appearance, known as dark cutting.

Between the two extremes of alkaline and acid rigor, represented by exhausted and well fed animals, respectively, lies the type characteristic of starved animals, that is, animals in which the muscle glycogen has been more or less reduced. If such animals are immobilized at death, the duration of the delay period will decrease, but by no means so drastically as by allowing a struggle to occur. For example, a 50% reduction of the initial glycogen content, that is, of the potential glycolysis, results in curtailment of the delay period by about one-third, but in a somewhat prolonged rapid phase (see Bate-Smith and Bendall, 1949).

We can thus distinguish the following types of rigor:

1. Acid rigor, characterized in immobilized animals by a long delay period and a fast rapid phase, and in struggling animals by drastic curtailment of the delay period. Stiffening is accompanied by shortening at body temperature only.

2. Alkaline rigor, characterized by very rapid onset of rigor and marked shortening, even at room temperature.

3. An intermediate type, characterized in starved animals by curtailment of the delay period, but not of the rapid phase. Some shortening occurs during rigor at room temperature.

As we shall see later, the type of rigor which will ensue at any temperature is strictly determined by the magnitude of the initial P-C, ATP, and glycogen contents, of which the initial and final pH values are an indirect measure (Marsh, 1954). The types described above are, of course, variations on what might be called a normal pattern. Abnormally curtailed patterns, as in some breeds of pig, will be discussed in Section V.

B. Concomitant Changes in "Texture"

Side by side with the changes of extensibility which occur during rigor, there is a marked change in the "texture" of the muscle, which is soft and sticky before rigor and later becomes hard and dry as stiffening sets in. This dry texture may later change to a moister condition often characterized by an actual exudation of fluid, the so-called weep. The extent of the weep is mainly dependent on the pH and the temperature, and becomes noticeable only when the final pH is well below 6.0. It is increased greatly by raising the temperature to 37°C, where the muscle may lose as much as 15% of its weight as weep (Bendall, 1970). This is a particular feature of the rapid rigor in some breeds of pigs; it will be discussed in Section V,B. Weep is entirely absent in cases of starvation or exhaustion, where the final pH of the muscles is above pH 6.3, and it cannot be induced under these conditions at 37°C, in spite of the considerable shortening which occurs (Bate-Smith and Bendall, 1956). It must be stressed that, weep or no weep, the extensibility remains low and constant at this stage.

The change during rigor from a soft and sticky texture to a dry and hard texture can be assessed in a semi-quantitative way by means of a sclerometer (Mangold, 1927), the principle of which is to measure the depth of indentation into the muscle substance of a weighted steel ball or plunger. Clearly, this method involves several parameters at once. The actual hardness and extensibility of the muscle bundles, the ease with which they may be pushed apart, and the indeterminate changes in form of the irregular solid mass are concerned in the measurement recorded. In spite of this, the pattern of change closely parallels the extensibility changes, so that a delay period, during which the hardness remains low but constant for several hours, is followed by a phase of rapid increase in hardness, which may in its turn be succeeded by a slower, but considerable, decrease. This latter phase is particularly noticeable in whale muscle (Marsh, 1952b), where it is accompanied by a low pH (5.3) and the extrusion of much weep, probably due to severe damage to the sarcolemma, without which the loss of fluid would not be likely to occur. This would have the effect, from the point of view of sclerometer measurements, of converting the muscle bundles from firm rods of intact fibers to flaccid tubes of easily expressible fluid.

C. Shortening during Rigor

The idea that rigor is a slow but irreversible contracture is mentioned frequently in the older literature (e.g., A. S. Taylor, 1910), and even in more recent times (Bendall, 1951; Davies, 1963) it has been suggested that shortening is an essential part of the process. Considerable shortening can, indeed, occur during rigor, both under alkaline conditions (e.g., Fig. 2d) and at higher temperatures (Fig. 2b), and coincides in time rather exactly with disappearance of ATP from the muscle. However, the shortening is always feeble by comparison with a living contraction, yields only a tiny fraction of the full power which can be developed and is overcome by comparatively light loads (Bendall, 1961, p. 237). Indeed, histological studies show that rigor shortening never involves more than a fraction of the fibers, the passive remainder being folded into more or less irregular S-shapes (Bendall, 1961, p. 237; Voyle, 1969).

The work done during rigor at various temperatures and pH values is plotted in Fig. 3 for rabbit psoas and beef LD muscles. It is seen that the work increases with temperature, and is greater the higher the ultimate pH of the muscle. The maximal work done by the rabbit psoas is about 15 gm cm/gm under optimal conditions at an ultimate pH of 7.2 and a temperature of 37°C. It takes approximately 30 min for the muscle to perform this work (= 0.5 gm cm/min per gram of muscle), whereas the work output by a glycerolated muscle fiber model, contracting in ATP solution at 20°C, is about 100 gm cm/sec/gm (= 6000

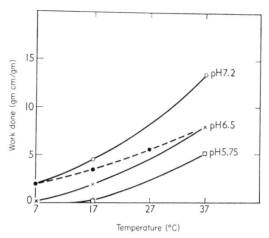

Fig. 3. Plot of work done against temperature during rigor in (a) rabbit psoa muscles at three different ultimate pH values (solid lines); (b) beef LD muscles (broken lines) at ultimate pH of about 5.7. (After Marsh, 1953.)

gm cm/min/gm) under optimal conditions (Bendall, 1969, p. 63). Expressed in this way, shortening during normal rigor is indeed feeble. On the other hand, the special conditions obtaining during thaw rigor (see Section V,A.) enable the muscle to develop much more power (about 400 gm cm/min/gm), but even this is only about 7% of the maximal.

From the point of view of tension development, the rigor muscle is also very feeble compared with a living muscle, the maximal tension developed being about 150 gm/cm^2 compared with about 4000 in a tetanus. Model glycerolated fiber systems can, however, develop tensions of about 1000 gm/cm^2 as they go into rigor (White, 1970).

D. Structural Basis of the Physical Change

Before the advent of the electron microscope, many models were put forward to explain the stretch-deformation curves of muscle, as measured before and after rigor. Most of these originated from theoretical considerations involving a series of damped and undamped springs and were not founded on any real structures within the muscle itself, apart from the well recognized sarcolemma on the outside of the fiber and the contractile elements within (see H. H. Weber, 1934; Houwink, 1937; Bate-Smith, 1939). Hill (1939) had already criticized theories based on this conception of free and damped springs. He suggested, on the other hand, that the main elements involved were a parallel elastic component, which took most of the strain when a resting muscle was stretched, and a series elastic component, which came into play only in the contracting muscle. He considered that the contractile elements themselves were too easily extensible under resting conditions to account for the deformation curves, which he attributed solely to the properties of the parallel component, perhaps identifiable with the endo-, peri-, and epimysium and the sarcoplasmic reticulum around the myofibrils. If this idea is considered in relation to rigor, it is clear that the main change during the process must be in the contractile elements themselves, since none of the concomitant chemical changes are likely to affect the extensibility of the other structures significantly. This, of course, has become even more likely since the discovery of actin as an essential part of the contractile mechanism and of the ability of this protein to combine with myosin in the absence of ATP (Szent-Györgv, 1945), and by the further significant discovery that myosin is located exclusively in the A bands of the muscle (Hanson and Huxley, 1955).

Out of the brilliant results of electron microscopy came the sliding filament model of contraction, developed by Hanson and Huxley (1955), (see Bendall, 1969, p. 20); this model explains satisfactorily all the facts of the rigor process. A diagramatic representation of it is given in Fig. 4. We note that the essential feature of the A band filaments is that they consist of packed longitudinal arrays of myosin molecules, the heads of which protrude from the filament, and are able to make contact

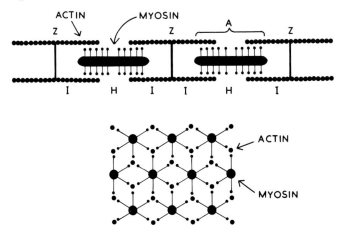

Fig. 4. Diagram of the sliding filament model of Hanson and Huxley (1955). Note thick filaments, with protruding heads of myosin molecules and the beaded structure of the thin filaments of actin. The sarcomere is shown at rest length (2.4 μm per sarcomere). The cross section, below, shows the typical hexagonal array of actin and myosin filaments in the region of overlap in the A band.

with the bead like actin molecules of the I filament. The heads of the myosin molecules contain an ATP hydrolase site, which is activated during the contractile process (Davies, 1963), and an actin-combining site which comes into play during rigor, when ATP has disappeared from the immediate neighborhood (Stracher, 1964). Thus, wherever actin filaments overlap myosin filaments there is a possibility of actin–myosin links being formed during rigor, locking the sliding mechanism in a rigid, inextensible condition from one end of the muscle to the other.

Huxley and Brown (1967) have been able to show by refined, low-angle X-ray studies that the characteristic layer lines of the muscle are seriously disturbed in the rigor muscle, indicating first that many more attachments are formed between actin and myosin than during a tetanus, and second that the characteristic 42.9 nm spacing along the backbone of the myosin filaments is replaced by layer lines which can be indexed on a 36 or 38 nm helical repeat. This latter change indicates that during rigor the myosin heads can swivel around, azimuthally, to find suitable actin partners (see transverse section in Fig. 4).

One obvious outcome of the sliding filament hypothesis is that fewer actin–myosin rigor bonds can be formed in a stretched muscle, where the overlap is less, than in a short muscle, where the overlap is greater. Thus the strength of a short muscle should be greater than that of a long one after rigor, just as it is during active tension development in a living muscle, where it can be shown that the tension is maximal at rest lengths but falls off sharply at greater lengths (Gordon *et al.*, 1966). It needs testing in rigor muscle, although the problem is a difficult one, because of the necessity of gripping the muscle firmly without damaging it during the measurement of the breaking strength. It has not yet been solved satisfactorily.

E. Resolution of Rigor

The term "resolution of rigor" carries the implication that the characteristically low elastic modulus of prerigor muscle, and the equally characteristic reversibility of its stretch-deformation–recovery curves, are somehow slowly reestablished in the postrigor period. However, in view of the firm linkages formed between the actin and myosin filaments in the absence of ATP, it is apparent that such a type of resolution could not occur without a supply of ATP becoming available *de novo*, as in the example of model rigor shown in Fig. 1. Such a miraculous occurrence is not highly probable in the intact muscle, at a time when its energy reserves are at their lowest ebb. Nevertheless there is considerable evidence that the series elastic component of the myofibrils (Hill, 1939, 1952) becomes weaker during the postrigor period and yields under breaking stresses which it could have easily borne during the onset of rigor, in spite of the earlier assertions of Marsh (1954) and Bate-Smith and Bendall (1956) to the contrary.

Reexamination of the experiment, quoted by Bate-Smith and Bendall (1956) to prove that resolution did not occur, does indeed show a marked decrease in the elastic modulus after 5 days storage of a psoas muscle at 3°C, as Fig. 5a illustrates. Up to that time however, there was no change in the very high rigor modulus, despite the fact that the muscle was alternately stretched and allowed to relax every 8 min. during the whole period. Thus we conclude that after about the fifth day at 3°C, some series linkages become more labile, with a reduction in their breaking stress.

The above type of resolution can be seen more clearly during rigor at 38°C, where frequently the muscle shortens and actively lifts the load on it, during the much curtailed rigor process at this temperature, but then soon afterwards begins to lengthen again with very marked changes in the deformation curves. An example is shown in Fig. 5b. We see that the half time for lengthening and decrease of modulus is here only about 1 hr compared with about 36 hr at 3°C (Fig. 5a), giving an approximate enthalpy of activation of 20 kcal/mole for the linkage-breaking process.

Davey and Gilbert (1967) carried out similar experiments with beef muscle, where they showed a marked decrease in the shearing stress (after cooking) during 3 days storage at 15°C. Davey and Gilbert (1969) also showed that homogenization of such stored muscle, at constant speed, results in the production of much shorter and more fragmented myofibrils than in the case of muscle homogenized immediately postrigor. They suggested, quite plausibly, that the main linkages affected by the aging process are in the Z disks, and indeed there is both light and electron microscope evidence to show that it is here that the most marked changes occur on storage, particularly loss of staining density (Voyle, 1970).

Although much more needs to be done to define the exact conditions under which resolution of rigor can occur, our own experiments on rabbit muscles of differing ultimate pH values, stored at 38°C, suggest that the process is highly pH dependent, occurring very rarely in exhausted muscle (ultimate pH 7.1) or muscle from starved animals (ultimate pH 6.4–6.8), but more or less markedly in all muscles from well fed animals with ultimate pH values between 5.3 and 5.8. As Scopes (1964)

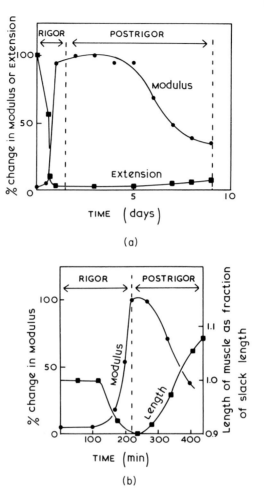

Fig. 5. (a) Rabbit psoas muscle, held at 3°C for 9 days, during which it was stretched and released alternately every 8 min. Load/cm² = 70 gm, ultimate pH = 5.78, ● = modulus change, ■ = extension change. (b) Rabbit psoas muscle held at 38°C for 7 hr. Load/cm² = 60 gm, ultimate pH = 5.74, ● = modulus change, ■ = length change.

and Penny (1967) have shown, these latter are precisely the pH conditions under which both the sarcoplasmic and myofibrillar proteins denature most rapidly, so it may well be that resolution is basically a denaturation process, perhaps of the α-actinin or tropomyosin which are thought to be the linking proteins between the ends of the actin filaments in the Z disk (Ebashi and Endo, 1968; Penny, 1972). Thus, there may be no further need to search about, apparently hopelessly, for a protease (cathepsin) which could break the rather specific linkages probably involved.

III. Chemical Changes Underlying the Rigor Process

A. The Pattern of Chemical Change

The main chemical changes which underly the anaerobic rigor process are the disappearance from the muscle of PC and glycogen, and later of ATP, and the concomitant formation of lactate, which is accompanied by a fall of pH (see Fig. 6a and b, and Fig. 7). As we see, the pattern is remarkably similar, whether the muscle contains a high initial level

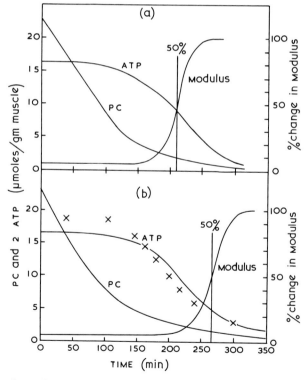

Fig. 6. Decline of PC and ATP levels, and increase of modulus, in rabbit psoas muscle at 38°C. Animals immobilized before death. Initial pH = 7.10; (a) ultimate pH = 6.58 (5 pairs); (b) ultimate pH = 5.70 (10 pairs). Crosses in (b) are the calculated ATP values from Eq. (15b).

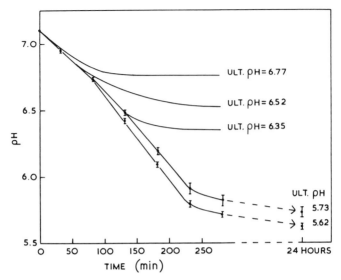

Fig. 7. Fall in pH during rigor in rabbit psoas muscles at 38°C. The pH/time curves are shown for five different ultimate pH values, as marked on the figure.

of glycogen, as in Fig. 6b, or this level has been considerably depleted by inanition of the animal, as in Fig. 6a. In our general description in this section, therefore, we shall mainly consider the changes character-istic of muscles from well fed animals (rabbit psoas) at 38°C, having a high initial pH (low initial lactate) and high levels of glycogen, PC and ATP (Fig. 7, lowest curve, and Fig. 6b). Such patterns of change apply generally to muscles from widely ranging species of mammals and to all types of rigor so far studied, including so-called thaw rigor (Bate-Smith and Bendall, 1947, 1948; Bendall, 1951, 1961, p. 246; Lawrie, 1953; Howard and Lawrie, 1956, 1957; Marsh, 1952a, 1954; Bendall *et al.*, 1963; Hallund and Bendall, 1965; Newbold and Scopes, 1967). The main differences are in time scale which can vary enormously according to temperature and other conditions. These will be discussed more fully later.

The analytical methods used for estimating various glycolytic inter-mediates were as follows: (1) in earlier work, the chemical methods for PC, ATP, P_i, total P, glycogen and lactate, described by Bate-Smith and Bendall (1949) and Bendall (1951); (2) the spectrophotometric methods for adenine and inosine nucleotides, described by Bendall and Davey (1957); (3) in recent work, particularly on beef and pig muscle, the more exact enzymatic methods for PC, ATP, lactate and other inter-mediates, described by Newbold and Scopes (1967).

Returning to Fig. 6b, we see that the main characteristic of the changes is that the ATP level remains constant for a considerable time, while the PC level and the pH (Fig. 7, lowest curve) are falling rapidly. Since the latter is a measure of lactate formation, this means that a great deal of resynthesis of ATP has occurred during this phase, both from this glycolytic source and from PC via the creatine kinase reaction. While the ATP level remains constant, it is clear that the resynthetic pathways are opposed by an equal and opposite breakdown of ATP, presumably by a hydrolase of some sort. We shall define this phase of constant ATP level as the *delay phase*, by analogy with the physical changes.

The overall reactions which occur during the delay phase can be formally represented by the following stoichemiometric equations:

(a) $\quad (3 + m)\,[\text{ATP}^{4-} + \text{H}_2\text{O} \rightarrow \text{ADP}^{3-} + \text{P}_i{}^{2-} + \text{H}^+]$ $\hfill$ (2)
$\qquad$ Hydrolases

(b) $\quad m\text{ADP}^{3-} + m\text{PC}^{2-} + m\text{H}^+ \rightarrow m\text{ATP}^{4-} + m(\text{creatine})$
$\qquad$ Creatine kinase

(c) $\quad 3\text{ADP}^{3-} + 3\text{P}_i{}^{2-} + (\text{glucose})_n + \text{H}^+ \rightarrow$
$\qquad\qquad\qquad\qquad 3\text{ATP}^{4-} + 2(\text{lactate})^- + (\text{glucose})_{n-1} + 2\text{H}_2\text{O}$
$\qquad\qquad$ Embden-Meyerhof glycolytic pathway

SUM: $\quad (\text{Glucose})_n + (1 + m)\text{H}_2\text{O} + m\text{PC}^{2-} \rightarrow (\text{glucose})_{n-1} + m\text{P}_i{}^{2-} + 2(\text{lactate})^-$
$\qquad\qquad\qquad\qquad\qquad\qquad\qquad\qquad + m(\text{creatine}) + 2\text{H}^+$

Note that ATP and ADP actually react as the Mg chelates, MgATP^{2-} and MgADP^-.

There is no way, a priori, of establishing what fraction $(m/3 + m)$ of the ADP formed by hydrolases is resynthesized to ATP by the creatine kinase pathway, but in practise the fraction is generally one-third initially, falling rapidly as the PC store is exhausted.

We note from the above scheme, that glycolysis can only be accompanied by acidification ($1\ \text{H}^+/\text{lactate}$), if ATP is the starting substance and is itself being continuously converted by hydrolases to ADP and P_i which act as the immediate substrates for glycolysis. If only ADP, P_i, glycogen and the glycolytic enzymes were present, then the result of the resynthesis of ATP and formation of lactate would, on the contrary, be an alkalinization to the extent of $-1\ \text{H}^+/2$ lactate, as Eq. (2c) shows. This is not always made clear in biochemical textbooks, e.g., Mahler and Cordes (1966).

Figure 6b shows that when the PC level has dropped to about 4 μmoles/gm muscle, the ATP level itself begins to fall in spite of vigorous glycolysis (see also Figs. 7 and 12a). We shall call this phase the fast phase of rigor. During this phase, the disappearance of ATP is accompanied by the appearance of ammonia in stoichiometric amounts, due to deamination of adenine nucleotides. As Bendall and Davey (1957) have shown, the overall reactions can then be formulated as:

(a) $(3 + 2q)$ [ATP^{4-} + H$_2$O → ADP^{3-} + P$_i$$^{2-}$ + H$^+$]

(b)   3ADP^{3-} + 3P$_i$$^{2-}$ + (glucose)$_n$ + H$^+$
$$\rightarrow 3\text{ATP}^{4-} + 2(\text{lactate})^- + (\text{glucose})_{n-1} + 2\text{H}_2\text{O}$$

(c) $2q$ADP^{-3} → qATP^{-4} + qAMP^{2-}
          myokinase

(d) qAMP^{2-} + qH$^+$ + qH$_2$O → qIMP^{2-} + qNH$_4$$^+$ (3)
          AMP-amino-hydrolase

SUM: (Glucose)$_n$ + $(1 + 3q)$H$_2$O + qATP^{4-} →
          (glucose)$_{n-1}$ + qIMP^{2-} + qNH$_4$$^{1+}$ + $2q$P$_i$$^{2-}$ + 2(lactate)$^-$ + $(2 + q)$H$^+$

Here again it is impossible to predict, a priori, what proportion $(2q/3 + 2q)$ of the ADP formed in reaction (3a) will be deaminated and dephosphorylated in reactions (3c) and (3d), although we can see that it will be critically dependent on the relative levels of the various adenine nucleotides, particulary ADP (for 3c) and AMP (for 3d). In practise, the proportion $(2q/3 + 2q)$ reaches a value of about 0.18 while glycolysis is still proceeding vigorously in the middle of the fast phase, but of course rises progressively higher as the glycogen store is later exhausted and resynthesis ceases.

We can see by inspection of Eqs. (2)–(3), that the total turnover rate of "energy-rich" phosphate ($= \sim$P) is given in general by:

$$d(\sim \text{P})/dt = -d(\text{PC})/dt + 1.5 \, d(\text{lactate})/dt - 2 \, d(\text{ATP})/dt \qquad (4)$$

The last term on the right is unimportant during the delay phase and the first term can be neglected as soon as the PC level falls below about 3 μmoles/gm. Equation (4) has to be modified if significant amounts of glycolytic intermediates accumulate during rigor, but such an occurrence is unusual under normal circumstances. As we shall see later, accumulation of small amounts of hexose 6-phosphate and α-glycerol phosphate does sometimes occur, however.

B. Relation between Chemical Changes and Modulus

Figures 6a and b show that the change most closely related to the increase in elastic modulus during the first phase of rigor is the disappearance of ATP from the muscle, as indeed we would expect from the model experiment shown in Fig. 1. The relationship between the two changes is not, however, quite as straightforward as it looks at first sight, because we see that 50% of the modulus change has already occurred in the case of the starved animal (Fig. 6a) before the ATP has fallen below 50% of the resting level, whereas in the well fed animal (Fig. 6b) the modulus has scarcely increased at this stage and reaches 50% of the final value only when the ATP level has declined to 2.5 μmoles/gm (i.e., 31% of the resting level). This feature is more clearly brought out in Fig. 8a, where the ATP level at 50% modulus change is plotted against the pH at this time, for muscles with varying initial glycogen contents (= glycolytic potential which varies inversely with

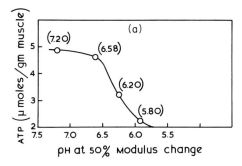

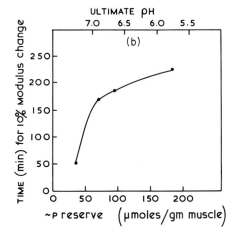

Fig. 8. (a) The ATP level at 50% modulus change plotted against ultimate pH in rabbit psoas muscles. The pH at the time of the modulus change is marked on the side of the curve in brackets. (b) The time for the beginning of modulus change (10%) plotted against the total $\sim$P reserve and the ultimate pH of rabbit psoas muscles at 38°C. Initial PC = 23 μmoles/gm; initial ATP = 8 μmoles/gm. Note that the total glycolytic $\sim$P reserve = 0 at pH 7.2 and 145 μmoles/gm at pH 5.6.

ultimate pH). We see that the lower the glycolytic potential, and the higher the ultimate pH and the pH at the time, the higher the ATP level is at 50% modulus change.

There is no really satisfactory explanation of this variation in ATP level at the time of rigor, particularly as we know from model systems that ATP concentrations as low as 0.5 mM are sufficient to bring about dissociation of the actomyosin complex under relaxing conditions, and also to relax glycerolated fiber preparations (White, 1970). Hence it is difficult to believe that any individual fiber of an intact muscle would go into rigor before the ATP level had declined below this concentration (= about 0.4 μmoles/gm of fresh muscle). However, since the modulus change is itself a direct measure of the proportion of fibers which are in rigor at any time, i.e., of those fibers where the free ATP level is below 0.4 μmoles/gm, we could argue that the modulus changes in Fig. 6 assume their particular form because of an uneven distribution of ATP, PC, and glycolytic potential between fibers during the fast phase.

Another way out of these difficulties over the ATP level at the time of rigor onset would be to assume that some other change was responsible for onset. Indeed, Kushmerick and Davies (1968) have suggested that the ATPase sites of myosin in resting, prerigor muscle are, in fact, occupied by ADP, and this has recently been confirmed by Marston and Tregear (1972). This "bound" ADP must, however, be in dynamic equilibrium with ATP (E. W. Taylor, 1970), so that when the latter disappears during rigor the bound ADP will also dissociate and be destroyed, as Table II demonstrates. It may well be that it is the latter change which actually determines rigor onset, though an extensive series of experiments under different conditions of ultimate pH would be required to prove it (see also Section IV,A,1).

C. Effect of Temperature on Rates of Change

The rigor process is dependent on a complex series of biochemical reactions and so it might be expected to have a high temperature coefficient ($Q_{10°C}$) and a high and positive enthalpy of activation (ΔH^*). This is indeed the case at temperatures of 20°C and above, as the Eyring plot in Fig. 9 shows. The parameters plotted in this figure are $\log_{10}(V_0/T)$ against $1/T$, with T in °K. The V_0 values are for $\Delta \sim P/\Delta t$ at pH 7, calculated from Eq. (4), at various temperatures. According to the Eyring theory of absolute reaction rates (Glasstone, 1947), the slope of this plot gives ΔH^*, because:

$$2.303 \log_{10}(V_0/T) = \log_e k/h + \Delta S^*/R - \Delta H^*/RT \tag{5a}$$

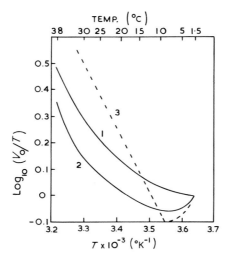

Fig. 9. Eyring plot of the rate of $\sim$P turnover (V_0) against temperature, for rabbit psoas and beef sterno muscles during initial stages of rigor mortis (pH = 7.0). Relative values of $\log_{10}(V_0/T)$ are plotted against $1/T$ (°K), taking $V_0/T = 1.0$ at 1.5°C. (1) Rabbit psoas from Bendall (1960); (2) beef sterno (intact) from Cassens and Newbold (1966); (3) beef sterno strips in paraffin, saturated with nitrogen, from Bendall and Taylor (1972).

where k = Boltzmann constant; h = Planck constant; S^* = entropy of activation, and since ΔS^* and ΔH^* are constant by definition:

$$-\Delta H^* = 4.6 \frac{d[\log_{10}(V_0/T)]}{d(1/T)} \tag{5b}$$

The three curves in Fig. 9, (1) for intact rabbit psoas, (2) for intact beef sternomandibularis, and (3) for the latter in the form of strips in paraffin oil, all show nearly linear portions from 38° down to 25°C, with similar slopes. The mean slope yields a value of 11 kcal/mole for ΔH^*. This is almost identical with ΔH^* for oxygen uptake by strips of respiring beef muscle (Bendall and Taylor, 1972), a process which is also driven by the ADP, liberated from ATP by the hydrolases of resting muscle (see Section IV,E,2). Thus, the chemical processes of rigor behave as an ideal chemical reaction in the higher temperature range, although they give a ΔH^* value considerably lower than that for splitting of ATP by the Mg/Ca-activated ATPase of contracting muscle (Bendall, 1969, pp. 50–56), which is of the order of 25 kcal/mole.

Ideally as the biochemical reactions may behave from 38° to 25°C, they depart more and more from the ideal as the temperature is reduced toward 0°C, as Fig. 9 clearly shows. In every case the rate (V_0) progres-

TABLE I

Time-scale Corrections for Changes in the Chemical and Physical
Parameters during Rigor at Various Temperatures (Rabbit
Psoas and Beef Sternomandibularis)

Temperature (°C)	Correction		
	Rabbit	Beef	Ideal
38	1.00×	1.00×	1.00×
35	1.20×	1.20×	1.19×
30	1.63×	1.63×	1.59×
25	2.03×	2.03×	2.16×
10	2.40×	2.26×	2.96×
15	2.75×	2.54×	4.10×
20	3.09×	2.79×	5.75×
2	3.50×	2.60×	10.20×

sively increases above the ideal value as the temperature is lowered.
In fact, intact beef muscle even shows an upward trend in the curve
at 1°C, the rate at that point being almost identical with that at 15°C
(see Cassens and Newbold, 1966; Newbold and Scopes, 1967). This extremely anomalous behavior at the lower temperatures can only be
accounted for by the increasing intervention of a new reaction, the nature of which may be surmised by analogy with the similar phenomenon
of thaw-contracture, as we shall see in Section V,A.

As far as normal rigor is concerned, the chemical and physical pattern
from 38°C to about 12°C is identical in form with that shown in Figs.
6a and b and Fig. 7. To convert it from one temperature to another
the time scale must be corrected according to the velocity plot in Fig.
9. The necessary corrections are shown in Table I.

D. Effects of Exhaustion, Inanition, and Struggling at Death

We have already briefly described the effects of the above factors
on the time for onset of the elastic modulus increase. Their effect on
the chemical changes runs a parallel time course, but some of the detailed changes are of general interest.

1. Exhaustion

Experimentally, animals can be exhausted either by massive injection
of insulin to induce a series of violent convulsions (Bate-Smith and

Bendall, 1956; Bendall, 1961, p. 234), or by the more gentle means of a subcutaneous injection of adrenalin (Howard and Lawrie, 1956; Bendall and Lawrie, 1962). Normally it is difficult to exhaust the larger animals by exercise alone, and in fact, with bullocks, Howard and Lawrie (1956; cf. Lawrie, 1962), found it almost impossible to do so, even under the hot and humid conditions of Australia. The more vigorous conditions of the hunt are, however, sufficient to exhaust most animals, particularly deer whose muscles then contain little or no glycogen reserve at death. The result of this type of exhaustion, especially that induced by large doses of insulin, is not only to reduce the glycogen reserve to zero, but often the PC reserves as well, so that the ATP level drops almost immediately after death and the modulus increases rapidly in step with this decline (see Section A,3). Thus, rigor is of very short duration, and often takes place at pH 7.2 or above, without any production of lactate or H^+ ion. This is the so-called alkaline rigor of Claude Bernard (1877), an observation which many years later led to the final abandonment of the lactic acid theory of rigor (Hoet and Marks, 1926) and the acceptance of the ATP hypothesis (Erdös, 1943; Bate-Smith and Bendall, 1947). With rabbit psoas muscle it is sometimes possible to catch the muscle while it still has a high level of PC and ATP, but no glycogen (Bate-Smith and Bendall, 1956). Such examples furnish a baseline for comparing the time for half change of modulus with the ultimate pH of the muscle and with the magnitude of its $\sim P$ reserve at death (see lowest point on Fig. 8b).

The use of subcutaneous injection of adrenaline, as a means of exhausting the muscles of glycogen, was first studied by Cori and Cori (1928) who showed that the blood lactate rose to high levels shortly after the injection and therefore suggested that the glycogen reserve disappeared through the glycolytic route, rather than via the oxidative (citric acid) cycle, as might have been expected in an aerobic muscle. Adrenaline is more useful experimentally than insulin, because it is possible to grade the effect and thereby obtain muscles with glycogen reserves ranging from zero to maximal (Bendall and Lawrie, 1962).

It is probable that slow adrenaline release is the main cause of the glycogen exhaustion which accompanies stress (Howard and Lawrie, 1956; Hedrick, 1958). Adrenaline treatment has distinctive effects on the chemical pattern, particularly: (1) very high initial levels of PC and ATP, in spite of the almost complete exhaustion of the glycogen reserves; (2) a significantly higher rate of $\sim P$ turnover in the postmorten period than in normal muscles over the same pH range (Bendall and Lawrie, 1962). Neither of these effects has yet found a satisfactory explanation, but the second tends to accelerate the rigor process even more than insulin convulsions, and may play a part in the very rapid glycolysis found in some pig muscles (see Section V,B.).

Adrenaline has been implicated in the so-called dark-cutting syndrome in beef (Hedrick, 1958), which is characterized by typical alkaline rigor and a very dark red appearance of freshly cut surfaces of muscle. Animals showing the syndrome are usually of extremely nervous disposition (Howard and Lawrie, 1956). In passing, we should note that this dark appearance of the muscle has probably nothing whatever to do with any change in muscle myoglobin, but rather with the fact that the translucence of muscle is much greater at high ultimate pH than at the low values found in normal well fed animals (MacDougall, 1969). The result is that one is looking deeper into the meat, which appears to deepen the characteristic, reddish-purple color of reduced myoglobin, because of reduced light scattering at the surface.

2. INANITION

Inanition has two contradictory effects on the muscle glycogen reserves: (1) In the short term it tends to reduce the reserves and thus raise the ultimate pH of the muscle (see Fig. 6a as an example in rabbit muscle). In rabbits it is rare to find ultimate pH values above 6.5 after 2 days or so inanition, though occasionally values as high as 6.8 occur (see Fig. 8b for example). In beef animals, Howard and Lawrie (1956) found only small increases in ultimate pH even after several days of inanition, and similar results have been reported for pigs (Gibbons and Rose, 1950). (2) The partial exhaustion of the glycogen reserves is frequently reversed after 1 or 2 days inanition. With rabbits and rats this is due to gluconeogenesis, mainly from amino acid and protein sources, but with bullocks it arises from the large amounts of fatty acids and of two-carbon compounds, mainly acetate, continuously being produced from cellulose in the gut, which serve to spare glucose and glycogen (Howard and Lawrie, 1956; Lawrie, 1962).

Thus inanition may or may not lead to curtailment of the rigor process, depending on the stage of gluconeogenesis at which one happens to catch the animal.

3. STRUGGLING AT DEATH

Most of the rabbits on which our arguments have so far been based had been immobilized with Myanesin for some time before death, so that when they were decapitated there was almost no struggle, and the psoas muscle had very high initial pH, PC, and ATP levels. If, however, the rabbit is killed by stunning and decapitation, the hindleg muscles go through spasmodic, but vigorous contractions, in which the psoas is also involved. The result is that the PC and glycogen reserves fall very rapidly, and the initial pH is low. Mean values for the psoas from rabbits killed in this way are: initial pH, 6.5; initial PC, 0–4 μmoles/gm, compared with resting values of 23 or higher; initial ATP, 5–8 μmoles/gm against resting values of 8. The ultimate pH is unaffected, because most of the lactate and H^+ produced in the struggle is locked up in the muscle, as the blood supply is cut off. The result is that the characteristic chemical and physical pattern of rigor is drastically foreshortened, as if the first 150 min or so of Fig. 6b had been cut off. The corresponding changes in extension at 17°C are shown in rigorgraph (c) of Fig. 2 [division of the time scale by 2.5 corrects this graph to 38°C, for comparison with graph (b) and with Fig. 6b for a rested psoas at 38°C].

The psoas muscles of beef animals, slaughtered by stunning, bleeding, and pithing, show very similar, low initial values of the various chemical parameters to those given for struggling rabbits, whereas the values in m. longissimus dorsi (LD), pectoralis superficialis, and sternomandibularis (sterno) are much higher and more similar to the initial values of the psoas from immobilized rabbits. With LD muscles from pigs, however, it is rare to find high values for any of the parameters, and with Landrace and Pietrain pigs in particular, pH values as low as 5.9 are often obtained within 5 min of slaughter. We shall return to these peculiarities of the various species in Section V,B.

4. Relation between Total ∼P Reserve, Ultimate pH, and Time for Half Modulus Change

As we saw in the last section, the initial reserves of glycogen, and to some extent of PC, can be varied within wide limits by exhaustion or inanition. Since the ultimate pH varies inversely with the glycogen reserve, it follows that this can be manipulated by such treatments to vary between 5.6 (high glycogen reserve) to 7.2 or above (zero glycogen reserve). A plot of such varying ultimate pH values against the time for half change of modulus is shown in Fig. 8b. The times have been corrected, where necessary, for small variations in initial PC level and represent an initial value of 23 μmoles/gm (as if all the rabbits had been immobilized before death). Also plotted on the ordinate are values for the total ∼P reserve at death [= 1.5× lactate units of glycogen + PC + 2×ATP, after Eq. (4)]. The latter is here entirely a function of the magnitude of the glycolytic reserve, since the other two terms, initial PC and $2\times$ATP, are constant, at 23 and 16 μmoles/gm for the whole series.

We see from Fig. 8b that increasing the $\sim$P reserve from 39μmoles/gm (the value in the absence of glycogen; ultimate pH = 7.3) to 70 μmoles/ gm (ultimate pH = 6.9) effects a large increase in the time for half modulus change from 35 to 170 min, whereas a further increase in the reserve from 70 to 185 μmoles/gm (ultimate pH = 5.8) only increases this time from 170 to 230 mins. The reason for this difference is that the total $\sim$P reserve is largely made up of PC in the higher range of ultimate pH values and of glycolytic potential in the lower. In other words, PC and glycolysis together are far more effective in maintaining the ATP level and prolonging the time for half change of modulus than either separately.

It is difficult, a priori, to see why glycolysis on its own account cannot maintain the ATP level at least until most of the glycogen reserve has been used up. It can perhaps be attributed to the fact that ATP is not only being split by hydrolases but is also used at the phosphofructo-kinase step of glycolysis (Eq. 16a), producing ADP and fructose diphosphate. In the presence of PC and creatine kinase, this ADP would be rapidly rephosphorlyated, but as soon as the PC reserve is almost exhausted three other enzymes are left to compete for it, myokinase, phosphoglycerate kinase and pyruvate kinase. Thus there is an increasing possibility that part of this ADP will be permanently lost through combined action of myokinase and AMP aminohydrolase [see Eq. (3c,d)], before it can be rephosphorylated by the phosphoglycerate kinase reaction (see Mahler and Cordes, 1966).

E. Other Chemical Changes

1. Glycolytic Intermediates and Coenzymes

Newbold and Scopes (1967) have made an extensive study of the changes in glycolytic intermediates which occur in beef sterno during rigor at various temperatures. They have shown that the main changes are in the levels of hexose 6-phosphate (mainly glucose 6-phosphate) and α-glycerol phosphate. These changes are summarized in Table II for beef sterno at 38°C. At lower temperatures the changes are, of course, slower but the levels of hexose 6-phosphate are higher throughout, particularly at 1° and 5°C. α-glycerol phosphate originates from dihydroxyacetone phosphate through the activity of α-glycerol-phosphate dehydrogenase (sarcoplasmic) and involves the formation of NAD^+ from the NADH

TABLE II

CHANGES IN ATP, ADP, HEXOSE-6-P AND α-GLYCEROL-P (IN μMOLES/GM) DURING THE COURSE OF RIGOR IN BEEF STERNO MUSCLE AT 38°C

Time post-mortem (min)	pH (H$_2$O-iodo)	ATP	ADP[a]	Hexose-6-P	α-Glycerol-P
6	7.20	5.1	1.15	4.0	0.6
30	7.13	5.0	"	—	1.2
50	7.08	4.8	"	3.4	2.1
100	6.86	4.4	"	3.4	2.8
150	6.55	3.8	"	3.2	3.8
200	6.35	2.8	1.09	2.3	3.6
250	6.17	1.6	0.85	2.1	3.8
300	6.05	0.7	0.61	2.5	4.5
400	5.91	0.4	0.55	5.0	3.1
440	5.85	0.2	0.50	4.3	2.8

[a] Newbold and Scopes (1967). Other results are by Bendall and Ketteridge (1970).

produced in the glyceraldehyde-phosphate-dehydrogenase reaction; hence while this side reaction proceeds, there is no NADH available for reduction of pyruvate to lactate by lactic dehydrogenase. Pyruvate and other intermediates might be expected therefore to pile up in proportion to the α-glycerol phosphate formed, but Newbold and Scopes (1967) were not able to show that this is so. Alternatively, NADH might be restored via oxidation of malate to oxalacetate by L-malate: NAD oxidoreductase.

The difficulties of accounting for the ADP level of about 1 μmole/gm, in resting muscle at pH 7 and above, are discussed in Section IV,A, where it is shown that not more than 0.5 to 0.7 μmole/gm are bound to structural proteins (mainly actin), yet the calculated level of free ADP, from the equilibrium of the creatine kinase reaction, is only 0.03 μmole/gm at pH 7.1. As the results in Table II show, the high resting level of ADP does, in fact, fall during rigor in beef muscle to a value of ~0.5 μmole/gm which is about the calculated value for binding to actin, assuming a 1:1 molar ratio (Perry, 1952; Kushmerick and Davies, 1968). Besides these changes, the NAD⁺ level falls slowly during rigor and in the postrigor period (Newbold and Scopes, 1967).

2. OTHER CHANGES IN ADENINE NUCLEOTIDES

We have already discussed the fate of ATP during the rigor process itself, and shown that as soon as the PC level falls below about 4

μmole/gm or so ATP is progressively dephosphorylated and deaminated through reactions c and d of Eq. (3). Bendall and Davey (1957) were also able to show that during this stage of rigor small amounts of inosine triphosphate (ITP) and inosine diphosphate (IDP) accumulated in the muscle. The origin of the IDP could be from ADP by the action of a specific ADP aminohydrolase (Webster, 1953; Deutsch and Nilsson, 1953). Alternatively, the following series of reactions could be catalyzed by the unspecific nucleotide diphosphate phosphokinase of H. A. Krebs and Hems (1953), while some ATP is still present:

$$
\begin{aligned}
&\text{(a)} \quad 2\text{ATP} + 2\text{IMP} \rightleftharpoons 2\text{ADP} + 2\text{IDP} \\
&\text{(b)} \qquad\qquad 2\text{IDP} \rightleftharpoons \text{ITP} + \text{IMP}
\end{aligned}
\tag{6}
$$

The lower the ATP level falls, so reaction (a) will be inhibited and reaction (b) will be pushed further and further to the left. Thus we should expect to find exactly what Bendall and Davey (1957) reported, more IDP than ITP in the later stages of rigor. As Bendall (1961, p. 258) pointed out, it is still a mystery how ITP can survive the attentions of the actin–myosin hydrolase which will accept ITP as a substrate.

The final fate of the adenine nucleotides is to be metabolized first to inosine and then to hypoxanthine (Bendall and Davey, 1957; Lee and Newbold, 1963; Rhodes, 1965). The time for half-change of IMP to hypoxanthine is $\sim$8 days at 2°C. Either of the two pathways in Eq. (7) is feasible (Lee and Newbold, 1963):

$$
\begin{array}{l}
\qquad\qquad\;\; \longrightarrow \text{hypoxanthine} + \text{ribose} \\
\text{IMP} \rightarrow \text{inosine} + \text{P}_i \\
\qquad\qquad\;\; \longrightarrow \text{hypoxanthine} + \text{ribose 1-phosphate} \\
\qquad + \text{P}_i
\end{array}
\tag{7}
$$

3. GLYCOGEN

Glycogen is a branched chain polysaccharide, containing straight chains of about 10 glucose units, linked 1,4, which then branch by 1,6 linkages. The 1,4 linkages are cleaved by phosphorylation with inorganic phosphate, catalyzed by phosphorylase A and B, giving rise to the first intermediate in the glycolytic chain, glucose 1-phosphate. The 1,6 links, however, can only be cleaved by an amylo-1,6-glucosidase, giving rise to free glucose (Sharp, 1957; Mahler and Cordes, 1966). During rigor we should therefore expect to find that one glucose molecule was produced for every nine or ten glucose units of glycogen, broken down by the phosphorylative pathway. Scopes (1970) has indeed shown that this is so.

TABLE IV

Calculated pH Changes during Rigor Mortis in Rabbit Psoas Muscle Due to Temperature or Iodoacetate Poisoning[a]

pH at 20°C	pH at 0°C	pH at 38°C	pH after iodo-acetate	pH before iodo-acetate
7.3	7.68	7.14	7.3	7.10
7.1	7.39	6.95	7.1	6.90
6.9	7.10	6.75	6.9	6.70
6.7	6.91	6.56	6.7	6.52
6.5	6.70	6.36	6.5	6.34
6.3	6.46	6.18	6.3	6.16
6.1	6.26	5.97	6.1	6.02
5.9	6.10	5.75	5.9	5.86
5.7	5.90	5.49	5.7	5.70

[a] Iodoacetate poisoning at 20°C. Ultimate pH of muscle = 5.7 at 20°C.

half-charged form at pH 6.6 (i.e., at its pK value). Hence there is a more or less marked alkalinization after poisoning in the range pH 7.3–6.0. Table IV summaries the results of the actual calculations in rabbit psoas muscle. The alkalinization amounts to about 0.2 pH units in the higher range of pH and about 0.15 in the lower, but falls to zero at the ultimate pH (= 5.7 in this case).

To arrive at the actual pH obtaining in the intact muscle before homogenization and poisoning, two other factors must be taken into account, first the temperature, and second the carbon dioxide tension (pCO_2).

Temperature affects the pK values of the muscle buffers, particularly carnosine and anserine (average pK at 20°C = 6.9) and the histidylimidazole residues of the proteins (pK = 6.4–6.8), and to a smaller extent the protein carboxyl residues (pK = 4.2–4.4) and the phosphate compounds (Bendall and Wismer-Pedersen, 1962). The precise effect can be calculated, if the enthalpies of ionization (ΔH^*) are known, by means of the van't Hoff equation, $-d(pK) = (\Delta H^*/2.303\ RT^2)dt$. ΔH^* for the imidazole residues of carnosine, anserine, and the proteins is 6900 cal/mole, and the change of the pKs from 20 to 38°C is thus approximately -0.0165 pH units/°C. The pKs of the carboxyl residues and of the phosphate compounds are much less affected, about -0.004 and -0.0016 pH units/°C, respectively, so that this effect can be ignored for all practical purposes, particularly as the pKs of the carboxyls are low and outside the experimental range of pH. Calculated values for the effect of temperature on the pH of intact muscle, as it goes through rigor, are given in Table IV. It is not possible to derive a simple relation between the parameters; we note, however, that in general the pH falls by about 0.15 units on raising the temperature from 20 to 38°C, and rises by about 0.2 units on lowering the temperature from 20 to 0°C. In the lower range of pH values the calculated effect agrees well with the observations of Bendall and Wismer-Pedersen (1962) in pig LD muscle, which gave an average value of —0.36 units on raising the temperature from 0 to 38°C. As we shall see in Section V,B, the fact that the pH at 38°C is lower than the measured pH at 20°C (i.e., the normal conditions of measurement) has to be taken into account at the low pH values frequently found within 20 min of death in some types of pig, because the protein denaturation which occurs at these low pH values and high temperatures is extremely sensitive to pH changes of about this order.

The effect of pCO_2 is important, if we wish to arrive at the real pH in living resting muscle. When a muscle is excised and homogenized in iodoacetate the pCO_2 falls from the *in vivo* value of about 50 mm Hg (Evans, 1956) to the value in air (= 0.304 mm Hg). This, of course, produces a considerable alkalinization, which can be calculated from the pH, the pK of HCO_3^- (= 6.33 at 38°C) and the solubility of carbon dioxide in muscle (Hill, 1965). Let us take the example of a truly resting mammalian muscle containing no lactate. The pH as measured after poisoning with H_2O/iodo at 20°C would be approximately 7.36 (see Table IV). After correction to the ionic strength of muscle [see Eq. (8)], this falls to 7.20, which is equivalent to 7.0 before poisoning (Table IV). Further correction to a body temperature of 38°C gives a resting pH, at the pCO_2 of air, of 6.85. At this pH, the HCO_3^- concentration at $pCO_2 = $ 50 mm Hg in the living muscle is 3.54 μmole/gm (from the Henderson-Hasselbach equation). The known buffering capacity of rabbit muscle is 65 μmole/pH/gm (see Table V), so the pH fall due to increasing the pCO_2 in this way is 0.055 units. Hence it follows that the pH of a living resting mammalian muscle at 38°C is approximately 6.80, a value of the order to be expected from the results of Caldwell (1958), who measured the internal pH of the large muscle fibers of the crab (*Carcinus maenas*) and found resting pH values of 7.0–7.25 at 20°C in the absence of carbon dioxide. We can summarize these calculations as: true initial pH at 38°C = measured pH at 20°C —0.2 units (iodo effect) —0.055 (CO_2 effect) —0.15 (temperature effect).

It might be thought that the problem of pH measurement could be solved by using probe electrodes inserted into the intact muscle. However, this introduces another set of corrections, quite as complex as those needed to correct the iodoacetate values. In particular there are the problems of the temperature gradient in the probe itself and of superimposed potentials which possibly arise from the membrane potential of the muscle (Bendall, 1970). In freshly excised beef sterno, for example, the probe electrode gives, initially, an impossibly high pH of about 7.8 which falls over 20 min or so to just below the value measured in iodoacetate. This artifact is not so marked in the pig LD, so that there the probe is a useful guide to the true fall of pH postmortem (Wismer-Pedersen, 1959; Bendall and Wismer-Pedersen, 1962; Bendall *et al.*, 1966).

B. Relation between Glycogen Disappearance, Lactate Formation, and Fall of pH

The relation between the initial glycogen and final lactate contents, on the one hand, and the ultimate pH on the other, is shown for rabbit psoas muscle in Fig. 10a, and that between lactate and pH for beef sterno in Fig. 10b. We see that there is an inverse linear relation between the pH (as measured in water or potassium chloride) and the lactate content, and that an identical relation obtains between initial glycogen content and ultimate pH down to pH 5.8 (note that the glycogen content is given in lactate units, for convenience; i.e. 1 lactate unit = $\frac{1}{2}$ glucose unit of glycogen). Below pH 5.8 increasing amounts of glycogen are seen to be left in the rabbit muscle after glycolysis is complete. This is the so-called residual glycogen described by Lawrie (1955) in horse muscle. The pH at which this residue accumulates appears to vary considerably from muscle to muscle of the same species. The main cause of the hold-up in glycolysis is probably the concomitant deamination of the adenine nucleotide which occurs progressively as the pH falls during rigor [see Eq. (3d)]. AMP is particularly needed as cofactor for phosphorylase B which is probably the operative form of phosphorylase during rigor, and for phosphofructokinase (Newbold and Scopes, 1967).

The relation between lactate formation and pH may be expressed in the form:

$$\text{Lactate} = B(\text{pH}_0 - \text{pH}_t) \tag{10}$$

where the pH is measured in H_2O/iodo or KCl/iodo, and pH_0 = the pH at zero lactate content, and pH_t = pH at any time, t.

The constant B may be called the buffering capacity and strictly speak-

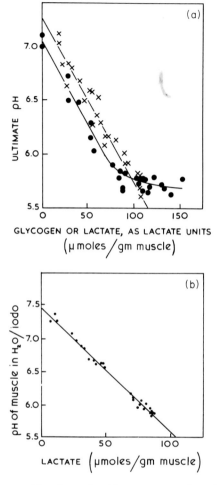

Fig. 10. (a) Ultimate pH values of rabbit psoas muscles plotted against initial glycogen content (in lactate units) and final lactate content. ● = initial glycogen content; ✗ = final lactate content. (b) Ultimate pH values plotted against final lactate content for beef sterno muscles. Correlation coefficient = 0.99 for $n = 27$. Note that pH in both figures was measured in H_2O/iodo solution (see text).

ing $= -\Delta[H^+]/\Delta pH/gm$ [see Eqs. (2)–(3)]. The buffering capacity is due mainly to the contents of P_i, ATP and other phosphate esters, and of the bases carnosine and anserine (Davey, 1960). A smaller part is due to buffering by the muscle proteins, which increases proportionately as the pH falls below 6.5. Values of the constant B and of pH_0 for muscles of various species are given in Table V.

The difference in the buffering capacities of the various muscles in

TABLE V

BUFFERING CAPACITY OF MUSCLES OF VARIOUS SPECIES
ESTIMATED FROM RELATION BETWEEN LACTATE FORMATION
AND pH FALL DURING RIGOR

Muscle type	Buffering capacity[a]	pH range	pH$_0$ in H$_2$O/iodo[b]	pH$_0$ in KCl/iodo[b]	Reference
Rabbit psoas	65.0	7.1–5.6	7.30	7.12	Bendall (1961, p. 237)
Beef sterno	54.0	7.3–5.9	7.47	7.29	Bendall and Ketteridge (1970)
Pig LD	71.7	7.2–5.4	7.34	7.16	Lister (1970)
Sheep pectoral	58.0	7.1–5.7	7.34	7.16	Scopes (1970)

[a] Buffering capacity $= \Delta[\mathrm{H}^{1+}]/\Delta\mathrm{pH}$.
[b] pH$_0$ = pH at zero lactate content.

Table V is attributable only in small part to the differing contents of P$_i$ and other phosphate esters. The remainder is almost certainly due to differing carnosine and anserine contents (Davey, 1960), though we have no exact data to confirm this for most of the species quoted in Table V (see Crust, 1970). One result of the differences is, for example, that 1.33 × the lactate must be produced in pig LD muscle as in beef sterno to produce the same fall of pH.

In the case of rabbit muscle, we know the exact content of buffering substances at the various stages of rigor (see Table III and Davey, 1960) and so it is possible to construct an approximate buffering curve over the pH range 7.3–5.7. In the intact muscle, the mean buffering capacity, calculated in this way is 62.9 μmoles H$^+$/pH/gm compared with the found value of 65 from the lactate/pH curve shown in Fig. 10a. After iodoacetate poisoning the calculated value would fall to about 60 μmoles/pH/gm in this range of pH.

C. Relation between pH and PC and ATP Contents

1. PC AND pH

The typical curve relating PC content to the pH during rigor is of exponential form, as shown in Fig. 11a for rabbit psoas and Fig. 11b for beef sterno muscle. The reason for this exponential relation probably lies in the fact that the creatine kinase reaction [Eq. (2b)] rapidly comes to equilibrium at each stage during rigor, the position of equi-

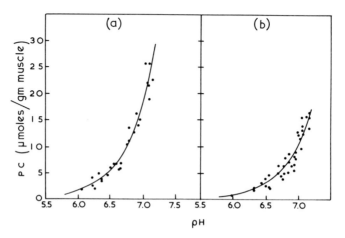

Fig. 11. Relation between PC content and pH of the muscle (a) for rabbit psoas muscles. (b) for beef sterno muscles. The lines through the points have been calculated from Eq. (12).

librium itself depending on the H^+ concentration:

(a) $K = (\text{creatine})(\text{MgATP}^{2-})/(\text{PC}^{2-})(\text{MgADP}^-)(\text{H}^+)$

(b) $(\text{PC}^{2-}) = 1/(\text{H}^+)[(\text{creatine})(\text{MgATP}^{2-})/K(\text{MgADP}^-)]$ (11)

where all the terms represent the concentrations of free substrates.

Providing that the term in the square bracket on the right of Eq. (11b) remains constant during rigor, it follows that (PC^{2-}) is inversely related to (H^+), and $\log_{10}(\text{PC}^{2-})$ is directly related to pH. Calculation of this term from the known value of $K/(\text{H}^+) = 130$ at pH 7.4 (Kuby and Noltmann, 1962) and from the experimental values of the other parameters, shows indeed that the ratio $(\text{creatine})(\text{MgATP}^{2-})/(\text{MgADP}^-)$ stays constant at 2500 ± 100 from pH 7.1 to 5.7 in rabbit muscle. (Note that (MgADP^-) must first be calculated from the creatine kinase equilibrium at the given pH). Thus a direct plot of $\log_{10}(\text{PC}^{2-})$ against pH (as measured) yields the following empirical equation:

$$\text{PC}_0/\text{PC}_t = \text{antilog } 1.06 \ (\text{pH}_0 - \text{pH}_t)$$ (12)

where PC_0 and pH_0 are the values at zero lactate content and PC_t and pH_t those at any time, t.

The curves through the points in Figs. 11a and b have been calculated in this way and are seen to give an excellent fit, particularly for rabbit psoas. The equation also applies to the curarized pigs, studied by Bendall (1966). Note that the constant on the right hand side of the equation probably represents an overall correction for the effect of iodoacetate poisoning and of the other factors mentioned in Section IV,A,2.

The equation applies to muscles with lowered glycogen reserves only to a limited extent, because then the term in the square bracket in Eq. (11b) changes as resynthesis of ATP via glycolysis begins to fail.

2. PC AND ATP

The relation between the PC and ATP levels is shown in Fig. 12a, for rabbit psoas and beef sterno muscles. We see that the ATP level stays virtually constant in both cases while the PC level is still above 10 μmole/gm. Below this PC level it begins to fall increasingly rapidly, until at PC = 2 μmole/gm about half of the ATP has disappeared. There are still traces of ATP left, even when all the PC has disappeared. It is possible that this ATP is within the sarcoplasmic reticular system and the mitochondria and thus immune from attack by hydrolases.

The relation between ATP and PC is of the type of a rectangular hyperbola, although no simple hyperbola can express it satisfactorily. A more exact and complex equation can be derived from the relation of pH to PC, already discussed, and that between pH and total turnover of $\sim$P, as we shall see in the next section.

The type of curve shown in Fig. 12a is also characteristic of pig and sheep muscles, and is indeed probably of general applicability.

D. Relation between Energy-Rich Phosphorus Potential and pH Fall and Lactate Formation

As we have already mentioned, the change during rigor of the energy-rich phosphate potential ($\sim$P) can be calculated from:

$$\Delta \sim P = -1.5\Delta \text{ lactate} + \Delta PC + 2\Delta ATP$$

$$= 1.5B\Delta pH + \Delta PC + 2\Delta ATP \qquad (13)$$

where B = the buffering capacity and the pH is measured in H_2O/iodo or KCl/iodo.

When this calculation is made for rabbit psoas, beef sterno, or pig LD muscles, it turns out that there is an exact linear relation between $\sim$P and pH, as shown for rabbit and beef muscles in Figs. 13a and b, respectively. The slope of the line, $\Delta \sim P/\Delta pH$, is 125 for rabbit psoas, 102 for beef sterno, and 129 for pig LD.

It follows from this relation that a knowledge of the pH, the PC content and the buffering capacity B of a muscle at any instant t

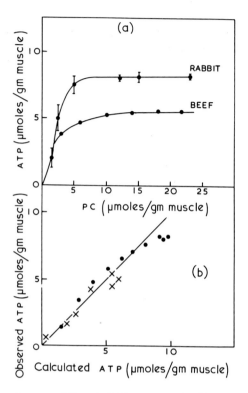

Fig. 12. Relation between PC and ATP content of rabbit psoas and beef sterno muscles: (a) ATP against PC content [vertical bars show SEMs of ATP values of rabbit psoas ($n = 6$)]. (b) ATP, calculated from Eq. (15b), plotted against observed ATP. The line is drawn with a slope of unity. ● = rabbit psoas; ✕ = beef sterno.

should enable us to calculate the ATP level, since Eq. (13) shows that:

$$2\Delta ATP = \Delta \sim P - 1.5\ B\Delta pH - \Delta PC$$

$$= a\Delta pH - 1.5\ B\Delta pH - \Delta PC$$

$$= (a - 1.5\ B)\Delta pH - \Delta PC \qquad (14)$$

where $a = \Delta \sim P/\Delta pH$.

We can derive the value of ΔPC from Eq. (12), since:

$$-\Delta PC = PC_0[1 - 1/\text{antilog } 1.06(pH_0 - pH_t)] \qquad (15a)$$

where PC_0 and pH_0 are the values at zero lactate content.

The full equation for the change in ATP content is terms of pH and PC_0 then becomes:

$$2\Delta ATP = (a - 1.5\ B)\Delta pH + PC_0[1 - 1/\text{antilog } 1.06(pH_0 - pH_t)] \qquad (15b)$$

$$ATP_t = ATP_0 - \Delta ATP \qquad (15c)$$

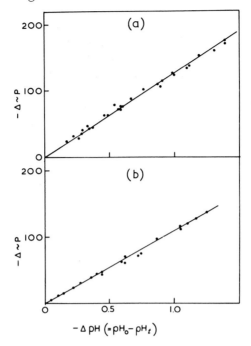

Fig. 13. Change in energy-rich P potential $(\Delta \sim P)$ plotted against change in pH (ΔpH): (a) for rabbit psoas, $r = 0.995$ for $n = 26$; (b) for beef sterno, $r = 0.998$ for $n = 53$.

If values of PC_0 and pH_0 are not available by extrapolation, it is possible to insert instead any pair of high values of these parameters into Eq. (15b), for example those given in Table III.

A plot of the calculated ATP against the observed is shown in Fig. 13b, for rabbit psoas and beef sterno muscles. The agreement is seen to be satisfactory, but the equation overestimates the initial ATP value, and underestimates the values as the ATP level falls. It is doubtful whether the accuracy could be improved further.

The calculated ATP values can be plotted against time, since they are dependent on the pH change for which the time can be read off from Fig. 7. The calculated values for the rabbit psoas of ultimate pH 5.7, illustrated in Fig. 6b, are shown on that figure as crosses, and are seen to agree remarkably well in time with the observed values, except initially, where they are too high. The observed time for half change of ATP is 228 min against the calculated time of 216 min. The pH values at these times would be 5.93 and 6.00, respectively.

Equation (15b) can be applied to a limited extent to muscles with reduced glycogen reserves, e.g., the muscle in Fig. 6a (ultimate pH =

6.58). At a pH of 6.60, 150 min after death, for example, the equation predicts that ATP should still be almost at the initial level of 8.1 μmole/gm. The found ATP level at this time is 7.0 μmole/gm (see Fig. 6a). In other words, the equation correctly predicts that there can still be high levels of ATP in this type of muscle when the glycogen reserves are almost completely exhausted, whereas in muscles with lower ultimate pH (Fig. 6b for example) the ATP has nearly disappeared by the time the ult. pH is reached at about 300 min after death.

It is of some interest to compare the resynthesis of ATP from ADP via the glycolytic pathway ($= 1.5$ B ΔpH) with that via the creatine kinase reaction ($= \Delta$PC). In rabbit psoas, ΔPC/1.5 B ΔpH $= 0.51$ at pH 7.05, 0.31 at pH 6.85 and only 0.08 at pH 6.3. The ratio, apart from being highly dependent in this way on pH and PC, must also be determined by the siting of the respective enzyme systems, by their relative V_M and K values, and by the rate of diffusion of ADP to them from the site of its production.

The first reaction in the glycolytic chain which converts ADP to ATP is the phosphoglycerate kinase reaction, in which 1:3 diphosphoglycerate is the $\sim$P donor. The concentration of the latter in relation to that of PC is probably one of the two most decisive factors in determining the magnitude of the PC kinase–glycolysis ratio. Unfortunately, this compound is unstable and is present at very low concentrations which makes accurate measurement difficult.

Another step in the glycolytic chain which is rate-determining is the phosphofructokinase reaction in which 1 mole of fructose 6-phosphate and 1 mole of ATP are converted to 1 mole of fructose diphosphate and 1 mole of ADP. The enzyme shows strong allosteric inhibition at the high ATP concentrations we find in the earlier stages of the rigor process, and this may slow down glycolysis sufficiently to give the PC kinase–glycolysis ratio actually observed at high PC levels (Newbold and Scopes, 1967).

E. Relation between Rate of $\sim$P Turnover and pH

The apparent rate of $\sim$P-turnover during rigor varies with pH in the manner shown in Figs. 14a and b, for rabbit, beef, and pig muscles. In the rabbit psoas and pig LD, the rate falls to a minimum as the pH reaches about 6.7 and then rises sharply to a new and higher maximum, at pH 6.25 in the rabbit psoas and at pH 5.9 in the pig LD. The plateau and maxima are not so obvious in the case of beef sterno.

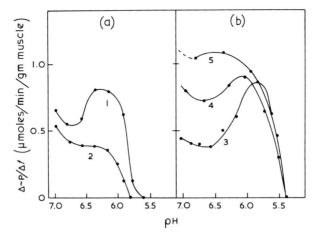

Fig. 14. $\Delta \sim P/\Delta t$ plotted against pH, at 38°C for: (a) 1. rabbit psoas, 2. beef sterno; (b) 3. LD of curarized Large White pig, 4. same, after tetanic stimulation, 5. LD of Pietrain pig, magnesium-injected.

The rates appear to fall off in all cases as the ultimate pH is reached, but this is because the glycogen reserve is then nearly exhausted, so that Eq. (4) no longer gives the true potential rate of turnover.

The form of the curve for pig LD is greatly modified when the muscle is given a series of tetani immediately after excision (Hallund and Bendall, 1965; Bendall, 1966), as curve 4 in Fig. 14b illustrates. No such marked and sustained increase in rate occurs after stimulation of rabbit psoas, however (Hallund and Bendall, 1965). The effect in the pig is to double the rate from pH 7.0 to 6.5, after which the curve gradually approaches that for the unstimulated muscle (curve 3), until it coincides with it at about pH 6.0, a point at which the glycogen reserves are becoming exhausted.

The fact that recovery from stimulation takes such a long time to wear off in pig muscle may be due to the fact that the calcium pump in the muscles of such highly bred and selected animals is not fully efficient, so that the calcium ions, which are released by a stimulus and act as the necessary activator of contraction, are incompletely pumped back during relaxation, allowing the actomyosin ATP hydrolase to remain slightly active over a prolonged period; we shall return to this question in Section VI.

When the glycogen reserves of the muscle are severely lowered by inanition, the ATP turnover rate falls off sharply as the ultimate pH is reached, in spite of the fact that the PC and ATP levels may be high.

This lowering of the rate, which in fact takes place rather slowly in time, helps to explain the apparently anomalous duration of rigor in muscles with lowered glycogen reserves, as illustrated in Figs. 6a and 8b. As we pointed out in Section III,D, it is also probable that the glycogen reserves, and thus the total $\sim$P potentials, are unevenly distributed in the fibers of muscles of this type.

V. Abnormal Types of Rigor

A. Thaw-Contracture and Cold-Shortening

1. THAW-CONTRACTURE

The phenomenon of thaw-contracture, that is the contracture which occurs on thawing out a muscle frozen in the prerigor state, has been known for many years (Chambers and Hale, 1932). The subject has been reviewed by Bendall (1961, p. 260) and will be discussed in detail in the present book by Dr. R. E. Davies. For these reasons we shall give here only an outline of its salient features.

When a strip of, say, rabbit psoas muscle is frozen quickly at -20°C in the prerigor state while the ATP and PC levels are high, it behaves on thawing out under load at 17°C, as shown in Fig. 15. We see that

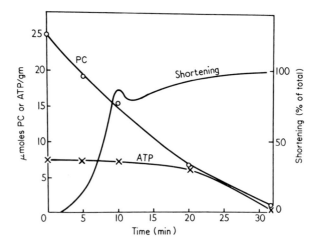

Fig. 15. Pattern of chemical and physical changes in rabbit psoas strip, frozen prerigor at -20°C and then allowed to thaw in air at 17°C. Dimensions of strip were $0.5 \times 1.5 \times 8.0$ cm; load $= 67$ gm/cm^2; total shortening $= 38\%$ of initial length; pH change 7.02 to 6.10.

release at low temperatures, which could perhaps be achieved by the elegant spectrophotometric method of Jöbsis and O'Connor (1966). It is also necessary to discover why fast muscles do not cold-shorten. Perhaps it is because they possess a more efficient and extensive calcium pump (Seidel *et al.*, 1964).

The cold-shortening phenomenon was first fully investigated by workers at the New Zealand Meat Research Institute, because customers had begun to complain of the toughness of New Zealand, chilled lamb. Marsh and Leet (1966) were able to show that it was precisely local shortening on the carcass which was the cause of the toughening. In an elegant series of papers they demonstrated that the toughness of meat increases exponentially with muscle shortening, reaching a maximum at about 0.6 rest length (40% shortening). It is a remarkable fact that this increasing toughness at lengths below rest length is inversely related to the fall of active tension which occurs at these short lengths, when a muscle is stimulated (see Aubert, 1956, quoted by Bendall, 1969, p. 128). In fact both phenomena may be due to the same basic change, the increasing double overlap of actin filaments as the muscle gets to lengths below rest length.

B. Rapid Rigor in Pig Muscle

A naturally occurring phenomenon of some interest is the so-called PSE (pale, soft, and exudative), or watery, condition found in some types of pig muscle. The condition is characterized by a very rapid pH fall postmortem, and by soft, pale, and watery musculature in the postrigor state (Ludvigsen, 1954; Briskey and Wismer-Pederson, 1961; Bendall and Lawrie, 1964; Briskey, 1964, 1968; McCloughlin and Tarrant, 1968; Lister, 1968). It was earlier classed as muscle degeneration (Ludvigsen, 1954) and was attributed to a lowering of the ACTH level in the living animal and also to thyroid insufficiency (Ludvigsen, 1968). It has since been shown conclusively that there are no signs of muscle degeneration, if a sample is taken by biopsy, or soon after the death of the animal (Bendall and Wismer-Pedersen, 1962); the muscle fibers then appear entirely normal, with clear cross-striations. On the other hand, if a sample is removed after the muscle has become watery, the cross-striations are obscured by what appear to be irregular bands of denatured protein. In cross-section, it is clearly seen that the dark-staining material is contained within the fiber boundaries and appears to be deposited around the individual fibrils (see Fig. 18). Scopes (1964) was able to show that the particular sarcoplasmic protein denatured under these conditions was creatine kinase which is present in sufficient amount to account for the dark-staining bands. It was later shown by

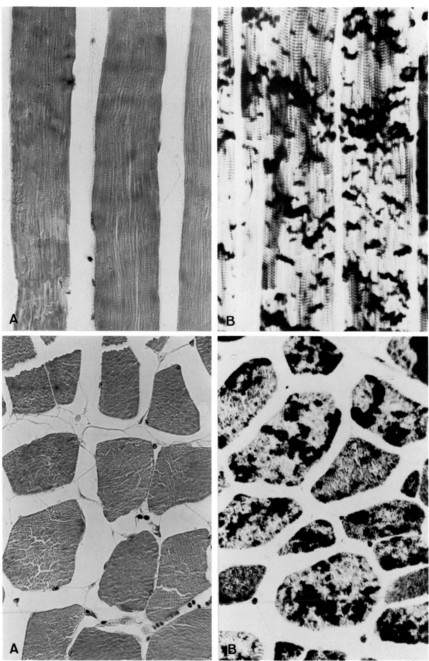

Fig. 18. Transverse sections of (A) normal muscle fiber, (B) fiber from a "watery" muscle of a Landrace pig. × 1000. Photo by courtesy of Mr. C. A. Voyle.

Penny (1967) that the actin and myosin filaments were also considerably denatured. These authors demonstrated, in fact, that both types of protein were extremely sensitive to temperature and pH in the range to be expected in a cooling carcass, i.e., 36–38°C and a pH of 6.0 or below within 1 hr of death in the most rapidly glycolyzing examples. We can, therefore, attribute the phenomenon entirely to protein denaturation at high temperature and low pH, combined with a partial disruption of the sarcolemma; the first factor reduces the water-holding capacity of the muscle proteins (Bendall and Wismer-Pedersen, 1962) and the second allows the solution of sarcoplasmic proteins to escape from the fiber and permeate the intrafiber space. Whale, beef, and rabbit muscles, subjected to high temperature rigor, also develop wateriness, so the condition is of general occurrence under these conditions and is not peculiar to pigs (Bendall and Wismer-Petersen, 1962; Bendall and Lawrie, 1964).

The problem of this type of muscle, therefore, resolves itself into determining the factors which bring about the exceptionally rapid glycolysis and ∼P turnover in the freshly excised muscle; but before discussing these, it is well to turn to the peculiar features of the chemical changes themselves. These have so far been observed in Danish and English Landrace, Pietrain, Poland China, and Hampshire breeds, but only to a very minor extent in English Large White or Middle White, which generally give a pattern and rate of postmortem change very similar to the rabbit (Bendall *et al.*, 1966). We shall mainly discuss here the results with Pietrain, obtained recently at the Meat Research Institute, Langford by Drs. Lister and Scopes; these will be compared with the normal rigor pattern shown by Large White pigs. Examples of the same phenomenon in Danish Landrace and Poland China breeds are given in publications by Bendall *et al.* (1933), Briskey (1964, 1968), Topel *et al.* (1968), Sybesma and Eikelenboom (1969), and Lister *et al.* (1970).

The observation on which much of the early work was based was the very rapid fall of pH in some types of pigs (Ludvigsen, 1954). This is illustrated in Fig. 19a, by comparing Pietrain pigs, treated in various ways, with English Large White pigs, curarized before death (Bendall, 1966). We see that all the Pietrain curves—(1) for control animals, (2) for curarized, and (3) for magnesium-injected) are far more rapid than the very slow curve for the Large White. Even in the lower range of pH, curarization of Pietrain pigs only slightly slows down the pH fall and, in the higher range from pH 6.9 to 6.6, the rate is quite exceptionally high. Magnesium injection, which in Large White pigs gives a pH–time curve similar to that for curarized animals, substantially decreases the high rate in the Pietrain from pH 6.9 to 6.3, but for some rea-

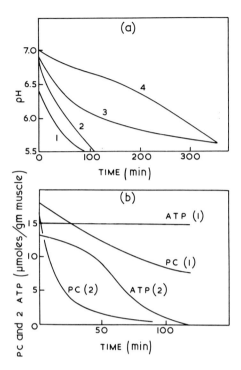

Fig. 19. Changes in chemical parameters during rigor in LD muscles of Large White and Pietrain pigs. (a) pH/time curves for: 1. control Pietrain, 2. curarized Pietrain, 3. magnesium-injected Pietrain, 4. magnesium-injected Large White. (b) PC and ATP/time curves for: 1. curarized Large White, 2. curarized Pietrain.

son has a more marked effect from pH 6.2 until the ultimate is reached at about pH 5.6.

In Fig. 18b, the ATP and PC–time curves are plotted for curarized Large White and Pietrain pigs. It is seen that there is a literally enormous difference between the two, the ATP, for instance, declining to zero in the Pietrain, i.e., the muscle is in full rigor, before it has even begun to decrease in the Large White. Similarly, the initial rate of PC decline in the Pietrain is more than 7 times that in the Large White (0.95 and 0.13 μmoles/min/gm, respectively).

Table VI summarizes some of the parameters, from which it is clear that the Pietrain, however treated, is much faster in every respect than the Large White. This is particularly obvious in the case of the rate of $\sim$P turnover which in the curarized Pietrain is ten times that in the Large White at pH 6.9, five times at pH 6.7, but only 1.5 times at pH 6.0. Magnesium injection appears to reduce the initial rate substantially in the Pietrain, although the rate is still very high (1.1 μmoles/min/gm) at

TABLE VI
Post-Mortem Changes in Large White and Pietrain Pigs

Type of pig and treatment	Half-life (min)		Rate of ∽P turnover (μmoles/min/gm)		
	PC	ATP	pH = 6.9	pH = 6.7	pH = 6.0
Curarized					
Large white	98	330	0.45	0.37	0.80
Pietrain	12	61	4.7	1.9	1.2
Mg-injected					
Large white	40	230	0.80	0.31	0.80
Pietrain	30	110	1.20	1.10	0.90

pH 6.7, a region in which the rate in the Large White is particularly low (0.37 μmoles/min/gm; see Fig. 14b).

Incidentally, it should be noted that the low initial pH of 6.4 in the control Pietrain muscles (see Fig. 19a) is in no way exceptional for animals killed by stunning and bleeding; it can occur quite frequently in Large White pigs, in beef animals, and in rabbits (see Fig. 1); it is due to struggling during slaughter and more or less severe muscle spasms. What is exceptional about the Pietrain curve is the maintenance of a high rate throughout the rigor process.

The absolute initial rate of ∽P turnover in the curarized Pietrain of 4.7 μmoles/min/gm is higher than in the rabbit thaw-rigor experiment in Fig. 15 which was 3.7 μmoles/min/gm. So, if we account for the high rates in thaw-rigor by calcium release and activation of the contractile ATPase, it is logical to attribute the same high rates in curarized Pietrains to the same cause. In that event, we should expect the muscle to shorten and go into contracture after excision, and indeed this is frequently observed (Scopes and Lister, 1970; Lister, 1968), the contracture often lasting for 2–5 min.

Release of calcium ions from the sarcoplasmic reticulum is brought about in intact living muscle by the inward movement of the action potential, generated by a direct or indirect stimulus to the plasmalemma. The depolarizing wave arrives at the triads of the sarcoplasmic reticulum via the transverse tubules which are continuous with the plasmalemma (Constantin and Podolsky, 1966). Normally this process is quickly reversed as the plasmalemma becomes repolarized in the wake of the action potential, but in some muscles longer lasting depolarizing agents, such as potassium chloride or caffeine, produce a more or less permanent contracture, i.e. the calcium pump continues to release calcium ions for a considerable time (Sandow, 1955). This is closely akin to what happens in Pietrain muscles when they are excised and then become anaerobic. However, such a state of affairs obviously does not exist in the living, aerobic muscle, otherwise Pietrain pigs would be permanently paralyzed.

A possible explanation of these apparently contradictory effects arises from measurement of the oxygen-uptake by strips of excised muscle. Under such aerobic conditions, the pyruvate produced in the glycolytic chain passes into the mitochondria, where it is oxidized to carbon dioxide and water via the citric acid cycle and the electron transport chain. During this process (including glycolysis) a total of 18.5 moles of ATP are re-synthesized from ADP for every mole of pyruvate oxidized, to give a P/O ration of 3 (Lehninger, 1967). Hence we would expect that for every mole of ADP produced during a contracture by the activated actomyosin ATP hydrolase, $\frac{1}{3}$ atom of oxygen would be taken up. The observed rate of ADP-production ($\equiv \sim$ P turnover), immediately after excision of the muscle, reaches a level of 4.7 μmoles/min/gm in the Pietrain, so that the expected oxygen uptake rate is about 1.6 μatoms/min/gm at 38°C. The observed rate, however, is only 0.33 μatoms/min/gm (Bendall and Taylor, 1972), equivalent to a $\sim$P turnover of 0.99 μmoles/min/gm; this is only slightly higher than the rate in the rabbit psoas (see Fig. 14a). Hence, we conclude that the high calcium levels, obtaining during the anaerobic contracture, are reduced sufficiently when the muscle is made aerobic to suppress the $\sim$P turnover to a more normal resting level. Indeed, Lehninger (1967) has shown that respiring mitochondria can accept free calcium ions and phosphorylate them to calcium phosphate, which is deposited on the inner side of the mitochondrial membrane; in other words, they can act as an additional calcium pump. Such a mechanism, operating in vivo, would control the resting metabolism, whenever the normal calcium pump was immobilized as a result of excessive stimulation. A similar phenomenon occurs during cold-shortening of beef muscle, if the experiment is performed in oxygen instead of nitrogen; the apparent rate of $\sim$P turnover then falls from 0.22 μmoles/min/gm in nitrogen to about 0.06 in oxygen (Bendall and Taylor, 1972).

Direct evidence of abnormality of the muscle membrane in stress-susceptible pigs is lacking at present, and no measurements appear to have been made of the fundamental parameters, the resting and action potentials and the time-course of de- and repolarization. We can summarize the indirect evidence for abnormality as follows:

1. Pietrain pigs are so susceptible to quite mild stress that they often die of heart failure after a chase of 5 min or so (see Sybesma and Eikelenboom, 1969). Boars frequently die from the exertion of mating.

2. The plasma K⁺ level rises sharply after the stress of slaughter, often doubling; a much smaller rise is observed in stressed Large White

pigs (Lister, 1972). Calculations show that the total K⁺ release is too large to be accounted for other than by substantial release of internal K⁺ from the musculature.

3. The failure of curare to prevent rapid $\sim$P turnover in anaerobic Pietrain muscles shows that the trouble lies on the membrane side of the neuromuscular junction and not in the end-plate region, since curare specifically inhibits the depolarizing effect of ACh release on the postsynaptic membrane (Dale *et al.*, 1936; Katz, 1966). The alleviating effect of magnesium injection must, therefore, be due to a side effect of the massive dose (22 mmoles/liter of blood) on the membrane itself, and not at its normal site of action at the motor end plate (Katz, 1966).

4. The neuromuscular-blocking drug succinylcholine and the common anesthetic halothane ($CF_3CHBrCl$) have lethal effects on stress-susceptible pigs (Wilson *et al.*, 1966; Sybesma and Eikelenboom, 1969; Berman *et al.*, 1970; Lister, 1970), causing the following symptoms to develop within 2 or 3 min of administration: doubling of the oxygen consumption rate, rapid pH fall in the musculature and the blood, a twelvefold rise in the blood lactate in about 20 min, increases in the plasma potassium, sodium, and calcium levels, and most dramatic of all, severe myotonia of the whole musculature and rapid hyperthermia, the muscle and body temperature rising at about 0.16°C/min to a limit of about 43°C. The myotonia and hyperthermia can certainly be attributed to stimulation of the contractile ATP hydrolase to a level of at least 10 μmoles/min/gm, judging by the rate of pH fall in the muscles (Sybesma and Eikelenboom, 1969), which would lead to complete exhaustion of the $\sim$P reserves and to full rigor in about 25 min. The heat generated by this accelerated anaerobic glycolysis would, in fact, account for most of the temperature rise in the muscles (Berman *et al.*, 1970), the increased oxygen consumption rate being mainly due to oxidation, in the liver, of the lactate produced. None of these symptoms develop in Large White pigs treated with these drugs, the body temperature tending if anything to fall.

The significant peculiarity of succinylcholine is that just before it blocks neuromuscular transmission, it frequently causes severe muscle fasciculations in normal subjects, evidently through a direct depolarizing effect on the muscle membrane (Goodman and Gilman, 1965). It is possibly intensification of this effect which causes the development of myotonia and hyperthermia in the stress-susceptible pigs.

VI. Nature of ATP Hydrolases Active in Resting Muscle

Although the nature of the ATP hydrolases, active in resting muscle, is of paramount importance in determining the pattern and time course of rigor mortis, present knowledge is too scanty to provide more than a tentative answer. The size of the problem can best be illustrated by comparing resting rates of ~P turnover with maximal rates of actin–myosin hydrolase activity (i.e., the contractile enzyme system) at various temperatures, as in Fig. 20. We see that at 38°C, the resting rate is less than 0.3% of maximal, rising to about 1% at 20°C and then more and more rapidly as the temperature is decreased, until at 1°C it reaches 13%. This is, of course, another way of expressing the fact that the resting rate has a low and anomalous enthalpy of activation (11 kcals/mole or less), whereas the maximal contractile rate has a high and constant one of 25 kcal/mole.

The known ATP hydrolases and other ATP-utilizing enzymes of muscle are of three types: (1) the myosin- and actin-modified myosin hydrolases, associated with the contractile apparatus; (2) the hydrolases associated with the mitochondria and the calcium pump of the sarcoplasmic reticulum; (3) sarcoplasmic enzyme systems which utilize ATP in a cyclic manner.

Taking these three types in reverse order, we know of two main cyclic systems among the sarcoplasmic enzymes, that involved in the phosphorylase B and A transformation, and the couple formed between PFK and fructose diphosphatase. In the complex cycle of the phosphorylase A to B transformation (E. G. Krebs et al., 1966), four ATP molecules are utilized by phosphorylase B kinase to convert two molecules of phosphorylase B to one of phosphorylase A, and four molecules of ADP are liberated. Phosphorylase A can be converted back to two molecules

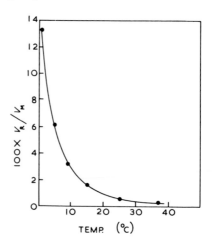

Fig. 20. The temperature dependence of the ratio of resting rate of ~P turnover (V_R) to the maximal rate of contractile ATP hydrolase activity (V_M). Beef LD muscle. V_M at 37°C = 180 μmoles/min/gm; V_M at 1°C = 1.3 μmoles/min/gm.

of phosphorylase B and four molecules of P_i by the action of phosphorylase phosphatase. This system might operate in the resting muscle, though there is no evidence to suggest that it does.

Similarly many muscles contain fructose-diphosphatase (H. A. Krebs and Woodford, 1965; Opie and Newsholme, 1967). This enzyme could operate after the phosphofructokinase (PFK) step of glycolysis to effect a cyclic utilization of ATP as follows:

(a) Fructose 6-phosphate^{2-} + ATP^{4-} $\xrightarrow{\text{PFK}}$ fructose diphosphate^{4-} + ADP^{3-} + H$^+$

$$(16)$$

(b) Fructose diphosphate^{4-} $\xrightarrow{\text{fructose-diphosphatase}}$ fructose 6-phosphate^{2-} + P$_i^{2-}$

SUM: ATP^{4-} → ADP^{3-} + P$_i^{2-}$ + H$^+$

However, the diphosphatase shows quite the wrong pH activity curve to account for the resting $\sim$P turnover, because its activity falls instead of rising at pH values below 6.7 (H. A. Krebs and Woodford, 1965); it is also strongly inhibited by AMP. Furthermore, rabbit white muscle contains more of it than pigeon breast (Opie and Newsholme, 1967), yet the $\sim$P turnover rate of the latter is about twice that of the former. Similarly, frog thigh muscles contain more than rabbit muscles, but the $\sim$P turnover rate in frog muscle is only half that in rabbit muscle.

The ATP hydrolase of the mitochondria is activated in the presence of calcium ions (Lehninger, 1967), but little is known of its potential activity in resting muscle, where the free calcium concentration is very low (about 10^{-8} M). Similarly, the calcium pump of the sarcoplasmic reticulum utilizes the free energy of ATP splitting to translocate calcium ions from the sarcoplasm (A. A. Weber *et al.*, 1966) and thus effectively acts as a hydrolase. It becomes active during the later stages of a twitch or tetanus, pumping back the calcium ions, released into the sarcoplasm by the stimulus (Jöbsis and O'Connor, 1966). Under resting conditions, however, its activity is likely to be very low, since it is only required to counteract the slow, passive diffusion of calcium ions out of the reticulum. Its contribution to the overall hydrolase activity cannot, however, be entirely neglected.

Of the two hydrolases associated with the contractile apparatus, the noncontractile myosin enzyme, although strongly inhibited at the high free Mg^{2+} concentrations of 2–5 mM found in muscle, is nevertheless capable of splitting ATP at a rate of 0.37 μmoles/min/gm muscle at 25°C, according to the results of Kanazawa and Tonomura (1965). This is close to the observed resting rate of $\sim$P turnover of 0.29 μmoles/min/gm in the rabbit psoas at this temperature (see Figs. 9 and

14a). Indeed if this enzyme has an enthalpy of activation of about 10.0 kcals/mole, similar to Ca^{2+}-activated myosin ATPase (Bendall, 1969, pp. 50–56), it would satisfactorily explain the resting rates of $\sim P$ turnover from 38 to about 20°C, shown in Fig. 9 (E. W. Taylor, 1970). We could then account for the anomalous rise in the rate, as the temperature is reduced toward 0°C, by increasing leakage of calcium ions from the sarcoplasmic reticulum and hence increasing activation of the contractile actin–myosin enzyme. As we have already pointed out, there is no doubt that this enzyme is progressively activated below 10°C in red muscles which then begin to shorten spontaneously (Section V,A,2).

Participation of the noncontractile myosin enzyme in the resting $\sim P$ turnover of muscle would also largely help to explain the finding of Bendall and Taylor (1972) that the oxygen uptake rates of strips of beef muscle show no anomalous temperature dependence, but follow a straight-forward Eyring plot from 38 to 0°C, to give a constant activation enthalpy of 10.2 kcals/mole; this would be expectable if the ADP required as acceptor in the ETC were produced by the myosin enzyme with a similar enthalpy of activation, but not if the actin–myosin enzyme, with an activation enthalpy of 25 kcals/mole, were involved. As we have mentioned, we would also have to assume in this case that active mitochondria are capable of removing the calcium ions released in the muscle as the temperature falls below 15°C or so, hence inhibiting the actin–myosin enzyme completely.

The other anomalous feature of resting $\sim P$ turnover which might perhaps be explained in terms of the two enzymes we have just discussed is its strange pH dependence, showing a minimum at pH 6.7 and maxima at pH 7.0 and 5.9 to 6.2 (see rabbit and pig curves in Fig. 14). The minimum might be produced by pH inhibition of the noncontractile myosin enzyme, whereas the lower maximum could be due to minute, but increasing leakage of calcium ions and activation of the actin–myosin enzyme at a subcontractile level, as the pH falls below 6.7. The pH profile is quite unlike that for calcium-activated myosin, for example, which is only slightly affected by pH, or that for the actin–myosin enzyme which is strongly inhibited by falling pH (Bendall, 1969, p. 37). Participation of the latter enzyme in resting turnover at lower pH values would not be excluded, however, if the pH inhibition were partly relieved by the increasing leakage of calcium ions we suggested above.

To summarize these arguments, we conclude that the noncontractile, myosin enzyme is the ATP hydrolase most likely to be responsible for the low $\sim P$ turnover in resting mammalian muscle under physiological conditions at a body temperature of 38°C. Under exceptional circumstances, however, as for example in the muscles of stress-susceptible pigs, it is clear that the contractile actin–myosin enzyme also becomes involved as shown by the continuing high $\sim$P turnover in such muscles after excision, and by their pronounced tendency to go into contracture. Similarly, it is obvious that the cold-shortening phenomenon in slow, red muscles is due to partial activation of the contractile enzyme, which almost certainly occurs also in fast, white muscles, such as rabbit psoas, when they are cooled below 10°C, though not at a level sufficient to cause contracture.

VII. Discussion

The constant theme running through this chapter has been the importance, to all the postmortem changes, of the dynamic equilibrium between ATP hydrolysis and its resynthesis via the creatine kinase reaction and anaerobic glycolysis. Ultimately the position of equilibrium, and its final displacement toward complete ATP destruction as the energy supplies fail, is determined by the primary reaction, the splitting of ATP to ADP and P_i. Without this reaction none of the postmortem changes we have described would take place, and glycolysis and the accompanying acidification would come to a stop very rapidly. Scopes (1970) has indeed shown in reconstituted enzyme systems, free of ATP hydrolases, but containing all the glycolytic enzymes plus creatine kinase, that the addition of glycogen, P_i, creatine and ATP results in the establishment of a quite different dynamic equilibrium:

(a) $\quad 3[ATP^{4-} + creatine \rightleftharpoons PC^{2-} + ADP^{3-} + H^{1+}]$

(b) $\quad (glucose)_n + 3ADP^{3-} + 3P_i^2 + H^+ \rightarrow$
$\qquad\qquad\qquad 3ATP^{4-} + 2(lactate)^- + (glucose)_{n-1} + H_2O$ $\qquad$ (17)

SUM: $\quad (glucose)_n + H_2O + 3(creatine) + 3P_i^{2-} \rightarrow 3PC^{2-} + 2(lactate)^- + 2H^+$

Since the equilibrium of reaction (17b) is far to the right, the final state of the system is that 3 moles of PC are formed per 2 equivalents each of lactate and H^+. Supposing that all the enzymes plus coenzymes and buffers were initially present in the amounts found in 1 gm of fresh rabbit muscle and that 42 μmoles of creatine, 8 μmoles of ATP and 50 μmoles of P_i were added to the system, the final analysis after an hour or so at 20°C would theoretically show: 8 μmoles ATP (no change), 42 μmoles PC, zero free creatine, 8 μmoles P_i and 28 μmoles lactate; the pH would have dropped by 28/65 units = 0.43 units. This is close to what was observed in an actual experiment where the changes at pH 7.3 occurred to the extent of 90% of theoretical. It is about as far removed as could be from any real situation in a muscle going through rigor.

This experiment has been described in detail to demonstrate first that the characteristic patterns of postmortem change illustrated in Section 3 could not occur in the absence of ATP hydrolysis, and second that it is unprofitable to seek for an explanation of the differing rates of postmortem glycolysis, between species and types of animal, in terms of quantitative changes in glycolytic enzymes, without first looking for differences in the rate determining step, ATP hydrolase activity. Indeed, much effort seems to have been wasted in the past on just such a vain search, with particular emphasis on stress-susceptible pigs (see Briskey, 1964, 1968, for reviews of the subject). Apart from the fact that many of the assays of glycolytic enzymes, and also of preparations of sarcoplasmic reticulum, were carried out on samples which had gone through rigor and had reached low and probably denaturing pH values, it is simple to demonstrate in quite normal muscles that the potential glycolytic rate is far greater than any encountered in even the fastest of stress-susceptible pigs (Sutherland and Cori, 1951). Damage to the muscle, due to mincing for example, also increases the rate of $\sim$P turnover; in a beef sterno muscle at 38°C it increases it from 0.5 to 3.0 μmoles/min/gm, with a corresponding increase in rate of pH fall from 0.0043 to 0.028 units/min (Bendall and Taylor, 1972). After homogenization, the rate doubles again to 6.0 μmoles/min/gm. These values compare with an average $\sim$P turnover rate of 4 μmoles/min/gm for stress-susceptible Pietrain pigs. It is obvious that the acceleration in the beef muscles has not been produced by any quantitative change in the glycolytic enzymes, but almost certainly by calcium release and stimulation of the contractile ATP hydrolase, consequent upon damage to the sarcoplasmic reticulum during mincing or homogenizing.

It is similarly unprofitable to try to relate the maximal ATP hydrolase activity of myosin or actomyosin to the differing rates of $\sim$P turnover found in normal as against stress-susceptible animals (see Briskey, 1968), because the rates in even the fastest of the latter are less than one-thirtieth of the maximal activity of these enzymes, as we have shown above. Indeed, the problem is far more difficult than the reasoning behind these quantitative assays would lead one to believe, and lies essentially in finding methods of detecting minute increases in the free Ca^{2+} level in intact muscles. It is clearly these which determine the increase in rate in all cases where contracture of the muscle occurs side by side with the other changes, as in Pietrain pigs, for example. The delicate spectrophotometric technique of Jöbsis and O'Connor (1966) offers most hope in this direction, though the calcium fluxes they measured during twitches and tetani were much higher than those likely to occur during rigor.

We are still largely ignorant of the threshold level of actin–myosin hydrolase activity at which a contracture will occur, although an approximate value can be deduced by comparing the ATPase activity of myofibrils with their contractile ability as measured by the decrease in centrifuged fibril volume after addition of ATP (Bendall, 1969, p. 42). For example at $0°C$, at an ionic strength of 0.15, beef myofibrils split ATP at a rate equivalent to 1.2 μmoles/min/gm intact muscle, but they are only just able to contract on addition of ATP under these conditions, as shown by the sensitive fibril volume test. Thus a value of 0.5–1 μmole/min/gm muscle can be suggested tentatively as the likely range of the threshold rate for contractile activity. At first sight, this contradicts the finding in the cold-shortening experiments described in Section V,A,2, where the muscle shortened when the splitting rate was only 0.22 μmoles/min/gm. However, as we pointed out there, less than half of the fibers had shortened to the full extent (Voyle, 1969).

Shortening during rigor itself, as described in Section II,C, must be sharply distinguished from the contractures mentioned above, since the former occurs only as the ATP level is beginning to fall, whereas the latter occur in the prerigor period while the ATP and PC levels are high (see Fig. 15, for example). One possible explanation of rigor shortening is that it begins in a few fibers, where the ATP content has fallen far below the overall value to a level insufficient for the calcium pump to work efficiently. This would result in release of calcium ions and stimulation of the contractile enzyme for the short time that traces of ATP remained available to it. Alternatively, it is also possible, as White (1970) has suggested, that the contractile proteins no longer require calcium ions to activate them, once the ATP level has fallen below 0.5 μmoles/gm and hence contract spontaneously until ATP is finally exhausted. In either case, by the time a few more fibers had reached the same state the first few would be in full rigor, thus limiting the rate and extent of the shortening to the low values actually observed. As we see from Fig. 3, pH has a very marked effect on this type of shortening, progressively increasing the work done as the ultimate pH rises from 5.6 (well fed animals) to 7.0 (exhausted animals).

Besides these questions, the outstanding problems which remain to be solved are, first and foremost, the exact nature of the ATP-hydrolases active in resting muscle and during the normal rigor process; second, whether the loss of the ADP, which is bound to the ATPase sites of myosin in resting muscle (Kushmerick and Davies, 1968; E. W. Taylor,

1970; Marston and Tregear, 1972), is the most immediate cause of rigor onset; and finally, the nature of the dynamic equilibrium between this ADP and ATP itself, which up to now has been considered as the only compound essential for maintaining muscle in prerigor state (see Bate-Smith and Bendall, 1947, 1949, 1956).

REFERENCES

Aubert, X. (1956). *J. Physiol. (Paris)* **48**, 105.
Bate-Smith, E. C. (1939). *J. Physiol. (London)* **96**, 176.
Bate-Smith, E. C., and Bendall, J. R. (1947). *J. Physiol. (London)* **106**, 177.
Bate-Smith, E. C., and Bendall, J. R. (1948). *J. Physiol. (London)* **107**, 2P.
Bate-Smith, E. C., and Bendall, J. R. (1949). *J. Physiol. (London)* **110**, 47.
Bate-Smith, E. C., and Bendall, J. R. (1956). *Brit. Med. Bull.* **12**, 230.
Bendall, J. R. (1951). *J. Physiol. (London)* **114**, 71.
Bendall, J. R. (1961). *In* "The Structure and Function of Muscle" (G. H. Bourne, ed.), Vol. 3, p. 264. Academic Press, New York.
Bendall, J. R. (1966). *J. Sci. Food Agr.* **17**, 333.
Bendall, J. R. (1969). "Muscle, Molecules and Movement," Heinemann, London.
Bendall, J. R. (1970). Unpublished observations.
Bendall, J. R., and Davey, C. L. (1957). *Biochim. Biophys. Acta* **26**, 93.
Bendall, J. R., and Ketteridge, C. C. (1970). Unpublished data.
Bendall, J. R., and Lawrie, R. A. (1962). *J. Comp. Pathol.* **72**, 118.
Bendall, J. R., and Lawrie, R. A. (1964). *Anim. Breed. Abstr.* **32**, 1.
Bendall, J. R., and Taylor, A. A. (1972). *J. Sci. Food Agric.* **23**, 61.
Bendall, J. R., and Wismer-Pederson, J. (1962). *J. Food Sci.* **27**, 144.
Bendall, J. R., Hallund, O., and Wismer-Pedersen, J. (1963). *J. Food Sci.* **28**, 156.
Bendall, J. R., Cuthbertson, A., and Gatherum, D. P. (1966). *J. Food Technol.* **1**, 201.
Berman, M. C., Harrison, G. G., Bull, A. B., and Kench, J. E. (1970). *Nature (London)* **225**, 653.
Bernard, C. (1877). "Leçons sur la diabète et la glycogenèse animale," p. 429. Baillière *et* Fils, Paris.
Briskey, E. J. (1964). *Advan. Food Res.* **13**, 89.
Briskey, E. J. (1968). *In* "Recent Points of View on the Condition and Meat Quality of Pigs for Slaughter (1968)" (W. Sybesma, P. G. van der Wal, and P. Walstra, eds.), p. 41. Res. Inst. Anim. Husb., 'Schoonoord,' Zeist, Holland.
Briskey, E. J., and Wismer-Pedersen, J. (1961). *J. Food Sci.* **26**, 207.
Briskey, E. J., Sayre, R. N., and Cassens, R. G. (1962). *J. Food Sci.* **27**, 560.
Caldwell, P. C. (1958). *J. Physiol. (London)* **142**, 22.
Carlson, F. D., Hardy, D., and Wilkie, D. R. (1967). *J. Physiol. (London)* **189**, 209.
Cassens, R. G., and Newbold, R. P. (1966). *J. Sci. Food Agr.* **17**, 254.
Chambers, R., and Hale, H. P. (1932). *Proc. Soc. Ser B* **110**, 336.
Cheah, K. S. (1970). Private communication.
Constantin, L. L., and Podolsky, R. J. (1966). *Nature (London)* **210**, 483.
Cori, C. F., and Cori, G. T. (1928). *J. Biol. Chem.* **79**, 309.
Crust, K. G. (1970). *Comp. Biochem. Physiol.* **34**, 3.

Dale, H. H., Feldberg, W., and Vogt, M. (1936). *J. Physiol. (London)* **86**, 353.

Davey, C. L. (1960). *Arch. Biochem. Biophys.* **89**, 303.

Davey, C. L., and Gilbert, K. V. (1967). *J. Food Technol.* **2**, 57.

Davey, C. L., and Gilbert, K. V. (1969). *J. Food Sci.* **34**, 69.

Davies, R. E. (1963). *Nature (London)* **199**, 1068.

Deutsch, A., and Nilsson, R. (1953). *Acta Chem. Scand.* **7**, 1288.

Ebashi, S., and Endo, M. (1968). *Progr. Biophys. Mol. Biol.* **18**, 123.

Erdös, T. (1943). *Stud. Inst. Med. Chem. Univ. Szeged* **3**, 51.

Evans, C. L. (1956). "Principles of Human Physiology," p. 753. Churchill, London.

Gibbons, N. E., and Rose, D. (1950). *Can. J. Res., Sect. F* **28**, 438.

Glasstone, S. (1947). "Textbook of Physical Chemistry," 2nd ed., p. 1098. Van Nostrand-Reinhold, Princeton, New Jersey.

Goodman, L. S., and Gilman, A., eds. (1965). "The Pharmacological Basis of Therapeutics," p. 603. Macmillan, New York.

Gordon, A. M., Huxley, A. F., and Julian, F. J. (1966). *J. Physiol. (London)* **184**, 170.

Hallund, O., and Bendall, J. R. (1965). *J. Food Sci.* **30**, 296.

Hanson, J., and Huxley, H. E. (1955). *Symp. Soc. Exp. Biol.* **9**, 228.

Hedrick, H. B. (1958). *Vet. Med.* **53**, 466.

Heldt, H. W., and Klingenberg, M. (1965). *Biochem. Z.* **343**, 433.

Hill, A. V. (1939). *Proc. Phys. Soc., London* **51**, 1.

Hill, A. V. (1952). *Proc. Roy. Soc., Ser. B* **139**, 468.

Hill, A. V. (1965). "Trails and Trials in Physiology." Arnold, London.

Hoet, J. P., and Marks, H. P. (1926). *Proc. Roy. Soc., Ser. B* **100**, 72.

Houwink, R. (1937). "Elasticity, Plasticity and Structure of Matter." Cambridge Univ. Press, London and New York.

Howard, A., and Lawrie, R. A. (1956). *Spec. Rep. No. 63, Food Invest. Dept. Sci. Ind. Res., London.*

Howard, A., and Lawrie, R. A. (1957). *Spec. Rep. No. 65, Food Invest. Dept. Sci. Ind. Res., London.*

Huxley, H. E., and Brown, W. (1967). *J. Mol. Biol.* **30**, 383.

Jöbsis, F. F., and O'Connor, M. J. (1966). *Biochem. Biophys. Res. Commun.* **25**, 246.

Kanazawa, T., and Tonomura, Y. (1965). *J. Biochem. (Tokyo)* **57**, 604.

Katz, B. (1966). "Nerve, Muscle and Synapse," p. 123. McGraw-Hill, New York.

Krebs, E. G., De Lange, R. J., Kemp, R. G., and Riley, W. D. (1966). *Pharmacol. Rev.* **18**, 163.

Krebs, H. A., and Hems, R. (1953). *Biochim. Biophys. Acta* **12**, 172.

Krebs, H. A., and Woodford, M. (1965). *Biochem. J.* **94**, 436.

Kuby, S. A., and Noltmann, E. A. (1962). *In* "The Enzymes" (P. D. Boyer, H. Lardy, and K. Myrbäck, eds.), 2nd ed., Vol. 6, p. 515. Academic Press, New York.

Kushmerick, M. J., and Davies, R. E. (1968). *Biochim. Biophys. Acta* **153**, 279.

Lawrie, R. A. (1953). *J. Physiol. (London)* **121**, 275.

Lawrie, R. A. (1955). *Biochim. Biophys. Acta* **17**, 282.

Lawrie, R. A. (1962). *Recent Advan. Food Sci.* **1**, 71.

Lee, C. A., and Newbold, R. P. (1963). *Biochim. Biophys. Acta* **72**, 349.

Lehninger, A. L. (1967). *In* "Modern Perspectives in Biology," p. 122. Harper, New York.

Lister, D. (1970). Unpublished observations.

Lister, D. (1968). *In* "Recent Points of View on the Condition and Meat Quality of Pigs for Slaughter (1968)" (W. Sybesma, P. G. van der Wal, and P. Walstra, eds.), p. 123. Res. Inst. Anim. Husb., 'Schoonoord,' Zeist, Holland.

Lister, D. (1972). In press.

Lister, D., Sair, R. A., Will, J. A., Schmidt, G. R., Cassens, R. G., Hoekstra, W. G., and Briskey, E. J. (1970). *Amer. J. Physiol.* **218**, 102–114.

Locker, R. H., and Hagyard, C. J (1963). *J. Sci. Food. Agr.* **14**, 787.

Ludvigsen, J. (1954). *Anim. Breed. Abstr.* **24**, 729.

Ludvigsen, J. (1968). *In* "Recent Points of View on the Condition and Meat Quality of Pigs for Slaughter (1968)" (W. Sybesma, P. G. van der Wal, and P. Walstra, eds.), p. 113. Res. Inst. Anim. Husb., 'Schoonoord,' Zeist, Holland.

MacDougall, D. B. (1969). Private communication.

McLoughlin, J. V., and Tarrant, P. J. V. (1968). *In* "Recent Points of View on the Condition and Meat Quality of Pigs for Slaughter (1968)" (W. Sybesma, P. G. van der Wal, and P. Walstra, eds.), p. 133. Rest. Inst. Anim. Husb., 'Schoonoord,' Zeist, Holland.

Mahler, H. R., and Cordes, E. H. (1966). "Biological Chemistry," pp. 410 *et seq.* Harper, New York.

Mangold, H. (1927). *Pfluegers Arch. Gesamte Physiol. Menschen Tiere* **196**, 200.

Marsh, B. B. (1952a). *Biochim. Biophys. Acta* **9**, 127.

Marsh, B. B. (1952b). *Biochim. Biophys. Acta* **9**, 247.

Marsh, B. B. (1953). *Biochim. Biophys. Acta* **9**, 478.

Marsh, B. B. (1954). *J. Sci. Food Agr.* **5**, 70.

Marsh, B. B., and Leet, N. G. (1966). *J. Food Res.* **31**, 450.

Marston, S. T., and Tregear, R. T. (1972). *Nature (London)* **235**, 23.

Martinosi, A. (1968). *Biochim. Biophys. Acta* **150**, 694.

Newbold, R. P. (1966). *In* "The Physiology and Biochemistry of Muscle as a Food" (E. J. Briskey, R. G. Cassens, and J. C. Trautman, eds.), p. 213. Univ. of Wisconsin Press, Madison.

Newbold, R. P., and Scopes, R. K. (1967). *Biochem. J.* **105**, 127.

Opie, L. H., and Newsholme, E. A. (1967). *Biochem. J.* **103**, 39.

Penny, I. F. (1967). *J. Food Technol.* **2**, 325.

Penny, I. F. (1972). In press.

Perry, S. V. (1952). *Biochem. J.* **51**, 495.

Rhodes, D. N. (1965). *J. Sci. Food Agr.* **16**, 447.

Sandow, A. (1955). *Amer. J. Phys. Med.* **34**, 155.

Scopes, R. A. (1970). Personal Communication.

Scopes, R. A. (1964). *Biochem. J.* **91**, 201.

Scopes, R. A., and Lister, D. (1970). Unpublished observations.

Seidel, J. C., Sreter, F. A., Thomason, M. M., and Gergely, J. (1964). *Biochem. Biophys. Res. Commun.* **17**, 662.

Sharp, J. G. (1957). *J. Sci. Food Agr.* **1**, 19.

Sreter, F. (1969). *Arch Biochem. Biophys.* **134**, 25.

Stracher, A. (1964). *J. Biol. Chem.* **239**, 1118.

Sutherland, E. W., and Cori, C. F. (1951). *J. Biol. Chem.* **188**, 531.

Sybesma, W., and Eikelenboom, G. (1969). *Neth. J. Vet. Sci.* **2**, 155.

Szent-Györgyi, A. G. (1945). *Acta Physiol. Scand.* **9**, Suppl. 25, 115.

Szent-Györgyi, A. G. (1953). "Chemical Physiology of Contraction in Body and Heart Muscle." Academic Press, New York.

Taylor, A. S. (1910). "Manual of Medical Jurisprudence," 6th ed. p. 101. Churchill, London.

Taylor, E. W. (1970). Personal communication.

Topel, D. G., Bicknell, E. J., Preston, K. G., Christian, L. L., and Matsushime, C. J. (1968). *Mod. Vet. Pract.* **49**, 40.

Voyle, C. A. (1970). Unpublished results.

Voyle, C. A. (1969). *J. Food Technol.*, **4**, 275.

Weber, A. A., Herz, R., and Reiss, I. (1966). *Biochem. Z.* **345**, 329.

Weber, H. H. (1934). *Pfluegers Arch. Gesamte Physiol. Menschen Tiere* **235**, 205.

Weber, H. H. (1952). *Proc. Roy. Soc. Ser. B* **139**, 512.

Webster, H. L. (1953). *Nature (London)* **172**, 453.

White, D. C. S. (1970). *J. Physiol. (London)* **208**, 583.

Wilson, R. D., Nichols, R. J., Dent, T. E., and Allen, C. R. (1966). *Anesthesiology* **27**, 231.

Wismer-Pedersen, J. (1959). *Food Res.* **24**, 711.

REGENERATION OF MUSCLE

P. HUDGSON and E. J. FIELD

I. Introduction

Early investigations of the repair and regeneration of muscle tissue date from the second half of the eighteenth and the beginning of the nineteenth century, when macroscopic changes were studied in dogs, cats, and rabbits. From these studies, it emerged that destructive lesions in muscle resulted in fibrous scarring, a conclusion that was generally accepted, despite the claims by Nannoni (1782), Zimmerman (1812), and Guensberg (1848) that skeletal muscle did in fact possess some power of regeneration.

The early literature has been summarized by Küttner and Landois (1913). During the second half of the nineteenth century, a number of well worked out papers appeared in which regenerative capacity of striated muscle was clearly recognized, though different views were taken of the details of the process. Allowing for the undeveloped state of biological technique at the time, the reader today is astonished at the accuracy of many of the observations made and the shrewdness with which they were interpreted. For example, as early as 1858, Bottscher noted the nuclear proliferation which occurred in the first 24 hr at the margins of a muscle wound in the frog, rabbit, and rat. Weber (1863, 1867), using dogs, cats, and rabbits, concluded that new muscle cells were derived from old surviving ones, and it is of considerable interest to find in Weber's later paper (1867, p. 236) a suggestion of the "budding" theory usually attributed to Neumann (1868). "*Zuweilen schiebt sich aus dem abgerissenen Ende eines Bündels innerhalb des konisch zulaufenden Sarkolemmschlauches oder vor das abgerissene Ende desselben hervortretend, ein schmaler, blasser, fein gestrefter Streifen kontraktilen Protoplasmas hervor, den man als neugebildet ansehen muss, da er von der alten quergestreiften Substanz sich deutlich abgrenzt.*"*

The correspondence of this description with the modern view, as will appear below, is striking. Another completely modern idea was brought forward by Aufrecht (1868) from his study of experimental wounds in rabbits and guinea pigs when he suggested that the condition of the sarcolemma was of paramount importance for regeneration. Waldeyer (1865) described the formation of *Sarkolemmschläuche* during muscle regeneration and Neumann (1868) introduced the term *Muskelknospen* to describe the buds he found growing from the ends of surviving muscle fibers.

* *Translation:* "Now and again there can be seen a pale, narrow, faintly striated process of contractile protoplasm issuing from a conically tapering sarcolemmal tube or from its torn off end. This protoplasmic process must be regarded as a new formation, since it is clearly distinct from the old transversely striated fiber substance."

About this time, a curious thesis was put forward (Maslowsky, 1868), namely, that new muscle cells could be formed from the leukocytes of the blood—a theory developed apparently under the all-pervading influence of the Cohnheim School.

Much experimental work was carried out, and in 1893 Volkmann's monumental paper appeared—"*Ueber die Regeneration des quergestreif-ten Muskelgewebes beim Menschen und Säugethier*"—in which the process of regeneration and repair was exhaustively examined in human, rabbit, and guinea pig material under a variety of conditions which he grouped into those which spared the sarcolemma and those in which it was destroyed. Meanwhile, Zenker (1864) had given his classic description of the waxy degeneration commonly occurring in typhoid fever (Typhus abdominalis of German authors) and this stimulated much study of the regeneration that was found to take place so completely in this condition (Hoffmann, 1867; Janowitsch-Tschainski, 1869, 1870). A further notable contribution was made by Küttner and Landois (1913), while the most recent detailed studies have been made by Le Gros Clark (1946) and Godman (1957).

In the past decade a great deal of new information has emerged on the problems associated with muscle development and regeneration. Much of it has confirmed long-held views, particularly those of Aufrecht (1868) and Volkmann (1893). The latter's insistence on the importance of sarcolemmal integrity in determining the outcome in muscle regeneration is established today, and his emphasis of the importance of continuous regeneration (budding) from surviving fiber segments in the higher vertebrates still commands much support. More modern work has, however, challenged conventional thought on the relative importance of this process as opposed to discontinuous regeneration; on the occurrence of mitosis in regenerating muscle cells; and on the origin of these cells. Possibly the most significant recent landmarks have been the satellite cell theory of muscle regeneration; the role of mitosis in myogenesis; and the recognition that regeneration, albeit ineffective, occurs in most forms of muscular dystrophy. This work will be reviewed in detail below.

II. Factors Affecting Degree of Muscle Fiber Regeneration

A. *General Factors*

Regeneration of skeletal muscle fibers after injury or illness varies considerably in extent and effectiveness. The degree to which a damaged fiber can repair itself is clearly influenced by many factors, e.g., the general condition and particularly the endocrine status of the subject, the local blood supply, the continuation or repetition of noxious stimuli (the repeated injection of irritant or myotoxic chemicals such as chloroquine or pethidine). In the latter case particularly, considerable fibrosis is likely to ensue, and this per se militates against effective regeneration. As indicated above, the integrity or otherwise of the sarcolemmal sheath is also an important factor in determining the degree to which a damaged fiber will regenerate. At one end of the scale, we have Zenker's waxy degeneration in which the sarcolemma is unscathed and which is followed by complete reconstitution of the muscle fiber, while at the other, destruction of both fiber and sarcolemma is followed by imperfect regeneration, usually associated with considerable interstitial fibrosis. In this context, it is of interest to speculate on the possible relationship between the supposedly abnormal (leaky) sarcolemmal membrane in muscular dystrophy and the abortive attempts at regeneration made by the damaged muscle fibers.

B. Species Variation

Skeletal muscle fibers have a considerable innate capacity to repair themselves after injury, although this varies significantly from species to species. Schminke (1907, 1908, 1909) made a comparative study of regeneration in different vertebrates. Ichthyopsida (fishes and amphibia) in general replaced damaged fibers by a budding process from surviving segments (continuous regeneration) with the exception of *Triton*, a tailed amphibian, where regenerating fibers derived from single myoblasts or spindle cells (discontinuous regeneration). A like process was recorded in the salamander (*Salamander maculata*) by Galeotti and Levi (1893). Schminke found that in Sauropsida (reptiles and birds) regeneration was continuous in most cases, though in the legless reptile *Anguis fragilis* (the blindworm) both continuous and discontinuous formation of new fibers was seen. The latter occurred particularly when the sarcoplasm of a fiber bud had disappeared completely leaving an empty "tube" with few isolated surviving nuclei from which regeneration might originate. In mammals, Schminke considered that regeneration was essentially continuous, although a considerable body of evidence has accumulated since then suggesting that this is not always or indeed commonly the case.

C. Constitutional Factors in Disease

In human or animal disease, the capacity of a fiber to regenerate may depend upon the genetic constitution of the affected individual.

In the days when infectious fevers were widespread, complete regeneration of necrotic muscle could be expected in an otherwise healthy individual who developed Zenker's degeneration during the course of typhoid fever, bacterial pneumonia, or cholera. These conditions are rare in Western European or North American practice now, but profuse and effective regeneration can be seen in some primary muscle disorders. It is, for example, the rule in early cases of hereditary paroxysmal myoglobinuria due to episodic massive muscle necrosis (Meyer–Betz disease) in which a progressive myopathy has not developed and in which there have been no irreversible renal complications. It may also be expected in those cases of polymyositis that remit spontaneously or that have been treated successfully with corticosteroids or other immunosuppressive agents. On the other hand regeneration, although readily demonstrable histologically in early progressive muscular dystrophy (particularly the Duchenne type of X-linked recessive disease), is clearly incomplete and ineffective in view of the ultimate fate of the muscle fiber in this situation. It appears probable that regeneration is ineffective because of an inherent defect in the protein synthetic mechanism of the muscle fiber, manifested both in abnormal development and subsequently in abnormal repair.

III. Morphology of Spontaneous Degeneration and Regeneration in Skeletal Muscle

In the previous edition of this chapter (Field, 1961), a detailed account was given of the morphologic changes occurring in skeletal muscle in Zenker's degeneration as representing an excellent model for regeneration in the human under ideal conditions. However, we feel that it is now so rarely encountered, that the emphasis of this section should be altered and propose therefore to discuss both degeneration and regeneration in more general terms and to consider how they may be modified by genetic factors and disease processes.

A. Macroscopic Appearance of Degenerating or Necrotic Muscle

Gendrin (1832), quoted by Hoffmann (1867), was the first to note the so-called fish flesh appearance of massive necrosis in muscle in patients dying with infectious fevers. Zenker found this, particularly in the adductors of the thigh and in rectus abdominis, in his necropsy studies of many of the patients who succumbed during the great typhoid epidemics in Dresden between 1859 and 1862. Nowadays the condition is extremely uncommon, the writers having seen it once only in a patient dying with massive *Clostridium welchii* septicemia. It might be noted in passing that the term "fish flesh" is now commonly applied to the appearance of skeletal muscle at autopsy in any severe muscle-wasting disease, e.g., the Duchenne type of progressive muscular dystrophy, where the appearance is due to widespread replacement of muscle fibers with fat and fibrous tissue, so that in its modern usage the term is historically, if not descriptively, inaccurate.

In the conditions in which the degeneration–regeneration sequence is seen nowadays, the loci are so patchy that they are rarely visible to the unaided eye. Indeed, it would probably help to scan muscle fasciculi under an operating microscope to detect foci of pathological changes before removing a biopsy sample, particularly in diseases like polymyositis.

B. Histopathology of Muscle Degeneration and Necrosis

Difficult problems beset the interpretation of histologic changes in muscle fibers. Apart from artifacts induced by fixation of fresh tissue and its embedding for section cutting, is the fact that in many diseases, e.g., polymyositis, abnormalities are extremely patchy. Nevertheless a reasonably clear-cut chain of events has been defined in the course of degeneration and these are essentially the same whatever the noxious stimulus inducing them. This sequence is summarized in Fig. 1.

The fundamental histopathologic change in most forms of muscular dystrophy and one of the cardinal abnormalities in polymyositis is segmental degeneration, or necrosis. This may take a number of forms, one of the commonest of which is hyaline change, the characteristic appearance seen in Zenker's degeneration. In this case the muscle fiber is rounded in transverse section and the sarcoplasm and its contents are aggregated into a dense, homogeneous, eosinophilic, and highly refractile mass, often retracted from the sarcolemma. While Adams (1969) has suggested that hyaline change may be an artifact of formalin fixation, it is, in our experience, seen in profusion only in Duchenne dystrophy and can virtually be regarded as a diagnostic histologic feature of this disease. Moreover, such hyalinized fibers can be seen in sections of biopsy material rapid frozen at −150°C and cut on a cryostat without exhibition to formalin (Fig. 2), so it is most unlikely that they result from formalin fixation. The time course of the development of hyaline

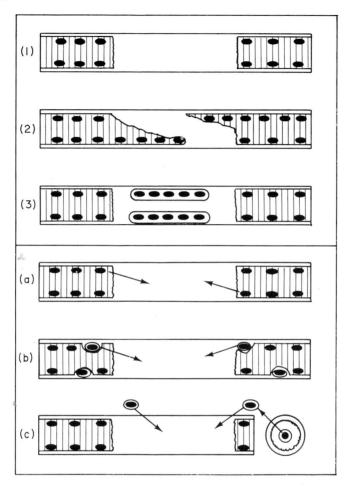

Fig. 1. Diagrams illustrating the cellular changes in a skeletal muscle fiber during the degeneration–regeneration at both the light microscopic and ultrastructural levels. Reproduced (with kind permission) from Dr. Michel Reznik's *Thèse pour l'agrégation*. University of Liège.

change is uncertain, but its appearance and the evidence gleaned from early studies (see Field, 1961) suggests that it evolves gradually (2–3 weeks in the case of Zenker's degeneration).

A muscle fiber may degenerate much more rapidly, in which case the appearance differs from that outlined above. The type of degeneration seen in this setting is variously described as granular or floccular and is most often encountered in acute and subacute polymyositis and may also be seen in the preclinical and early clinical phases of Duchenne dystrophy. Occasionally, numerous tiny vacuoles may be found in the

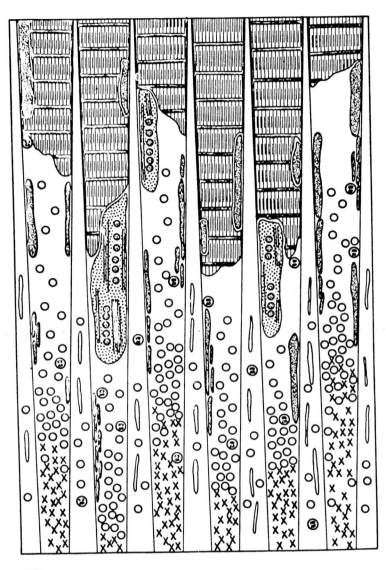

Fig. 1 (Continued)

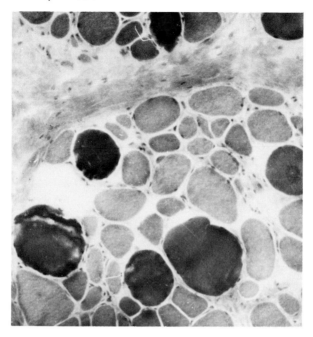

Fig. 2. Biopsy from an early clinical case of Duchenne type X-linked recessive dystrophy showing numbers of enlarged, hyalinized fibers in a cryostat section. Hematoxylin–eosin; ×150.

substance of such a fiber, an appearance claimed by Pearson (1969) to be most often seen in the myositis associated with systemic lupus erythematosus.

According to Volkmann (1893) the earliest change in a degenerating segment is proliferation of the sarcolemmal nuclei. It is unlikely, however, that there is a real increase in the number of these nuclei at this time, the impression probably being given by their central migration and by their tendency to form chains. Field (1961) suggested that increase may be brought about by amitotic division, but it is now considered that this is unlikely (see below). This is followed by a granular clouding and swelling of the sarcoplasm with loss of cross striations and sometimes of myofibrils. It seems likely on the basis of ultrastructural studies that the I band suffers first in this process (see below), and this may be the reason for Bowman's "discoidal" or "conchoidal" degeneration (Adams *et al.*, 1962; Adams, 1969). In Bowman's degeneration, the fibers in formalin-fixed muscle have a tendency to shatter or fragment across the Z lines, sometimes in a zigzag manner. This appearance was first described by Bowman (1840) in what was probably Zenker's de-

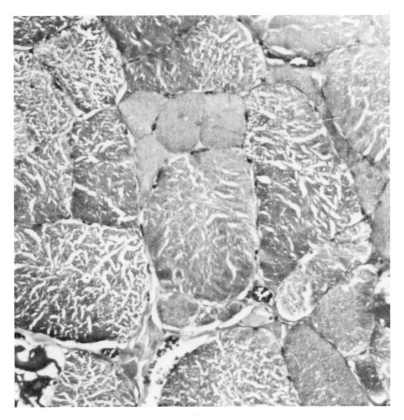

Fig. 3. Biopsy from a case of spinal muscular atrophy with considerable secondary myopathic change. This section shows splitting of both large- and small-diameter fibers. Hematoxylin–eosin; ×242.

generation but may occur in any type of muscle necrosis. Sometimes well marked longitudinal splitting of a degenerate fiber is found with connective tissue between the two parts (Fig. 3). The explanation of this phenomenon is uncertain, but it occurs most often in hyalinized fibers.

The final phase of degeneration in a muscle fiber is invasion by phago-cytes (Fig. 4). According to Adams (1969), this may take place within 24–48 hr of injury and is a valuable criterion for establishing the ante mortem nature of degenerative changes seen in the fiber.

Walton (1973) has recently described an unusual finding in a case of the Becker type of X-linked recessive dystrophy. The patient demon-strated gross pseudohypertrophy in the absence of clinical weakness, but pathologically, his quadriceps contained only hypertrophied fibers

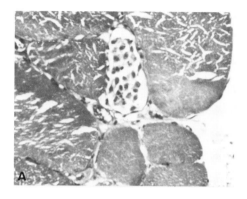

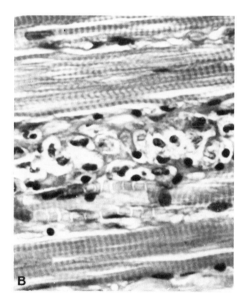

Fig. 4. Same biopsy as in Fig. 2 with transverse (top) and longitudinal (bottom) sections showing phagocytosis of a necrotic fiber.

(average diameter greater than 100 μ) with no evidence of degenerative change. This presumably compensatory process had evidently developed in the absence of its usual stimulus, i.e., fiber atrophy or death.

C. Ultrastructural Changes in Necrotic Fibers

Pearce (1964, 1966) claimed that at the ultrastructural level dilatation of the longitudinal reticulum, and particularly of its terminal cisterns, was the first degenerative change observed in muscle in the dystrophies and in other myopathies. There now seems to be general agreement, however, that the Z band and the segments of the actin filaments immediately adjacent to it are the parts of the muscle fiber most susceptible to damage whatever the noxious stimulus. Dissolution of the Z band and actin filaments (streaming) (Hudgson and Pearce, 1969) is the earliest degenerative change seen in many naturally occurring myopathies and in denervation atrophy (Fig. 5). It can also be induced in the experimental animal by ischaemia (Moore et al., 1956) and by poisoning with plasmocid (Price et al., 1962) and chloroquine (Aguayo

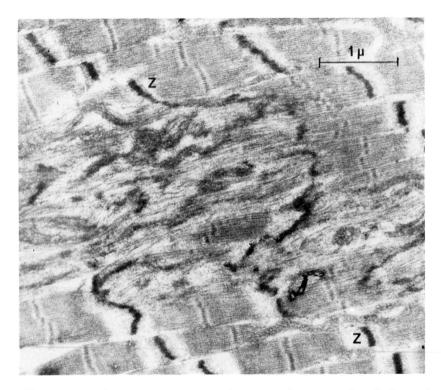

Fig. 5. Biopsy from a young woman with a myopathy associated with abnormal lipid storage in type I fibers. The electron micrograph shows a gross example of Z band and actin filament degeneration (Z band streaming). (×33,000.)

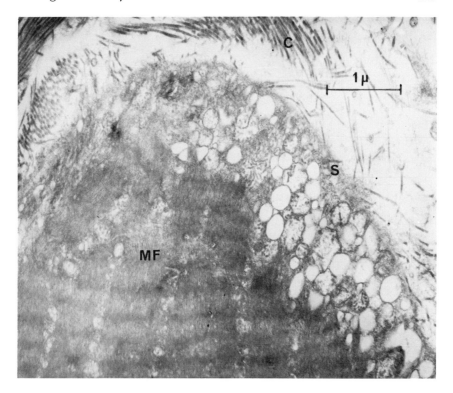

Fig. 6. Biopsy from a clinical case of Duchenne type X-linked recessive dystrophy. The electron micrograph shows coagulative necrosis of the myofibrils with intense vacuolar activity among the mitochondria lying beneath the thickened and blurred sarcolemma. There is a considerable increase in the amount of collagen near the sarcolemma. C = collagen, S = sarcolemma, MF = myofibrils. (×30,750.)

and Hudgson, 1970). It is possible that this selective vulnerability of the Z band region is the basis for Bowman's discoidal change described above. In a completely necrotic segment, the ultrastructural appearance of the fiber is determined by the rate at which necrosis developed. In an acute process, the fiber may appear as a gray, granular, and relatively homogeneous mass enclosed in a sarcolemma of which the two laminae are fused. In this situation, it may be impossible to distinguish any of the contractile apparatus and sarcoplasmic organelles apart from a few unduly electron-dense and clearly degenerate mitochondria. In more gradually developing necrosis, myofibrils may still be discernible, though thickened, fused, and abnormally electron-dense (Fig. 6). Scattered vacuoles (possibly the remnants of dilated sarcoreticulum) are often seen scattered in these electron-dense masses.

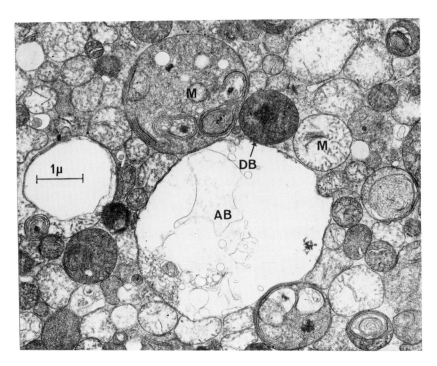

Fig. 7. Biopsy from a patient with a mitochondrial myopathy resembling facio-scapulohumeral dystrophy. The electron micrograph shows a large autophagic vacuole with a number of dense bodies and myelin whorls lying in an aggregate of apparently degenerate mitochondria. M = mitochondrion, DB = dense body, AV = autophagic vacuole. (×21,000.)

The fused fibrils together with any surviving organelles may be re-tracted from the sarcolemma, and the whole picture closely corresponds to hyaline change at the light microscopic level. The classic descriptive term coagulative necrosis has been applied to this appearance by Mas-taglia *et al.* (1969).

In any necrotic fiber, intense lysosomal activity may be seen as judged by the presence of numbers of small dense bodies, autophagic vacuoles containing degenerate remnants of the contractile apparatus and or-ganelles of the muscle cell, and complex lipid figures of all kinds (Fig. 7). In either form of necrosis, invasion of a segment by phagocytic cells may be found (Fig. 8), although this is less frequently seen than in the light microscope studies because of the relatively small samples studied.

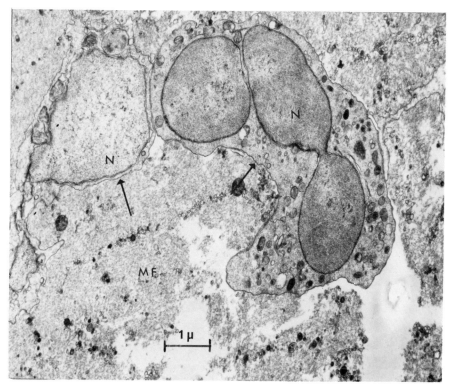

Fig. 8. Electron micrograph of dystrophic mouse muscle showing phagocytic cells—with their plasma membranes (arrows)—one possibly a polymorph, invading a necrotic fiber. N = nucleus, MF = myofilament. (×14,500.)

D. Histological Changes in a Regenerating Muscle Fiber

Considerable uncertainty still exists as to the origin and nature of the cells in the loci from which regeneration begins. This problem will be discussed in detail later, suffice it to say at this stage that in most cases both continuous and discontinuous regeneration occur. In the first case, regeneration commences with the outgrowth of one or more sprouts, or buds, from a normal segment into an adjacent necrotic one after the latter has been cleared of necrotic sarcoplasm, myofibrils, etc. by phagocytic cells (Fig. 9). The efficacy of this type of regeneration clearly depends upon the integrity of the sarcolemmal tube into which the sprouts are growing. In discontinuous regeneration, the process depends upon the presence of viable myoblasts, or at least viable myoblast

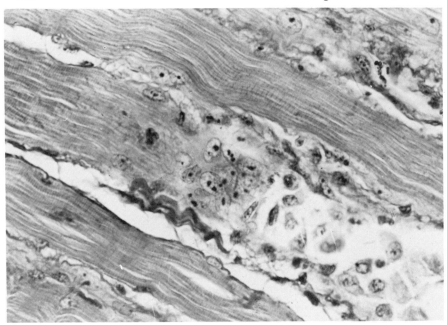

Fig. 9. Biopsy from a patient with polymyositis showing budding from the surviving part of a mature muscle fiber, which has undergone segmental necrosis. Hematoxylin–eosin; ×640. From Mastaglia and Kakulas (1970).

nuclei within the empty sarcolemmal tube, which may act as seeds from which regeneration can occur. In transverse sections of end-stage necrotic muscle cells, such nuclei may be seen surrounded by a thin rim of basophilic cytoplasm and lying immediately under the plasma membrane of an otherwise empty sarcolemmal tube (Fig. 10). These cells may originate from satellite cells (see below). These so-called spindle cells multiply and fuse to form short straps and eventually a myotube (Fig. 11).

It is clearly difficult to be dogmatic about the relative importance of these processes, but it seems reasonable that when gross destruction has occurred in muscle tissue, e.g., after a crush lesion, discontinuous regeneration is the only possible means whereby a fiber can reconstitute itself. On the other hand, in those conditions where one might expect the sarcolemma of a necrotic segment to be intact, e.g., in Meyer–Betz disease, continuous regeneration is probably the more important process. The position in muscular dystrophy and polymyositis is uncertain, for while an apparently intact sarcolemma may be seen at the ultrastructural level, its functional integrity might be abnormal, because of a possibly fundamental leaky defect in the case of dystrophy or as a result of

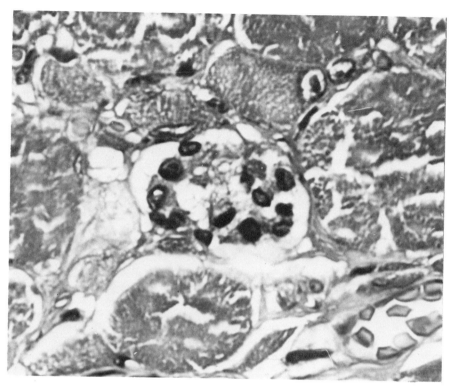

Fig. 10. Biopsy from a clinical case of Duchenne dystrophy showing a sarcolemmal tube in transverse section with a ring of myoblasts lying just under the sarcolemma. Hematoxylin–eosin; ×1224.

the ravages of sensitized lymphocytes in polymyositis. Budding in polymyositis has recently been demonstrated by Mastaglia and Kakulas (1970) in their study of regeneration in this condition. In conclusion we should point out that there is no incompatibility between continuous and discontinuous regeneration and that both probably go on side by side in most diseases. Indeed, Volkmann (1893) stated this with remarkable clarity in view of the rigid positions taken up by later authors on one side or the other.

In spindle cells (myoblasts) and even in immature muscle straps, it is impossible to distinguish myofibrils or their cross striations. By the time the myotube has formed, however, it becomes possible to distinguish myofibrils, and as maturation proceeds, cross striations can be seen as well. The appearance of the regenerating segment at this stage is quite characteristic in both longitudinal and transverse sections, regardless of its origin. The segment is usually smaller than its surviving neighbors, with a basophilic sarcoplasm and a few, coarse myofibrils distributed around the periphery of the muscle cell. The muscle nuclei are invariably large and vesicular, with one or more prominent nucleoli, and are usually centrally situated. They may be aggregated into clumps or chains. If discrete, they tend to lie in the hollow center of the segment, an appearance which gave rise to the term myotube (Fig. 11). The basophilia of the sarcoplasm is associated with its high concentrations of ribonucleic acid (RNA), no doubt responsible for the rapid synthesis of contractile and other structural proteins at that time. The presence of RNA can be demonstrated by staining with any basic dye (we use Azur B) controlled by predigestion of serial sections with ribonuclease

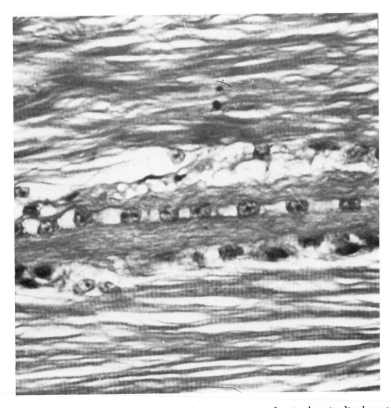

Fig. 11. Same biopsy as in Fig. 3 showing a myotube in longitudinal section. This immature fiber contains coarse fibrils in which the cross striations can barely be discerned, as compared with adjacent fibers, and a row of vesicular central nuclei with prominent nucleoli apparently lying in the hollow center of the tube. Hematoxylin–eosin; ×612.

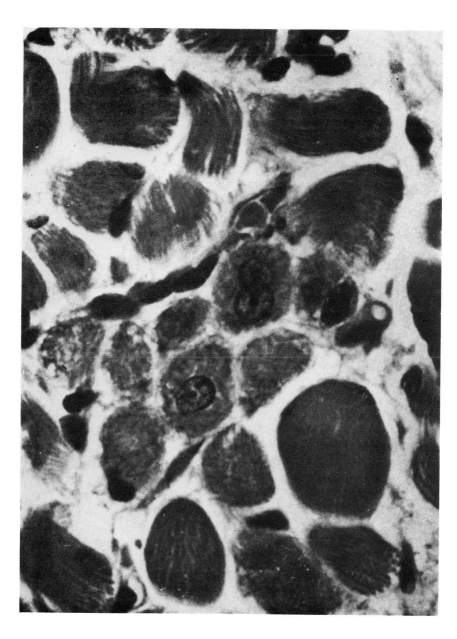

Fig. 12a

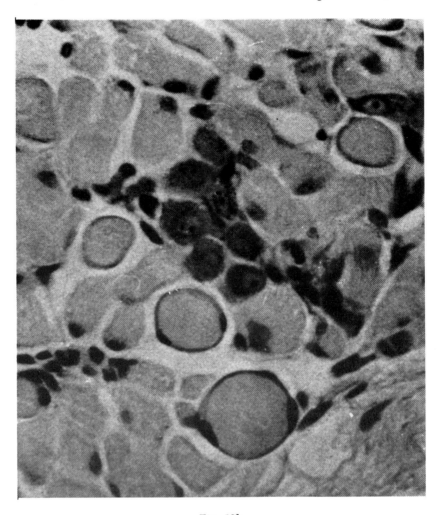

Fig. 12b

(Fig. 12). The nature and distribution of the RNA is further discussed below.

The ultimate fate of a regenerating segment depends upon the nature of the noxious stimulus inducing degeneration in the first instance and on the fiber's inherent capacity for regeneration. In Zenker's original degeneration, restoration to normality was the rule if the subject survived the underlying illness. The same is true in Meyer–Betz disease, and complete regeneration may also occur in polymyositis if the immunologic mechanisms underlying the disease are effectively suppressed. In muscu-

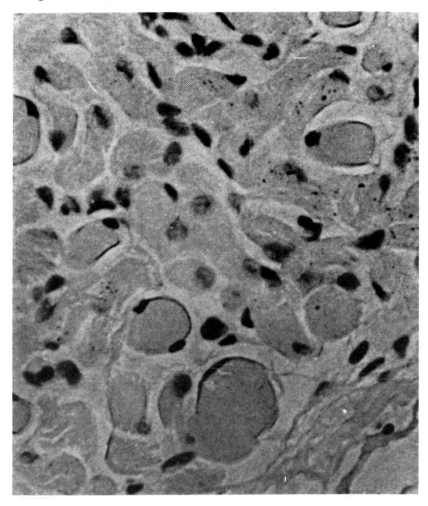

Fig. 12c

Fig. 12. (a) Biopsy from a preclinical case of Duchenne type X-linked recessive dystrophy showing a clump of apparently normal regenerating fibers. Note the coarse myofibrils, the plump vesicular central nuclei containing prominent nucleoli, the obvious difference in staining reaction between these basophilic fibers and their neighbors. Hematoxylin–eosin; ×1230. (b–c) Serial sections to that illustrated in (a) stained with Azur B before and after predigestion with ribonuclease. These show clearly that the regenerating fibers contain high concentrations of ribonucleic acid. (×490). From Hudgson *et al.* (1967).

lar dystrophy, particularly of the Duchenne type, a very different situation obtains. In the preclinical phase of Duchenne dystrophy, profuse regeneration can be found in muscle biopsy samples, and this will often appear to be completely normal histologically (Fig. 12). It is clearly ineffective, however, in view of the almost total devastation that ensues in affected muscles. According to Hudgson *et al.* (1967), Mastaglia *et al.* (1969), and Mastaglia and Kakulas (1969), regeneration in this situation becomes abnormal so that the fiber begins degenerating again before it is fully reconstituted. This sequence repeated many times will eventually produce extreme atrophy if not death of the fiber, and its consequence in terms of muscle malfunction are clear. Certainly a careful search of biopsy sections from such cases will reveal foci of regeneration in which the regenerating segments appear to be abnormal (Fig. 13).

Histochemical studies of regeneration, spontaneous or experimental, have produced some interesting information about the metabolism of developing muscle fibers. Smith (1965) showed that pentose phosphate shunt enzyme activity was increased in regeneration, possibly providing pentose sugars for nucleic acid synthesis. This finding was confirmed by Susheela *et al.* (1968), who also found that the deeply situated type I fibers in the triceps of dystrophic mice were more dystrophic than the superficially placed intermediate and type II fibers. They also

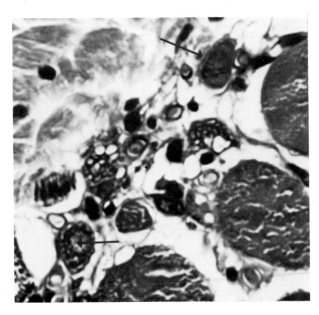

Fig. 13. Same biopsy as in Fig. 2. Transverse section showing apparently abnormal regenerating fibers. The sarcoplasm of some of the regenerating fibers is retracted from the sarcolemma and is often vacuolated. One of these fibers (upper arrow) appears to be hyalinized and another (lower arrow) in undergoing central necrosis. The area is being invaded by phagocytic cells. Hematoxylin–eosin; ×1224.

demonstrated that these type I fibers contained abnormally high concentrations of oxidative enzymes and sudanophilic lipids (Fig. 14). On the basis of Nile blue sulfate staining, they suggested that this may have been due to the presence of high concentrations of free fatty acids. They also suggested that the increased pentose phosphate shunt activity noted by Smith (1965) may have been related to this by furnishing $NADPH_2$ for fatty acid synthesis. In a later study (Susheela et al., 1969), they produced crush lesions in the superficial part of dystrophic mouse triceps. They showed that these lesions regenerated more rapidly than similarly placed lesions in normal control material, but the regenerated muscle was more obviously dystrophic than the corresponding area in the opposite limb. They also found that its histochemical profile had changed to type I, with abnormally high oxidative enzyme activity and lipid concentrations in the regenerated fibers. Reznik and Engel (1970) recently studied the histochemical changes in normal mouse gastrocnemius muscle, where they found that necrotic fibers lost both oxidative and glycolytic enzyme activities, an observation also noted by Susheela et al. (1968). They also showed that early regenerating fibers stained intensely for both oxidative enzymes and myofibrillar ATPase. Later in the course of regeneration, the fibers developed a phosphorylase reaction.

E. Ultrastructural Changes in Regenerating Muscle

It is only during the past 5 years that there has been any coherent account of ultrastructural changes occurring in regenerating human muscle. The technical problems involved in studying patchy changes in biopsy material under the electron microscope are easy to understand, but this problem has been solved in part at least by the practice of studying "thick" (0.25–1.0 μ) plastic sections cut from blocks prepared for electron microscopy and stained with toluidine blue. This permits relatively rapid identification of those samples containing foci of degeneration and/or regeneration.

The first studies of the ultrastructure of regenerating skeletal muscle in the dystrophies and in polymyositis came from workers at the Institute of Muscle Disease, Inc., New York (Milhorat et al., 1966; Shafiq et al., 1967), who reported their findings in biopsy samples from patients with the above diseases. In particular, they noted the presence of large numbers of ribosomes in the sarcoplasm of regenerating muscle cells and suggested this was the basis for sarcoplasmic basophilia in regenerating muscle at the light microscopic level. They also noted that in some

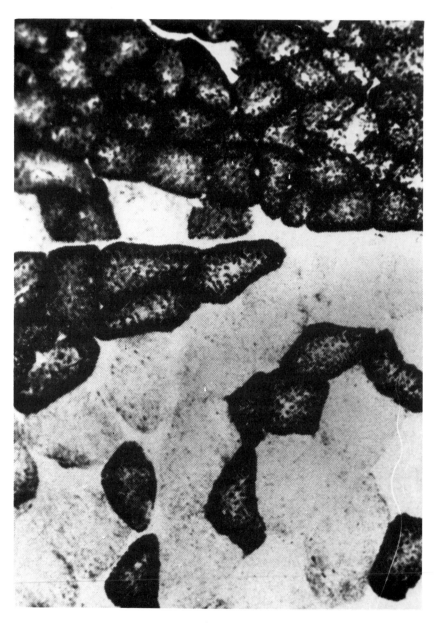

Fig. 14a

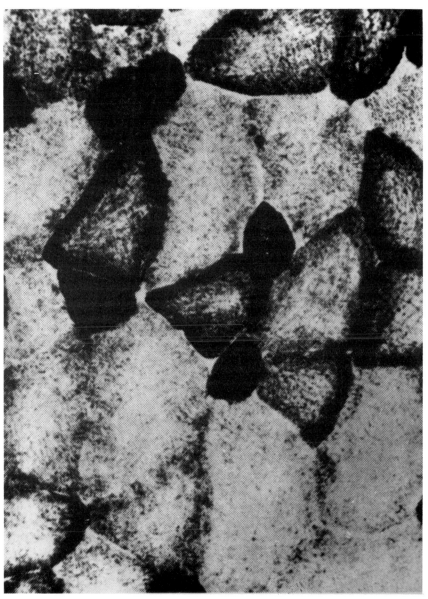

Fig. 14b

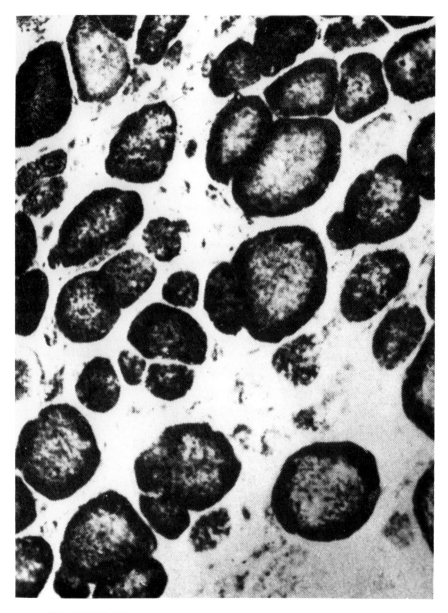

Fig. 14c.

Fig. 14. (a) Normal mouse muscle showing the clear demarcation between the
type I and type II fiber zones in triceps. Cryostat, succinate dehydrogenase; ×92.
(b) Dystrophic mouse muscle showing the relatively normal type II fiber area.
Cryostat, succinate dehydrogenase; ×192. (c) Dystrophic mouse muscle illustrating
the more severely affected type I fiber area. Cryostat, succinate dehydrogenase;
×192.

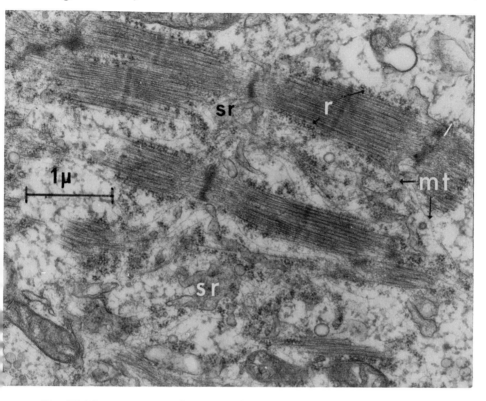

Fig. 15. Electron micrograph of 15-week fetal human muscle showing alignment of ribosomes with myosin filaments. (×23,500.)

instances ribosomes appeared to be associated with developing myofibrils in these cells (Shafiq *et al.*, 1967, note their Fig. 6 particularly), without drawing any particular inferences from this observation. Workers in these laboratories (Larson *et al.*, 1969, 1970a; Hudgson *et al.*, 1970) have been impressed by the close alignment of some of the ribosomes with myofibrils in regenerating muscle and indeed in myoblasts growing in tissue culture and in the human fetus (Fig. 15). Because this relationship is so constant, they have suggested that these ribosomes may be involved in the synthesis of myosin subunits. Colleagues (Ross *et al.*, 1970) have correlated these morphological findings with their quantitative studies of ribosomal RNA in fusing myoblasts. Milhorat *et al.* (1966) have commented on the apparent lack of orientation of filaments in developing dystrophic myoblasts in striking contrast to their appearance in "normal" regeneration. Hudgson (1968) also found that the myofilaments in regenerating cells from dystrophic subjects appeared to be disorien-

tated, though due allowance must clearly be made for the state of contraction of the muscle cells and other possible artifacts in assessing these
observations.

Shafiq *et al.* (1967) found numerous satellite cells in regenerating
muscle from patients with dystrophy and with polymyositis. These cells,
first described by Mauro (1961) in frog muscle, lie between the basement and plasma membranes of the sarcolemma. Their role in muscle
physiology and in disease is contentious (see below), but many workers
regard them as the source of myoblasts in discontinuous regeneration.
Certainly Shafiq and colleagues considered that there were more satellite
cells in foci of regeneration than in control normal muscle. They also
felt that satellite cells in regenerating muscle contained more cytoplasm
and more rough endoplasmic reticulum than the occasional satellite cells
found in normal muscle (Shafiq *et al.*, 1967, C.f. their Figs. 9, 10, and
11). Similar findings have been reported by Mastaglia *et al.* (1969).

Shafiq *et al.* (1967) did not make any fundamental distinction between
regeneration in dystrophy and polymyositis, though they suggested that
the laying down of myofilaments was more orderly in the latter. Mastaglia (1970), quoted by Walton and Hudgson (1973), has recently
completed an ultrastructural study of regenerative phenomena in polymyositis. He found that regeneration proceeded in an orderly manner,
with myofibrils being laid down parallel to the long axis of the fiber.
He also found large numbers of ribosomes in the developing cells, some
free in the sarcoplasm, in rosettes or chains, and others arranged in close
proximity to developing myosin filaments. In common with Shafiq and
colleagues (1967), he identified numerous satellite cells in regenerating
foci although they suggested that their appearance was not necessarily
indicative of a role in myogenesis. The ultrastructural appearance of
regenerating muscle in polymyositis and in dystrophy is illustrated in
Figs. 16 and 17.

Although the late appearance of the cross striations in both spontaneous and experimental regeneration at the light microscopic level
has been stressed (see above), myosin filaments arranged in an orderly
fibrillar fashion, and Z bodies (Z disk precursors) can readily be identified in the electron microscope at an early stage of regeneration, even
in unfused myoblasts (Fig. 18) (Hudgson *et al.*, 1970; Larson *et al.*,
1970a).

IV. Experimental Studies of Regeneration

As indicated previously, the concept that skeletal muscle can effectively regenerate itself after injury has only recently been generally ac-

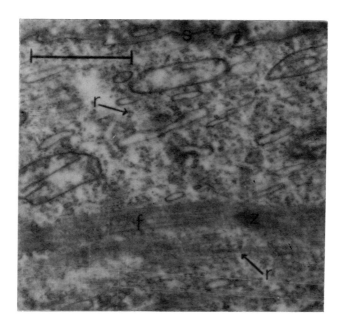

Fig. 16. Biopsy from a young woman with polymyositis. Electron micrograph of a regenerating fiber in which numerous sarcoplasmic ribosomes are seen. Two well orientated fibrils can be seen in the lower half of the figure, and in both, ribosomes are aligned with the developing myosin filaments. (×37,500.)

cepted. As late as 1958, Payling Wright stated that skeletal muscle had only a limited capacity for regeneration, and Denny-Brown (1951, 1952, 1962) maintained that an inherent incapacity for regeneration was one of the cardinal pathologic stigmata of muscular dystrophy. It is now clear, however, that regeneration, effective or otherwise, is a common phenomenon in primary muscle diseases of many kinds, though the mechanism(s) involved remain to be elucidated. Dawson (1909) wrote that "No changes in relation to a healing of wounds have been more difficult to interpret than those in muscle." The difficulties have arisen for two reasons, (1) because of the close association between degenerative and regenerative phenomena; and (2) because of the cytologic similarities of cellular elements derived from the sarcolemmal nuclei,

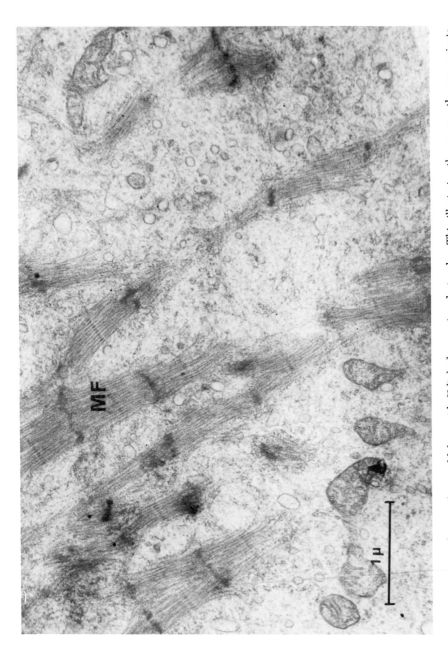

Fig. 17. Biopsy from a 5-year-old boy with X-linked recessive dystrophy. This illustrates the apparently poor orientation of the developing myofibrils. Ribosomes can be seen in alignment with some myosin filaments. MF = myofilament. (×26,500.)

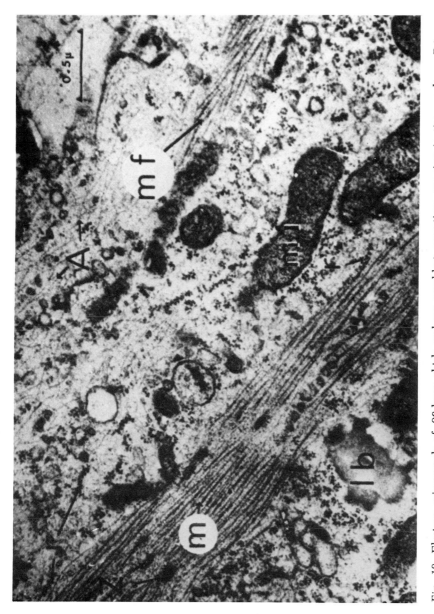

Fig. 18. Electron micrograph of 96-hour chick embryo myoblast syncytium growing in tissue culture. Primordial Z bodies can be seen lying across a developing myofibril. A = actin filament, lb = lipid body, m = M band, mf = myofibril, mi = mitochondrion, r = ribosome, Z = Z body. (×29,500.)

endomysium, and the circulation. The most important problem complicated by these difficulties is the source of the nuclei from which regeneration originates.

A. *Origin of Regenerating Muscle Cells*

This problem in fact dates back to the years immediately following
Zenker's description of waxy change. Waldeyer (1865), who first emphasized the importance of regenerative changes in this condition and described the sarcolemmal tubes crammed with phagocytic cells as
Muskelzellenschläuche, was led to suggest this name because he thought
that many cells were produced by amitotic multiplication of sarcolemmal
nuclei, giving rise to *Muskelkörperchen,* whose cytoplasm, too, was derived from surviving sarcoplasm. He emphasized the production of cells
within the sarcolemmal tube, while Zenker had attached more importance to changes in the endomysium. He also suggested that the *Muskelkörperchen,* in addition to forming free cells, might develop into a
syncytium from which differentiation could occur in the direction of
muscle fibers or connective tissue (Waldeyer, 1865, pp. 506, 507). This
speculation is interesting to compare with Levander's suggestion, discussed below p. 349, that mesenchyme may be "induced" to form muscle
fibers. Waldeyer's ideas were elaborated by several workers over the next
few years. Thus, Weber (1863, 1867) maintained that cells derived from
muscle cells could become transformed into "pus cells," contravening
the idea of specificity of adult tissues then being canvassed. Since his
time, many observers have agreed that Waldeyer's interpretation is reasonable but extremely difficult to prove or disprove (Le Gros Clark,
1946). The opposite view, that the phagocytic cells seen within
Waldeyer's muscle cell tubes were not really formed there but were
cells which had migrated in from outside, was maintained by Güssenbauer (1871).

In an attempt to distinguish the different sorts of phagocytic cells
that are obviously at work within the sarcolemmal tubes, some workers
have employed vital staining techniques. A remarkable early attempt
to do so was made by Janowitsch-Tschainski (1870), who injected
aniline dyes intravenously at various stages during degeneration and
regeneration in experimental wounding. He noted the capacity of some
cells to take up vital dyes but was unable to draw any firm conclusions
from his experience. Much later, Kiyono (1914) confirmed that some
of the cells in the sarcolemmal tubes would take up dye while others
would not, and he regarded these latter as of truly muscular origin.

However, he could not be sure that they, too, did not help in the phago-cytosis of degenerate contents of the tubes.

Lymphocytes, plasma cells, and polymorphs present within and around tubes are obviously of vascular origin and are not usually numerous except in polymyositis. Their role in this situation is not clear, but the prelimi-nary studies of Saunders *et al.* (1969) and of Currie (1970) suggest that the mononuclear cells may be specifically sensitized to muscle anti-gen(s) and may be primarily responsible for damaging muscle cells.

Most attention has been paid to the various types of phagocytic cell within the sarcolemmal tubes. Adams *et al.* (1962) maintain that it is possible to distinguish phagocytic cells of muscular origin by their compact and rather heavy chromatin, together with the absence of a prominant nucleolus, but the distinction is difficult and unreliable (Forbus, 1926a). The same author (Forbus, 1926b) used the method of vital staining to analyze the phagocytic cells that appeared after intramuscular injection of irritants such as alcohol, phenol, or boiling water. In order to distinguish preexisting local histiocytes, which may wander into the sarcolemmal tubes subsequent to the damage from non-phagocytic cells, he stained up the experimental animals as a preliminary with trypan blue, thus labeling all histiocytes by the blue granules in their cytoplasm. After allowing an interval for the disappearance of the dye from the plasma, lesions were made. Under these conditions, he found that some of the phagocytic cells within the tubes were dye-labeled, while others, morphologically identical with them, were not. Presumably, the stained cells were histiocytes that had wandered in, while the unstained cells were produced after the damage, though the author adduces reasons for setting aside this conclusion. Thus, in further experiments, he showed that cells in the sarcolemmal tubes that were unquestionably of muscle origin could be induced to take up dye. It is recognized that the interpretation of seemingly simple vital dye injec-tion experiments may be difficult, and that cells that are obviously phago-cytic may nevertheless fail to stain (Field, 1956). In spite of the intro-duction of sophisticated techniques such as the radioisotopic tagging of nuclei acid precursors, e.g., tritiated thymidine (thymidine-^{3}H) and radioautography, little progress has been in solving the problems dis-cussed above. Walker (1963) studied the site of origin myoblast nuclei in an interesting series of experiments. He induced lesions in the connec-tive tissue of an experimental animal adjacent to a muscle fiber and tagged the regenerating nuclei in that situation with thymidine-^{3}H. After several weeks, he injured the muscle itself, and the regenerating nuclei in the myoblasts appearing in this lesion failed to take up the label. The experiment was then repeated when the muscle instead of the connective tissue had been injured in the first experiment. In this case, myoblast nuclei were strongly labeled, clearly indicating that they were of muscle origin. Rebeiz (1968) has recently repeated these experiments and confirmed Walker's observations.

The next problem to be faced is that if regenerating myoblasts are of muscle origin, from what part of the intact muscle fiber do they come? The satellite cell of Mauro (1961) is a possibility that has received support from Church *et al.* (1966), Shafiq *et al.* (1967), Reznik (1969), and, Reznik and Engel, (1970) in both spontaneous and experimental regeneration. Furthermore, Larson *et al.* (1970b) in these laboratories have identified numerous satellite cells in fetal human muscle at a time when active myogenesis, a process closely resembling regeneration, is in progress (Fig. 19). One might fairly ask at this point of whence do satellite cells come and what is their nature? Reznik (1969) subscribes to the view that they are single myoblasts sequestrated by folds

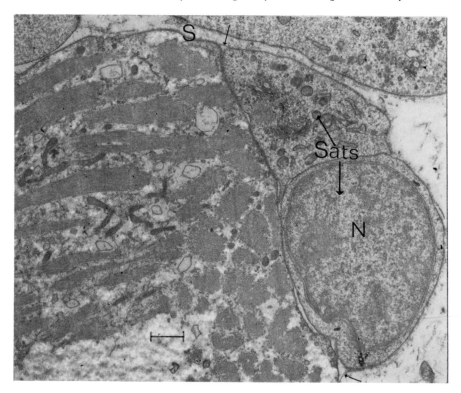

Fig. 19. Electron micrograph of 18-week fetal human muscle showing two adjacent satellite cells. The basement membrane of the sarcolemma (arrows) underlies these two cells. S = sarcolemma, Sats = satellite cells, N = nucleus. (×12,000.)

of the plasma membrane during development, but Shafiq *et al.* (1968) deny this. In a series of radioautographic studies using thymidine-³H tag, Larson and colleagues (1970c) have been unable to confirm that myoblast nuclei are derived from satellite cells in experimental degeneration. In addition, Mastaglia (1970), in studying regeneration in polymyositis, noted that many satellite cells more closely resembled macrophages than myoblasts. In particular, he commented upon the presence of large amounts of rough endoplasmic reticulum in their cytoplasm (see also Shafiq *et al.*, 1967, Fig. 11 particularly), a feature rarely seen in immature muscle cells whether developing or regenerating in our experience, and suggested that they may in fact have a phagocytic potential. It is clear, therefore, that the question must remain open for the time being.

There is, however, a remote possibility that might be worth examining. Field (1961) suggested that the lymphocytes that appear at the site of muscle damage may be replacement nuclei and might colonize fragments of surviving sarcoplasm. It is interesting that in a similar situation where the number of mitoses seem inadequate to account for the amount of cell renewal apparently going on, namely in the epidermis, Andrew and Andrew (1949) have suggested that lymphocytes may become transformed and incorporated into the epithelium. Although this idea has been rejected (Andreasen, 1952), the suggestion that lymphocytes may constitute a store of nonspecialized nuclei which may be called upon from time to time when "regeneration" must take place is an attractive one and deserving of further attention, especially since it is now realized that (under special conditions, it is true) nuclei may be transferred from one cell to another and yet maintain their vital activity (Lorch and Danielli, 1953; Danielli *et al.*, 1955). Some support has been lent to this concept by the work of Bateson *et al.* (1967), who claimed to show on the basis of thymidine-³H uptake studies that myoblast nuclei in regeneration muscle could be derived from circulating leukocytes. This study is of considerable interest but remains unconfirmed to date.

B. Mode of Proliferation of Myoblasts in Regenerating Fibers

For many years there have been two schools of thought on the way in which immature muscle cells proliferate during regeneration. Volkmann (1893) claimed to have observed amitotic division of sarcolemmal nuclei in the earliest stages of degeneration followed by mitotic divisions in the young muscle fibers, and he illustrated these with serial camera lucida drawings. Le Gros Clark (1946) noted occasional mitotic figures in the sarcolemmal sheath cells of tubes occupied by regenerating fibers, but could not be sure that many did not in fact belong to histiocytes still remaining within them. In common with the majority of previous workers, he ascribed the striking increase in numbers of nuclei to amitosis. After colchicine, there were, for example, many arrested mitoses in connective tissue cells, but none could be assigned definitely to regenerating muscle cells. Altschul (1947) made similar observations with colchicine, and Godman (1955) found that the drug was apparently without effect on the number or appearance of nuclei in rapidly growing muscle sprouts in tissue culture. More recently, Godman (1957) has reported the finding of only one instance of a mitotic figure in a muscle strap either *in vivo* or *in vitro*. However, the increase in number of nuclei is enormous and, in the absence of direct evidence, has been assumed to take place by amitosis. The grouping of nuclei in doublets or triplets, their arrangement in columns, and the frequent presence of indentations certainly support this view. Lash *et al.* (1957) believe that mitosis occurs only in the mononucleated cells, which are prominent in the early stages of regeneration, and not in multinucleate masses within sarcolemmal tubes. They did not, indeed, find any convincing evidence of mitosis either and suggest that the accumulation of centrally placed nuclei is the result of mobilization rather than proliferation, a conclusion they support by measurements of DNA content of "regenerating" nuclei. Konjetzny (1953) was of the opinion that the great number of nuclei was due to degeneration of sarcoplasm with consequent crowding rather than to actual multiplication. It is thus apparent that the source of the very numerous nuclei seen in histological preparations of regenerating muscle is by no means clear.

In the past 5 years, there has, however, been a growing body of evidence to suggest that mitosis is much commoner in the early stages of regeneration than was once supposed, and there has been a corresponding fall in the popularity of the theory of amitotic division. Gilbert and Hazard (1965) in a beautifully illustrated study of regeneration in human muscle disease reported frequent mitotic figures in spindle cells arranged circumferentially immediately beneath the sarcolemma of necrotic fibers that did not appear to be undergoing phagocytosis (see their Figs. 1–6). Certainly, these cells lie in the expected position for myoblasts in the earlier stages of the degeneration–regeneration sequence (see above) and may well have been of muscle origin.

In experimental studies on muscle development about this time, MacConnachie *et al.* (1964) claimed that the increase in the number of muscle nuclei in early postanatal life was due to mitotic division of myoblast nuclei, and Johnson (1966) found that up to 6.5% of the myoblasts in regenerating muscle from suckling mice infected with coxsackie A23 virus contained mitotic figures at any given time, and she concluded that this was due to the extent of muscle necrosis and not to any mitogenic effect of the virus. Various workers (Chiakulas and Pauly, 1965; Petersen and Baserga, 1965; Gordon *et al.*, 1966) have shown an increase in DNA synthesis in muscle cell nuclei during postnatal muscle growth. Shafiq *et al.* (1968) studied both skeletal and cardiac muscle in 3-week-old rats treated with colchicine a few hours before sacrifice. They found numbers of mitotic figures in tibialis and in cardiac muscle. In both tissues these figures were seen in undifferentiated myoblasts (they do not say how they characterized these cells as such) and in other free cells, and in tibialis they were also seen in satellite cells. No mitoses were seen in formed fibers or in myoblasts in which myofibrils could be distinguished.

Perhaps part of the explanation for the two opposed views on the mechanism underlying the proliferation of muscle cells has been supplied by Okazaki and Holtzer (1966). Studying chick embryo myoblasts in tissue culture, using thymidine-^{3}H incorporation into muscle nuclei as an index of DNA synthesis rates, and the binding of fluorescein-labeled antimyosin and antiactin antibodies to detect the laying down of myofilaments, they concluded that mitotic activity on the one hand and fusion with myosin synthesis on the other were mutually exclusive. They suggest that presumptive myoblasts (i.e., those containing no myofibrils) repress DNA synthesis and withdraw from the mitotic cycle prior to adapting themselves for myosin synthesis. Holtzer (1970) has elaborated this concept and suggests that proliferating myoblasts undergo a set number of mitotic divisions after which mitotic activity ceases and they begin to fuse and lay down myosin filaments. Certainly, Shafiq *et al.* (1968) were unable to find any myofibrils in cells displaying mitotic activity and, conversely, found no mitoses in muscle straps or blast cells with developing fibrils. Interestingly, Ishikawa *et al.* (1968) described filaments intermediate in size between actin and myosin (approximately 100 Å in diameter) in blast cells undergoing mitosis and concluded that they were not uniquely associated with myogenesis. Similar filaments can be found in other cells of mesodermal origin, e.g., fibroblasts, chondrocytes.

In conclusion, we feel the following points should be mentioned in brief:

1. While there is no absolute certainty that studies of myogenesis *in vitro* can be applied to regenerative phenomena *in vivo*, there are many similarities linking embryonic muscle development on the one

hand and regeneration and myogenesis *in vitro* on the other. Indeed, workers in these laboratories (Ross and Hudgson, 1969) have suggested that the latter is probably an acceptable laboratory model for both physiologic and pathologic processes.

2. Another problem is the nature of the cells emanating from a muscle explant growing in tissue culture. Konigsberg (1963) believes that this can be satisfactorily solved by cloning myoblasts from established cultures and subculturing them, and colleagues here have had some success with this technique using trypsinized chick embryo and human fetal myoblasts. Certainly, in many hundreds of tissue culture sections examined in the electron microscope, the vast majority of the single cells seen have been undoubtedly myogenic, implying either incredible good luck or (more likely) the absence of a sampling problem.

3. Finally, it should be stated that there is as yet no absolute proof that amitotic division does not occur, though this now seems highly unlikely.

C. Continuous and Discontinuous Regeneration

This problem has occupied the attention of very many workers and presents considerable difficulties. On the face of it, it would seem to be a simple matter to examine areas of regenerating muscle by serial sectioning and to establish clearly whether individual isolated myoblasts are first developed which later join up with one another to form strands, or whether regeneration is always by a sprouting from the ends of surviving fibers. Actually, it is found in practice to be virtually impossible to follow individual muscle fibers or sprouts through serial sections, especially in the highly cellular fields found during muscle regeneration. Fortunately, the problem seems largely an academic one, and Volkmann's (1893) synthesis of the two views, so far as mammalian muscle is concerned, has been referred to above and has much to commend it.

Among more recent writers, Le Gros Clark (1946), from his study of the regeneration of crushed and grafted muscle in the rat and rabbit, has found "no evidence that regenerating muscle fibers arise otherwise than as continuous outgrowths from the stumps of old, pre-existing fibers." He found in the rabbit the maximum rate of such outgrowth to be about 1.0–1.5 mm per day and described the advancing tips of the muscle protoplasmic strands as labile, multinucleated structures. On the other hand, Chu (1956a,b), after studying regeneration in white rats following transverse surgical division of the gluteus maximus muscle, is of the opinion that "regeneration of striated muscle fibers is only carried out by the development of . . . embryonic single myoblasts derived from the old muscle nuclei" and concluded that the "opinion of . . . terminal budding seems incorrect." He thus disagrees with both Le Gros Clark (1946) and Levander (1941, 1945, 1955).

The latter worked with muscle transplants in rabbits and failed to find evidence of budding from muscle cells but noted the occurrence of myoblasts in the connective tissue distinctly separate from the implant (*"in deutlichen Entfernung von dem Implant"*). He likened the collections of small spindle-shaped muscle fibers distinctly separate from the main mass to "shoals of fish," a term adopted by Ross and Hudgson (1969) to describe the configuration of myoblasts growing in tissue cultures. Since they had apparently no continuity with preformed muscle fibers, Levander concluded that muscle regeneration can take place from nonspecific mesenchymal tissue by a process of induction. The difficulty of establishing such connection has, however, been remarked upon, and Le Gros Clark thinks there may be an error in interpretation. Moreover, to underline the difficulty, he himself (Le Gros Clark and Blomfield, 1945) considers it "possible also that fibroblastic elements may contribute to myogenesis." Konjetzny (1953), in his study of regeneration in human material, found no evidence of Levander's "induction" but thought it took place always by the continuous sprouting process. While Levander (1955) has recently adduced fresh evidence in favor of his views, Oettgen (1955) has found no evidence of "induction" under well controlled experimental conditions.

In concluding this section, we suggest that the work of Walker (1963) and Rebeiz (1968) leaves little room for doubting the muscle origin of the cells participating in regeneration. There is equally little doubt that Levander's contention that discontinuous regeneration plays an important role in the reconstitution of damaged fibers is valid. Indeed, Johnson (1966), in her study of experimental viral myositis in suckling mice, considered that discontinuous regeneration played the major part in muscle repair.

D. Joining of Muscle Sprouts

When regeneration takes place by the discontinuous process—i.e., from isolated myoblasts, however produced—the discrete foci join up to form a continuous muscular strand, and this process has been described as myogenesis (Godlewski, 1902). The question arises whether muscular sprouts arising by extension from the ends of surviving muscle fibers may approach, join, and so bridge a gap. Despite much study, the point remains difficult, if not impossible, to decide in histological preparations. Lalonde (1950) thought that "fibers from one side of the wound appear to join, but it is granted that this may be an illusion as it is very difficult to follow the fibers in serial section." Jones (1949) found that when the rectus muscle of the dog had been allowed to heal after clean division, stimulation of one part of the muscle led to contraction in the other portion beyond the scar. Gay and Hunt (1954), working with the rectus abdominus and tibialis anterior muscles of the rat, thought that direct union of muscle sprouts coming from the ends of the divided muscle fibers could take place. Indeed their "impression was that the majority of the transected fibers had reunited." More recently, Cotte and Inglesakis (1956) removed segments of the esophagus of the dog (composed of striated muscle) and traced the process of healing. Despite the inevitable sepsis and the size of gap left, they found a surprisingly full regeneration of muscle fibers, beginning with multinucleated outgrowths from the divided fiber ends as early as the second day. They thought the buddings from either side met and fused, and that the rate of repair was of the same order as that estimated by Le Gros Clark (1946) in the rabbit. A curious finding was the persistence of muscle buddings even as late as 226 days, when some degenerated fibers, too, were still present.

Such complete muscle regeneration has sometimes been claimed by surgeons examining human material at operation, and sometimes on the basis of animal experiments. Thus, Bier (1917), for example, observed complete restoration of muscle fibers after large excisions, though Martin (1919) and Bundschuh (1923) were unable to confirm this. More recently, Horn and Sevitt (1951) have reported well marked regeneration in the tibialis anterior muscle after degeneration consequent upon rupture of the popliteal artery.

In the process of healing across a defect in muscle fibers, the greatest importance attaches to the endomysial sheaths, which act as directing planes for muscle sproutings in much the same way as do neurilemmal tubes during regeneration of peripheral nerve. Where injury is such that sarcolemmal tubes remain virtually intact, for example in Zenker's degeneration or in localized experimental freezing of muscle, regeneration is perfect. When, however, sarcolemmal tubes are destroyed, the degree of reconstitution of muscle fibers depends (among other things) on the extent to which the more resistant endomysial connective tissue planes have survived. The pronounced influence of these was clearly shown by Le Gros Clark (1946), who found in ischemic lesions that fibroblasts invade the tissue in advance of muscle sprouts and form new endomysial tubes to reinforce or reconstruct the old ones and so reproduce the original pattern of the muscle when the muscle sprouts follow them in. Moreover, if a strip of muscle be excised and reimplanted after rotation through an angle of 90°, then the ingrowing muscle sprouts may be seen to bend round and follow the new line of endomysial planes.

E. Maturation of Myoblasts

Individual myofibrils usually appear in the peripheral cytoplasm of young muscle cells and are concentrated under the sarcolemma. They may make their appearance quite early and are sometimes seen in isolated muscle cells that have not yet joined to make a continuous fiber, though in general they are not prominent at this stage (Adams *et al.*, 1962). Longitudinal striations appear before transverse. While Forbus (1926a), reported the latter to appear very late, Le Gros Clark (1946) found them as early as the sixth day of growth and very distinct after 2 or 3 weeks. Fishback and Fishback (1932) found them to be well advanced by the twelfth day. Great variations apparently occur under different natural and experimental conditions, and much depends upon the vitality and vigor of the regeneration process, so that generalization has but limited interest for an individual case. It seems, however, that in light microscopic preparations, longitudinal striation precedes transverse, and the general sequence of appearance of the latter follows that in embryogenesis, the A bands appearing before the Z disks. This is not borne out by ultrastructural studies in which it is clear that myosin filaments and Z bodies (Z disk precursors) appear quite early in the course of myogenesis both in *in vivo* and in *in vitro* situations (Hudgson *et al.*, 1970; Larson *et al.*, 1970a) (Fig. 18).

F. Time Scale of Changes

Only a very general indication can be given of the time sequence of changes in regeneration. Von Meyenburg (1929) has summarized the main times reported in the literature as follows:

Sarcolemmal tubes and giant cells appear on the second to fifth days after injury; spindle-shaped muscle cells on the third to fifth days; budding of surviving muscle fibers on the sixth day onward and lasting 6–8 weeks [though Cotte and Ingelsakis (1956) found them much later (see Section III, D.3)]; young muscle fibers on the sixth to eighth days;

and cross striation during the course of the second to third week. The rate of muscle develpment in tissue culture is greatly accelerated, and it may even behave differently from the *in vivo* situation, e.g., myotubes may branch (Okazaki and Holtzer, 1966).

V. Experiments with Human Muscle

A. *Normal Muscle*

Volkmann (1893) carried out intramuscular carbol–glycerin injection experiments after preliminary cocainization in limbs doomed to amputation for some condition such as tumor growth. Cases were selected in which the muscle was otherwise normal, and examination was carried out between 6 hr and 43 days after injection. The changes were essentially similar to those in his rabbit injection experiments, although the injected material tended to spread more readily in the coarsely fasciculated human muscle, producing a very banded lesion.

Walton and Adams (1956) used a sterile suspension of carbon particles in a mixture of oil and alcohol as a means of producing a readily identifiable lesion in human muscle that was later removed at an operation undertaken for some other reason. As did Volkmann, they found a close parallel between the sequence of changes in the rabbit and in the human, the process being perhaps a little less rapid in the latter in the earlier stages, but reaching approximately the same stage by the tenth day. They found both buds and isolated spindle cells to take part in the regeneration process.

B. *Pathological Muscle*

Walton and Adams (1956) also examined the regenerative capacity of muscle in clinical cases of muscular dystrophy, denervation atrophy, and polymyositis, using volunteers who were willing to undergo muscle biopsy. They concluded that impairment of the regenerative potential of the muscle-cell is not a primary effect of either dystrophy or denervation. But a muscle fiber in an advanced stage of atrophy or degeneration, as a result of either denervation of muscular dystrophy, loses its ability to regenerate. However, more recent studies have indicated that this view may need modification (see Section VI).

VI. Effect of Denervation and Muscle Tension on Regeneration

Kirby (1892), working with rabbits, studied the effect of preliminary denervation on regeneration in a damaged muscle. The calf muscles were ligated in their upper third until the lower part had assumed a dark blue appearance. Usually, the ligature was *in situ* for about 3–3½ hr. It was then removed and the wound was sewn up. In other cases, preliminary division of the sciatic nerve was carried out. Histological investigation showed that, as an immediate result of the ischemia, muscle sarcoplasm was disrupted, but connective tissue planes remained intact. On comparing the course of the very good regeneration that took place, Kirby was surprised to find that nerve section 5–10 days prior to the infliction of injury did not in any way hinder muscle regeneration. He also found that nerve section did not lead to muscle fiber degeneration for some weeks.

Denny-Brown (1951), on the other hand, found that while the regenerative ability of muscle was unimpaired for the first 2 weeks after denervation, it was but feeble and abortive after 3 weeks. He pointed out that Kirby had limited his observations to the few days following nerve section. After 3 months, he found trauma to cause "first a proliferation of the nuclei and then fragmentation" (Adams *et al.*, 1962). Walton and Adams (1956), however, found regenerative capacity to be retained for much greater periods in humans, recording active regeneration in an injured muscle that had been weak and atrophic for over 2 years and that had all the histological appearances of long-standing denervation. They concluded that it is only when denervation has produced profound atrophy of muscle cells that they fail to respond to injury. Similarly, Saunders and Sissons (1953) found that recovery from a crush took place in exactly the same way in denervated as in intact muscle. They point out that because of the nature of the injury inflicted, repair must in any case be taking place in many denervated muscle fibers even in an intact muscle. They extended the interval between nerve section and the injury to 3 weeks, using the rat as an experimental animal, and took precautions that reinnervation should not occur.

The influence of tension in the muscle during regeneration following a burn has been examined by Denny-Brown (1951). He showed that division of the tendon at the time of making the lesion, even though it prevented separation of the ends of the muscle fibers, retarded regeneration. With an intact tendon, there was more distortion during repair, but the tension appeared to stimulate all phases of the process, both actual formation of muscle fibers and their maturation.

The observations discussed above are reflected in the ease with which a muscle explant in tissue culture can produce new muscle fibers in the absence of the normal trophic influence of motor neurons and without the physiological tension associated with attachment to the skeleton. Shimada et al. (1967) and Larson et al. (1970a) found that the ultrastructure of chick embryo myoblasts growing in tissue culture developed normally in every particular. However, the myotubes that form in this situation showed no tendency to develop into intact muscles. Nakai (1965) showed that a chick embryo muscle explant would develop into a normal whole muscle if subjected to a very slow constant pull (imparted by a geared down clockwork mechanism) when growing in vitro.

On the other hand it has become apparent that chronic denervation may itself be followed by regeneration. Adams et al. (1962) noted that experimental denervation in the dog and cat was followed by complete reconstitution of the affected fibers if they were reinnervated within a few weeks. Longer periods of denervation resulted in secondary myopathic degeneration—a phenomenon subsequently recognized in chronic denervation atrophy in the human (Drachman et al., 1967; Gardner-Medwin et al., 1967). The former authors described occasional foci of regenerative activity in the myopathic areas of the muscle they studied (see their Fig. 4).

In a later study, Drachman and colleagues (1969) were able to induce profuse regeneration in the extraocular muscles of cats by experimental denervation, and they found histologically similar changes in biopsy material from human subjects suffering from ocular myopathy. Because of this, they questioned the concept of this disease as a form of muscle dystrophy (Kiloh and Nevin, 1951; Magora and Zauberman, 1969), which seems reasonable because of the general lack of association of muscular dystrophy with pigmentary degenerations of the retina and system degenerations within the neuraxis.

That motor neurons may have a trophic effect on skeletal muscle fibers, a deficiency of which may produce changes in the fiber other than those of simple denervation atrophy has recently been suggested by McComas et al., (1970). They studied the motor unit population of extensor digitorum brevis in human subjects suffering from various forms of muscular dystrophy and found a significant reduction in population in all these. On this basis, they have invoked a neurogenic hypothesis for muscular dystrophy, but this work is clearly in its earliest stages and requires careful evaluation. In this context it is also interesting that Caspary et al. (1972) have demonstrated that lymphocytes from a group of patients with Duchenne dystrophy (including one preclinical case) are sensitized to sciatic nerve basic protein (SNBP) as well as to muscle.

VII. Effect of Corticosteroids on Muscle Regeneration

The effect of cortisone on repair of rabbit skeletal muscle has been investigated by Sissons and Hadfield (1953) and by Ellis (1955). The former used the crush technique of Le Gros Clark and found that 10–20 mg per kilogram body weight per day retarded the onset and rate of regeneration but did not alter its course of eventual outcome. Similar findings have recently been published by Sloper and Pegrum (1967) in a study of regeneration in mouse muscle. The effect of cortisone was much less pronounced than on bone repair. Ellis found that healing was rapid once cortisone treatment was discontinued and that rabbits on potassium-deficient diet showed changes similar to those found in the cortisone animals. It seems possible that the delaying action of cortisone may be due to its depressant action upon mobility and phagocytic activities, which are so important in the process of preliminary scavenging associated with muscle regeneration. Thus, Paff and Stewart (1953) found cortisone to diminish ameboid activity of phagocytes in tissue culture. These factors have been discussed in another connection (Field, 1957).

It is now clear that the various corticosteroid drugs can themselves induce a severe proximal myopathy in subjects taking large doses for prolonged periods, the 9α-fluorinated compounds being particularly dangerous in this respect (Golding *et al.*, 1961). Recent studies of steroid myopathy in humans and experimental animals (Afifi *et al.*, 1968; Afifi and Bergman, 1969) make it clear that regeneration is not prominent in this disease. This is not surprising in view of the generalized catabolic effects of steroids, but in addition, they may well specifically repress regeneration, perhaps by inhibiting enzyme activity in the direct oxidative pathway and so restricting the availability of pentose sugars for nucleic acid synthesis. The case of polymyositis, however, deserves special mention. The vast majority of patients improve rapidly on treatment with steroids and, in one case subjected to quadriceps biopsy (opposite sides) before and after a course of treatment, no regeneration was seen in the first biopsy and profuse, widespread activity in the second. This presumably means that its immunosuppressive effects in this situation outweigh any direct effect on the muscle fiber.

VIII. Regeneration of Cardiac Muscle

Many authors have examined the regenerative capacity of heart muscle and expressed varying opinions, although the majority have found little

(Fleischer and Loeb, 1910; Christian *et al.*, 1911; Heller, 1914; Karsner and Dwyer, 1916; Collier, 1922; Warthin, 1924; Bright and Beck, 1935; King, 1941; Harrison, 1947; Walls, 1949; Ring, 1950). Just as in the case of skeletal muscle, this variation is in some measure due to the different nature of the primary toxic influence after which regeneration has been studied. Thus some (e.g., Harrison, Walls, Ring) deny that regeneration may take place at all, while others (Heller, Warthin, King, for example) believe that it may.

From what has been said above of the regeneration of skeletal muscle, it will be apparent that the character of the primary damage will be of importance in determining the vigor and success of any subsequent attempts at regeneration. Thus, an insult that leaves the general architecture of the heart muscle undisturbed should present the best chances for successful repair. Such injury is best seen in the myocarditis, which may accompany diphtheria, and has been well described in detail by Heller (1914) and Warthin (1924). The essential lesion here is a toxic, parenchymatous, hyaline degeneration or necrosis associated frequently with fatty degeneration or with cloudy swelling. Waxy degeneration also occurs, and as in the case of the degeneration of skeletal muscle of typhoid fever, the lesions are patchy in their distribution. It is followed by a reparative inflammatory process with foreign body giant cell formation as necrotic material is removed. Warthin speaks of "perimysial tubes" becoming filled with detritus and inflammatory and regenerating cells in much the same way as occurs in skeletal muscle. Regenerating takes place later by an ingrowth of muscle sprouts into the tubes from adjacent surviving fibers. Bulbous swelling stuffed with myoblastic nuclei may be seen at the living ends of the muscle defects. Warthin was able to confirm Heller's (1914) findings in all respects and both support the view that heart muscle fibers do have a sarcolemma.

Associated with heart muscle fiber regeneration is some fibrosis, and there is no doubt that this is a common result of most forms of cardiac injury, just as it is in skeletal muscle injury. Thus, Martinotti (1888) noted only a slight proliferation of cardiac muscle nuclei (with mitotic figures) in the early stages after a stab wound of the rat heart, but did not regard this as of significance in healing, which took place largely by fibrous tissue. Anitschkow (1913), studying wounds of the rabbit heart, found no true regeneration of muscle cells. He thought, however, that damaged heart muscle cells could be converted through the loss of contractile substance first into myocytes, with characteristic elongated nuclei showing serrated chromatin (Anitschkow cells), and then later into fibrocytes. Similar views with respect to skeletal muscle have been discussed in Section III.

Among more recent workers, Harrison (1947) concluded that rabbit cardiac muscle has not the same regenerative capacity as general somatic muscle, and Walls (1949), too, failed to find evidence of regeneration after burning of the rabbit ventricular myocardium. However, such thermal lesions are inevitably associated with some destruction of all the elements of the myocardium. Ring (1950) produced ischemic lesions in the cat and rabbit heart by ligation of branches of the left coronary artery and found revascularization of the infarct to take place within 14 days without regeneration of muscle fibers. This he was inclined to attribute to the absence of a sarcolemma (but see Heller and Warthin, above). Törö (1939), however, has described a marked improvement in the healing process when an extract of embryonic heart muscle is administered. According to this work (which remains to be confirmed), such treatment favors formation of elastic tissue rather than dense fibrous tissue and has a beneficial effect upon the regenerative efforts of the heart muscle fibers themselves.

Destruction of heart muscle in virus infections, notably those with coxsackievirus, has been considered in detail by Bell and Field (1970) in Volume III, Chapter 3, Section IX of this treatise. Regeneration after coxsackievirus infection is inconspicuous (Field, 1961), but the degeneration produced by reovirus types 1 and 2 may be followed by well marked regeneration in cardiac muscle (Walters *et al.*, 1965).

Mention must be made of the theory propounded by Hofmann (1902) and supported by Retzer (1920) that the atrioventricular connecting system (conducting system of the heart) is really a source for wear-and-tear replacement of the myocardium. Van der Stricht and Todd (1919) describe Purkinje cells (in the human heart) as giving rise to ordinary heart muscle cells by longitudinal fission and claim to have seen this process particularly clearly in diseased hearts where greater replacement was necessary. Todd (1932) has brought further evidence for this, and the question is considered in more detail by Field (1951).

IX. Summary

Skeletal muscle fibers have a considerable inherent capacity for regeneration, but the extent to which this is successfully exercised depends upon many factors, local and general. Among the more important of these are (a) the presence of viable muscle elements from which regeneration may begin either at the extremes of or within a segment of necrotic muscle; (b) the integrity of the sarcolemmal sheath and the endomysial framework of the necrotic segment so that the growth of the new fiber can be directed along anatomic pathways; and (c) the constitutional capacity of the individual fiber to regenerate itself.

A number of other factors, such as blood supply and tissue oxygenation, the endocrine status of the subject, and the presence of irritants or foreign bodies, may influence the course of regeneration. We believe, however, that the single most important factor is the fiber's constitutional capacity for regeneration, the importance of this factor being illustrated by the differences in regenerative efficiency between Duchenne dystrophy and a disease like polymyositis, for example. There is no doubt that virtually complete restoration of normal fiber architecture has been observed in patients with typhoid fever and Zenker's degeneration who survived their original illness, but optimal conditions for regeneration are rarely encountered nowadays. The nearest approach to complete regeneration is seen in the early stages of Meyer–Betz disease (although chronic degenerative changes become more prominent in a stepwise fashion with successive attacks of muscle necrosis), and regeneration may also be effective in uncomplicated polymyositis. In these conditions, there is no known inherent defect in the muscle fiber's ability to repair itself after injury, but in dystrophy (particularly Duchenne), there may well be a genetically determined malfunction of protein synthesizing processes leading to abnormal development and later abnormal regeneration. Necrosis and regeneration may be going on *pari passu* in adjoining fibers and even in adjacent segments of a particular fiber, so that eventually, regeneration in this situation becomes histologically abnormal.

Orthodox histological studies have firmly established the various cytological changes in regenerating muscle, and it is interesting to see that latter day accounts vary only in a few minor particulars from those of nearly a century ago. The balance of evidence available indicates that the cells from which regenerating fibers originate are of muscle origin, but their location in the normal and necrotic fiber is uncertain. Certainly, the role of satellite cells in this respect has not yet been established and requires clarification.

The widespread use of tissue culture studies of myogenesis has provided us with a useful model for both muscle development and regeneration. Ultrastructural studies of muscle cells grown in this environment and from foci of regeneration in biopsy material have provided a wealth of new information about the mechanism involved in myogenesis. It seems clear that there is intense ribosomal activity in myoblasts and that the ribosomes are not concentrated on endoplasmic reticulum, but lie free in the sarcoplasm, sometimes in close relationship to developing myofilaments. These ribosomes may be involved in the laying down of myosin subunits. Unfortunately, these studies have not yet produced any clear evidence to suggest that there are any fundamental differences between myogenesis in the normal subject and in subjects with genetically determined myopathies. The results of further work along these lines is awaited with interest and guarded optimism.

ACKNOWLEDGMENTS

The writers wish to thank various colleagues for their advice and criticism during the preparation of this chapter, and they are particularly grateful to Professor J. N. Walton and to Drs. Paul Larson and Frank Mastaglia. The technical work of Mr. J. Fulthorpe, Misses M. Jenkison and A. Brown, and Mrs. V. Giles, which has formed the basis of many of the observations made in the Muscular Dystrophy Research Laboratories, is acknowledged with gratitude. Dr. Hudgson's work has been supported by the Muscular Dystrophy Associations of America, Inc., the Muscular Dystrophy Group of Great Britain and the Medical Research Council.

REFERENCES

Adams, R. D. (1969). *In* "Disorders of Voluntary Muscle" (J. N. Walton, ed.), 2nd ed., p. 143. Churchill, London.

Adams, R. D., Denny-Brown, D., and Pearson, C. M. (1962). "Diseases of Muscle. A Study in Pathology," 2nd ed. Harper (Hoeber), New York.

Afifi, A. K., and Bergman, R. A. (1969). *Johns Hopkins Med. J.* **124,** 66.

Afifi, A. K., Bergman, R. A., and Harvey, J. C. (1968). *Johns Hopkins Med. J.* **123,** 158.

Aguayo, A. J., and Hudgson, P. (1970). *J. Neurol. Sci.* **11,** 301.

Altschul, R. (1947). *Rev. Can. Biol.* **6,** 485.

Andreasen, E. (1952). *Acta Dermato-Venereol.* **32,** 17.

Andrew, W., and Andrew, N. V. (1949). *Anat. Rec.* **104,** 217.

Anitschkow, W. (1913). *Beitr. Pathol. Anat. Allg. Pathol.* **55,** 373.

Aufrecht, E. (1868). *Arch. Pathol. Anat. Physiol. Klin. Med.* **44,** 180.

Bateson, R. G., Woodrow, D. F., and Sloper, J. C. (1967). *Nature (London)* **213,** 1035.

Bell, T. M., and Field, E. J. (1970). *In* "The Structure and Function of Muscle" (G. H. Bourne, ed.), 2nd ed., Vol. 3. Academic Press, New York. (In press.)

Bier, A. (1917). *Deut. Med. Wochenschr.* **43,** 285.

Bottscher, A. (1858). *Arch. Pathol. Anat. Physiol. Klin. Med.* **13,** 227.

Bowman, W. (1840). *Phil. Trans. Roy. Soc. London* **130,** Part 2, 457.

Bright, E. F., and Beck, C. S. (1935). *Amer. Heart J.* **10,** 293.

Bundschuh, E. (1923). *Beitr. Pathol. Anat. Allg. Pathol.* **71,** 674.

Caspary, E. A., Currie, S., and Field, E. J. (1971). *J. Neurol., Neurosurg. Psychiat.* **34,** 353.

Chiakulas, J. J., and Pauly, J. E. (1965). *Anat. Rec.* **152,** 55.

Christian, H. A., Smith, R. M., and Walker, J. C. (1911). *AMA Arch. Intern. Med.* **8,** 468.

Chu, J. (1956a). *Acta Exp. Biol. Sinica* **5,** 199 (English Abstr.).

Chu, J. (1956b). *Acta Exp. Biol. Sinica* **5**, 371 (English Abstr.).

Church, J. C. T., Noronha, R. F. X., and Allbrook, D. B. (1966). *Brit. J. Surg.* **53**, 638.

Collier, W. D. (1922). *J. Med. Res.* **34**, 21.

Cotte, G., and Inglesakis, J. A. (1956). *C. R. Soc. Biol.* **150**, 212.

Currie, S. (1970). *Acta Neuropathol.* **15**, 11.

Danielli, J. F., Lorch, I. J., Ord, M. J., and Wilson, E. G. (1955). *Nature (London)* **176**, 1114.

Dawson, J. W. (1909). *J. Pathol. Bacteriol.* **13**, 174.

Denny-Brown, D. (1951). *J. Neuropathol. Exp. Neurol.* **10**, 94.

Denny-Brown, D. (1952). *Can. Med. Ass. J.* **67**, 1.

Denny-Brown, D. (1962). *Rev. Can. Biol.* **21**, 507.

Drachman, D. B., Murphy, S. R., Nigam, M. P., and Hills, J. R. (1967). *Arch. Neurol. (Chicago)* **16**, 14.

Drachman, D. B., Wetzel, N., Wasserman, M., and Naito, H. (1969). *Arch. Neurol. (Chicago)* **21**, 170.

Ellis, J. T. (1955). *Amer. J. Phys. Med.* **34**, 240.

Field, E. J. (1951). *Brit. Heart J.*. **13**, 129.

Field, E. J. (1956). *J. Anat.* **90**, 428.

Field, E. J. (1957). *J. Neuropathol. Exp. Neurol.* **16**, 48.

Field, E. J. (1961). *In* "Structure and Function of Muscle" (G. H. Bourne, ed.), 1st ed., Vol. 3, p. 139. Academic Press, New York.

Fishback, D. K., and Fishback, H. R. (1932). *Amer. J. Pathol.* **8**, 193.

Fleischer, M. S., and Loeb, L. (1910). *AMA Arch. Intern. Med.* **6**, 427.

Forbus, W. D. (1926a). *Arch. Pathol.* **2**, 318.

Forbus, W. D. (1926b). *Arch. Pathol.* **2**, 486.

Galeotti, G., and Levi, G. (1893). *Beitr. Pathol. Anat. Allg. Pathol.* **14**, 272.

Gardner-Medwin, D., Hudgson, P., and Walton, J. N. (1967). *J. Neurol. Sci.* **5**, 121.

Gay, A. J., and Hunt, T. E. (1954). *Anat. Rec.* **120**, 853.

Gendrin (1832). Cited by Hoffmann (1867).

Gilbert, R. K., and Hazard, J. B. (1965). *J. Pathol. Bacteriol.* **89**, 503.

Godlewski, E. (1902). *Arch. Mikrosk.-Anat. Entwicklungsmech.* **60**, 111.

Godman, G. C. (1955). *Exp. Cell Res.* **8**, 488.

Godman, G. C. (1957). *J. Morphol.* **100**, 27.

Golding, D. N., Murray, S. M., Pearce, G. W., and Thompson, M. (1961). *Ann. Phys. Med.* **6**, 171.

Gordon, E. E., Kowalski, K., and Fritts, M. (1966). *Amer. J. Physiol.* **210**, 1033.

Guensburg, A. (1848). Cited by Küttner and Landois (1913).

Güssenbauer, C. (1871). *Arch. Klin. Chir.* **12**, 1010.

Harrison, R. G. (1947). *J. Anat.* **81**, 365.

Heller, A. (1914). *Beitr. Pathol. Anat. Allg. Pathol.* **57**, 223.

Hoffmann, C. E. E. (1867). *Arch. Pathol. Anat. Physiol. Klin. Med.* **40**, 505.

Hofmann, H. K. (1902). *Z. Wiss. Zool.* **5**, 189.

Holtzer, H. (1970). Personal communication.

Horn, J. S., and Sevitt, S. (1951). *J. Bone Joint Surg., Brit. Vol.* **33**, 348.

Hudgson, P. (1968). *In* "Research in Muscular Dystrophy. Proceedings of the Fourth Symposium of the Muscular Dystrophy Group of Great Britain" (Members of the Research Committee of the Muscular Dystrophy Group, eds.), p. 207. Pitman, London.

Hudgson, P., and Pearce, G. W. (1969). *In* "Disorders of Voluntary Muscle" (J. N. Walton, ed.), 2nd ed., p. 277. Churchill, London.

Hudgson, P., Pearce, G. W., and Walton, J. N. (1967). *Brain* **90**, 565.

Hudgson, P., Jenkison, M., and Larson, P. F. (1970). *In* "Muscle Diseases" (J. N. Walton, N. Canal, and G. Scarlato, eds.), p. 90. Excerpta Med. Found., Amsterdam.

Ishikawa, H., Bischoff, R., and Holtzer, H. (1968). *J. Cell Biol.* **38**, 538.

Janowitsch-Tschainski, S. (1869). *Med. Vestn., St. Petersburg* **9**, 371.

Janowitsch-Tschainski, S. (1870). *Inst. Exp. Pathol. Wien, Stud.* p. 86.

Johnson, M. A. (1966). Ph.D. Thesis, University of Newcastle upon Tyne.

Jones, D. S. (1949). *Anat. Rec.* **103**, 473.

Karsner, H. I., and Dwyer, J. E. (1916). *J. Med. Res.* **34**, 21.

Kiloh, L. G., and Nevin, S. (1951). *Brain* **74**, 115.

King, E. S. J. (1941). "Surgery of the Heart," p. 728. Williams & Wilkins, Baltimore, Maryland.

Kirby, E. (1892). *Beitr. Pathol. Anat. Allg. Pathol.* **11**, 302.

Kiyono, K. (1914). "Die vitale Karminspeicherung." Fischer, Jena.

Konigsberg, I. (1963). *Science* **140**, 1213.

Konjetzny, G. E. (1953). *Chirurg.* **24**, 49.

Kraske, P. (1878). Habilitationsschrift, Halle.

Küttner, H., and Landois, F. (1913). "Die Chirurgie der quergestreiften Muskulatur," Deut. Chir., Part 1, p. 303. Enke, Stuttgart.

Lalonde, I. L. (1950). Ph.D. Thesis, Loyola University, Chicago (cited by Gay and Hunt, 1954.

Larson, P. F., Hudgson, P., and Walton, J. N. (1969). *Nature (London)* **222**, 1168.

Larson, P. F., Jenkison, M., and Hudgson, P. (1970a). *J. Neurol. Sci.* **10**, 385.

Larson, P. F., Jenkison, M., and Hudgson, P. (1970b). Unpublished observations.

Larson, P. F., Jenkison, M., and Fulthorpe, J. J. (1970c). Unpublished observations.

Lash, J. W., Holtzer, H., and Swift, H. (1957). *Anat. Rec.* **128**, 679.

Le Gros Clark, W. E. (1946). *J. Anat.* **80**, 24.

Le Gros Clark, W. E., and Blomfield, L. B. (1945). *J. Anat.* **79**, 15.

Levander, G. (1941). *Arch. Klin. Chir.* **202**, 667.

Levander, G. (1945). *Nature (London)* **155**, 148.

Levander, G. (1955). *Ark. Zool.* **8**, 565.

Lorch, I. J., and Danielli, J. F. (1953). *Quart. J. Microsc. Sci.* **94**, 445.

McComas, A. J., Sica, R. E. P., and Currie, S. (1970). *Nature (London)* **226**, 1263.

MacConnachie, H. F., Enesco, M., and Leblond, C. P. (1964). *Amer. J. Anat.*, **114**, 245.

Magora, A., and Zauberman, H. (1969). *Arch. Neurol. (Chicago)* **20**, 1.

Martin, B. (1919). *Arch. Klin. Chir.* **111**, 673.

Martinotti, G. (1888). *G. Accad. Med. Torino* **7**, 348.

Maslowsky, J. (1868). *Wien. Med. Wochenschr.* **18**, 192.

Mastaglia, F. L. (1970). Unpublished observations.

Mastaglia, F. L., and Kakulas, B. A. (1969). *Brain* **92**, 809.

Mastaglia, F. L., and Kakulas, B. A. (1970). *J. Neurol. Sci.* **10**, 471.

Mastaglia, F. L., Papadimitriou, J. M., and Kakulas, B. A. (1969). *Proc. Aust. Ass. Neurol.* **6**, 93.

Mauro, A. (1961). *J. Biophys. Biochem. Cytol.* **9**, 493.

Milhorat, A. T., Shafiq, S. A., and Goldstone, L. (1966). *Ann. N.Y. Acad. Sci.* **138**, 246.

Moore, D. H., Ruska, H., and Copenhaver, W. M. (1956). *J. Biophys. Biochem. Cytol.* **2**, 755.

Nakai, J. (1965). *Exp. Cell Res.* **40**, 307.

Nannoni (1782). "Diss. de simularium partium h.c. constituentium regeneratione." Mediolanum (cited by Küttner and Landois, 1913).

Neumann, E. (1868). *Arch. Mikrosk. Anat. Entwicklungsmech.* **4**, 323.

Oettgen, H. F. (1955). *Frankfurt. Z. Pathol.* **66**, 48.

Okazaki, K., and Holtzer, H. (1966). *Proc. Nat. Acad. Sci. U.S.* **56**, 1484.

Paff, G. H., and Stewart, R. (1953). *Proc. Soc. Exp. Biol. Med.* **83**, 591.

Payling, Wright, G. (1958). "Introduction to Pathology," 3rd ed. Longmans Green, New York.

Pearce, G. W. (1964). *In* "Disorders of Voluntary Muscle" (J. N. Walton, ed.), 1st ed., p. 220. Churchill, London.

Pearce, G. W. (1966). *Ann. N.Y. Acad. Sci.* **138**, 138.

Pearson, C. M. (1969). *In* "Disorders of Voluntary Muscle" (J. N. Walton, ed.), 2nd ed., p. 501. Churchill, London.

Petersen, R. O., and Baserga, R. (1965). *Exp. Cell Res.* **40**, 340.

Price, H. M., Pease, D. C., and Pearson, C. M. (1962). *Lab. Invest.* **11**, 549.

Rebeiz, J. J. (1968). Personal communication to Ross and Hudgson (1969).

Retzer, R. (1920). *Carnegie Inst. Wash., Contrib. Embryol.* **9**, 145.

Reznik, M. (1969). *Lab. Invest.* **20**, 353.

Reznik, M., and Engel, W. K. (1970). *J. Neurol. Sci.* **11**, 167.

Ring, P. A. (1950). *J. Pathol. Bacteriol.* **62**, 21.

Ross, K. F. A., and Hudgson, P. (1969). *In* "Disorders of Voluntary Muscle" (J. N. Walton, ed.), 2nd ed., p. 319. Churchill, London.

Ross, K. F. A., Jans, D. E., Larson, P. F., Mastaglia, F. L., Parsons, R., Fulthorpe, J. J., Jenkison, M., and Walton, J. N. (1970). *Nature (London)* **226**, 545.

Saunders, M., Knowles, M., and Currie, S. (1969). *J. Neurol., Neurosurg. Psychiat.* **32**, 569.

Saunders, J. H., and Sissons, H. A. (1953). *J. Bone Joint Surg.* **35B**, 113.

Schminke, A. (1907). *Verhandl. physik. med. Ges. Würtzburg* **39**, 15.

Schminke, A. (1908). *Beitr. pathol. Anat. allgem. Pathol.* **43**, 519.

Schminke, A. (1909). *Beitr. pathol. Anat. allgem. Pathol.* **45**, 424.

Shafiq, S. A., Gorycki, M. A., and Milhorat, A. T. (1967). *Neurology* **17**, 567.

Shafiq, S. A., Gorycki, M. A., and Mauro, A. (1968). *J. Anat.* **103**, 135.

Shimada, Y., Fischman, D. A., and Moscona, A. A. (1967). *J. Cell Biol.* **35**, 445.

Sissons, H. A., and Hadfield, G. J. (1953). *J. Bone and Joint Surg.* **35B**, 125.

Sloper, J. C., and Pegrum, G. D. (1967). *J. Pathol. Bacteriol.* **93**, 47.

Smith, B. (1965). *J. Pathol. Bacteriol.* **89**, 139.

Susheela, A. K., Hudgson, P., and Walton, J. N. (1968). *J. Neurol. Sci.* **7**, 437.

Susheela, A. K., Hudgson, P., and Walton, J. N. (1969). *J. Neurol. Sci.* **9**, 423.

Todd, T. W. (1932). *In* "Special Cytology" (E. V. Cowdry, ed.), Vol. II, Section 29. Hoeber, New York.

Törö, I. (1939). *Z. Zellforsch, u. mikroskop. Anat.* **22**, 304.

Van der Stricht, O., and Todd, T. W. (1919). *Johns Hopkins Hosp. Rep.* **19**, 1.

Volkmann, R. (1893). *Beitr. pathol. Anat. allgem. Pathol.* **12**, 233.

Von Meyenburg, H. (1929). *In* "Handbuch der speziellen pathologischen Anatomie" (O. Lubarsch, F. Henke, and R. Rössle, eds.), Vol. IX, Part I. Springer, Berlin.

Waldeyer, W. (1865). *Arch. pathol. Anat. Physiol. Virchow's* **34**, 473.

Walker, B. E. (1963). *Exp. Cell Res.* **30**, 80.

Walls, E. W. (1949). *J. Anat.* **83**, 66.

Walters, M. N.-I., Leak, P. J., Joske, R. A., Stanley, N. F., and Perrett, D. H. (1965). *Brit. J. Exp. Pathol.* **46**, 200.

Walton, J. N. (1973). *Proc. Int. Acad. Pathol.*, *1970* (in press).

Walton, J. N., and Adams, R. D. (1956). *J. Pathol. Bacteriol.* **72**, 273.

Walton, J. N., and Hudgson, P. (1973). *In* "The Molecular Basis of Neurology" (S. Appel and E. Goldensohn, eds.). Lea & Febiger, Philadelphia, Pennsylvania (in press).

Warthin, A. S. (1924). *J. Infectious Diseases* **35**, 32.

Weber, O. (1863). *Zentr. med. Wiss.* **34**, 529.

Weber, O. (1867). *Arch. pathol. Anat. Physiol. Virchow's* **39**, 216.

Zenker, F. A. (1864). "Über die Veränderungen der wirkurlichen Muskeln im Typhus abdominalis." C. W. von Vogel, Leipzig.

Zimmerman (1812). *Arch. Physiol.* **11**, 131.

7

MUSCLE SPINDLE

ZBIGNIEW L. OLKOWSKI AND SOHAN L. MANOCHA

I. Historical Perspective

The study of muscle spindles and related nerve endings began in 1861 when Weismann (1861a, b) described the small embryonic-type striated muscle cells called Weismann bundles. Kölliker (1863), in the belief that they are the centers for the muscle cell reproduction, named them *Muskelknospen*. In the same year, Kuhne (1863) introduced the name muscle spindle because of the presence of the usual fusiform enlargement of these endings in mammals. In 1886, Babinski reported the presence of pathological structures" consisting of "circular rings with several atrophied muscle fibers" in the cross-striated muscles in the course of chronic myelitis. Later on, in 1886, he described the bundles consisting of three to seven small muscle fibers with blood vessels and nerves in the space surrounded by laminated connective tissue. The detailed morphological structure of the muscle spindles was described by Kölliker (1889) in frog, as Kuhne (1863) had described it earlier in mammals. Kölliker (1862) had assumed that the *Muskelknospen* were the result of the longitudinal cleavage of dichotomously dividing muscle fibers. Bremer, in 1883, working on the muscle spindles of lizards and mice, concluded likewise that they were young muscle fibers. He observed that they are innervated by large myelinated fibers as well as by fine ones having either a thin myelin sheath or none at all; the large fibers were interpreted as motor and the fine ones as sensory. Kerschner, in 1888, considered that the innervation of the muscle spindles is sensory and that it is effected either by large nerve fibers surrounding the muscle bundle or the individual fibers as a spiral.

Ruffini (1893, 1898) later gave a detailed description of the innervation of the muscle spindles in cat and human muscles. He observed that the nerves entering the muscle spindle broke up into branches forming *"terminaisons annulospiralles"* surrounding the intrafusal fibers in the middle third of the muscle spindle. As a second type of innervation, he described the *"terminaisons à la fleur"* coming from the fine fibers and appearing in the shape of ball-like enlargements on the intrafusal fibers. The different nerve endings in the muscle spindles were described by Ramon y Cajal (1897) in the frog and by Ruffini (1892, 1893) in the cat. Cajal had found two types of endings in the muscle spindles, located on the polar region of the intrafusal fibers and in the equatorial region. Ruffini (1892, 1893) was the first, however, to describe different types of endings as simple, intermediate, and complex, according to the presence of secondary endings on the basis of staining procedures using gold chloride. The sensory nature of the muscle spindle innervation, suspected by Cajal, was experimentally proved by Sherrington (1895). Sherrington found that the sensory fibers of the muscle spindles ramifying at the equatorial region of the muscle spindle measure 7–18 μ. He did not show at that time the motor fibers innervating the muscle spindles, although Huber and DeWitt (1897) believed that such fibers are present. Horsley (1897), studying these organs in the dog and in the cat, discovered that they remained unchanged even after sectioning the sciatic nerve. Dogiel in 1902 showed that the coarse fibers form the sensory endings and the fine fibers create the motor endings, thereby confirming the results of Ruffini and Kerschner. By 1910 the histology of the muscle spindles was known in detail, as is evident by a review article written by Climbaris, but it was not until 1928 that the presence of the motor endings on the intrafusal fibers was experimentally proved by Hines and Tower. They studied the nerve degeneration in the muscle spindles of the cat and described in the polar region grapelike endings that degenerated following the cutting of the anterior spinal roots. Simultaneously, the equatorial innervation disappeared after the removal of the spinal ganglia.

Excellent review articles have been written during the last few decades on the morphology, innervation and physiology of muscle spindles. Some of the important articles to which the reader is referred are written by Boyd (1962) and Barker *et al.* (1966), who described in detail the morphology and innervation of the muscle spindles. The problems of the physiology and pathology of the muscle receptors were discussed by Bach-y-Rita (1959), Cooper (1960), Buchwald *et al.* (1963), Mathews (1960), Cohen (1964, 1965), Eccles and Lundberg (1958), Eldred (1960, 1965), Eyzaguirre (1964), Ferrari and Sonna (1965), Granit (1957, 1962a,b,c, 1964), Homma (1963), Hunt and Perl (1960), Inman (1962), Adams *et al.* (1962), Kuffler and Hunt (1952), Laporte (1962), Rushworth (1962, 1964a,b), C. M. Smith (1963), Stewart *et al.* (1963), and Mathews (1971).

II. Development of Muscle Spindles

A. *Intra- and Extrafusal Fibers*

The embryological studies of Felix (1888), Christomanos and Strossner (1891), Forster (1902), and others, demonstrating the presence of muscle spindles as early as the sixteenth week of fetal life, stimulated

systematic investigations on the development of neuromuscular structures. Felix (1888) characterized the "primitive muscle bundles" and
differentiated them from the extrafusal muscle fibers in 4–9-month-old
human fetuses. The main difference between the extrafusal fibers and his
"primitive muscle bundles" is that the latter are covered by the capsule of
connective tissue, show an enlargement of the middle part, show abundant nuclei, and show a wider striation compared to the extrafusal fibers.
Felix did not appreciate the connection between nerves and his primitive
muscle bundles because of a lack of good staining method at that time.
This was delayed until Ruffini (1893) discovered the nerve endings.
Christomanos and Strossner (1891) and Baum (1900) described the
development of capsule, the time of appearance of intrafusal fibers,
and their structure, and Forster (1902) stained the muscle spindles in
4-, 5-, and 9-month-old human fetuses quite adequately so as to differentiate between the intrafusal and extrafusal fibers. She also showed that
the intrafusal fibers were formed earlier as compared to the ordinary
cross-striated fibers and concluded that the muscle spindles are the important components of the adult muscles.

Tello (1922) was the first to describe the innervation of the fetal muscle
spindles and showed the nerve terminations in the muscle spindles from
6-month-old fetuses. Cuajunco (1940) later brought out an elaborate
account of the develoment of human muscle spindles, by taking the
specimens from the musculus biceps brachii of 10–39-week-old fetuses
and staining them with silver and with alum carmine. Such a staining
procedure showed the nerves as well as the muscle fibers and capsule.

The myoblasts in the 10-week-old embryo appear as tubelike structures
containing sarcolemma, minute myofibrils without the cross striation and
nuclei in the central part of the fiber. During the eleventh week of
fetal life, the picture of the myofibrils is altered; they are stained more
strongly, and the cross striation with a typical iso- and anisotropic band
appears. At this stage of development, there is no evident difference
in the structure of the extrafusal and intrafusal fibers (Cuajunco, 1940).
The differentiation of intrafusal fibers starts in the muscles of 12-week-
old fetuses when the equatorial region of the muscle spindles enlarges
and a bundle of slightly enlarged muscles with typical fusiform shape
appears in this region. The muscle spindles are easily recognized, although they do not show clearly the characteristic difference between
organization of particular parts of the muscle spindle. At the end of
the twelfth week of fetal life, the existence of two poles and the equatorial region is apparent. The nuclei in the axial bundle of the equatorial
region of the muscle spindle remain larger than those newly formed
bundles at the periphery.

The diameter of the muscle spindle in the equatorial region varies from muscle to muscle and from species to species (Cooper, 1960). In the case of human musculus brachii, the size of the equator remains unchanged during the period from the thirty-first week of fetal life to birth, whereas the number of intrafusal muscle fibers gradually increases from the fourteenth to the thirtieth weeks of fetal life, being at least eleven in the muscle spindle belonging to a 31-week-old fetus (Cuajunco, 1940). The increase in the number of the intrafusal fibers after birth has been extensively studied. Bravo-Rey *et al.* (1969) have shown that intrafusal fibers (nuclear bag and nuclear chain fibers) in rat leg muscles increase in number from two at birth to four fibers by the sixth day. At birth, one of each of these fibers is present in every muscle spindle in the rat (Marchand and Eldred, 1969) as well as in mouse (Wirsen and Larsson, 1964). Several authors have also described the presence of extrafusal fibers within the muscle spindles (Swett and Eldred, 1960; Cooper and Daniel, 1963). Latyshev (1958), studied developing human and cat muscle spindles, and concluded that secondarily acquired intrafusal fibers were newly formed from tissue elements in the inner layer of the capsule. Shanthaveerappa and Bourne (1962) presented evidence that the capsule elements are derived from the ectodermal elements.

The question of whether or not multiplication of the intrafusal fibers involves a mitotic division was investigated by Marchand and Eldred (1969). It was observed that a dose of 1200 R of X irradiation or an injection of colchicine blocked the mitotic process. The autoradiographic investigation of these workers also showed a negligible effect on the increasing number of intrafusal fibers proportional to the age of the fetus and did not show the incorporation of this thymidine-^{3}H into the nuclei of the intrafusal fiber of new born rats, whereas the other components of the muscle spindle were labeled. Marchand and Eldred concluded that the increase in the number of intrafusal fibers after birth takes place in rat muscle spindles by simple longitudinal divisions of the existing fibers at birth. There is also a possibility of adult intrafusal fibers bifurcating into two; Weismann (1861b) described the splitting of the intrafusal fibers in the muscle spindle of the frog. Batten (1897) and Forster (1894) have found that the intrafusal fibers could separate at the equatorial level and reunite at the pole. Haggquist (1960) held a similar opinion. Cooper and Daniel (1956, 1963), Boyd (1958a,b), and Swett and Eldred (1960) have emphasized, however, that intrafusal fibers do not branch so easily. On the other hand, Barker (1959) and Barker and Gidumal (1961) described the branching of intrafusal fibers in the cat among the nuclear chain fibers, and Boyd (1960, 1962) believed that the nuclear bag as well as nuclear chain fibers could divide,

the latter doing so more often. Marchand and Eldred (1969) resolved
the problem to a great extent by their autoradiographic studies using
thymidine-^{3}H as a DNA precursor. They showed clearly that the postnatal increase in the number of intrafusal fibers arises through the split
of existing fibers with a simple redistribution of nuclei.

B. Capsule

The capsule covering the muscle spindle develops from mesenchyme.
It is evident that mesenchymal cells have great potentiality in the developing embryo, and they differentiate into the many kinds of cells in
the connective tissue. The first part of the capsule, containing one layer
of cells, appears in the twelfth week of fetal life. The cells divide continuously, and at the fifteenth week, two layers of these cells may be
observed. An increase of the cell layers in the capsule continues until
the thirty-first week of fetal life and finally emerges a variable number
of concentric, tubular sheaths of cytoplasm of specialized fibrocytes. At
this stage, the capsule covers the equatorial region of the muscle spindle.
The periaxial space makes its initial appearance at the fourteenth week
of fetal life, increases in diameter, and extends to half the length of
the poles adjacent to the equatorial region in the latter stages of development. The thickness of the capsule is not uniform throughout the muscle
spindle. The main differences are in the place of entering of the nerve
fibers where several more layers of the sheath cells may be found.

C. Nerve Endings

At the eleventh week of fetal life, the nerve trunks are observed ramifying within the developing muscles. At this particular stage, the terminal
twigs form the network of nerve fibrils around some myoblasts, and
at their intersections, they form a delicate network of dots and ringlets
which come into contact with the muscle fibers. According to Cuajunco
(1940), it is the only reliable evidence of the earliest formation of the
muscle spindle. Since all myoblasts in this particular stage of development are alike, it is not easy to differentiate between the future extra-
and intrafusal fibers. This network may be considered the fundamental
sensory innervation of the human muscle spindles. Cuajunco noted that
one or more nerve branches may contribute to this network, which is
formed in the equatorial region about the twelfth week of fetal life.

He also described the distinctly separated nerves, far apart from those forming the sensory endings, in the polar regions at the late stages of muscle spindle development. These may form other types of endings. Also observed are typical motor endings in the shape of knob termination on the future intrafusal fibers, and at the eleventh week of fetal development, a distinct morphological difference between the sensory and motor endings in the future muscle spindle is evident. At the twelfth week of embryonic life, the sensory endings become more prominent because of the increasing size of the nerve fibers. At the thirteenth week, the distinctly defined capsule covers the equatorial region of the muscle spindle and the nerve fibers along with their branches are seen traversing the intrafusal fibers. Myelinization of large nerves in the human fetus starts at the fourteenth week, and well myelinated fibers might be seen after the twenty-fourth week of development. Myelinated nerve fibers enter the capsule of the muscle spindles in the equator or in the polar regions. Sometimes they retain their myelin sheaths in the periaxial space, but their ramifications before terminating are naked. This picture is mostly observed in the case of equatorial region fibers. In most cases, the fibers entering the polar region lose their myelin sheaths immediately beneath the capsule.

The number of muscle spindles keeps increasing until the fifteenth week of human embryonic life, although the immature tubular myoblast might also be seen in older fetuses. The network formation of the sensory nerve endings on the intrafusal muscle fibers was, however, not seen in the muscles of embryos older than 15 weeks (Cuajunco, 1940). Cuajunco concluded that most of these sensory endings are formed up to the fifteenth week of human fetal life, and in the later stages of fetal development, the delicate network (supposed to be the earliest type of the sensory endings) covers the larger area of the fibers. It may be changed in some intrafusal fibers into the larger and denser anastomosis of the nerve terminal fibers, although in some intrafusal fibers, it remains as a network of tenuous nerve fibers. These changes are observed up to the thirty-first week or till the time the fetal sensory nerve endings have been developed.

As the elongation of muscle spindle takes place, it becomes apparent that their polar regions are supplied by different nerve branches than those that form the sensory endings. They show the characteristics of motor end-plates and as described by Boeke (1921) enter the Doyer hillock in order to terminate the knoblike masses of nerve reticulum among the nuclei of the differentiated sarcoplasm of the end-plate.

Recent studies in the anatomy and physiology of muscle spindles have resulted in a detailed description of the nerve endings in the muscle spindles. In humans, a single spindle receives one large afferent nerve fiber to the primary endings on the nuclear bags and chains, one or more nerve fibers to secondary afferent endings, a group of small to medium sized nerve fibers, and very fine nerve fibers going to blood vessels and capsule (Cooper and Daniel, 1963). For a detailed description of the nerve endings on the intrafusal fibers in the muscle spindles, the reader is referred to the following sections, which deal with morphology and ultrastructure.

D. Histochemistry of Muscle Spindle during Development

Not many histochemical studies on the muscle spindles during development exist. Several enzymes, carbohydrates, lipids, and proteins have been investigated in the chicken embryo by Anda and Rebollo (1968). They have shown that starting from the fourteenth day of embryonic life, the equatorial region of the muscle spindle is PAS (periodic acid–Schiff) positive. The PAS-stained material is resistant to digestion with diastase, being at the same time positive with Sudan black B. The external capsule is sudanophilic, and polar areas of the muscle spindle contain glycogen from the sixteenth day of development. The Barnett and Seligman method for proteins showed more intensely in the intrafusal fiber as compared with extrafusal ones at all stages of development. The succinate dehydrogenase and cytochrome oxidase activity appeared at about the twelfth day in the central parts of the myotube and persisted up to the tenth postnatal day of life. The phosphorylase reaction is first observed on the sixteenth day of development only in the large-diameter fibers (probably nuclear bag). Different types of enzymic reactions in the intrafusal muscle fibers within the muscle spindles have been described by Wirsen and Larsson (1964). On the seventeenth and eighteenth day of embryonic life, these authors observed a moderately positive reaction for phosphorylases in the primary fibers and a weak reaction in the secondary fibers.

The cholinesterase activity as shown by Mummethaler and Engel (1961) appeared in the intrafusal fibers of the muscle spindle on the eleventh day of the chicken embryo development. The enzyme activity in the equator is first observed on the twelfth day of development, and the earliest observations of the polar motor end-plates on the intrafusal fibers comes from 18-days-old embryo. Mavrinskaya (1967) described AChE activity in the periequatorial and polar regions of the embryonal muscle spindle from 20-week-old human fetuses. The enzyme activity was located in clusters as well as small or large platelike structures. Anda and Rebollo (1968) showed cholinesterase activity in the motor end-plates on the fourteenth day of development. These studies also point out that the muscle spindles are functionally active during the development as was earlier believed by Cuajunco (1940).

III. Morphology of the Muscle Spindle

A. Species Differences—Invertebrates and Vertebrates

A review of the muscle spindles in different animals clearly shows that there is a simpler structure in the lower animals compared to the mammals. In invertebrates, the muscle receptors have been studied in crustaceans: Alexandrowicz (1951) in lobster, Kuffler (1954) in lobster and crayfish, Florey and Florey (1955) in crayfish, Wiersma *et al.* (1953) in crayfish, Eyzaguirre and Kuffler (1961) in lobster and crayfish, Dudel and Kuffler (1961) and Dudel (1963, 1965a,b,c) also in the lobster and crayfish. The ultrastructure of the stretch receptor of the crayfish muscle has been elaborately described by Kosaka (1969).

In vertebrates, the picture is more complicated and the muscle receptors differ greatly from species to species and from muscle to muscle. Beginning in 1899, Poloumordwinoff described "pencil shaped endings" in the fish, and Allen (1917) described the muscle spindles in the cordis caudalis muscle of the hagfish, consisting of the capsule and three intrafusal muscle fibers having the spiral of annular nerve endings. Fish muscle receptors have also been described by Barets (1961) who reported a "myosepted pencil" shaped ending in the dogfish, and by Sabussow *et al.* (1964). In the tailed amphibians, Mather and Hines (1934) located three kinds of sensory endings in the muscle of *Triturus forosus*. In the tailless amphibians, e.g., frogs and toads, the muscle spindles show typical motor end brushes and small grape endings as well as encapsulated sensory regions (E. G. Gray, 1957). Robertson (1957a, 1960), J. A. B. Gray (1959), and Uehara and Hama (1965) gave an excellent description of the ultrastructure of frog muscle spindles.

The reptilian muscle spindles consist of a single intrafusal fiber that is usually innervated by one sensory nerve fiber and a few motor fibers (Hines, 1932; Szepsenwol, 1960; Hunt and Wylie, 1970). Cooper (1960) distinguished two types of reptilian muscle spindles, one with a short capsule and the other with a large sensory ending in the shape of a spiral surrounding the fiber beneath the sarcolemma, where the accumulation of nuclei appears. The motor end-plates are distributed on both sides of the intrafusal fiber outside of the capsule. The second type of spindle is characterized by a large capsule with much wider sensory region and with a different arrangement of the nuclei. The intrafusal fiber in this kind of muscle spindle has a single row of nuclei. Most of the motor endings are in the shape of end-plates.

The fine structure of the snake spindle from the costocutaneous muscle has been recently described by Fukami and Hunt (1970). They distinguished two types of muscle spindles which they suggest correspond to the tonic and phasic physiological types. The tonic type of muscle spindle is more slender in the sensory region and myofilaments run through this zone. The phasic muscle spindle shows a more marked enlargement, but the intrafusal fibers in the sensory region are either absent or very minute. Tortoise muscles have a similar spindle, but it is multifascicular (Cooper, 1960).

Birds are said to have well developed encapsulated muscle spindles, having sensory as well as motor endings. The primary and secondary endings in bird muscle spindles were studied by Cooper (1960). The muscle spindles in the gastrocnemius muscle of the adult chicken were studied by Anda and Rebollo (1967), who have found single as well as complex muscle spindles having three types of intrafusal fibers with sensory and motor endings.

B. Mammalian Muscle Spindles

Voss (1937) and Cooper and Daniel (1949) counted the number of spindles in the lumbrical muscles of man and monkey. A number of detailed studies were conducted later and the results, especially of Voss (1956, 1959), Freimann (1954), and Schulze (1955) led to the conclusion that different human muscles have different amounts of muscle spindles (Table I). In general, the muscle spindles are located all over the belly of the muscle, being concentrated near the nerve ending (Cooper, 1960). It becomes apparent, however, that the muscles from the various regions of the body can be classified as spindle-poor or spindle-rich. The muscles of the shoulder and thighs belong to the spindle-poor group, the number of muscles spindles increases in the distal muscles of the arm and leg, and the highest number of muscle spindles in man appears in the muscles of the upper part of the vertebral column and the head.

TABLE I

Total Number of Muscle Spindles in Certain Human Muscles[a]

Muscle	Absolute number of spindles	Weight of muscle (gm)	Relative number of spindles per gram	Motor nerve supply	Investigator
Limb muscles					
1st lumbrical, hand	51	3.1	16.5		Voss (1937)
2nd lumbrical, hand	36	1.8	19.7		Voss (1937)
3rd lumbrical, hand	20	1.6	12.2		Voss (1937)
4th lumbrical, hand	23	1.3	17.5		Voss (1937)
Abductor pollicus brevis	80	2.7	29.3		Schulze (1955)
Opponens pollicis	44	2.5	17.3		Schulze (1955)
Flexor pollicis brevis					
superficial head	27	2.3	11.5		Schulze (1955)
deep head	27	1.5	18.6		Schulze (1955)
Adductor pollicis					
oblique head	63 }75	6.5	11.6		Schulze (1955)
transverse head	12 }				
Pronator quadratus	120	12.1	10		Voss (1956)
Latissimus dorsi	368	246	1.4		Voss (1956)
Cranial muscles					
Geniohyoid	15	3.3	5.4	XII cranial	Voss (1956)
Trapezius	437	201	2.2	XI cranial	Voss (1956)
Stylohyoid	6	1.2	5.0	VII cranial	Voss (1956)
Digastric posterior belly	0	5.4	0	VII cranial	Voss (1956)
Digastric anterior belly	6	5.4	approx. 2	V cranial	Voss (1956)
Mylohyoid	0	5.4	0	V cranial	Voss (1956)
Masseter					
deep part	42	3.8	11.2	V cranial	Freimann (1954)
superficial part	118	16.0	7.4	V cranial	Freimann (1954)
Temporal	217	14.7	14.7	V cranial	Freimann (1954)
Medial Pterygoid	155	7.6	20.3	V cranial	Freimann (1954)
Lateral Pterygoid	0	4.4	0	V cranial	Freimann (1954)

[a] Reproduced from Cooper (1960).

Cooper and Daniel (1963) showed that in the deep neck muscles in man, a large number of muscle spindles appear. Barker (1962) and his group (Chin *et al.*, 1962) studied the cat hindlimb muscles, basing the count of muscle spindles on the single muscle of a particular kind of animal, and expressed the results as a proportion of muscle spindle content to the weight of volume of the particular muscle. Another method used was the estimation of the proportion of the total length of the muscle spindles to the weight of the muscle. Barker and Chin (1960) and Barker *et al.* (1962) based their observations on twenty pairs of each cat's rectus femoris muscle, tibialis anticus and posticus, semitendineous, soleus, the medical and lateral heads of the gastrocnemius muscle, flexor digitorum longus, and the fifth interosseous muscle of the forelimb, in order to determine whether there was any correlation between the number of muscle spindles in the left and right sides of a particular muscle. They concluded that the muscle spindle-capsule content in a particular muscle varies in different animals and that the content of the capsules of the pair of individual muscles is closely equivalent. The distribution of the muscle spindles bears a constant relation to the entry and distribution of the nerve supply, being located in the neighborhood of the major intramuscular nerve trunks. Steg (1962), Zelena (1963), and Mellstrom and Skoglund (1965) counted the muscle spindles in rat muscles and found an average of fifty-six spindles per muscle. Almost similar results were obtained by Hunt (1959) and Swett and Eldred (1960), their count being fifty-eight and fifty-three, respectively.

1. Muscle Spindles in Extremities and Trunk

Muscle spindles are found in most human muscles (Cooper and Daniel, 1963) as well as in most muscles of other mammals studied. In general, the vertebrate muscle spindle consists of various numbers of intrafusal muscle fibers of small diameter enclosed in a fusiform capsule. The average number of intrafusal fibers in the rabbit is said to be four, in the cat six, and ten in man and other primates, with great variability in the latter group. The capsule covering the muscle spindle consists of several concentric sheets of cellular and connective tissue elements and is thought to contain circularly disposed fibrils. It is continuous with the intra- and extrafusal endomysium as well as with the perineurium of the nerve fibers passing into the muscle spindle. A number of blood vessels appear between the sheets of the capsule; but there is no network of capillaries such as is typical of the extrafusal fibers.

The blood vessels found in the muscle spindles may go through the capsule into the periatid space and have been observed to travel close to the intrafusal fibers (Cooper 1960).

The width of the muscle spindle in the equatorial region, where the capsule stands away from the intrafusal fibers, depends on the amount of stretch in the muscle at the time of fixation. In the human muscle spindles coming from the lumbrical muscles of the hand, the width of these organs rarely exceeds 200 μ (Cooper and Daniel, 1963). The diameter of the capsule decreases at the polar region, so that the lamellae are closely wrapped around the intrafusal fibers. The number of layers in the capsule decreases so that the polar region of the muscle spindle is covered by a single layer of connective tissue. Cooper and Daniel (1963) pointed out that elastic tissue is particularly marked at the poles of the muscle spindle. The elastic tissue, found over the bags, near the secondary endings and the poles counteracts the stretch and brings the muscle spindle back to its resting position (Cooper and Daniel, 1963). The elastic tissue was also observed by Amersbach (1911) and described at the ultrastructural level by Merrillees (1960). A small number of intrafusal fibers also continues for a considerable distance beyond the capsule. In the equatorial region they may divide, giving off branches, and join again inside the capsule. The lateral extension of the capsule in the lumbrical muscles as well as in the deep neck muscles in man have been described by Cooper and Daniel (1963). According to them, the band of connective tissue coming from the equatorial region of the muscle spindle surrounds various numbers of extrafusal fibers for a distance of 750 μ or more. Sometimes the capsule as well as the longer intrafusal fibers are bound to the extrafusal fibers ending in their perimysium.

The appearance of so-called tandem spindles, described by Barker (1962), is in close relation to the capsules covering these muscle spindles. The capsule itself is attached to the perimysium, or to fascial septae, or it may be that threads of connective tissue pass on in the intramuscular spaces to join a slip of the tendon. The capsule usually is not unbroken; it surrounds the intrafusal fibers and entering blood vessels. It is also continuous with a sheet of Henle of the incoming nerve fibers and with the perineural epithelium (Fig. 1) (Shantha *et al.*, 1968). It may be stressed here that the perineural epithelium, which is the continuation of the leptomeninges of the central nervous system, covers the nerve fasciculi and extends along the terminal nerve fibers to their terminations. At the terminations of the nerve fibers, the perineural epithelium forms the capsule of the end organs such as muscle spindles (Fig. 1). The structure of the perineural epithelium has been studied

Perineural epithelium
reflected showing
two layers

Nerve fibers with
Schwann sheaths
and endoneurium

Epithelial capsule of muscle
spindle, showing two layers
[continuous with perineural
epithelium]

Perineural
connective
tissue

Intrafusal
muscle
fibers

Connective tissue
covering muscle
spindle

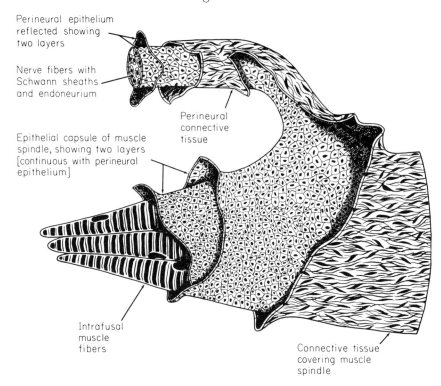

Fig. 1. Diagrammatic representation of a part of a neuromuscular spindle showing in detail the relationship between the coverings of the extrafusal nerve, which supplies the muscle spindle as well as the covering of the muscle spindle itself. The perineural epithelium covering the nerve of supply continues to cover the muscle spindle as its capsule. This covering is reflected and shown to be made up of multiple layers of cells lying one on top of the other. Immediately surrounding the perineural epithelium is the perineural connective tissue covering which extends onto the capsule of the muscle spindle. This layer of connective tissue is cut and reflected back on the nerve as well as on the muscle spindle to show the continuity of the perineural epithelium of the nerve fasciculus as a capsule of the muscle spindle as seen in isolation studies. (After Shantha *et al.,* 1968. Karger Publ.)

by Shanthaveerappa and Bourne (1962) and Shantha *et al.* (1968) and consists of a few layers of squamous epithelial cells having serrated borders. In rat, guinea pig, and cat sciatic nerve the perineural epithelium consists of five layers of these cells, whereas in frog only two layers are discernible. The nuclei of these cells are pale, and a sex chromatin has been observed in the adult female mammals (Shanthaveerappa and Bourne, 1962). A detailed cytochemical analysis of the muscle spindles reveals that the perineural epithelium shows the activity of several dephosphorylating enzymes (Figs. 59–64) (Shantha *et al.*, 1968). It is of an ectodermal origin and is formed by elements that surround the nerve fibers and follow them wherever they go. It has also been shown to cover the sensory as well as the motor endings (Shantha and Bourne, 1968).

2. Muscle Spindles in the Head Region

The greatest concentration of muscle spindles in most of the muscles of the body is at the region of the belly of the muscle. The muscle spindles are largest in number at the place of nerve entry. The trunk muscles are richly endowed with sensory receptors, varying in number according to the particular muscle. This is not the case in the muscles of the head region. Cooper has reviewed in detail the muscle spindles and other receptors in the mammalian cranial muscles, and a brief summary is given here. The jaw muscles resemble the limb muscles in the structure of the extrafusal muscle fibers and the muscle spindles. The jaw muscles are supplied by the fifth cranial nerve, and both the motor and sensory pathways are located in its motor root, and the cells of the origin of the sensory nerves from the muscle spindles of the jaw muscles are located in the mesencephalic nucleus of this cranial nerve (Cooper, 1960). Karlsen (1969) showed the muscle spindle in the lateral pterygoid muscle of *Macaca foscicularis*, and identified thirteen muscle spindles in the middle region of the muscle. Eight occurred singly, and five were grouped in a larger muscle spindle unit. Muscle spindles have not been located in the pterygoid muscles of a larger number of animals, such as goat (Cooper, 1960), rat (Franks, 1964; Karlsen, 1965), cat (Cooper, 1960; Franks, 1964; Karlsen, 1969), and guinea pig and rabbit (Franks, 1969). However, Honee (1966) observed one to fifteen muscle spindles in the pterygoid muscles of man and along with the investigations of Karlsen (1969) in monkeys, it may be assumed that muscle spindles are present in this muscle in the primates.

The extrinsic muscles of the eye are another group of cranial muscles primarily meant to give a stable field of vision. The muscle spindles in the extrinsic eye muscles have been located only in man, higher primates, and certain ungulates (Broccoli, 1970). The eyes in human are moved constantly both reflexively and voluntarily; about fifty muscle spindles may be observed in the inferior rectus muscle (Cooper and Daniel, 1963). Muscle spindles of the human eye muscles are short and are best observed in the cross sections. The impulses from these muscle spindles travel to the brain stem in the nerves that join the opthalmic branch of the fifth cranial nerve (Cooper, 1960).

The facial muscles are supplied by the seventh cranial nerve and are believed by a number of workers to contain no sensory receptors. However, Kadanoff (1956) identified a few muscle spindles and other endings in the human mimetic muscles that did not show any characteristics of motor endings that could be sensory in nature. The lack of sensory supply has also been a matter of doubt and speculation in the intrinsic muscle of the larynx, and a number of workers have failed to locate muscle spindles in these muscles. Keene (1961) observed a few muscle spindles in the laryngeal muscles of man.

The intrinsic muscles of the tongue are supplied by the twelfth cranial nerve, and it is easy to locate the muscle spindles in these muscles in a number of animals including man. A detailed study of the muscle spindles in the intrinsic and extrinsic muscles of the rhesus monkey's tongue has been made by Bowman (1968), who showed that muscle spindles are distributed in all the intrinsic as well as some extrinsic muscles as previously reported by Cooper (1953) and L. B. Walker and Rajagopal (1959). An analogy of the tongue muscle spindles with those of the limbs may explain the rate and extent of tongue movement and shows that the touch endings of the tongue are involved in a primary capacity in signalling the direction of lingual movement (Bowman, 1968). Bowman believed that if "the spindle in the human tongue function in the same manner as do those in the monkey's tongue, the afferent pattern derived from these end organs in response to movement may be significantly discriminative to play a leading role in the monitoring of the motor patterns characterizing speech."

C. Muscle Spindle—Extrafusal Muscle Relationship

The studies of Yellin (1969b) carried out on the medial gastrocnemius, plantaris, and tibialis anterior muscles of the rat have shown that prominent segregation of the extrafusal fiber types appeared. Axial fascicles contained mainly type B and C fibers (classification of Stein and Padykula, 1962), whereas the superficial fascicles were made up of type A fibers and with only a single C fiber. Muscle spindles in the muscle studied by Yellin (1969b) were confined to the axial region, composed with B and C fibers, and were never observed among the type A fibers. The extensor digitorum longus muscle exhibited less complete segregation of fiber types and the distribution of muscle spindles was less apparent. Yellin (1969b) did not observe any particular distribution of muscle spindle with respect to extrafusal fibers for soleus and lumbrical muscles. He concluded that in rat muscles exhibiting prominent segregation of extrafusal muscle types, muscle spindles are less frequent in the region with the predominance of type A fibers.

D. Tandem Muscle Spindles

During the Symposium on Muscle Receptors, held at the University of Hong Kong in 1962, the problem of the nomenclature of muscle spindles was discussed by Barker, Boyd, and others. Barker suggested the term "single spindle" to indicate one encapsulated sensory region with "just the ordinary polar region." Boyd proposed the term "spindle unit" for what is termed "tandem spindles." The term "tandem spindle" was proposed for the arrangement of two or more sensory regions along the bundle of intrafusal muscle fibers. Such arrangement of the sensory regions was first observed by Vihvelin (1932) in amphibian muscle spindles and Cooper and Daniel (1960) described the tandem spindles in man. E. G. Gray (1957) defined a compound muscle spindle in the frog toe extensor muscle, and a histological study of the tandem spindles in mammals was carried out by Boyd (1960), Barker and Ip (1961), Barker and Cope (1962), and Swett and Eldred (1960) in the soleus and medial gastrocnemius muscles from young cats and the cat soleus and interosseus muscles. According to Barker (1962), tandem spindles in the cat muscle typically consist of a large capsule with ten to twelve intrafusal fibers bearing one primary and two or three secondary endings, and one or more capsules with three or four muscle fibers bearing a primary ending only. In general, one or two intrafusal fibers go right through the tandem linkage, in most cases. The quantitative studies of Barker (1962) showed that the most common tandem is the double kind in which two capsules are linked along the long axis of the intrafusal fibers. Out of the total number of tandem spindles in the rectus femoris muscle of the cat, 88% were double, 9% were triple, and 3% quintuple.

The role of the tandem spindles in the function of the muscles is not quite clear, but it has been suggested that a fraction of the population of tandem spindles in mammalian muscles may serve to collect more generalized information on the length of the muscle fibers (Swett and Eldred, 1960). Barker and Cope (1962) also emphasized the functional significance of the tandem spindles. The neurophysiological studies of Ito (1969) showed that the static and dynamic indices of responses of single fibers innervating one capsule in tandem or compound spindles were apparently higher than those from single fibers innervating a capsule in the single muscle spindle. Mathews (1960) and Toyama (1966) suggested that the viscoelastic properties of the underlying tissues embedded with the sensory terminals may be an important factor for the determination of the functional properties of their spindle receptors. Large numbers of intrafusal fibers may result in an enhancement of viscosity in the polar regions and in relative decrement of viscosity in the sensory regions. The number of contractile elements, which are believed to cause the viscosity, is decreased in the sensory regions (Karlsson *et al.*, 1966). Ito (1969) suggested that if the intrafusal fibers in a bundle are combined with each other, the sensory region may be more extended during the stretch of the muscle when a large number of intrafusal fibers are present, contrary to the situation when only a few intrafusal fibers are found. Ito concluded that the tandem muscle spindles are provided to respond sensitively to minute changes of the muscle length.

E. Type and Number of Intrafusal Fibers

1. NUCLEAR BAG AND NUCLEAR CHAIN FIBERS

As far as the intrafusal fibers of the muscle spindles are concerned, Cooper and Daniel (1956) in man and Boyd (1958a,b, 1960) in the cat muscles were the first to describe the two morphologically different intrafusal fibers. The longer fibers (nuclear bag fibers) have a larger diameter in most regions of the spindle and pass out of the capsule at the poles to lie among the extrafusal muscle fibers, whereas the shorter ones with a smaller diameter, referred to as nuclear chain fibers, rarely extend beyond the poles of the spindle. A study of transverse serial sections of the spindle reveals that these fibers differ significantly, in width as well as in length. The nuclear bag intrafusal fibers in the human muscles are more than 15 μ wide. They pass out of the capsule at the polar region, start tapering, and finally attach to the perimysium of the extrafusal fibers. Sometimes the nuclear bag fibers are attached to the muscle tendons or to the capsule of the tendon organ. Cooper and Daniel (1963) described the reencapsulation of some nuclear bag fibers, which previously left the capsule so that two or more muscle spindles in tandem appeared. They also calculated the length of the nuclear bag fibers beyond the capsule in the lumbrical muscle of man and found that they continue for about 0.8 mm, their total length being approxi-

TABLE II

NUMBER AND DIMENSIONS OF INTRAFUSAL FIBERS IN MUSCLE SPINDLE FROM
FIRST LUMBRICAL HUMAN MUSCLE[a]

Spindle	Length of capsule (mm)	Extracapsular length of nuclear bag fibers		Total length of spindle (mm)	No. of i.f. fibers	No. of nuclear bag fibers	No. of nuclear chain fibers
		Proximal (mm)	Distal (mm)				
1	2.7	0.7	0.5	3.9	8	3	5
2	3.1	0.4	1.0	4.5	7	3	4
3	3.3	0.6	0.7	4.6	10	4	6
4	3.4	0.4	1.5	5.3	8	3	5
5	3.6	1.2	0.5	5.3	9	3	6
6	3.7	0.7	0.8	5.2	11	4	7
7	4.2	0.7	1.5	6.4	11	4	7
8	3.8	1.0	—	11.3 in tandem	8	3	5
		1.0[b]					
9	4.9	—	0.6		13	4	9

[a] Reproduced from Cooper and Daniel (1963). Data from a group of spindles in the belly of a first lumbrical muscle.

[b] 1.0 mm = intercapsular distance.

mately 5 mm (Table II). The nuclear chain fibers are less than 15 μ in width, and as soon as they reach the polar region, they get smaller in diameter and finally end on the sides of nuclear bag fibers or at the internal wall of the capsule inside or just a little outside the capsule. Porayko and Smith (1968), however, observed in the rat lumbrical muscles, that both nuclear bag and nuclear chain fibers were extended beyond both poles of the capsule. Cooper and Daniel (1963) studied the lumbrical muscles of man and observed certain muscle spindles which contained only nuclear bag fibers. They did not observe any muscle spindles that contained only the nuclear chain type fibers. In contrast, in the rabbit, most of the intrafusal fibers are the nuclear bag type (Barker and Hunt, 1964). There is no evidence of splitting the nuclear bag fibers into nuclear chain type in man (Cooper and Daniel, 1963), but such a possibility cannot be excluded in pathological muscles. Gutman and Zelena (1962) described the increased number of intrafusal fibers in the muscle spindles of experimentally denervated muscles of rat up to 50% more than the control value, and interpreted this fact as being due to a division of intrafusal fibers in the equatorial zone.

TABLE III

NUMBER OF NUCLEAR BAG AND NUCLEAR CHAIN MUSCLE FIBERS PER
SPINDLE IN THREE CAT HINDLIMB MUSCLES[a]

Muscle	\multicolumn{10}{c}{Fibers per Spindle}

Muscle	1	2	3	4	5	6	7	8	9	Mean
Nuclear bag fibers										
Ext. dig. brevis	6	15	4							1.9
Soleus	3	20	2							2.0
Med. gastroc.	4	17	4							2.0
Nuclear chain fibers										
Ext. dig. brevis		2	2	6	10	2	1	2		4.8
Soleus			2	6	14	3				4.7
Med. gastroc.				4	6	4	6	4	1	6.1

[a] Reproduced from Eldred *et al.* (1962). Data based on samples of 25 nontandem spindle units from each muscle. Average number of intrafusal fibers in this sample of spindles from EDB and soleus was 6.7, and from gastrocnemius 8.1. Means for all spindles in these muscles were: EDB 7.2, soleus 6.5, and gastrocnemius 7.4.

The muscle spindles show a varying number of intrafusal fibers; from four in the case of the rabbit, to six in the cat (Table III), and an average of ten in the primates and man (Cooper, 1960; Cooper and Daniel, 1963). Porayko and Smith (1968) have found that in the rat spindle coming from plantar lumbrical muscles, there are usually two nuclear bag and two nuclear chain fibers. The spindles in the human muscles in transverse serial sections taken at 4 mm intervals were studied by Cooper and Daniel (1963) in order to estimate the average number of intrafusal fibers in these muscle spindles. According to their observations, the number of intrafusal fibers varied from two to fourteen in each muscle spindle; among them, one to four nuclear bag fibers were observed in a single spindle (Fig. 2). The number of nuclear chain fibers is comparatively larger and frequently up to ten fibers were shown in a single spindle. The number of intrafusal fibers on both poles in the spindle is sometimes different and invariably this difference is in the number of nuclear chain fibers.

Nuclear bag fibers are long, wide, and rich in myofibrils, and as Boyd (1962) observed, do not atrophy easily. At the equatorial region, the thin layer of myofibrils surrounds the large number of big round nuclei. The nuclei show a small nucleolus, very light chromatin network, and are surrounded by a prominent nuclear membrane. In the large bags of the human muscle spindles, about one hundred nuclei can be observed. The bag is tapered at both ends and one or two nuclei can be observed in this area. The shape and the size of the bag, however, depends on the stretch of the muscle during the fixation procedure (Cooper and Daniel, 1963). Bags observed in well stretched muscles are long, whereas in the shrunken muscles, very short swollen bags are observed. The dimensions of these bags, therefore, vary from 100 μ to more than 250 μ in length and from 20 to 50 μ in width in the widest region (Cooper and Daniel, 1963). The nuclear bag fibers may lie close together in the bag region, but in the spindle they may lay a little distance apart due to the extension of the nuclear bag. If such distance is more than 0.5 mm and two groups of nuclear bags are observed, the tandem spindle shares a common capsule. The myotube region is much shorter in the case of the human muscle spindle as compared to that of rabbit (Barker, 1948). The strands of elastic tissue are distributed on the surface of the muscle at the ends of the nuclear bag region (Cooper and Daniel, 1963).

Nuclear chain fibers are shorter and smaller in diameter, having a smaller amount of myofibrils and showing a more rapid rate of atrophy. Nuclear chain fibers sometimes branch into two parts (Barker, 1962). The myofibrils in the equator region of the nuclear chain fibers are reduced to a thin layer and beneath the striated fibrils, a single row of the various numbers of nuclei can be observed. The chain is usually about 100 μ long and about 10 μ wide, having ten to twenty nuclei with the same morphological characteristics as observed in the case of nuclear bag intrafusal fibers. Secondary nerve endings of the nuclear chain fibers differ from the nuclear chain fibers due to accumulation of myofibrils under the secondary ending in the distinct narrow tube, containing elongated nuclei. Cooper and Daniel (1963) observed an accumulation of argentophilic substance in the sarcolemma, between the nuclei and related it to the osmophilic substance described earlier by Gruner (1961) in human spindles.

2. Myofibril Density in Intrafusal Fibers

The myofibril density varies greatly in the nuclear bag and nuclear chain fibers and appears to change when the intrafusal fibers are traced in transverse sections from end to end. According to Barker (1962), through most of their length, nearly all fibers have a fibrillar pattern (Barker, 1962), whereas the areal pattern occurs sporadically in the polar regions and is typical for the juxtaglomerular region of the muscle spindle. Each intrafusal fiber is usually areal in some particular length and in large fibers, the areal component is smaller, whereas in small

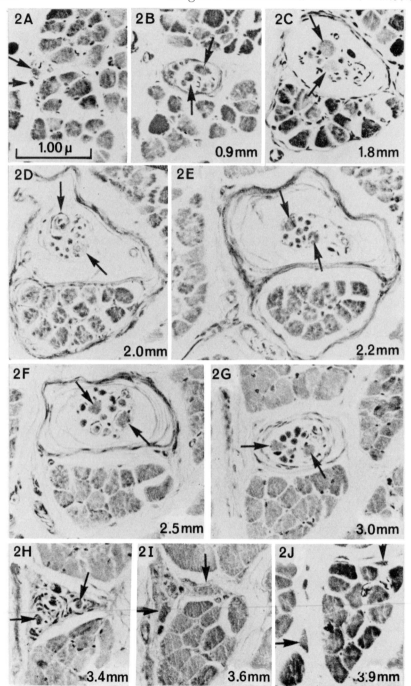

Fig. 2.

fibers the ratio of areal length to fibrillar length is higher (Barker, 1962). Barker suggested that these two patterns of myofibril density are the expression of waves of contraction passing along the intrafusal muscle fiber. More details have been discussed in a later section on the ultra-structure of muscle spindles.

3. OTHER FIBERS

Cooper and Daniel (1963) observed a single muscle fiber that is smaller than the extrafusal ones and goes through the entire length of the muscle spindle. This fiber has neither a nuclear bag nor a nuclear chain at the equatorial region, nor does it appear to be the same as the intermediate intrafusal fiber described by Barker (1962), because of the difference in the morphological structure. The nature of this muscle fiber is not clear at this time. The intermediate fibers are of the medium size as compared to other intrafusal fibers and they possess a nuclear bag in the myotube region. They differ from the nuclear bag fiber because their bags have a smaller number of nuclei and the diam-

---

Fig. 2. Transverse 15 μ sections taken through different regions of a spindle in a lumbrical muscle. Holmes' stain. The distance from A is shown on each photomicrograph. The spindle is associated with fourteen extrafusal muscle fibers; and extension of connective tissue from the spindle capsule encloses these fibers for 1.14 mm and can be seen in C, D, and E. There are two nuclear bag fibers, marked with arrows, and nine nuclear chain fibers. A small blood vessel and nerve fibers are also seen in the spindle in many of the sections. In A and nuclear-bag fibers are seen in the distal extracapsular region together with a small nerve fiber. The 2 muscle fibers are about to end against the extrafusal fibers in the lower central part of the figure; these are the same fibers that are seen enclosed by the extension of the capsule later. In B the capsule has just formed, and the nuclear chain fibers are starting. In C the capsule has greatly increased in diameter, and its extension can be traced round the 14 extrafusal fibers. The extension continues in D and E past the equatorial region of the spindle, and in D some fine nerve fibers can be seen inside the capsule extension about to supply the motor end plates of the extrafusal fibers. In both D and E parts of the nuclear bags and the primary endings can be seen; the nuclear chain fibers remain very small in diameter. The nerve ring in F may be part of the secondary ending on a nuclear chain fiber. In F the capsule is closing in again, and it is ending in H and I. In H, note the many nuclei and nerve fibers which probable supply the motor end plates on the nuclear bag fibers. In J the capsule has ended. The 2 nuclear bag fibers are widely separated; 3 of the tapered ends of nuclear-chain fibers are seen just about to finish. The group of 14 extrafusal fibers is clearly seen; in succeeding sections the nuclear bag fibers can be traced for another 1 mm, and they end among the group of extrafusal fibers. The whole spindle is about 5 mm long. The scale for A to J is given in Fig. 2A. (Reproduced from Cooper and Daniel, 1963. St. Martin's Press, Inc., New York).

eter of myotube region is smaller. They sometimes undergo bifurcation
and their diameter in the polar region is not constant. Barker and
Gidumal (1961) observed the bifurcation of intermediate fibers in the
polar region. Anda and Rebollo (1967) described the intermediate fibers
in the muscle spindles of the chicken. The muscle spindles were stained
with ferric hematoxylin and intrafusal fibers of intermediate diameter
with thin myofibrils aggregated in the Conhaim's fields were observed.
These fibers generally have a relatively smaller amount of sarcoplasm
as compared to the other intrafusal fibers.

F. Innervation of Muscle Spindles

1. Sensory Innervation

The sensory endings of the muscle spindles derive from two groups
of afferent nerve fibers: (1) larger fibers (12–20 μ in diameter) from
group Ia afferent fibers of the muscle nerves (Rexed and Therman,
1948), and (2) a group of smaller fibers (4–12 μ in diameter) belonging
to group II afferent fibers. The fibers of group I terminate as "primary"
or annulospiral endings, surrounding the center of the nuclear region,
whereas the fibers of group II terminate as secondary or flower-spray
endings in the juxtaequatorial region (Ruffini, 1898). Barker (1967)
divided the nerve fibers supplying the sensory innervation of the muscle
spindles into four classes—large, medium, small, and unmyelinated c
fibers. Each muscle spindle has a primary sensory ending, consisting
of a number of spiral terminations (Boyd, 1972; Barker, 1967) which
surround the nuclear region of the nuclear bag as well as the nuclear-
chain intrafusal fibers (Fig. 3). Boyd (1962) and Barker (1967) observed that the large spirals surrounding the nuclear-bag fibers and small
spirals are located on the nuclear-chain fibers. Primary sensory terminals
and their group Ia afferent fibers provide an afferent pathway common
to both nuclear bag and nuclear chain intrafusal systems (Barker, 1967).
The fibers from the Ib group provide the innervation of the Golgi tendon
organ. Secondary sensory endings are found predominantly on the nuclear-chain fibers in the juxtaequatorial and/or myotube regions (Figs.
4–6).

In the mammals, the different localization of these endings reveals
that the muscle spindles of the rabbit are located mainly in the region
of myotube of the nuclear bag intrafusal fibers (Barker and Hunt, 1964).
Apart from small accessory arborization on the nuclear bag fibers, the
secondary sensory endings rarely assume the spray form. Muscle spindles

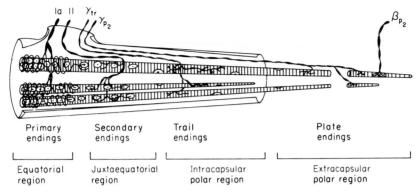

Primary endings   Secondary endings   Trail endings   Plate endings

Equatorial region   Juxtaequatorial region   Intracapsular polar region   Extracapsular polar region

Fig. 3. Schematic diagram of the distribution of sensory and motor fibers in a cat hindlimb muscle spindle. One nuclear bag (*upper*) and two nuclear chain (*lower*) fibers have been drawn. At the equatorial region of the spindle there are primary sensory endings supplied by group Ia nerve fiber (Ia). At the juxtaequatorial region secondary sensory endings are supplied by group II nerve fiber (II). Trail endings originating from fusimotor fiber innervate both nuclear bag and nuclear chain particularly, but not exclusively, at the intracapsular polar region of the spindle. Plate endings originating from both skeletofusimotor (p_1) and fusimotor (p_2) fibers innervate intrafusal muscle fibers at extracapsular polar region on the spindle. Type p_1 plate endings have been described so far only on nuclear-bag fibers, while p_2 plate endings have been found on both nuclear bag and nuclear chain fibers. Reproduced from Corvaja *et al.* (1969), University of Pisa Press, Pisa.

of the cat may contain up to four (Barker and Ip, 1961) or five (Boyd, 1959) secondary endings, and the region they occupy on both sides of the primary endings has been designated as S_1 to S (Boyd, 1962). In the case of cat's hindlimb muscle spindles, the secondary innervation consists usually of one ending (Barker and Ip, 1961). These endings, which are distributed in the nearest region to the primaries, consist mainly of rings and spirals and are supplied by thicker axons than those located farther away, having a more irregular form. According to Barker and Ip (1961), the annulospiral secondaries are about twice as common as flower-spray secondary endings in the cat. In muscle spindles with a secondary innervation, it is believed that the primary endings are usually supplied by a large-diameter Ia nerve fiber (Adal and Barker, 1965b). The quantitative studies of Barker *et al.* (1962) have also shown that about 29.2% of the total number of sensory fibers in the cat soleus muscle remained unallocated, comprising over one-third of Group II and three-quarters of Group III. Similar results were obtained by Zelena and Hník (1963) on the rat soleus muscle. Porayko and Smith (1968) studied the sensory endings in the muscle spindles from the lumbrical

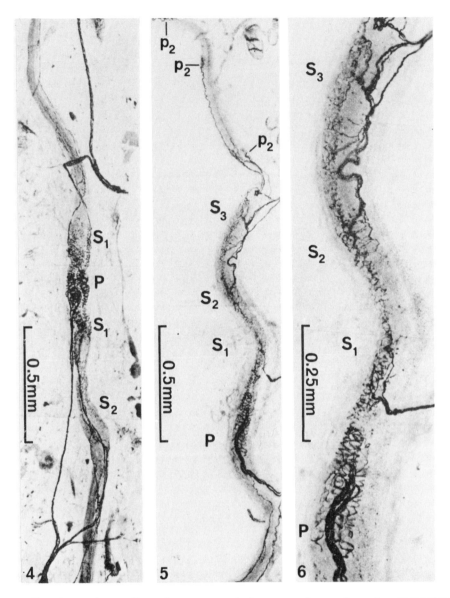

Fig. 4. Muscle spindle with primary and three secondary endings from normal flexor digitorum longus. From Barker (1967).

Fig. 5. Muscle spindle with primary and three secondary endings from normal tenuissimus. Note clarity of motor innervation in this silver preparation as compared with that in the preceding gold chloride preparation. From Barker (1967).

Fig. 6. Enlargement of the sensory innervation of the spindle shown in Fig. 4. The secondary endings are located mainly on the nuclear chain muscle fibers

muscles of rat in silver-stained specimens of normal and deefferented muscles. They found that 50% of the studied muscle spindles contained only one primary ending. This occupied the central 130–140 μ of the intrafusal muscle bundle and usually appeared in the form of tightly wound spirals on the nuclear bag as well as the nuclear chain fibers. The remaining 50% of the preparations contained only one primary sensory ending and one or two secondaries in the shape of fine sprays of nerve terminals, usually on the juxtaequatorial regions of nuclear chain fibers. Gladden (1969), however, obtained different results on the intertransverse caudal muscle of the rat. He found that only 10% of muscle spindles have a single primary, while 10% have one secondary, 60% have two secondaries, and 20% have three secondary endings.

2. MOTOR INNERVATION

Motor innervation of the muscle spindles has been the subject of controversy over the past decade. The types and distribution of motor endings was a subject of prominent discussion at the symposium in Hong Kong in 1962 to be continued later at the Nobel Symposium in Stockholm in 1966. Boyd (1962) believed that end-plates terminate only on the nuclear bag intrafusal fibers, whereas Barker maintained that they terminate on both nuclear bag and on nuclear chain fibers. At the symposia, disagreement was also expressed on the distribution of diffuse type of motor endings on the different end-plates. These diffuse endings were first detected by Coërs and Durand (1956) and Coërs (1962), who used a cytochemical technique for acetylcholinesterase for their demonstrations. They were named trail endings by Barker and Ip (1965). Boyd (1962) suggested that the diffuse endings innervated the nuclear chain fibers but not the nuclear bag ones, whereas Barker (1962) maintained that there is no selective innervation of this kind. Some important investigations between 1945 and 1952 indicated that the intrafusal muscle fibers are innervated mainly by small γ fusimotor nerve fibers (Leksell, 1945; Hunt and Kuffler, 1951; Kuffler and Hunt, 1952). Later, Granit *et al.* (1959a,b), Bessou *et al.* (1963a,b,c), Kidd

---

(coursing through on the left hand side of the spindle) and only to a small extent on the nuclear bag muscle fibers. Figures 4–6 are teased gold chloride preparations of cat muscle receptors; P = primary ending, P_2 = end-plate of the P_2 type, S_1, S_2 and S_3 = secondary endings located in positions increasingly distant from the primary ending. Photographs of plate and trail motor endings in teased, silver preparations (modified de Castro method) of deaffarentated cat muscle spindles; b.b. = blood vessels, e = end-plate, pf = plate ending fusimotor fiber, tf = trail ending fusimotor fiber, t = trail ending. From Barker (1967).

(1964, 1966, 1967), Mathews (1960), Adal and Barker (1965a), Barker
and Ip (1965), Brown *et al.* (1965), Barrios *et al.* (1966, 1967), Carli
et al. (1966), Emonet-Denand and Laporte (1966), and Haase and
Schlegel (1966) showed that some of the extrafusal and intrafusal fibers
do have common innervation. Hagbarth and Wolfhart (1952) and Feinstein *et al.* (1955) estimated the number and diameter of the fibers
in the whole muscle nerve in the cat muscle. Hagbarth and Wolfhart
concluded that about 60% of large nerves were motor type, an observation later confirmed in the muscles of newborn humans by E. Christensen (1959). Boyd and Davey (1962) also estimated around 60% of the
nerves in the large muscles of the cat to be motor fibers. In 1970, Barker
and his colleagues carried out comprehensive studies embracing morphology, histochemistry, and ultrastructure of the motor endings in the
muscle spindles and shed more light on these controversial problems.
They studied a fusimotor innervation of the muscle spindles from the
muscle peroneus longus, peroneus brevis, peronus digiti quinti, soleus,
flexor hallucis longus, flexor digitorum longus, tenuissimus, tibialis
anterior and posterior, and the interosseous and lumbrical muscles of
the hindfoot of the cat. The motor endings are mainly located intracap-
sullarly, on the poles and in the juxtaequatorial region of the intrafusal
fibers. They recognized three types of motor endings: p_1 end-plates,
being the collaterals of the α nerve fibers; p_2 end plates and trail endings,
originating from γ nerve fibers (Barker *et al.*, 1970). The diameter of
the nerve fibers supplying the muscle spindles was found to be different
in different motor endings. Barker *et al.* estimated that in the peroneal
muscles, the diameter of p_1 fibers is 1.2–3.0 μ, and they are similar to
those of the terminal branches of extrafusal motor fibers. The diameter
of the p_2 fibers varied from 1.4 to 3.8 μ. The trail endings are supplied
by myelinated, 1.0–3.8 μ in diameter, and nonmyelinated fibers have
a smaller diameter approximately 1.0 μ. The p_1 nerve fibers enter the
muscle spindle in the polar region, and they terminate usually in one
end plate. According to Barker *et al.*, the interesting feature of the
p_1 fiber is that they derive from large intramuscular nerves located near
the muscle spindle and rather less frequently from the spindle nerve
trunks. The p_1 end plates are located in the polar region of the intrafusal
fibers (Figs. 7, 8). They are distributed on nuclear bag as well as on
the nuclear chain fibers. Barker *et al.* (1970) estimated that on the
sample of one hundred p_1 plates in peroneal and flexor hallucis longus
muscles, seventy-five p_1 end-plates were located on the nuclear bag and
twenty-five on the nuclear chain intrafusal fibers. The p_1 plates resemble
the extrafusal motor end-plates. The picture of these endings is characterized by the different forms, beginning from simple plates having a

few axon terminals to many axons coming from several branches of the supplying α nerve fibers. The axon terminals appear in various shapes, i.e., spindles, knobs, or rings, being distributed on the nucleated sole plate with a Doyer's eminence. No significant differences were observed by Barker between p_1 end-plates on nuclear chain and nuclear bag intrafusal fibers.

The γ innervation terminates in the p_2 plates and in trail endings. The trail endings—first shown in the histochemical preparations for cholinesterase by Coërs and Durand (1956) in rat, cat, and men; Kupfer (1960) in man; Hess (1961a,b) in mouse and rabbit; Coërs (1962) in rat, cat, and man—were later shown by Boyd (1962) in gold chloride preparations as a network of fine axons and small elongated nerve endings. The considerable distance occurring between the terminal node and the final ramification creating the synaptic contact as well as the presence of many branches given off from preterminal nodes are one of the characteristic features of these endings (Fig. 9), as shown by silver preparation of Barker and Ip (1965), Barker (1966a, 1967, 1968b), and Jones (1966). The trail endings are supplied by myelinated and nonmyelinated nerve fibers. As Barker *et al.* (1970) pointed out, the nonmyelinated nerve fibers may become intermittently myelinated, and the primarily myelinated fibers may lose their myelin sheath after entering the muscle spindle. In a sample of eighty-seven peroneal spindle poles, Barker *et al.* (1970) found thirty-two of them innervated by nonmyelinated trial fibers in addition to myelinated axons. They also described the recurrent preterminal trail axons (Fig. 9), which come from the site of trail endings, leave the muscle spindle, the finally reach the trail ending area of another spindle, occurring as a nonmyelinated trail axon. The shape of trail endings differs according to their location. In the juxtaequatorial region, they are usually simple in form, whereas those located on the poles have knoblike terminals (Barker *et al.*, 1970). The length of the ramifications of trail endings vary from 10 μ in the case of the simple endings, up to 210 μ in the case of complex ones. Among the ramifications, the nuclei was observed by Barker *et al.* (1970); but in contrast to the p_1 endings there was no evident sole plate. Trail endings have also been observed on nuclear bag as well as on nuclear chain fibers, although no quantitative data, or their proportional distribution is available. The electron microscope studies of Adal and Barker (1967) confirmed the presence of trail endings on both kinds of intrafusal fibers. Histological preparations of muscle spindles (Barker and Ip, 1965) reveal the previously inaccessible nature of muscle nerve endings. However, Barker (1966b; 1967) pointed out that a correct interpretation of the morphological criteria of the fusimotor in-

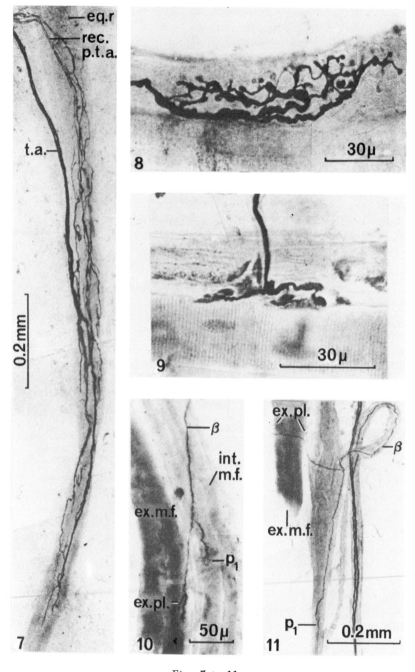

Figs. 7 to 11.

nervation in the muscle spindles is very difficult, because the motor endings do not maintain their fixed morphological identity. The collateral and terminal branches of the motor axons that effect renewal in cat hindlimb muscles has been shown to be between 30 and 40% of fusimotor plate endings axons and in about 20% skeletomotor axons (Barker and Ip, 1965). It is also interesting that some trail endings coming from the polar region of the spindle insert into the tendon, and according to Barker *et al.* (1970), seem to form "presumable abortive contacts."

The second kind of the endings in the muscle spindles, supplied by the γ fibers are p_2 motor end plates. According to Barker *et al.* (1970), in the peroneal muscles of the cat, they are usually located in midpolar region and in extreme end of the poles (Figs. 10, 11). The p_2 plates innervate mostly nuclear bag intrafusal fibers. Among the fifty peroneal muscles of the cat, 90% of the p_2 plates were located on the nuclear bag and 10% on the nuclear chain intrafusal fibers. Barker *et al.* have shown most of the p_2 nerve fibers ending in only one motor end plate, located among the trail endings as well as outside the capsule. Then quantitative analysis revealed that out of one hundred peroneal p_2 nerve fibers, seventy have ended in only one plate, twenty-two in two, five in three, and three in four end-plates. The mean number of p_2 fibers supplying the polar region of the muscle spindles in the peroneal muscles of the cat is 1.6 per pole, whereas the mean number of trail fibers supplying the same region is 4.2 per pole. In rat, Gladden (1969) described three forms of fusimotor endings in the intertransverse caudal muscle. His results showed that the plate endings confined to the polar region of the intrafusal fibers, have a nucleated sole plate and a discrete

---

Fig. 7. A nerve fiber that innervates a p_1 plate, located on the pole of the muscle spindle of the interosseous muscle of the cat, through one of its branches. The extrafusal motor end plate (ex. pl.) is innervated through a second branch of the nerve fiber; ex. m. f. = extrafusal muscle fibers, int. m. f. = intrafusal muscle fibers.

Fig. 8. A preparation similar to that one illustrated in Fig. 7, but taken from an *interosseous* muscle of the rabbit. Figures 7 and 8 were obtained by Dr. Barker from "L'innervation motrice du fuseau musculaire de mammiferes" by M. C. Ip, Theses Ph.D., Durham, 1965. Figures 7–11 are reproduced from Barker, 1968b, Director: A. M. Monier, Masson et Cie, Paris.

Fig. 9. The polar region of the muscle spindle of the m. peroneous of the cat innervated exclusively by trail endings from myelinated axons (t.a. = trail axons) and by nonmyelinated ascending preterminal (rec. p.t.a.) nerve fibers. A part of the equatorial region (eq. r.) of the spindle is visible at the top.

Fig. 10. Surface view of the p_2 plate. Note the bouton nerve endings in the muscle spindle of the m. peroneous of the cat.

Fig. 11. Side view of a p_2 plate. Muscle spindle of the m. peroneous of the cat.

subneural apparatus, structurally similar to the extrafusal motor end-plates. The second type of ending is formed by the plate endings located on the juxtaequatorial region and are twice the size of the plates located on the poles. These endings show no presence of nucleated sole plate and are supplied by the axon, which has twice the diameter of the axon supplying the polar end-plates. The third type of motor innervation is in the form of multiterminal endings similar to the trail endings of Barker. The end-plates, however, correspond to p_1 and p_2 end-plates of Barker (1966b).

The fusimotor innervation is different in the muscles of different animals. For instance, the rabbit muscle spindles, where nearly all intrafusal muscle fibers are of the nuclear bag type, received p_1 plates as well as trail endings (Barker and Hunt, 1964). The rat lumbrical muscle spindles were studied by Meier (1969) using a cytochemical technique for cholinesterase combined with a Sudan black-staining in order to visualize the motor endings on the intrafusal fibers. He distinguished two types of end-plates localized on the extracapsular polar region similar to p_1 and p_2 plates of Barker in the peroneal muscle spindles of the cat. Meier (1969) assumed that p_1 plates are located on the nuclear chain fibers, whereas p_2 plates are distributed on the nuclear bag fibers. The nuclear bag fibers are also innervated by the most of the trail endings. The origin of the fusimotor innervation of the cat muscle spindles has been discussed at length by Barker et al. (1970) in an excellent manner, but it must be emphasized that the problem requires further investigations to find some satisfactory answers concerning the fusimotor innervation.

IV. Ultrastructural Studies of the Muscle Spindle and Golgi Tendon Organ

A. Muscle Spindle

Robertson (1957a,b, 1960) was the first to study the ultrastructure of the muscle spindle in the frog, to be followed by investigations of Katz (1960, 1961) and Karlson et al. (1966). Cat muscle spindles from the lumbrical muscles were studied by Merrillees (1960), Adal and Barker (1967), Adal (1969), and Corvaja et al. (1967, 1969). In 1966, Landon presented an excellent description of the muscle spindles from rat muscle in the symposium "Control and Innervation of Skeletal Muscle" held at Queen's College in Dundee. The fine structure of human muscle spindles in biopsied material has been given by Gruner in 1961 and Hennig (1969), and Düring and Andress (1969) described the sensory and motor endings in the muscle spindles of a number of mammalian species.

1. CAPSULE

The fine structure of the capsule surrounding the central part of the muscle spindle is similar in all the animals studied. The outer portion of the capsule is composed of a changeable number of layers of flattened cells (Fig. 12), which Merrillees (1960) described as capsular sheath cells. These capsular cells contain flattened nuclei, a small number of mitochondria, occasional ribosomes, cisternae of endoplasmic reticulum, and glycogen granules. The presence of numerous pinocytic vesicles suggests active transport within these cells (Bennett, 1956, 1960). On the external surface of the capsule and between its constituent layers there are a number of collagen fibers (Landon, 1966) running both transversally and longitudinally (Corvaja et al., 1969). Merrillees (1960) and Landon (1966) indicated that collagen fibrils are oriented transversely or circumferentially. The inner sheath of the capsule is formed by one or two thick layers of cells, which, according to Merrillees, are endomysial cells. They resemble the capsular sheath cells, but do not have a basement membrane (Corvaja et al., 1969). Also, there are no pores or fenestrations in the capsule except where it is penetrated by the nerve fibers of intrafusal muscle fibers. The outer sheath of the capsule does not cover the polar region of the nuclear bag as well as the nuclear chain fibers, which enter the endomysial tissue. Although Corvaja et al. (1969) found a close apposition between the cells of inner sheath, they did not observe the tight junctions between these cells as earlier described by Landon (1966) in the rat spindles. In the transverse sections of the spindles, the existence of the subcompartments between the particular layers of the external and internal sheaths has often been described by a number of workers even under the light microscope. Blood vessels are generally found in the outer sheath of the capsule and have been described by Merrillees (1960), Landon (1966), and Corvaja et al. (1969). The latter also found some capillaries placed centrally to the most internal layer of capsular cells.

2. INTRAFUSAL FIBERS

A. NUCLEAR BAG FIBERS. Three regions are usually differentiated within the nuclear bag fibers: the poles, myotube, and the nuclear bag, and

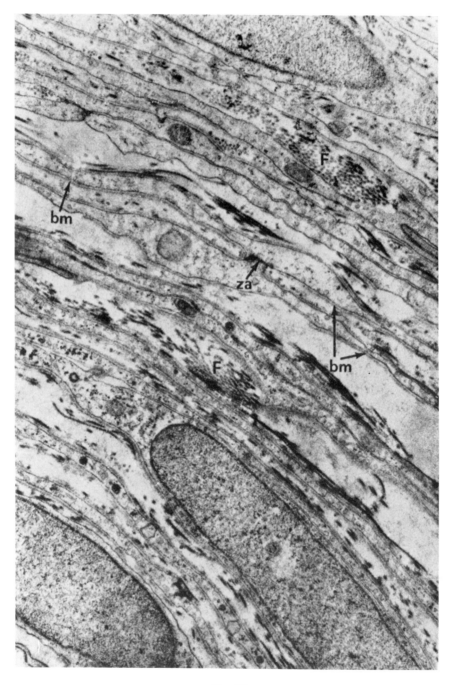

Fig. 12.

their ultrastructure is similar in nearly all mammals. The polar region of the nuclear bag fibers contains numerous tightly packed myofibrils and a small amount of sarcoplasm (Merrillees, 1960; Landon, 1966; Corvaja *et al.*, 1969). The length of sarcomeres in the myofibrils in the nuclear bag fibers is similar to those of extrafusal fibers from the same preparation; however, Landon (1966) and Corvaja *et al.* (1969) described an ill defined H band and a lack of M line. The polar region of the nuclear bag fibers contains elongated nuclei with one or two nucleoli distributed along the extracapsular part of the fiber (Landon, 1966). The nuclei become round and form a column within the region of the myotube. A progressive increase in the volume of sarcoplasm and a decrease in the myofibrils in the myotube as compared to the polar region was observed by Merrillees (1960) in the rat muscle spindle as well as by Corvaja *et al.* (1969) in the cat muscle spindle. It is not known how the myofilaments terminate in the myofibrils which fail to pass from the pole to the equator. Merrillees (1960) and Corvaja *et al.* (1969) did not observe the continuity of myofibrills between the myotube regions separated by the nuclear bag. Landon (1966), however, described the continuity of these myofibrils and their different distribution in the myotube and nuclear bag regions.

The sarcoplasmic reticulum in the nuclear bag fibers is poorly developed. The triades of this reticulum are found at the junctions of I and A bands (Merrillees, 1960; Landon, 1966; Corvaja *et al.*, 1969). Corvaja *et al.* also found approximately 100 Å spaces between the opposed membranes of the T system and the sarcoplasmic reticulum, filled with electron-dense material. Similar observations were made by Franzini-Armstrong and Porter (1964), Huxley (1964), M. Walker and Schrodt (1965), D. S. Smith (1966), and Hagopian and Spiro (1967), and a relationship between the tubules of the T system and a plasma membrane in the nuclear bag fibers was established. In the nuclear bag region, the number of nuclei is greatly increased (Figs. 13, 14), whereas the number of myofibrils is decreased, and they split into a number of distinct bands (Landon, 1966).

B. Nuclear Chain Fibers. Besides different size and shape, as revealed by light microscopic studies, there are other differences between these

---

Fig. 12. Muscle spindle in the lumbrical muscle of the adult cat. Cross section through the equatorial region of the spindle capsule. The capsular sheath is formed by concentric layers of these cells, separated by collagen fibrils (F). A thin basement membrane (bm) surrounds the cells. It disappears at the level of a "zonula adherenses." ×16,800. Reproduced from Corvaja *et al.* (1969) Pisa University Press.

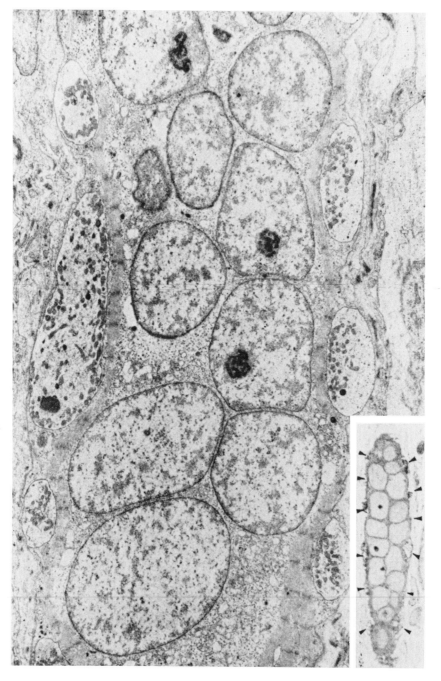

Fig. 13.

two kinds of intrafusal fibers. In contrast to the nuclear bag fibers, the nuclear chain fibers have well defined H bands and prominent M lines (Fig. 14), although Z lines are thicker in the nuclear bag myofibrils as compared to nuclear chain ones (Corvaja *et al.*, 1969). The sarcoplasm in the nuclear chain fibers is more abundant, showing a well developed sarcoplasmic reticulum, numerous glycogen granules, and numerous large mitochondria. The triades are more numerous than in the nuclear bag fibers (Fig. 14). However, there are certain differences in the description of Landon (1966) and Corvaja *et al.* (1969) regarding the distribution of T systems. Landon found the triade of tubules occurring most frequently at the junction of the A and I bands in the nuclear chain fibers. Corvaja *et al.* (1969) attributed this difference to the greater degree of contraction in the nuclear chain fibers.

The nuclei in the nuclear chain fibers have a lenticular outline at the ends of the fiber and rectangular shape in the equatorial region, and they appear less granular than those in the nuclear bag fibers (Landon, 1966). The perinuclear cytoplasm and the nuclei are surrounded by a peripheral sheath of myofibrils. Corvaja *et al.* (1969) described the nuclei in the equatorial region of the nuclear chain fibers in cat muscles as often surrounded by two concentric layers of myofibrils, an internal longitudinal and an external circular or helical layer. They found similar orientation among the peripheral subsarcolemmal myofibrils along the juxtaequatorial region and under the spirals of the sensory endings. However, such an orientation of myofibrils was never observed in the nuclear bag fibers.

It has been known from the light microscopic studies that the intrafusal muscle fibers appear to branch and reunite throughout the long axis of the spindle. Barker and Gidumal (1961) described it in the small nuclear chain fibers. Corvaja *et al.* (1969) discussed it in the light of their observations on the fine structure of the muscle spindle in the cat. They observed that individual fibers retain their integrity throughout, being separated from each other by about 180 Å of extracellular space. Only in some instances do the thin cytoplasmic prolongations of satellite cells separate the membranes of two adjoining intrafusal fibers.

Fig. 13. Muscle spindle in the lumbrical muscle of the adult cat. Longitudinal section through the "bag region" of a nuclear bag fiber. Note the great number of closely packed nuclei. The contractile elements are disposed in a thin shell at the periphery of the fiber, while the central region contains the nuclei. The bag region of the fiber is surrounded by the sensory coil of the primary ending. The inset shows a light micrograph from a semi-thin section of the same intrafusal fiber in which the sensory primary endings are indicated by arrows. Magn. ×5,000 (Inset ×750). Reproduced from Corvaja *et al.* (1969), Pisa University Press.

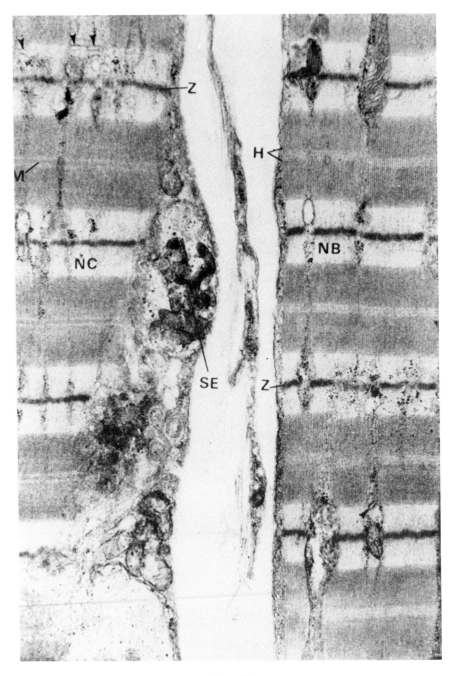

Fig. 14.

Light microscopic studies have shown the existence of junctions between the intrafusal fibers. Adal (1969) followed the sections through
the equatorial region of the cat muscle spindle in the electron microscope
and described the so-called pairing of the nuclear chain fibers. This
is not observed in the nuclear bag fibers. In the former, the borders
of the two chain fibers approach very closely to one another at two
points with no basement membrane between them. The nuclear chain
fibers are separated by their own plasma membranes with about 80
mμ space between them. Similar observations have been made by Corvaja *et al.* (1969) in the muscle spindles of the cat. They observed
that the intrafusal fibers come in close apposition with their cell membranes but never fuse, as indicated by the persistence of the intracellular
space of about 200 Å across. The subjacent sarcoplasm of these two
neighboring fibers often show increased electron density and an increased amount of fibrillar and amorphous material. The significance
of these junctions is not clear. They resemble *"zonulae adherentes"* described by Farquahr and Pallade (1963) in the epithelial cells.

The intrafusal fibers also show tapering as observed by Merrillees
(1960) in the rat muscle spindle and by Corvaja *et al.* (1969) in the
cat muscle spindle. In transverse sections, tapering of nuclear chain
fibers is marked by a progressive loss in diameter, by a loss of their
circular contours, and by the decrease of myofibril content. The terminal
endomysium usually merges with the endomysium of a neighboring intrafusal fiber. So far, no detailed investigations have been made on the
changes appearing during the tapering of the nuclear bag fibers.

The analysis of the fine structure of the intrafusal muscle fibers indicates that great structural differences exist between nuclear bag and
nuclear chain fibers. These differences concern the myofibrils as well
as the triades of the sarcoplasmic reticulum. The nuclear bag fibers
have no M line, ill defined H bands, and thick Z lines (Landon, 1966;
Corvaja *et al.*, 1969). The sarcoplasm of these fibers is poorly endowed
with different organelles, and finally, the sarcoplasmic reticulum is less
developed as compared to that of the nuclear chain fibers. In the latter,
triades are also less numerous in the nuclear bag fibers. One can compare
the differences in the fine structure of the nuclear chain and nuclear

---

Fig. 14. Muscle spindle in the lumbrical muscle of the adult cat. Longitudinal
section through the juxtaequatorial region of a nuclear chain (NC) and nuclear
bag (NB) fiber. The M line can be detected only in the nuclear bag fiber. The
arrows indicate some triads. A segment of a sensory nerve ending (SE) lies on
the surface of the nuclear-chain fiber. H-H band; Z-Z line. ×16,000. Reproduced
from Corvaja *et al.* (1969), Pisa University Press.

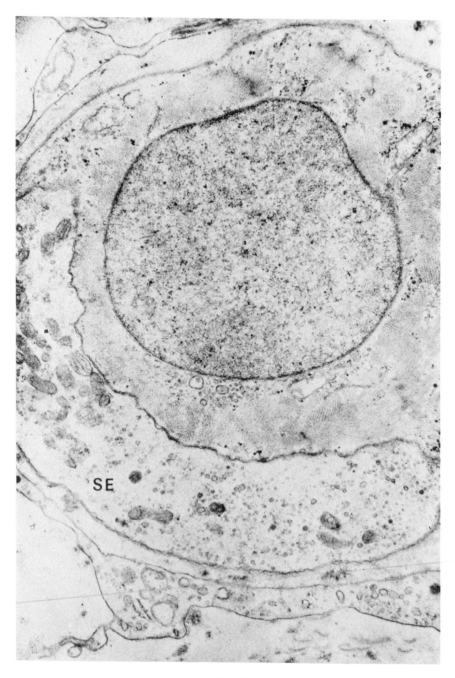

Fig. 15.

bag fibers with those between the twitch and the slow skeletal muscle fibers. The slow fibers generally have no M line, ill defined H bands, and thick Z lines, in contrast to the twitch fibers. Based on these observations, the nuclear bag fibers resemble to some extent the frog's slow muscles, although the nuclear chain fibers are more similar to the frog twitch fibers (Landon, 1966; Corvaja *et al.*, 1969).

3. OTHER INTRAFUSAL STRUCTURES

A. LEPTOFIBRIL-LIKE STRUCTUES. In 1961, Katz described a particular kind of structure named microladders in the muscle spindles in the frog. They are distributed longitudinally among the myofibrils of the intrafusal fibers. Karlsson and Andersson-Cedergren (1968; Karlsson *et al.*, 1966) later confirmed the observations of Katz in the frog muscle spindles. Gruner (1961), in comparable sites of human intrafusal fibers, found similar structures, except their binding to the underlying myofibrils appeared at the long axis of the fiber. Landon (1966) found them in the rat muscle spindle in nuclear chain fibers, being banded around the myofibrils at the level of the Z disk of the dark and light bands at intervals of about 120 mμ. Corvaja *et al.* (1969) found these structures in the muscle spindles of the cat in the nuclear bag (Fig. 15) as well as in the nuclear chain fibers and named them leptofibrils, because they resemble the identical structures described in skeletal and cardiac muscle by Ruska and Edwards (1957), Viragh (1968), and many others. The leptofibrils are more or less miniature myofibrils, and have alternate dark (300 Å) and light (1200 Å) periodic bands. The light bands are filamentous. Leptofibrils are most often located under the sarcolemma (Figs. 15, 16), connecting the myofibrils with a plasma membrane, but these fibrils may also be observed occasionally in the vicinity of sensory endings. The functional significance and the chemical composition of these structures is unknown. Since similar fibrils have been described in growing fibers of the skeletal and heart muscle (Ruska and Edwards, 1957), one may speculate on the similarity between the intrafusal fibers and immature embryonic extrafusal fibers as far as the development of contractile material in the fibrils is concerned.

Fig. 15. Muscle spindle in the lumbrical muscle of the adult cat. Leptofibrils within nuclear bag and nuclear chain fibers. Longitudinal section of a nuclear bag fiber. Surrounded by a rectangle in a leptofibril, which seems to originate from a Z line of a myofibril and proceeds peripherally towards the sarcolemma. Another leptofibril is indicated by the arrow. ×16,000; the region in the frames magnified to ×35,000. Reproduced from Corvaja *et al.* (1969), Pisa University Press.

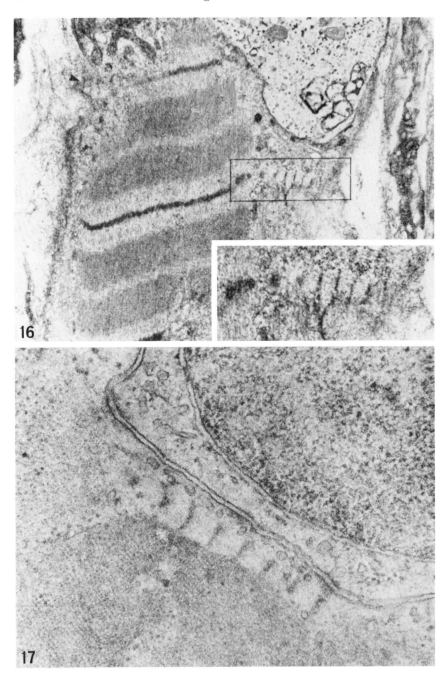

Figs. 16 and 17.

B. SATELLITE CELLS. Landon (1966) described in rat muscle spindles the fine structure of small fusiform cells, which have a close morphological relationship to the intrafusal fibers in the spindles. Mauro (1961) described similar cells in extrafusal skeletal muscle fibers of the frog and named them satellite cells. The satellite cells contain a strongly stained nucleus and a small amount of cytoplasm almost devoid of organelles other than ribosomes and pinocytic vacuoles (Fig. 16, 17). Corvaja *et al.* (1969) also found mitochondria and centrioles. They are generally located in a trough on the surface of the intrafusal fiber, included within its basement membrane in such a manner that their presence does not alter the contour of the fiber (Landon, 1966). Corvaja *et al.* (1969) observed the satellite cells in the polar regions in the intrafusal fibers lying under the basement membrane with a main axis being parallel to the longitudinal axis of the fiber. The role of the satellite cells is not clear at this time. Embryological studies indicate that they may be resting myoblasts, which could be activated by local injury, or may be a source of the efflux of small cells from explants of striated muscle in the early stages of their tissue culture (Muir *et al.*, 1965; Landon, 1966).

4. FINE STRUCTURE OF MUSCLE SPINDLE IN DIFFERENT SPECIES

Crustacean stretch receptors and vertebrate muscle spindles both possess muscular elements that are supplied with sensory input fibers, and when these elements are pulled or stretched, the afferent fibers are excited. They both receive efferent motor innervation, which in vertebrates supplies the intrafusal muscle fibers with the muscle spindle and in crustacea the fine muscle strands of the receptor units (Kosaka, 1969). In crayfish stretch receptors, Kosaka investigated the morphological difference between inhibitory and excitatory synapses. Excitatory synapses contain spherical vesicles, whereas ellipsoidal vesicles appearing in the inhibitory synapse on the stretch receptor muscles are found in

Fig. 16. Cross section of a nuclear chain fiber, showing a leptofibril in a peripheral region of the sarcoplasm, near the sarcolemma. Thin filaments are visible across the clear bands. The leptofibril contacts a myofibril cut through the Z line. A satellite cell is present at the top right. ×60,000. Reproduced from Corvaja *et al.* (1969), Pisa University Press.

Fig. 17. Muscle spindle in the lumbrical muscle of the adult cat. Cross section near the equatorial region of a nuclear chain fiber. A satellite cell (SC) underlying the basement membrane (BM) of the muscle fiber is shown at low and medium power. EC = endomysial cell. ×64,000 (Inset ×7500). Reproduced from Corvaja *et al.* (1969), Pisa University Press.

the vicinity of receptor neurons. Many synapses containing the ellipsoidal vesicles are found at the muscular part of the stretch receptor. Kosaka suggested that the inhibitory synapses of the stretch receptor muscles are branched from the inhibitory axon to the dendrite of the stretch receptor neuron. The muscle elements are innervated by motor nerves, which cause contraction and then raise efferent discharge in the sensory region. Kosaka believed that these muscles are presumably analogous to intrafusal muscles of the vertebrates, and he identified two different types of synapses in the fast-stretch receptor muscles. One contained spherical synaptic vesicles, the second type contained ellipsoidal ones. Based on the electrophysiological studies of Uehara and Hama (1965), Kosaka concluded that synapses with ellipsoidal synaptic vesicles are characteristic for inhibitory neuromuscular junctions.

The reptilian spindle is known to consist of only a single intrafusal fiber which is usually innervated by one sensory nerve fiber ending and by one of the new motor fibers (Fukami and Hunt, 1970; Szepsenwol, 1960). The fine structure of muscle spindles in the costocutaneous muscle of the gartner snake was studied by Fukami and Hunt (1970), who showed that the spindles are present through the entire length of this muscle, being parallel to the extrafusal fibers. The spindles are generally of two types—a slender-type spindle, which exhibits a more tonic type of behavior, whereas the other exhibits a more phasic behavior in response to the external stretch. Both these types show significant difference in their fine structure.

The sensory terminations in the reptilian spindle are different from the frog or mammalian sensory endings. Fukami and Hunt (1970) showed that in the snake spindle, the sensory nerve endings appear crescent-shaped in cross section. They are closely applied to the intrafusal fibers, and as in mammals, they are separated by a gap of 200 Å. These terminals contain varying number of mitochondria and electron-dense glycogen granules as well as fine fibers and vesicles. These endings come in contact with an intrafusal fiber on the large area, but no membrane fusion on the right junctions was observed. Morphologically two types of muscle spindles may be pointed out in the snake. One has a thick fusiform sensory region; the other has slender and longer sensory terminals. The main difference between them is that myofilaments do not run through the central region of the thicker sensory area, whereas in the slender type, they do run through the sensory area. The motor nerve terminations are usually located on one side of the sensory endings, although occasionally, they can occur on both sides of these terminals. Fukami and Hunt described a large number of vesicles 400 Å in diameter as well as numerous mitochondria located in these endings. In general, these motor endings are similar to those observed on the extrafusal fibers.

The fine structure of frog muscle spindles has been described by Robertson (1957a), Katz (1961), and Uehara and Hama (1965). In the sensory terminals, the myelinated nerve fibers enter the spindle spaces at the equatorial zone, lose their myelin sheath and break up into numerous terminal arborizations. They are characterized by swellings that give them the appearance of beaded chains. These terminal endings lie along the spindle fiber. Most of them are distributed in the equator, whereas a decreased number of these terminals may be observed at the polar region. They are never observed on the extracapsular portions of the intrafusal fibers. Most of the terminal arborizations are not associated with Schwann cells. The nerve terminals show different number of mitochondria, vesicles, and glycogen granules. The plasma membrane of the nerve ending and the sarcolemma of the intrafusal muscle fiber are separated from each other by a gap of 150–200 Å. At the equatorial zone, the sensory endings are covered by the thin cytoplasmic processes of the endomysial cells.

The motor endings are always observed on the extracapsular portions of the spindle in the amphibians (Uehara and Hama, 1965). However, the basic structure of the motor endings remains like that described in the mammalian skeletal muscle; although in frog, the endings are smaller and simpler than the other species' extrafusal ones. Between the sarcolemma of the intrafusal muscle fiber and nerve ending, a gap of 500–600 Å appears to be filled with basement membrane material. The axoplasm of the nerve ending shows the presence of neurofilaments, granules of glycogen, mitochondria, and clusters of synaptic vesicles of 450–600 Å in diameter (Uehara and Hama, 1965). Andersson-Cedergren and Karlsson (1967) described a different arrangement of polyribosomes in the equatorial region of the intrafusal muscle fibers in the frog muscle spindles. They interpreted it as a site of the active protein synthesis going on in this area of the muscle spindle.

B. Nerve Endings

The fine structure of the nerve endings on the intrafusal fibers of muscle spindles in the cat has been described by Corvaja *et al.* (1969) and Adal (1969); in rat by Merrillees (1960), Teravainen (1968), and Rumpelt and Schmalbruch (1969); in snake by Fukami and Hunt (1970); in frog by Uehara and Hama (1965); Robertson (1957a), Katz (1960, 1961); and in rabbit by Corvaja and Pompeiano (1970). The neuromuscular junctions have been studied in crayfish stretch receptors by Kosaka (1969) and in human by Gruner (1961) and Rumpelt and Schmalbruch (1969). Several prominent differences exist between the ultrastructure of the particular endings of the mammals and lower animals and even between the reptilians and the frog (Fukami and Hunt, 1970). In general, there are two particular types of nerve endings in the intrafusal fibers. The motor end-plates are mostly located in the juxtaequatorial region in the intracapsular polar region and in the extra-capsular polar region. The sensory endings are mostly found in the equatorial region. Each spindle receives a primary sensory ending which consists of a number of spiral terminations (Barker, 1967; Corvaja *et al.*, 1969). Each spiral surrounds the nuclear region of each individual intrafusal fiber. The primary sensory endings provide an afferent pathway which is common to the nuclear bag as well as nuclear chain intrafusal fibers (Boyd, 1962; Barker, 1967). Secondary sensory endings are located mainly in the juxtaequatorial region. There are three kinds of motor end-ings on the intrafusal fibers (Barker, 1967), the diffuse multiterminal trail endings and two types of plate endings (p_1 and p_2). The trail endings, first detected by the cholinesterase technique (Coërs and Durand, 1956; Coërs, 1962), originate from γ-fibers and are mainly lo-cated on the nuclear chain fibers in both sides of the sensory endings. The p_1 and p_2 plates originate respectively from β- and γ-fibers. While p_1 plates are found on the extreme polar region of the nuclear bag fibers, the p_2 plates are observed in the mid-polar region of the nuclear chain as well as in the nuclear bag region (Barker, 1966b, 1967).

1. Sensory Endings

The myelinated nerve fibers enter the spindle space at the equatorial region, lose their myelin sheaths, and break up into numerous terminal arborizations. These terminal aborizations lie along the spindle fiber throughout its equatorial intracapsular course. They are numerous in this region, and their number seems to decrease at the polar zone. They have not been observed at the extracapsular polar area of the intrafusal fibers. In the transverse or longitudinal sections at the sensory region of the muscle spindle of the cat, no differences were observed between the sensory terminals on the nuclear chain and the nuclear bag intrafusal fibers (Adal, 1969). Landon (1966) pointed out that the sensory coil on the nuclear chain fibers was finer and without any equatorial enlarge-ment as compared to the nuclear bag fibers in the rat muscle spindles. Corvaja *et al.* (1969) also described the same difference between the sensory endings in the cat, having two types of sensory endings on the nuclear bag fibers. One type measures less than 1 μ in diameter and extends along the axis of the muscle fiber and gives rise to transverse or oblique branches lying in slightly raised gutters or muscle substance. The second one is in a flattened form and is observed on the same nuclear bag fiber. The presence of the sensory endings outside on the polar region of the intrafusal nuclear bag fiber region has not been observed. Contrary to the raised and rounded shape of the primary endings on the nuclear bag fibers, those present on the nuclear chain fibers are low and flattened (Corvaja *et al.*, 1969). The secondary endings on the nuclear chain fibers have the same size as the primary endings surrounding the same intrafusal fiber. The main difference between the primary and secondary endings is the absence in the secondary endings of the regular spiral arrangement and the presence of elongated branches parallel to the axis of the intrafusal fiber. This represents connecting links between rings, which only partially surround the fiber (Fig. 18) (Corvaja *et al.*, 1969). The fine structure of the internal parts of both types of endings appear to be similar. They both contain numerous mitochondria, varying in size from 0.25 to 0.60 μ in diameter and up to 3 μ length, and neurofibrils and diffuse vesicular structures varying in density and size (Figs. 18, 22, 23). Adal (1969) described vesicles containing dense bodies about 20–50 mμ in diameter, as well as multivesicular bodies about 60–400 mμ in diameter. It is interesting that a continuity between the neuroplasm of the sensory endings and the sarcoplasm of the intrafusal fibers has not been observed in any muscle spindles, although occasionally cross terminals between the nuclear chain fibers in the cat muscle spindles have been reported (Adal, 1969). The fine structure of the cross-terminal sensory endings linking the two nuclear chain fibers is similar to those lying on the individual intrafusal fibers and contain mitochondria, neurofibrils, vesicles and dense bodies (Figs. 19–21). These cross terminals have not been observed among the nuclear bag fibers and between the nuclear chain and nuclear bag fibers (Adal, 1969). On either side of the sensory ending, the sarcoplasm of the particular intrafusal fiber often extends to form lips of sarcolemma (Fig. 22) lying on a portion of the sensory terminal (Adal, 1969). The adjacent cell membranes of the nerve endings and the intrafusal fiber are always separated by a gap of about 15 mμ (Fig. 23). The sensory terminals and its sarcolemma are bounded by the basement membrane which surrounds the intrafusal fiber. The amount, size, and distribution of the cell organelles in the different sensory endings differ greatly from ending to ending. No evidence of continuity between the neurolemma of the sensory ending and the sarcolemma of the intrafusal fibers

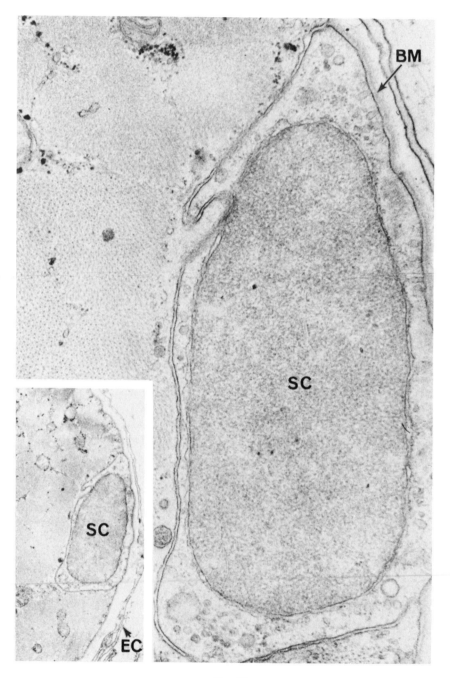

Fig. 18.

has been found, which is contrary to the findings of Katz (1961), who suggested the existence of such a continuity in the frog muscle spindle.

2. Motor Endings

There are two types of motor endings in mammalian muscle spindles: the trail endings, being distributed in the form of fine diffuse terminals near the secondary sensory endings on the nuclear chain fibers, and the plate endings (Landon, 1966). Corvaja *et al.* (1969) observed two types of end-plates in the case of the muscle spindles of the cat. The nerve terminals lie on the depressions of the sarcolemma of the intrafusal fibers. The endings contain varying number of mitochondria and a large number of 35–45 mμ thick vesicles (Fig. 24). Unlike the sensory endings, there is an amorphous basement membrane of the muscle fiber between the neuronal ending and the sarcolemma (Fig. 24). Junctional folds characteristic of the extrafusal motor end-plates are also observed in the motor endings on the intrafusal fibers. In nearly all intrafusal motor endings, there appeared a thin layer of sarcoplasm containing varying amounts of myofilaments. Occasionally one may observe the nucleus of the intrafusal fiber just beneath the the motor ending (Corvaja *et al.*, 1969). The processes of a Schwann cell often form a cap covering these motor endings, and there is no basement membrane between the Schwann cells and the nerve endings. These endings, being located outside the sensory terminals at the end of the spindle capsule as well as in the extracapsular region, seem to be trail endings, as indicated by the observations of Adal and Barker (1967). Corvaja *et al.* (1969) described the fine structure of another kind of motor ending in the cat muscle spindles. This motor ending is located mainly on the extra polar region of a nuclear bag fiber. It appears to penetrate deeply into the intrafusal fiber and shows the presence of a basement membrane between the nerve ending and the sarcolemma. Similar endings have been described by Merrillees (1960, 1962) in the rat muscle and by Adal and Barker (1967) in the cat spindles. The accumulation of nuclei and mitochondria in the sarcoplasm below these endings has also been observed. These endings probably come from the same motor axons as the trail endings.

Fig. 18. Muscle spindle in the lumbrical muscle of the adult cat. Cross section of a nuclear chain fiber at the equatorial region of the spindle. An annulospiral sensory ending (SE) covers the fiber for almost half of its circumference. A great number of vesicles occupy the region of the axoplasm, particularly in the lowest part of the ending where mitochondria are absent. Note the oblique orientation of the external myofibrils near the sensory ending. ×15,000. Reproduced from Corvaja *et al.* (1969), Pisa University Press.

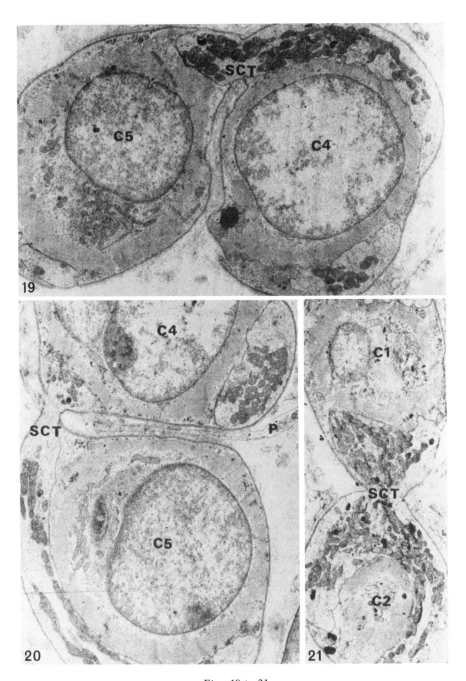

Figs. 19 to 21.

3. Sympathetic Innervation

There is no positive evidence in the literature of the existence of sympathetic innervation of the intrafusal fibers (Norberg, 1967). Hinsey (1928) suggested that the sympathetic nerves end in the walls of the blood vessels. The evidence for the lack of sympathetic innervation of the spindles was also indicated by the cytochemical investigations of Coërs and Durand (1956). They showed cholinesterase activity in those endings that were suspected to be sympathetic in nature. Enzyme activity disappeared after cutting the ventral roots (Boyd, 1962). Barker (1967) believed that the sympathetic innervation of the muscle spindles is concerned with (or limited to) the walls of the blood vessels.

4. Nerves to the Blood Vessels and Capsule

The nerves supplying the capsule often enter the muscle spindle with a nerve fiber to the primary ending (Cooper and Daniel, 1963). They go around through the sheaths of the capsule and occasionally they give out short branches. Another nerve fiber might enter the capsule with the afferent nerves and then divide near the primary endings inside the spindle (Cooper and Daniel, 1963). The nerves to the blood vessels may enter the capsule together with the vessels or with a main nerve fiber as traced by Cooper and Daniel (1963) in the human muscle spindles.

C. Golgi's Tendon Organ

The Golgi tendon organ, the nerve-spray receptors around the joints, and Ruffini receptors in the subcutaneous tissues presumably react to mechanical distortion of the connective tissue and have been collectively

---

Fig. 19. Cat muscle spindle from m. tenuissimus. "Sensory cross terminals" between neighboring nuclear-chain muscle fibers from a primary ending. Transverse section (A). The sensory terminal on chain fiber C_4 is linked to the terminal on the neighboring chain fiber C_5, forming a sensory cross terminal (SCT) between two intrafusal fibers. ×5100. From Adal (1969).

Fig. 20. At slightly different level toward the pole, the sensory cross terminal (SCT) continues to cover nearly half the perimeter of chain fiber C_5. A process of an inner capsule cell (P) is seen between the two chain fibers. It appears that this sensory cross terminal consists of a linkage of two sensory half spirals from neighboring nuclear-chain muscle fibers. ×5100. From Adal (1969).

Fig. 21. Another sensory cross terminal (SCT) between chain fibers C_1 and C_2. ×5100. From Adal (1969).

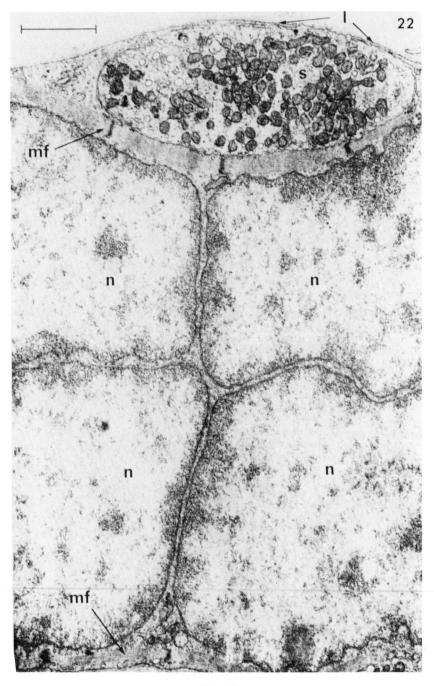

Fig. 22.

called Ruffini endings (Stilwell, 1957). Boyd (1954), comparing the Ruffini sensory unit in the joint capsule and the Golgi sensory unit in the tendon, pointed out that Golgi sensory unit is three to four times larger than the Ruffini unit and that it is supplied by 12 mμ afferent nerve fibers, whereas the Ruffini unit is innervated by approximately 6 mμ thick nerve fiber. Neurotendinous organs or tendon organs of Golgi are found at the junctions of muscles and tendons and in the tendinous aponeuroses of the muscles. The classic name "nervous musculotendinous endoorgan" was given by Camillo Golgi to this specialized nerve ending in 1880, and he described it along with associated sensory nerve endings in skeletal muscles of amphibia, reptiles, birds, and mammals. The tendon organ was also found in mammalian skeletal muscles by Cattaneo (1888), and in a series of degeneration experiments, he demonstrated the sensory function of this organ. Since the tendon organ has not been associated with muscle fibers all the time, the name introduced by Golgi appeared somewhat inappropriate, and in 1891 Ciaccio proposed the name "Golgi's tendon spindle." Since then, the tendon organ has been studied by a number of workers. Huber and DeWitt (1900) studied the nerve terminations in the tendon organs and greatly contributed to an understanding of their physiology. A fuller understanding of the physiology of the tendon organs was, however, achieved by the investigations of Barker (1959), Swett and Eldred (1960), and Wohlfart and Henriksson (1960).

Neurotendinous organs are composed of specialized encapsulated fascicles of collagen, which are offshoots from the primary tendon of origin or insertion of the muscle. In a study on the structure of the tendon organ in the cat, Bridgman (1968) showed that the total length of this organ in cat hindlimb muscles is about 900–1200 μ, and they have a fusiform shape. The width in the midsection is 100–200 μ, and it tapers toward the ends up to 50–60 μ at the point of its origin and up to 80–100 μ at the place where it attaches to the extrafusal muscle fiber. The capsule is lamellar, similar to the one observed in the muscle spindle, and it covers nearly all the tendon organ. Since the nerve fibers enter the tendon organs, it probably contains the perineural epithelium

Fig. 22. The muscle spindle from the plantar lumbrical muscle of the rat. An axial longitudinal section through the equator of a nuclear bag intrafusal fiber. The sensory nerve ending (S) is transversely sectioned as it lies in a trough in the surface sarcoplasm and is almost completely covered by lips of sarcolemma (L). The ending contains large numbers of mitochondria together with vesicles of varying sizes. The muscle fiber nuclei (n) are packed closely together, and the myofibrils (mf) are reduced to thin peripheral bands. Scale = 1 μ, $\times 20,400$. Reproduced from Landon (1966), Williams and Wilkins, Baltimore.

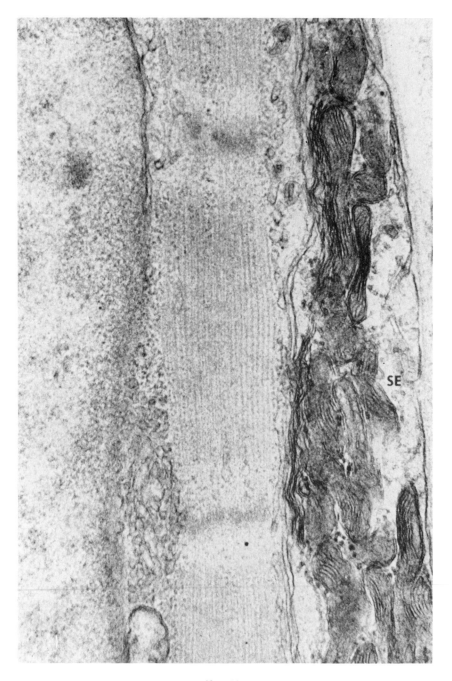

Fig. 23.

(Shanthaveerappa and Bourne, 1962). Adjoining the muscle aponeurosis, the capsule, built with a connective tissue, surrounds the collagen fascicles about 250–300 μ from the place where they originate from the main aponeurosis. The collagen fascicles break up into bundles, separated by the layers of capsule, forming the compartments. Sometimes the tendon organ forms a so-called dyad where it is associated with a muscle spindle (Marchand *et al.*, 1966). As Barker *et al.* (1962) described it, the innervation of tendon organs comes from Ib fibers. Bridgman (1968) suggested, however, that the size of the large axons resembles the Ia fibers. The nerve fibers, after losing its myelin sheath, enters the capsule in the central region of the tendon organ and immediately branches into a thin threadlike fiber 3–4 μ in diameter, which enter the space between fine bundles of collagen fibers. A compression of these endings by the collagen bundles, as they straighten under tension, could be a source of graded generator potentials (Bridgman, 1968). At the distal end of the tendon organ, projecting into muscle fibers, the collagen bundle form dense fascicles attaching the extrafusal muscle fibers. The capsule is no longer discernible beyond this point. No nerve endings are observed in the part of the tendon organ which are not covered by the capsule.

An electron microscopic study of the tendon organ from the white rat lumbrical muscles shows that the capsule, having the same fine structure as that of muscle spindle, extends into the organ and divides it into several compartments containing small bundles of collagen lying among the fibroblasts and Schwann cells. The outer part of the capsule is covered by the zone of collagen, and myelinated as well as unmyelinated nerve endings are identified in the tendon organ (Merrillees, 1962). The nerve endings show a large number of mitochondria, a few vesicles, some fine filaments, and some amorphous electron-dense material. The nerve processes are often deeply indented by the bundles of collagen fibers. Merrillees (1962) observed blood vessels in the form of capillaries inside as well as outside the tendon organ. According to this author, the nerve processes appear as "sprays of the nerve ending," which are thin, nonmyelinated fibers studded with granular leaf- or clasplike enlargements that partly wrap around the tendon fasciculi. Bridgman (1968) discussed the possible mechanism of the function of this nerve

---

Fig. 23. Muscle spindle in the lumbrical muscle of the adult cat. Longitudinal section of the equatorial region of the nuclear chain fiber showing a primary sensory ending (SE). Note the parallel course of the plasma membrane of both the nerve ending and the muscle fiber. The ending contains a greater number of elongated mitochondria with a dense matrix and longitudinally oriented cristae. There is also a series of parallel membranes on the deep surface of the nerve ending. ×48,000. Reproduced from Corvaja *et al.* (1969), Pisa University Press.

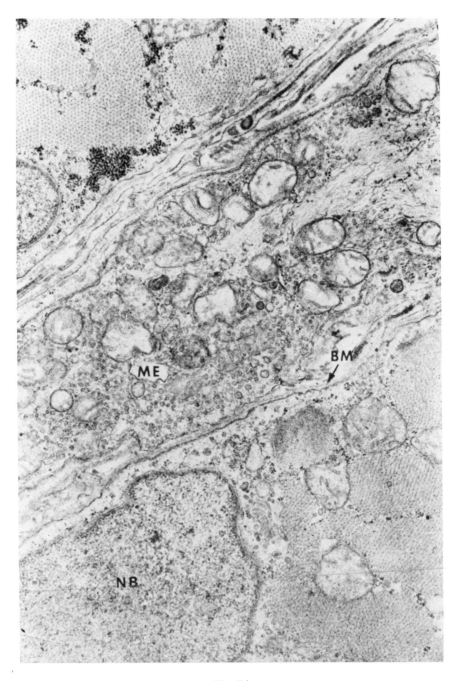

Fig. 24.

ending. Since the collagen fibers are not very extensible, the possibility of the lengthening or bending of nerve fibers located within collagen fine bundles is somewhat questionable, and a longitudinal stress could exert a lateral compression on the nerve fibers, thereby displacing the endings.

V. Cytochemistry of the Muscle Spindle

The following tabulation contains a list of the abbreviations of enzymes and substrates used in this section.

AChE	Acetylcholinesterase (specific)
AD	Aldolase
3',5'-AMPase	Cyclic 3',5'-nucleotide phosphodiesterase
AP	Amylophosphorylase
ATPase	Adenosinetriphosphatase
BChE	Butyrylcholinesterase (nonspecific cholinesterase)
LDH	Lactic dehydrogenase
MAO	Monoamine oxidase
NAD	Nicotinamide adenine dinucleotide (DPN)
NADH	Reduced nicotinamide adenine dinucleotide
NADH$_2$-TR	NADH$_2$-tetrazolium reductase
NADPH$_2$-TR	NADPH$_2$-tetrazolium reductase
NADP	Nicotinamide adenine dinucleotide phosphate (TPN)
SDH	Succinic dehydrogenase
SE	Simple esterase
TPPase	Thiamine pyrophosphatase
UMP	Uridine monophosphate
UTP	Uridine triphosphate
CMP	Cytidine monophosphate
CTP	Cytidine triphosphate
DFP	Diisopropyl fluorophosphate

A. Polysaccharides

Von Brzezinski (1961a), in his studies on the muscle spindles of the guinea pig, demonstrated various reaction for polysaccharides in the

Fig. 24. Muscle spindle in the lumbrical muscle of the adult cat. Cross section through the polar end of the spindle capsule showing a motor nerve ending on a nuclear bag fiber (NB). The motor ending (ME) contains mitochondria, microtubules, and a great number of neurofilaments. It contains also a large number of small vesicles intermingled with some vesicles of large size. The basement membrane (BM) lies between the adjacent nerve and muscle cell membranes. ×25,000. Reproduced from Corvaja *et al.* (1969), Pisa University Press.

different parts of these endings. The capsule is strongly positive for PAS and PAS after digestion with diastase. Alcian blue staining indicated a positive reaction for acid mucopolysaccharides in the capsule, which also shows β-metachromasia after staining with toluidine blue. The peripheral parts of the intrafusal fibers are also strongly positive after PAS and PAS–diastase staining as well as after staining with Alcian blue. β-Metachromasia is also observed at the poles of the intrafusal fibers. The lymph spaces are almost negative after PAS and PAS-diastase staining, but a strong reaction for acid mucopolyaccharides is evident. Toluidine blue staining also shows γ-metachromasia in the "lymph" space of the muscle spindle. The cytoplasm of the intrafusal fibers shows α-metachromasia after toluidine blue staining, and no glycogen or acid mucopolysaccharides are present. Studies of Rebollo and Anda (1967) on the muscle spindles of the chicken revealed the presence of PAS-positive reaction in the capsule and in the cytoplasm of the intrafusal fibers; however, it is only in the capsule that the reaction is resistant to digestion with ptyaline, indicating that the positive PAS reaction is not due to the presence of glycogen. The greatest accumulation of glycogen appeared at the polar and interpolar zones. Thin fibers are reported to contain more glycogen compared to the large and intermediate fibers. Rebollo and Anda (1967) emphasized that the increase in the reaction for glycogen in the intrafusal fibers appeared in the equatorial zone. Wirsen (1964) observed the PAS reaction in the muscle spindles of the dog, and the distribution of glycogen was found to be similar to that observed by Rebollo and Anda. They reported the appearance of the metachromasia in the sarcolemma of the intrafusal fibers, whereas the external capsule and the equatorial zone were negative. Contrary to observations of Brzezinski (1961a), Rebollo and Anda (1967) did not observe a positive reaction for mucopolysaccharides in the muscle spindles, with the exception of the positive Alcian blue reaction in the walls of the blood vessels. The muscle spindle lymph space contains a high amount of hyaluronic acid, probably formed by the capsule compounds (von Brzezinski, 1961a). Spiro and Beilin (1969b), in their investigations of the human muscle spindle, found no differences in the PAS reaction between nuclear bag and nuclear chain fibers. Both types contain small amounts of PAS-positive substances.

B. Lipids

Rebollo and Anda (1967) stained muscle spindles of the chicken for phospholipids (Baker test with pyridine control), for carbonyl groups, unsaturated lipids, for fatty acids and neutral fats. Sudan black B staining revealed a different pattern of reaction in the muscle spindle. In the equator region, large amounts of lipids were observed, as opposed to a very weak reaction in the polar regions. The capsule and the nerve fibers showed positive staining. The reaction for phospholipids was similar to that of Sudan black B. The capsule was negative, whereas the nerve fibers showed a positive reaction. Neither fatty acids nor neutral fats were observed in the chicken muscle spindle. A study of muscle spindle from cat gastrocnemius and toad iliofibularis muscles did not show any Sudan black B or other fat-staining material, and Spiro and Beilin (1969a,b) did not observe any difference in the lipid contents of the nuclear bag or nuclear chain fibers of the human muscle spindles (Spiro and Beilin, 1969a,b).

C. Nucleic Acids

DNA and RNA have been studied in muscle spindles from musculi lumbricales of different laboratory mammals by von Brzezinski (1965) by using Feulgen method for DNA and methyl-green pyronine with a subsequent RNase digestion for RNA. The nuclear bag intrafusal fibers showed high DNA content, being in close relation with a large amount of nuclei in the bag region. von Brzezinski also observed a small amount of RNA in the cytoplasm of intrafusal fibers. No attempt to quantitate the results was done by this author.

D. Proteins and Myoglobin

Von Brzezinski (1965) observed a distinct perinuclear concentration of sulfhydryl groups in the muscle spindles of different laboratory animals and man. He considered it as a sign of the higher metabolism of the equatorial region. Tryptophan is also present in the same region and may be related to the role of this amino acid in 5-hydroxytryptamine synthesis. Rebollo and Anda (1967) studied the protein content of the muscle spindles of the chicken and showed the presence of the proteins with disulfide and sulfhydryl groups in the intrafusal fibers as well as in the external capsule. They also observed a moderately positive Millon reaction for tyrosine and tryptophan in the large intrafusal fibers in the muscle spindle of the chicken, which seems to confirm earlier observations of von Brzezinski (1965) in the muscle spindles of the guinea pig.

Myoglobin is probably bound to the myofibrils and has been demonstrated in the intrafusal fibers of the mouse, rat, guinea pig, cat, rabbit, macaque monkey, and man (N. T. James, 1968). The intensity and the pattern of staining appeared different in various kinds of intrafusal fibers. The most common pattern of staining in rat and in mouse muscle spindle was that the intrafusal fibers were stained stronger than the extrafusal fibers, whereas the nuclear bag fibers showed slightly stronger reaction than nuclear chain fibers. In the cat and in the guinea pig, the nuclear chain fibers showed a negative to negligible reaction, whereas the nuclear bag fibers showed a moderate to moderately strong staining reaction. These differences in the staining pattern between the two types of intrafusal fibers were very prominent in the guinea pig as compared to other investigated species. In the case of rabbit muscle spindles, almost all intrafusal fibers were stained equally well, which may be attributed to the uniformity of the intrafusal muscle fibers in this species. From the localization and amount of myoglobin in the intrafusal fibers, it may be concluded that myoglobin-rich fibers contract more slowly and for longer periods than myoglobin-poor fibers (N. T. James, 1968); similarly, the nuclear chain fibers would contract rapidly and nuclear bag fibers slowly. James emphasized that the present nomenclature of muscle fibers is misleading because the terms red and white muscles should refer specifically to myoglobin-rich and myoglobin-poor fibers demonstrated cytochemically and not to a specific fiber type.

E. Enzymes

Based on their enzyme content, the extrafusal muscle fibers may be placed in three broad categories. Muscle fibers with low oxidative and high glycolytic enzyme activity are designated as type A. They are usually observed in white fast-twitch muscles. Muscles with moderate oxidative and low glycolytic enzyme activity are designated as type B. They are dominant in the red slow-twitch muscles. Type C has high oxidative and low glycolytic enzyme activity and are observed in red as well as in white muscles (Stein and Padykula, 1962). The cytochemical characteristics of the extrafusal and intrafusal fibers have been studied by a number of workers. Ogata and Mori (1962), Wirsen (1964), Wirsen and Larsson (1964), and Nyström (1967) found similar enzyme activity in the two types of fibers, whereas some other studies showed significant differences in the activity of extrafusal and intrafusal fibers (Nachmias and Padykula, 1958; Henneman and Olson, 1965; Rebollo and Anda, 1967). Yellin (1969a,b) showed a number of different types of intrafusal fibers; some of them resembled the extrafusal fibers, while the others showed an enzyme pattern unlike those commonly described in the extrafusal fibers. The large intrafusal fibers showed particularly low phosphorylase and low succinate dehydrogenase activity (Figs. 25 and 26). Yellin related this kind of fibers to the amphibian slow muscle fiber described earlier by Peachey and Huxley (1962) and R. S. Smith (1966). Additional investigations of the amphibian muscles seem to confirm the assumption that histochemically non reactive fibers are slow (tonic) contracting, while the fibers with a moderate to high oxidative and glycolytic activity are fast (twitch) contracting (Engel and Irwin, 1967). Romanul and Van der Meulen (1967) showed in the normal reinnervated and cross-innervated mammalian fast and slow muscles that faster contracting muscles possess greater amounts of fibers rich in glycolytic enzyme activity.

1. OXIDATIVE AND GLYCOLYTIC ENZYMES

Several types of intrafusal fibers according to their enzymic pattern have been described in the rat muscle spindles (Yellin, 1969a). Some showed the same pattern of enzyme activity as the A, B, C types of intrafusal fibers, whereas some others showed low succinate dehydrogenase (SDH) as well as phosphorylase activity. This is particularly true of large intrafusal fibers. The small intrafusal fibers always showed appreciable amounts of SDH and phosphorylase activity (Yellin, 1967). The phosphorylase reaction in the intrafusal fibers of the rhesus and squirrel monkeys' muscles (Figs. 27 and 28) also differ greatly (Nakajima et al., 1968); the large intrafusal fibers show stronger enzyme reaction compared to the smaller ones. Spiro and Beilin (1969a,b) studied rabbit intrafusal fibers and found weak amylophosphorylase activity in all the fibers. Based on phosphorylase reaction, Rebollo and Anda (1967) categorized three types of intrafusal fibers in chicken muscle spindles, such as fibers with intense enzyme activity, thin fibers with a weak reaction, and fibers of intermediate diameter with no enzyme reaction at all. They also studied the activity of UDPG transferase and found that the intermediate fibers showed the strongest reaction, whereas the thin fibers were negative and the thick ones showed only very weak enzyme activity. Strong aldolase activity (Fig. 29) in the intrafusal fibers was shown by Nakajima et al. (1968). In the spindles, the different intensity of the SDH reaction has been described in different zones by Germino and D'Albora (1965) and Yellin (1969b). In chicken muscle spindles, Germino and D'Albora reported that SDH activity in well differentiated motor zones of intrafusal fibers and in the myofibrils remained like that observed in the extrafusal fibers. In the second zone, located between the equator and polar region, the enzyme activity was distributed around the nuclei, and in the third zone, with a single row

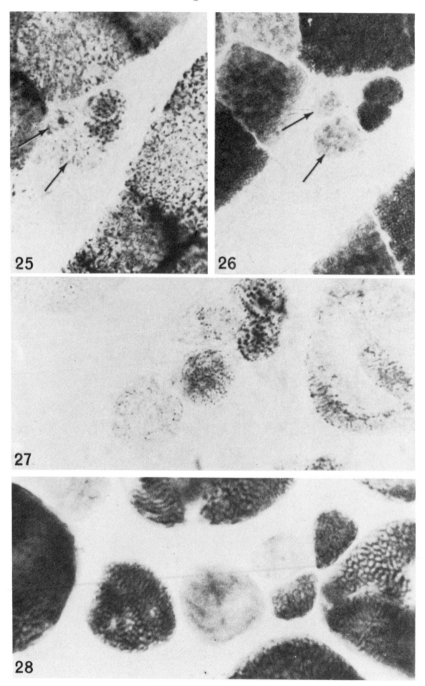

Figs. 25 to 28.

of nuclei, intense SDH reaction was observed among the nuclei. Yellin (1969b) observed significant enzymic differences between the polar contractile region of the intrafusal fibers and the equatorial area of the fibers.

Ogata and Mori (1962), studying the succinate, lactate, malate, and α-glycerophosphate dehydrogenases, as well as DPNH and TPNH diaphorases in the muscle spindles of mouse muscles, found three types of intrafusal fibers with different intensity of oxidative enzyme activity, starting from a strong reaction in the smaller intrafusal fibers to a weak enzyme activity in the large fibers. In the medium-sized fiber, SDH activity was intermediate between large and smaller intrafusal fibers. Ogata and Mori (1962) related these particular types of intrafusal fibers to the type A, B, and C of extrafusal fibers based on their enzymic reaction. Nyström (1967), using cat gastrocnemius and toad iliofibularis muscles, studied $NADH_2$-TR and found that of all examined muscle spindles in the cat muscle, only one type of fiber has high $NADH_2$-TR activity. The toad muscle spindles showed a weaker activity of $NADH_2$-TR as compared with the intensity of reaction in the cat gastrocnemius. The intrafusal fibers also showed varying enzymic reactions. The large fibers showed a weak $NADH_2$-TR activity, whereas the reaction in small fibers was weaker. Rebollo and Anda (1967) studied succinate and malate dehydrogenases, cytochrome oxidase, and TPNH diaphorase ($NADPH_2$-TR) in the muscle spindles of the chicken and also found three different types of intrafusal fibers according to their enzymic pattern in the polar region. They described thin fibers with thick subsarcolemmal clusters and positive central clusters, thick fibers with uniformly arranged formazan deposits, and intermediate fibers with an

---

Fig. 25. Succinate dehydrogenase activity (using nitro-BT) of extrafusal and intrafusal fibers in the medial gastrocnemius muscle of the rat. ×600. From Yellin (1969b).

Fig. 26. A serial section of the muscle fibers depicted in Fig. 33 demonstrates their phosphorylase activity. The intrafusal fibers are of two enzyme profiles: type C fibers and those nonreactive for either enzyme (arrows). The intrafusal fibers are viewed at their polar regions. The size (diameter) potential of one of the nonreactive fibers is not fully expressed at this level, probably as a result of the common out-of-register origin of intrafusal fibers. ×600. From Yellin (1969b).

Fig. 27. Interosseous muscle of the hand of rhesus monkey. There is one large, strongly amylophosphorylase-positive intrafusal fiber and five moderately positive ones. ×480. From Nakajima *et al.* (1968).

Fig. 28. Interosseous muscle of the hand of squirrel monkey. Two intrafusal fibers show a negligible amylophosphorylase reaction, and three intrafusal fibers show a mild reaction. Note two moderately positive extrafusal fibers (m) and the strongly positive one (s). ×480 From Nakajima *et al.* (1968).

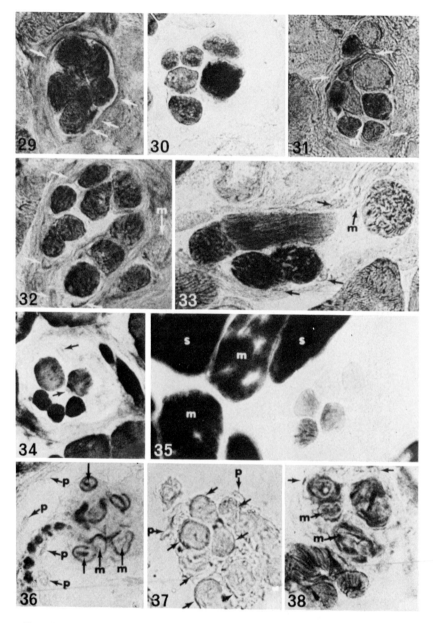

Fig. 29. Interosseous muscle of the hand of rhesus monkey. Aldolase (AD) test. Note the very strongly stained rims and the moderately (arrows m) and very strongly positive internuclear areas. The pericapsular epithelial cells are moderately positive (arrows). ×480. From Nakajima et al. (1968).

Fig. 30. Interosseous muscle of the hand of rhesus monkey. The spindle contains

average amount of formazan. The enzymic reaction for all oxidative enzymes was increased in all intrafusal fibers at the equatorial zone, where the formazan granules occupy the whole sarcoplasm. Nakajima *et al.* (1968) carried out a detailed study on the distribution of several oxidative enzymes in the muscle spindles of the hand and foot interosseous muscles and m. longissimus dorsi of the rhesus monkey as well as in those of the hand interosseous muscles of the squirrel monkey. The polar region of the intrafusal fibers in the muscle spindles of the rhesus monkey showed two types of reaction intensity for oxidative enzymes. The moderate and strong reaction for SDH and LDH (Figs. 30 and 31) was observed in the small intrafusal fibers, whereas the reaction was weaker in the large intrafusal fibers. In the equatorial region the intranuclear area of sarcoplasm showed a moderate to very strong reaction for SDH (Figs. 32 and 33), LDH (Figs. 30, 31, 34, and 35), and negligible for MAO, whereas a moderate reaction for LDH and MAO (Fig. 36) was found in the rim just beneath the sarcolemma

---

only one type of (very strongly LDH positive) intrafusal fibers. The pericapsular epithelial cells are moderately stained (arrows). ×480. From Nakajima *et al.* (1968).

Fig. 31. Interosseous muscle of the hand of rhesus monkey. One intrafusal fiber is moderately lactate dehydrogenase (LDH) positive (arrow m), and the others show very strong activity. Note moderately stained pericapsular epithelial cells (arrows). ×480. From Nakajima *et al.* (1968).

Fig. 32. Longissimus dorsi muscle of rhesus monkey. Note one large, moderately succinate dehydrogenase (SDH) positive intrafusal fiber and four small, strongly stained ones. The pericapsular epithelial cells are mildly (arrow m) and moderately positive (arrows). ×300. From Nakajima *et al.* (1968).

Fig. 33. Interosseous muscle of the hand of squirrel monkey. The internuclear areas (AR) are moderately (arrows m) and very strongly succinate dehydrogenase (SDH) positive (arrows). Note the rims also show moderate and very strong reactions. An intrafusal fiber, supposedly longitudinally cut, reveals very strong internuclear activity. The pericapsular epithelial cells (arrows p) are mildly and moderately stained. ×380. From Nakajima *et al.* (1968).

Fig. 34. Interosseous muscle of the hand of rhesus monkey. Lactate dehydrogenase (LDH) test. The internuclear areas are stained very strongly and moderately (arrow m). Note moderately stained pericapsular epithelial cells (PE) (arrows) and very strongly stained rims. ×480. From Nakajima *et al.* (1968).

Fig. 35. As in Fig. 65. Infant, 3 months. From Coërs (1962).

Fig. 36. Interosseous muscle of the hand of rhesus monkey. The rim of the equator shows moderately positive monamine oxidase (MAO) activity (arrows). The pericapsular epithelial cells are mildly positive (arrows p). ×480. From Nakajima *et al.* (1968).

Fig. 37. Succinate dehydrogenase activity, using MTT of intrafusal fibers in a spindle of a hindlimb (middle) lumbrical muscle of the rat. ×920. From Yellin (1969b).

Fig. 38. Phosphorylase activity of the same intrafusal fibers appearing in Figure 29. Four fiber types are readily apparent. ×920. From Yellin (1969b).

of the intrafusal fibers. The glucose-6-phosphate dehydrogenase activity was negligible in the intrafusal fibers. Different results were obtained by Spiro and Beilin (1969a,b). They studied NAD-dependent dehydrogenase, cytochrome oxidase and SDH in the human muscle spindle. Their study demonstrated a similarity in the histochemical reaction for oxidative enzymes in both the nuclear bag and nuclear chain intrafusal fibers and in none of the muscle spindles the cytochemical reactions in the intrafusal fibers duplicated the findings in either the type I or type II extrafusal fibers. The most common enzyme profile observed by Yellin (1969b) among the small intrafusal fibers is that of high SDH and high phosphorylase activity (Figs. 37 and 38). He suggested that it indicates a qualitatively similar innervation for this type of fibers. According to Engel (1965), Romanul and Van der Meulen (1967), and Yellin (1967), in the mammalian motor unit, all muscle fibers innervated by a particular motoneuron are alike as far as their metabolic characteristics are concerned. But the existence of small intrafusal fibers with SDH as well as low phosphorylase activity makes it more complicated.

2. Hydrolytic Enzymes

Nakajima *et al.* (1968) studied several hydrolytic enzymes in muscle spindles of interosseous and latissimus dorsi muscles of rhesus and interosseous muscles of squirrel monkeys (Figs. 39–43). The muscle spindles showed butyrocholinesterase activity (Figs. 41 and 42) in the walls of blood vessels which were also alkaline phosphotase positive (Fig. 43), but a negative acid phosphatase reaction. Lipase is located in the small as well as in the large intrafusal fibers of the chicken muscle spindle, being more abundant in the small fibers (Rebollo and Anda, 1967), and its activity is increased in the equatorial zone in both nuclear chain and nuclear bag fibers.

Magnesium-activated (Fig. 44) and calcium-activated ATPases (Figs. 39 and 45) are located in the walls of the blood vessels. These enzymes showed strong activity in the polar region of the intrafusal fibers as well as in the equator area, where magnesium-activated ATPase was located in the intranuclear cytoplasm as well as in the rim just beneath the sarcolemma. The intranuclear sarcoplasm, except beneath the sarcolemma, also showed calcium activated ATPase activity. Two types of intrafusal fibers have been described in the cat with Mg-ATPase (Henneman and Olson, 1965) and with Ca-ATPase activity (Nyström, 1967). Rebollo and Anda (1967) found that the activity of calcium-activated ATPase in chicken muscle spindles varies from fiber to fiber, from very small to large amounts.

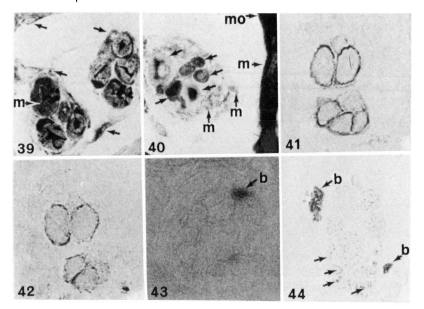

Fig. 39. Interosseous muscle of the hand of rhesus monkey. CaATPase test. The internuclear areas are moderately positive to strongly positive. The rims are moderately positive (arrows). The pericapsular epithelial cells show mild (arrows m) and moderate (arrows mo) staining. The capsular connective tissue is moderately stained. ×480. From Nakajima *et al.* (1968).

Fig. 40. Interosseous muscle of the hand of rhesus monkey. The strong acetylcholinesterase (AChE) reaction is found around the polar region of intrafusal fibers. ×480. From Nakajima *et al.* (1968).

Fig. 41. Interosseous muscle of the hand of rhesus monkey. Butyrocholinesterase (BuChE) test. Moderate activity is observed around the polar region. ×480. From Nakajima *et al.* (1968).

Fig. 42. Butyrocholinesterase (BuChE) test. The equator shows no activity. Note the moderately positive small blood vessel in the muscle spindle (arrow b). ×480 From Nakajima *et al.* (1968).

Fig. 43. Alkaline phosphatase test. The reaction of intrafusal fibers is negligible. The blood vessel in the spindle shows strong staining (arrows b). The pericapsular epithelial cells are mildly positive (arrows). ×300. From Nakajima *et al.* (1968).

Fig. 44. Longissimus dorsi muscle of rhesus monkey. The polar region of intrafusal fibers shows moderate and strong MgATPase reactions. Pericapsular epithelial cells are mildly (arrow m) and moderately positive (arrows). (D) ×300. From Nakajima *et al.* (1968).

Spiro and Beilin (1969a) described two distinctly different cytochemical reactions in the intrafusal fibers of each spindle from rabbit (Fig. 46), despite their morphological homogeneity (Barker, 1967). One of them has a very strong reaction, whereas the other is lightly stained. They related the difference to the contractile elements of the intrafusal

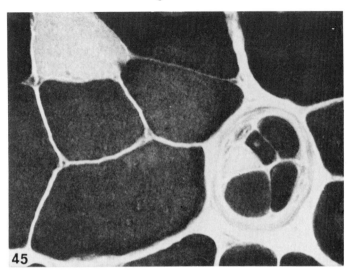

Fig. 45. Interosseous muscle of the hand of rhesus monkey. CaATPase test. Two large intrafusal fibers show a moderate reaction, and the small ones are stained very strongly. There are mildly positive pericapsular epithelial cells (arrows). ×300. From Nakajima *et al.* (1968).

fibers and to the different innervation. Spiro and Beilin (1969b), also showed in human muscle spindles different activity of myofibrillar ATPase in different intrafusal fibers. The nuclear chain fibers show a relatively higher level of this enzyme. The nuclear bag intrafusal fibers are generally of two types. Some of them are stained as type I of extrafusal fibers (weak activity of ATPase) and approximately an equal number of nuclear bag fibers are more lightly stained. Yellin (1969a) studied actomysin ATPase in the intrafusal and extraocular muscle fibers and found at least two forms of actomysin ATPase positive mammalian skeletal muscles. One was relatively alkali-stable, acid-labile localized in the fibers of fast-twitch muscles; the second form was alkali-labile, acid-stable and was located in the slow-twitch muscles (Figs. 47–52). Each showed only one of the two ATPase forms (Yellin, 1969a). However, the large intrafusal fibers of nearly every muscle spindle from hind leg muscles of rat have shown high levels of both alkali-stable and alkali-labile actomysin ATPase activity. Yellin observed this dual activity along the intrafusal fibers from their juxtaequatorial region up to the poles. This is in contrast to the nonreactive equatorial area. The nuclear chain fibers show only one form of ATPase. Since mammalian muscles are composed mainly of single efferented twitch muscle fibers, each of which exhibits only one form of ATPase, the dual enzyme activity in the same type of intrafusal fibers would suggest their multiaxonally innervation and capability for tonic, graded contraction (Barker, 1966b; Boyd, 1962, 1966; R. S. Smith, 1966). Experiments with the deinnervation and subsequent reinnervation of muscles seem to confirm this idea (Karpati and Engel, 1967; Romanul and Van der Meulen, 1967).

It may be concluded that the cytochemical pattern reflecting the metabolism of the fibers is different in the extrafusal and intrafusal fibers. Significant cytochemical differences in the nuclear bag and nuclear chain fibers are also evident. However, the preponderance of ATPase activity in the intrafusal fibers points to a higher level of energy metabolism.

The capsule of the muscle spindles contain varied amounts of dephosphorylating enzymes (Figs. 53–58), as was shown by Shantha *et al.* (1968). They concluded that the capsule may act as a metabolically active diffusion barrier.

Since AChE activity is essential for the transmitter function of acetylcholine at neuromuscular junctions, its localization would indicate the sites of such endings in those neuromuscular junctions, which are believed to be cholinergic. The cytochemical demonstration of AChE has provided a new approach to the study of polymorphic nerve endings on the muscle spindles. Motor endings contain a high level of cholinesterase (Fig. 40) and are located in the subneural apparatus. Coërs and Durand (1956) were the first to study AChE activity in the muscle spindles and described it in the motor endings on the intrafusal fibers of the muscle spindles (Figs. 35, 59–64) (Coërs, 1959, 1962; Coërs and Woolf, 1959). They described the AChE reaction in the muscle spindles from m. rectus abdominis of the rat, cat, and man, and stained the nerve endings vitally by methylene blue, in animal as well as human material, in order to compare the dimensions and the shape of the nerve terminals in the spindles (Table IV).

The detailed studies of Coërs indicate that in the equatorial region of the rat muscle spindle no AChE activity could be observed over a distance of 300–600 μ in young animals (10 days to 2 months old) and 420–660 μ in young animals (10 days to 2 months old) and 420–660 μ in 3-month-old rats. On both sides of the juxtaequatorial region, a strong reaction for AChE appeared in the nerve endings. The enzyme reaction was somewhat diminished in the extreme polar regions as compared to the equatorial area. In the cat muscle spindles from m. rectus abdominis, the equatorial region showed no AChE activity in an area of 370–660 mμ. The enzyme activity, however, appears on both sides of the intrafusal fibers, where the reaction product is located in well delimited subneural apparatus. As in the rat, the density of the AChE reaction in the nerve terminals of cat muscle spindles tends to diminish

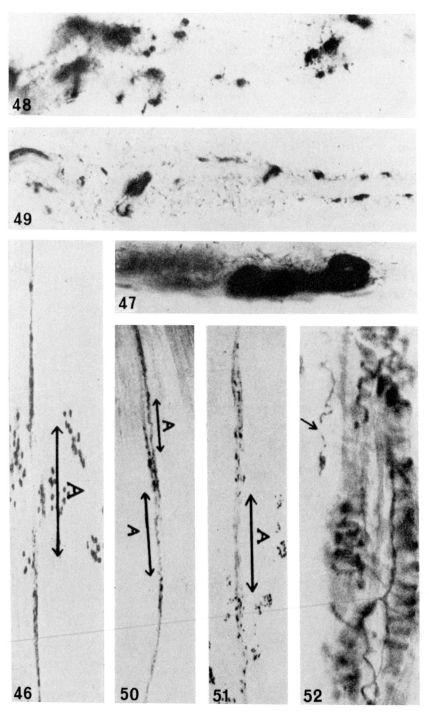

Figs. 46 to 52.

from equator to polar regions. In muscle spindles of man, AChE activity is observed in the cuplike structures, which according to Coërs resemble the extrafusal subneural apparatus. The equator area of intrafusal fibers in human muscle spindles is also devoid of AChE activity over a distance of 400–500 μ. A comparison of AChE activity with the picture obtained after vital staining of muscle spindles suggests a motor function of several types of nerve endings, being in close relation to the areas of AChE activity. Coërs concluded that the juxtaequatorial endings, generally described as flower-spray, include a motor ending, and the variety of nerve terminals suggests a heterogeneity of γ-motor system that supports two types of discharge in γ-fibers (Coërs, 1962). Häggquist (1960) studied the musculi lumbricales of the cynomolgus monkey and revealed that the equatorial region of the muscle spindle, in which the afferent innervation is concentrated, is free of AChE activity. The reaction appeared on the polar regions of the intrafusal fibers and was more marked in the proximal regions of the muscle spindle as compared to the distal area. Kupfer (1960) investigated cytochemically the AChE content of the human ocular muscle spindles and found the diffuse terminals along the large intrafusal fibers and typical motor end-plates containing strong AChE activity. He concluded that the innervation of the thin fibers comes from γ-nerves, whereas the large intrafusal fibers additionally received the endings from the α-motor fibers. Hess (1962) studies revealed two types of intrafusal endings in mammalian muscle

Fig. 46. Cross section of portion of rabbit muscle containing spindle. One light and four dark intrafusal fibers within laminated capsule. Two types of extrafusal fibers. ATPase reaction. $\times 520$. Reproduced from Spiro and Beilin (1969b).

Figs. 47–52. Actomyosin ATPase in medial gastrocnemius muscles of the rat. Figs. 47, 49, and 51 are prepared after alkaline preincubation, and Figs. 48, 50, and 52 after acid preincubation. The tissue in Fig. 52 was incubated in the presence of excess potassium (0.65 M potassium chloride). Muscle fibers labeled I exhibit the acid-stable, alkali-labile form of the enzyme, whereas fibers labeled II possess the alkali-stable, acid-labile variety. Figures 48 and 49 are closely adjacent sections illustrating the long muscle fiber, a spindle intrafusal fiber (arrow), exhibiting dual enzyme activity. Figures 49 and 50 illustrate the dual enzyme activity of an intrafusal fiber (arrow) in a second spindle, the latter sectioned at the juxtaequatorial region. In Fig. 49, all of the intrafusal fibers have clear central areas, indicating the presence of nuclei. In Fig. 46, at the equatorial region to the latter spindle, enzyme activity of the intrafusal fibers is minimal due to a reduction in the number of myofibrils and the aggregation of nuclei into nuclear bags and nuclear chains. In Fig. 52, an additional section of the latter spindle illustrates the inhibition by potassium of the acid-stable form of the enzyme, with the exception of that of the lone intrafusal fiber (arrow) which had exhibited dual enzyme activity. Nerve bundles in and about the spindles are relatively unreactive after either alkaline or acid preincubation. $\times 265$. Figures 47–52 reproduced from Yellin (1969b).

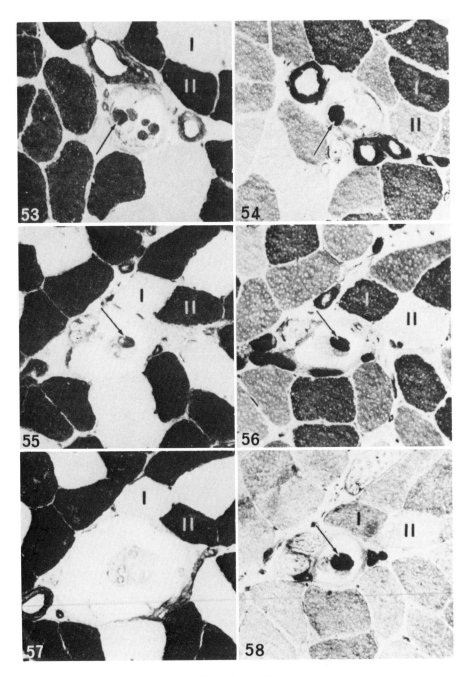

Figs. 53 to 58.

spindles. One group of the endings was located on the polar region, and the second one in the neighborhood of the equatorial area. Hess suggested that the diffuse multiterminal endings come from thin motor fibers, whereas the polar motor end plates come from large motor fibers. Rebollo and Anda (1967) studied cholinesterase in the muscle spindles of chicken and found the enzyme activity at the polar and equatorial zone of the large as well as thin intrafusal fibers. The reaction persisted in the control slides incubated with DFP.

VI. Pathology of the Muscle Spindle

A. Approach to the Problem

Because of the complicated etiopathology of most of the neuromuscular diseases, studies on alterations in the microscopic and fine structure of muscle spindles in these diseases are of great importance, and many problems related to muscle tone may be clarified on the basis of improved knowledge of the structure and behavior of muscle spindles during experimental and naturally occurring muscular diseases. The early investigations of Sherrington (1895), Batten (1897), Horsley (1897), and others indicate that muscle spindles remain unchanged in many muscular diseases, whereas the extrafusal muscle fibers show most of the pathological changes. Eisenlohr (1876) examined muscles from a case of infantile paralysis and found some pathological changes in the

---

Fig. 53. A 5′MP substrate. Note the strong positive activity in the nuclei as well as in the cytoplasm of the capsular cells (arrows). Note also that the intrafusal muscle fiber gives stronger reaction than the extrafusal muscle fiber. ×370. Figures 53–58 (showing the guinea pig thigh muscle) reproduced from Shantha *et al.* (1968).

Fig. 54. ATP substrate. The capsule of the muscle spindle (arrows) shows moderately strong positive reaction for the test. The intrafusal muscle fibers also show stronger reaction than the extrafusal muscle fibers. ×592.

Fig. 55. UMP substrate. Note the strong positive activity in the capsule (arrows) as well as in the intrafusal muscle fibers of the spindle. ×370.

Fig. 56. UTP substrate. The capsule (arrows) as well as the intrafusal muscle fibers show moderately strong positive activity. ×148.

Fig. 57. CMP substrate. The capsule (arrows) and the intrafusal muscle fibers show very strong positive activity for the test. The blood vessel (arrow = bv) in between the perineural epithelial layers is also strongly positive. ×370.

Fig. 58. CTP substrate. Moderate positive reaction in the capsule (arrows) of the muscle spindle as well as in the blood vessel found in between the capsular cells (arrows = bv). ×370.

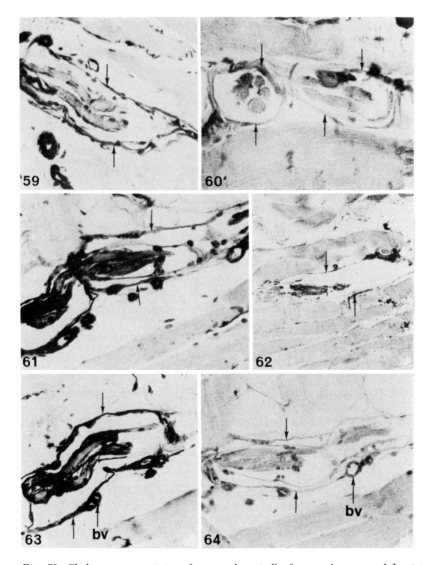

Fig. 59. Cholinesterase activity of a muscle spindle from rat's rectus abdominis muscle. (A) equatorial region devoid of cholinesterase activity. From Coërs (1962).

Fig. 60. High magnification of an intrafusal subneural apparatus of the usual (extrafusal) type. Rat, rectus abdominis muscle. From Coërs (1962).

Fig. 61. Scattered subneural apparatuses on intrafusal muscle fibers. Adult human palmaris longus muscle. From Coërs (1962).

Fig. 62. Compound muscle spindle at low magnification. Infant, 3 months; deltoid muscle. (A) region devoid of cholinesterase activity. From Coërs (1962).

Fig. 63. Muscle spindle at low magnification. Infant, 8 months; vastus medialis muscle. (A) equatorial region. From Coërs (1962).

Fig. 64. Annulospiral endings in human muscle spindle. Arrow indicates unbranched ending coming from small nerve fiber. From Coërs (1962).

TABLE IV

Length of Areas in Rat and Cat Spindles Devoid of Cholinesterase and Occupied by Annulospiral (a-s) and Flower-Spray (f-s) Endings[a]

		Histochemical method		Vital staining method	
Age	Muscle	Total spindle length (mm)	Length without choline- sterase (μ)	Length (μ) annulo- spiral ending	Length (μ) flower- spray ending
Rat					
10 days	Rect. abd.	2.0	360	—	—
		2.6	600	—	—
2 months	Rect. abd.	2.2	420	240	—
		2.0	420	360	—
		2.2	360	480	—
		2.1	480	540	—
3 months	Rect. abd.	2.2	420	620	—
		2.2	600	—	—
		2.6	600	—	—
		2.2	620	—	—
		2.4	660	—	—
2 months	Vast. med.	2.1	420	—	—
		2.1	540	—	—
Cat					
1 month	Rect. abd.	—	370	340	400[b]
		—	420	—	480[c]
		—	480	400	440
		—	480	420	240[b]
		3.6	500	—	350[c]
		4.5	500	500	—
		4.0	560	550	—
		4.0	660	—	—
Adult	Sartorius	—	250	—	—
		—	430	—	—
		—	460	—	—
		—	600	—	—

[a] Reproduced from Coërs (1962).
[b] Devoid of cholinesterase.
[c] Occupied by annulospiral and flower-spray endings.

fibers. Fraenkel (1878) studied the muscles of the tuberculosis patients and described "*unschnurte Bundel*" located usually near the nerves and blood vessels. Babinski (1886) reported an existence of several atrophied muscle fibers surrounded by a circular ring in the case of chronic myelitis. In 1886, Babinski discovered bundles consisting of three to seven

TABLE V

QUALITATIVE AND QUANTITATIVE FEATURES OF MUSCLE SPINDLES AND
INTRAFUSAL CONTENTS[a]

Serial number	Condition of extrafusal muscle	Number of cases	Number of muscles	Number of spindles	Size of capsule (μ)
1	Normal	25	25	32	50–320
2	Denervation atrophy	56	65	140	29–180
3	Nonspecific atrophy (including nutritional)	13	13	21	54–200 (one of 615 μ)
4	Muscular dystrophy	19	20	34	50–215
5	Myotonia dystrophica	5	5	11	72–145
6	Chronic polymyositis	7	7	12	61–215
7	No opinion	5	5	7	36–180
	Total	130	140	257	
Post mortem muscles					
8A	Fetal	2	8	50	30–108
8B	Others	4	9	42	—
	Total	6	17	92	

[a] Reproduced from Patel *et al.* (1968).

small muscle fibers with blood vessels and nerves covered by laminated connective tissue capsules. Greenfeld *et al.* (1957) and Adams *et al.* (1962) in their monographs on the muscle diseases confirmed the existing view among neuropathologists about the immunity of neuromuscular spindles in the course of various muscular diseases, and the draft report of the Research Group on Neuromuscular Diseases of the World (1967) (Federation of Neurology on the quantitation of muscle biopsy findings) pointed out that "Very little evidence is available on the pathological changes in muscle spindles."

The evaluation of morphological alterations in the muscle spindles in the course of muscular diseases is difficult for many reasons. Regarding methodology, biopsies are taken by different methods; the muscle might be in different functional states, and the fixation of biopsied material in different fixatives also changes the properties of the tissue. The ideal procedure would be to freeze the specimen in liquid nitrogen, which is not performed in most laboratories. Another problem is the small size of the spindle and the difficulty of finding these neuromuscular junctions in the small pieces of the biopsied material. Lapresle and

Diameter of intrafusal fibers (μ)	Number of intrafusal fibers	Capsule thickening	Intracapsular connective tissue	Intrafusal eosinophilic material	Intrafusal fibers structural change	Central nuclei	Nuclear clumping
4–32	2–10	1	1	0	0	2	1
4–40	0–19	7	4	7	2	1	0
4–25	2–12	1	0	1	0	1	0
4–29	2–14	1	2	2	0	1	2
4–22	2–10	1	1	0	0	4	2
4–25	2–10	0	1	3	0	0	1
4–32	2–9	1	0	0	0	0	0
3–18	1–10	0	0	0	0	0	5
—	—	—	—	—	—	—	—

Milhaud (1969) found muscle spindles in 150 cases out of 1200 biopsies and finally only 52 were selected for a detailed study. Patel *et al.* (1968) found the muscle spindles in only 130 of 1,000 biopsies (Table V).

Nor do we know the absolute and relative sizes of the nuclear bag and nuclear chain fibers, the degree to which the intrafusal fibers decrease in diameter reaching the poles of the spindles, and the extent of the variations in size of the intrafusal fibers in relation to the muscles to which they belong. Cooper and Daniel (1963) stated that between the equator and the poles of the spindle, the nuclear bag fibers are 15 μ or more in width, as compared with 10–15 μ for the nuclear chain fibers. On the other hand, Lapresle and Milhaud (1969) described the large intrafusal fibers in the m. peroneus longus lateralis having 30–40 μ diameter and small fibers with 20–25 μ diameter. Barker and Ip (1965) and Barker (1967) developed the theory regarding the cyclic renewal and decay of peripheral endings and this makes the interpretation of the morphology of the spindles more difficult. It is, however, extremely important to learn of the duration of a particular neuromuscular disease, because the behavior of the spindles is closely conditioned with the

duration of the illness. Also, the interpretations of the end changes
are not easy because of the complex structure of the muscle spindle.
Long after some of the extrafusal muscle fibers have ceased to function
and are in the state of degeneration, the intrafusal fibers may appear
intact. These fibers have nerve endings coming from sensory as well
as motor systems. Even if one ending remains intact, it may have the
property of greatly prolonging the existence of the muscle fiber. Thus,
in a section of a muscle stained by the standard tissue stains, there
may be very little to suggest abnormality of the muscle spindle. A very
different picture may be obtained when the specific stains for nerves
and their endings are employed.

B. Experimentally Induced Neuromuscular Disorders

1. EXPERIMENTAL DENERVATION OF MUSCLES

In order to investigate the behavior of muscle spindles in the course
of neuromuscular diseases, the experimental denervation procedure has
been carried out in many laboratories, giving the advantage of studying
the dynamic changes in the sensory and motor nerve endings in the
intrafusal fibers due to damage of a particular part of the nervous system.
Sherrington (1895) and Horsley (1897) believed that the spindles remain unchanged following denervation. Sherrington (1895) examined
muscles in which all motor fibers had been divided and the muscle
fully denervated. He observed that the spindles retained their characteristic features, i.e., the intrafusal fibers were normal and the nerve fibers
were well preserved and could be traced to the sensory roots.

Lesions of both motor and sensory fibers during the division of anterior
and posterior roots in the dog cause marked changes in the intrafusal
muscle fibers (Onanoff, 1890). In contrast, Horsley (1897) did not find
any changes in the intrafusal fibers of muscle spindles in the gastrocnemius and soleus muscles of the cat during experimental dividing
of the sciatic nerve in a period up to 3 years. The only alteration was
the shrinkage of the spindle on the seventieth day after the operation.
Batten (1897) studied muscle spindles in cats 3 weeks and 1, 2, and
3 months after section of the sciatic nerve and did not find any sign
of atrophy of the intrafusal muscle fibers. Another point of view was
represented by Chor *et al.* (1937), who described atrophic changes in
intrafusal as well as extrafusal fibers in monkeys after section of the
sciatic nerve.

Tower (1932) carried out a series of elegant experiments in which
she cut motor or sensory roots to muscles in the cat and, after appropriate

time for degeneration, examined the muscles using Bielschowsky's technique. She observed that if the dorsal root ganglia were removed, only the equatorial regions of the spindles were affected; the nuclei of the nuclear bags were reduced in number and volume, and the cross striation of the intrafusal fibers were more clearly seen over the central regions. The capsules invested the muscle fibers closely, the large nerve fibers degenerated, and there was no thickening of the capsule. If, however, the ventral roots were cut, the intrafusal muscle fibers and their nerves in the polar regions of the spindles atrophied. The capsule also thickened, but the equatorial regions remained intact, and the spindles were very conspicuous. When the whole nerve was cut, the early changes in the spindles were less obvious than in the rest of the muscle. Later the intrafusal fibers became smaller, their nuclei swelled and degenerated, the cross striations faded, the longitudinal striations became more obvious, and the amount of the fibrous connective tissue increased.

Sunderland and Ray (1950) later showed that the atrophic changes appearing up to 485 days after the experimental denervation of muscles in Australian opossum if the intrafusal fiber of the muscle spindles are not as developed as in the extrafusal fibers. Boyd (1962) also showed that 3 months after an appropriate ventral root section, the mean diameter of the nuclear bag fibers, nuclear chain fibers, and extrafusal fibers were reduced from normal values to 90%, 65%, and 50%, respectively. The equatorial region of the spindle remained unchanged. During experimental lesion of the posterior roots, Boyd found gradual atrophy and disappearance of the central nuclei at the equatorial region of the spindle, first in the nuclear chain fibers and later in the nuclear bag fibers. Gutman and Zelena (1962) found a gradual decrease in the number of nuclei in the nuclear bag fibers as early as 1 month after complete denervation. The atrophy of intrafusal fibers during their experiments was less prominent as compared to the changes in extrafusal fibers. Gutman and Zelena (1962) concluded that the denervation altered mainly the equatorial zone of the nuclear bag fibers, and no decrease in the number of spindles in the experimental muscle takes place compared to the control material. The spindles from experimentally denervated muscle showed a 50% decrease in diameter, chiefly due to a reduction in size of "lymphatic space." The degree of the degeneration and the extent of the atrophy in individual muscle spindles has always been difficult to estimate.

2. Muscle Atrophy

Esaki (1966) immobilized the hind limb of a rabbit by covering it with stockinette sleeve and gypsum bandage for a period of 1–9 weeks

TABLE VI

Muscle Spindle in Denervation Atrophy[a]

Case	Age	Muscle	Disease	Stage of the disease	Overall (outer) diameter of the spindle (μ)	Number of intrafusal muscle fibers
1	1 year	Quadriceps	Werdnig–Hoffmann disease	Moderately advanced	115	10
					115	10
2	6 months	Quadriceps	Werdnig–Hoffmann disease	Early	126	11
					182	12
					98	5
					91	4
					42	6
					42	4
3	18 months	Quadriceps	Werdnig–Hoffmann disease	Advanced	98	5
					140	4
					35	4
					56	4
					115	3
4	4 months	Quadriceps	Werdnig–Hoffmann disease	Advanced	126	15
					112	11
5	7 months	Quadriceps	Werdnig–Hoffmann disease	Early	98	8
					70	6
					98	4
6	2 years	Quadriceps	Werdnig–Hoffmann disease	Early	98	10
7	9 weeks	Quadriceps	Werdnig–Hoffmann disease	Advanced	154	6
8	4 years	Biceps brachii	Kugelberg–Welander syndrome	Moderately early	140	3
9	20 years	Deltoid	Kugelberg–Welander syndrome	Moderately advanced	98	3
10	3 years	Deltoid	Kugelberg–Welander syndrome	Advanced	84	3
11	25 years	Deltoid	Kugelberg–Welander syndrome	Early	84	9
					63	4
12	48 years	Quadriceps	Motor neurone disease	Advanced	112	4
13	55 years	Biceps brachii	Diabetic neuropathy	Early	84	6
14	60 years	Tibialis anterior	Chronic poly-neuropathy	Moderately advanced	154	6
					140	5
					126	4

[a] Reproduced from Cazzato and Walton (1968).
f Fibers not clearly recognisable as nuclear bag or nuclear chain.

in order to establish the relationship between muscle atrophy and the degeneration of muscle spindles. The muscle spindles from plantaris, soleus, and tibialis anterior muscles were studied from both the immobilized and free hind limb. It was observed that degeneration of muscle spindles started in the immobilized muscles as early as 1 week after

Number of nuclear bag fibers	Number of nuclear chain fibers	Mean diameter of nuclear bag fibers (μ)	Mean diameter of nuclear chain fibers (μ)	Mean diameter of 100 normal extrafusal fibers (μ)	Mean diameter of 100 moderately atrophic extrafusal fibers (μ)	Mean diameter of 100 severely atrophied extrafusal fibers (μ)
4	6	8.4	4.2	—	16.8	5
3	7	9.8	5.6	—		
1	10	21	11.2			
1	11	28	14			
1	4	28	14	38		
2	2	21	11.2		—	
2	4	21	11.2			
4	—	18.2	—			7
—	—	—	5.6[b]			
—	—	—	4.2[b]			
—	—	—	7.3[b]		—	—
1	3	21	9.8			
1	2	21	11.2			
5	10	11.2	7			
—	—	—	8.4	24	—	4.2
1	7	28	12.6			
—	—	—	12.6[b]	31	20	9.8
—	—	—	12.6[b]			
4	6	8.4	4.2	51.8	—	7
3	3	12.3	7	—	14	6
2	1	28	14	28	—	12.6
3	—	21	—	51	—	11.2
2	1	16	8.4	35	—	5.6
3	6	28	9			
2	2	20	10	35	—	9
2	2	7	4.2	32	14	6
3	3	12.3	7	35	—	12.6
1	5	21	12.6			
3	2	21	12	42	—	8.4
3	1	21	8			

the experimental procedure, especially within the sensory nerve endings, which became irregular in shape, and its spiral structure as well as its ending disappeared. In addition, the fusimotor nerve fiber endings disappeared, and the capsule became swollen, especially in the equator of the spindle. The number of intrafusal fibers in the experimental muscle

TABLE VII

THE MUSCLE SPINDLE IN PROGRESSIVE MUSCULAR DYSTROPHY[a]

Case	Age	Muscle	Type of dystrophy	Stage of the disease	Capsule	Periaxial space	Septa	Blood vessels
1	11 months	Quadriceps	Duchenne	Preclinical	Thickened	Enlarged	Thickened	Thickening of the walls
					Slightly thickened	Normal	Normal	Normal
2	2 years	Quadriceps	Duchenne	Preclinical	Normal	Enlarged	Thickened	Normal
					Slightly thickened	Enlarged	Thickened	Thickening of the walls
					Thickened	Normal	Slightly thickened	—
					Thickened	Enlarged	Slightly thickened	—
					Normal	Enlarged	Normal	—
					Normal	Enlarged	Normal	—
3	2 years	Quadriceps	Duchenne	Preclinical	Normal	Enlarged	Normal	—
					Normal	Slightly enlarged	Normal	Normal
4	6 years	Quadriceps	Duchenne	Early	Thickened	Normal	Normal	Normal
5	7 years	Rectus abdominis	Duchenne	Early	Normal	Enlarged	Slightly thickened	Normal
6	8 years	Biceps brachii	Duchenne	Moderately advanced	Normal	Enlarged	Normal	—
7	13 years	Quadriceps	Duchenne	Advanced	Thickened	Absent	—	—
					Thickened	Enlarged	Slightly thickened	Thickening of the walls
					Normal	Enlarged	Normal	Thickening of the walls
8	32 years	Biceps brachii	Limb–girdle	Early	Slightly thickened	Enlarged	Normal	Normal
					Normal	Normal	Normal	—
9	41 years	Pectoralis major	Limb–girdle	Moderately early	Edematous swelling	Enlarged	Slightly thickened	—
10	4 years	Quadriceps	Limb–girdle	Moderately early	Normal	Normal	Normal	Normal
					Normal	Slightly enlarged	Normal	—
11	59 years	Quadriceps	Limb–girdle	Moderately advanced	Thickened	Reduced	Thickened	Thickening of the walls
					Thickened	Absent	Thickened	Thickening of the walls
					Thickened	Reduced	Thickened	—
					—	Filled with connective tissue	—	—
					Normal	Reduced	Thickened	—
12	24 years	Triceps brachii	Limb–girdle	Moderately advanced	Edematous swelling	Enlarged	Normal	—

[a] Reproduced from Cazzato and Walton (1968). In this table, only those sections containing one or more abnormal muscle spindles have been listed. In addition, in certain sections nine other spindles were found that were regarded as completely normal in every respect.

[b] Fibers not recognizable as nuclear bag or nuclear chain.

Intrafusal muscle fibers	Sarcolemmal nuclei (nuclear bag fibers)	Overall diameter of spindle (μ)	Number intrafusal muscle fibers	Number nuclear bag fibers	Number nuclear chain fibers	Mean diameter nuclear bag fibers (μ)	Mean diameter nuclear chain fibers (μ)
Normal	Normal	154	3	3	—	21	—
Normal	Normal	136	11	2	9	25.2	7
Normal	Normal	140	5	2	3	21	11.2
Atrophied	Normal	182	2	1	1	14	9.8
Atrophied	Normal	91	6	3	3	15.4	7
Atrophied	Two pyknotic	210	10	—	—	—	12.6[b]
Atrophied; splitting of a fiber	All pyknotic	140	6	3	3	12.6	7
Normal	Clumps of dark-stained	154	12	3	9	12.2	11.2
Atrophied	Normal	119	8	3	5	12.6	7
Normal	Normal	84	8	4	4	18.2	8.4
Normal	Normal	168	8	3	5	21	9.8
Atrophied	One pyknotic	126	2	2	—	16.8	—
Atrophied	Normal	98	2	2	—	14	—
Only fragments	Few and pyknotic	154	—	—	—	—	—
Atrophied	Pyknotic and shrunken	119	4	—	—	—	11.2[b]
Some empty sarcolemmal sheaths	—	140	1	—	—	—	14[b]
Two atrophied	Two dark-stained	160	7	3	4	21	9.2
Normal	Normal	140	8	5	3	22	7
Atrophied	—	252	8	—	—	—	14[b]
Normal	Normal	182	7	2	5	21	12.6
Atrophic and eosinophilic	Dark-stained	84	6	4	2	15.4	9
Normal	Normal	154	9	5	4	25.2	7
Atrophied	Normal	140	3	2	1	16.8	7
Degenerating with replacement fibrosis	Few and shrunken	168	5	—	—	—	19.6[b]
No fibers	Some normal amid strands of connective tissue	140	—	—	—	—	—
Normal	Normal	160	9	5	4	28	7
Atrophied	Some pyknotic	252	5	1	4	16.8	8.4

spindles was also reduced. Esaki (1966) concluded that the reduction of impulses from the spindles due to the immobilization of the muscle might be the reason for muscle atrophy.

3. Experimental Demyelination

McDonald and Gilman (1968) showed that in cats with experimental demyelination in the region of dorsal root ganglia caused by experimental diphtheric polyneuritis, the stretchability of many muscle spindles' primary receptors in the early ataxic phase is raised.

4. Experimental Studies of the Muscle Spindle in Hereditary Neuromuscular Diseases

Since the hereditary neuromuscular disorders in mutants of mice are potentially analogous to the same type of disease of man, experimental studies on the animals are of great importance and may help to reveal the pathogenesis of these illnesses. Meier (1969) studied the spindles from psoas, iliopsoas, semitendinosus, and gastrocnemius muscles in normal and mutant mice suffering from various muscular and neuromuscular diseases. The muscles were taken from shambling, ducky, spastic, lethargic, disoriented, rabbit, and dystrophic mice and were compared to the clinically normal homozygous or heterozygous littermates. The pathological changes included increased diameter of the intrafusal fibers, vacuolization, fiber necrosis, and central rowing of nuclei in the shambling animals compared to no such abnormalities in the mutant mice. Also, no pathological alterations were observed in the muscle spindles of spastic and dystrophic mice. In teetering and ducky mice the muscle spindles were small; there was no qualitative difference between the infants and the controls. In the rabbit mutants, occasional thickening of the intrafusal fibers was detected, whereas in others a discrete thickening of the capsule was observed. Similar changes were found in the lethargic mutants. In most cases, the extrafusal fibers showed severe changes in the form of lesions. Meier (1969) concluded that muscle spindles in most of the cases under investigation showed only minor abnormalities as compared to the severe changes in the extrafusal fibers.

C. Naturally Occurring Neuromuscular Diseases

In most instances of human neuromuscular diseases, etiopathogenesis is very complicated. First, both sensory and motor fibers are often involved, and second, the material obtained from different patients is difficult to compare because of the varying degree of development of the particular illness. Also, the different parts of the nervous system may be involved, which further complicates the picture of the changes in the muscles.

Forster (1894), in a case of transverse myelitis, found complete degeneration of the extrafusal muscle fibers, but muscle spindles remained intact. Siemerling (1889) found a normal spindle in the case of alcoholic neuritis, whereas Gudden (1896), who also studied alcoholic neuritis, observed that the muscle fibers in the spindles were degenerated and the nerve in the spindles showed different kinds of affinity to the stain as compared to the normal ones. The degeneration of the muscle in advanced cases of polyneuritis was also observed by Wohlfart (1949). Dastur (1967) found disorganization and degeneration of the sensory nerves in the muscle spindle in the case of leprous neuritis; however, in other cases of the same illness, the normal axons were seen within the spindle. Adams *et al.* (1962) in their book "Diseases of Muscle" described a case of chronic polyneuritis showing great distention and capsular space with investion of mononuclear cells.

A series of systematic studies of muscle spindles in the neuromuscular disorders were carried out on twelve hundred biopsies by Lapresle and Milhaud (1969). They described in eighteen out of twenty-three cases of sensory motor polyneuropathy, changes affecting intrafusal muscle fibers and nerve fibers supplying the spindle. In three cases, the intrafusal fibers were markedly atrophic, and in five cases the proliferation of their nuclei was observed. In the case of denervation atrophy, Patel *et al.* (1968) showed extrafusal denervation changes in fifty-six cases. Eleven were of spinal origin; twelve were peripheral neuropathies; and in thirty-three cases, the nature of denervation was not known. Out of the fifty-six cases, seven specimens showed capsular thickening, four showed increased intracapsular connective tissue, and seven showed an accumulation of eosinophilic material. Only two cases have shown the structural changes in the intrafusal fibers, and one showed no intrafusal fibers at all (Table V).

Patel *et al.* observed, in cases of denervation atrophy, that three of the five intrafusal fibers come out of the capsule for a short distance and then reenter it. This may be related to the developmental disorder of this muscle spindle. Dastur (1967) described how in some cases of leprous neuritis degeneration and disorganization of the sensory nerves in the muscular spindle appear, whereas in others, the normal axons were seen in the muscle spindle. Greenfeld *et al.* (1957) studied one hundred twenty-one cases of various neuromuscular diseases and

TABLE VIII

THE MUSCLE SPINDLE IN CONGENITAL MUSCULAR DYSTROPHY
AND MYOTONIC DYSTROPHY[a]

Case	Age	Muscle	Disease and stage of the disease	Capsule	Periaxial space	Septa	Blood vessels
1	12 years, 8 months	Quadriceps	Congenital dystrophy, advanced	Slightly thickened	Enlarged	Thickened	—
				Slightly thickened	Enlarged	Thickened	—
				Thickened	Enlarged	Thickened	Normal
2	18 months	Deltoid	Congenital dystrophy, early	Thickened	Slightly enlarged	Normal	—
3	16 years	Biceps brachii	Congenital dystrophy, moderately advanced	Normal	Normal	Normal	Normal
4	1 year	Quadriceps	Congenital dystrophy, moderately advanced	Normal	Enlarged	Normal	Normal
5	4 years	Biceps brachii	Congenital dystrophy, advanced	Thickened	—	Thickened	Normal
				Thickened	Reduced?	Thickened	—
				Thickened	—	Thickened	Thickening of the walls
				Edematous swelling	Enlarged	Thickened	Normal
				Thickened	Enlarged	Normal	Thickening of the walls
				Normal	Normal	Normal	Normal
				Normal	Normal	Normal	—
				Normal	Normal	Normal	Normal
				Thickened	Enlarged	Thickened	—
				Normal	Normal	Normal	Normal
				Normal	Enlarged	Normal	Thickening of the walls
				Normal	Normal	Slightly thickened	Normal
				Slightly thickened	Enlarged	Slightly thickened	Thickening of the walls
				Slightly thickened	—	Thickened	—
				Normal	Enlarged	Slightly thickened	—
				Normal	Enlarged	Normal	—
				Thickened	Slightly enlarged	Normal	Normal
6	26 years	Deltoid	Myotonic dystrophy, moderately advanced	Thickened	Enlarged	Thickened	Thickening of the wall
7	4 years	Deltoid	Myotonic dystrophy, early	Slightly thickened	Enlarged	Thickened	—

[a] Reproduced from Cazzato and Walton (1968).
[b] Fibers not recognisable as nuclear bag or nuclear chain.

found only eight spindles in which some changes could be observed. Cazzato and Walton (1968) studied several hundred biopsies and selected fifty-six cases suitable for the study of pathological changes of

Intrafusal fibers	Sarcolemmal nuclei (nuclear bag fibers)	Overall diameter of spindle (μ)	Number intrafusal fibers	Number nuclear bag fibers	Number nuclear chain fibers	Mean diameter nuclear bag fibers (μ)	Mean diameter nuclear chain fibers (μ)
Fragments of fibers; replacement fibrosis	—	140	—	—	—	—	—
Fragments of fibers; replacement fibrosis	—	145	—	—	—	—	—
Atrophied but normal in appearance	Some pyknotic	210	6	2	4	14	7
Atrophied	Normal	84	8	3	5	14	5.6
Normal	Normal	126	8	3	5	23.8	11.2
Atrophied	Normal	49	4	2	2	12.6	8.4
No fibers	—	180	—	—	—	—	—
Atrophied; two eosinophilic	Some shrunken	140	6	1	5	14	7
No fibers	—	140	—	—	—	—	—
Eosinophilic	Pyknotic and shrunken	260	4	—	—	—	11.2[b]
Normal	Some pyknotic	140	5	1	4	21	12.6
Normal	Normal	126	6	2	4	19.6	11.2
Normal	One pyknotic	98	7	3	4	21	11.2
Normal	Normal	98	9	3	6	21	10.6
Eosinophilia and fragmentation	Pyknotic and shrunken	126	5	—	—	—	9.8[b]
Normal	One pyknotic	105	6	4	2	21	12.6
Atrophic and eosinophilic	Pyknotic	105	6	3	3	14	5.6
Normal	Normal	130	5	2	3	28	14
Atrophied	Some pyknotic	140	5	3	2	12.6	9.8
No fibers	—	140	—	—	—	—	—
Normal	One pyknotic	168	6	1	5	21	7
Two degenerating	Three pyknotic	182	6	1	5	14	9.8
Atrophic and eosinophilic	One pyknotic; two shrunken	105	6	3	3	14	7
Atrophy and degeneration	Some pyknotic	350	9	4	5	14	7
Atrophy and degeneration	Some pyknotic	250	5	—	—	—	11.2[b]

the muscle spindles in various neuromuscular and muscular disorders and compared them with a number of normal autopsies. The selected cases included severe infantile spinal muscular atrophy (Werdnig-

TABLE IX

MUSCLE SPINDLE IN BENIGN CONGENITAL HYPOTONIA AND NEMALINE MYOPATHY[a]

Case	Age	Muscle	Disease	Overall diameter of spindle (μ)	Number intrafusal fibers	Number nuclear bag fibers	Number nuclear chain fibers	Mean diameter nuclear bag fibers (μ)	Mean diameter nuclear chain fibers (μ)	Mean diameter of 500 extrafusal fibers (μ)
1	6 years	Quadriceps	Benign congenital hypotonia	112	10	—	—	—	9.8[b]	29
2	2 years	Quadriceps	Benign congenital hypotonia	140	6	3	3	14	7	15.8
				140	8	5	3	16.8	9.8	
				84	8	3	5	14	5.6	
									12.6[b]	
3	2 years, 8 months	Biceps brachii	Benign congenital hypotonia	126	3	—	2	—	8.4	32.6
				149	4	2		18.6	8.4	
4	11 years	Quadriceps	Benign congenital hypotonia	63	4	2	2	16.8	8.4	46
				84	6	2	4	19.6	9.8	
						2	8	14	8.4	
5	15 months	Quadriceps	Benign congenital hypotonia	112	10				11.2[b]	15.7
				112	12	2	11	16.8	11.2[b]	
				112	13	2	6	21	14	
6	3 years	Quadriceps	Benign congenital hypotonia	120	8	1	4	21	12.6	15.1
				210	2	1		19.6	—	
				119	5	2	6	20	12.6	
7	4 months	Quadriceps	Benign congenital hypotonia	48	1		8	15.4	5.6	10.3
				84	8	4	6	7	5.6[b]	
				84	28	2	5	7	4.2	
				84	12	2	12	11.2	4.2	
				49	8	5	9	9.8	4.2	
				49	7	4		7	5.6	
				112	17				4.2	
				49	13				5.6[b]	
				49	6				5.6	
8	16 years	Quadriceps	Nemaline myopathy	280	9	4	5	19.6	5.6	

[a] Reproduced from Cazzato and Walton (1968).
[b] Fibers not recognisable as nuclear bag or nuclear chain.

Hoffman) (Figs. 65 and 66), congenital dystrophy (Fig. 67), adult motor nervous disease, diabetic *neuropathy*, peripheral *neuropathy*, congenital and progressive muscular dystrophy (Figs. 68–70), *dystrophia myotonica, polymyositis, sarcoid myopathy, nemaline myopathy* and *benign congenital hypotonia*. The results are shown in the Tables VI–X. Table XI outlines the variations in intrafusal fiber diameter in pathological conditions.

In their evaluation, Cazzato and Walton (1968) used the following criteria: (1) the site where the muscle spindles are found; (2) the site of muscle spindle capsule; (3) the septa inside the capsule; (4) the nerve supplying the muscle spindle (normal, atrophied, endoneural, or perineural fibrosis); (5) the total diameter of the muscle spindle; (6) the relative number of nuclear bag and nuclear chain fibers; (7) the morphology of nuclei; and (8) intrafusal fibers and vessels inside the capsule and the presence and the degree of the atrophy of the intrafusal fibers. In the case of denervation atrophy, the atrophy of the intrafusal fibers affecting mainly the nuclear chain fibers was observed simultaneously with an enlargement of periaxial space and a slight thickening of the capsule. Progressive muscular dystrophy was characterized by thickening of the capsule and the connective tissue inside the muscle spindle as well as by the atrophy of nuclear bag and nuclear chain fibers. Lapresle and Milhaud (1969) observed degenerative changes and atrophy of intrafusal fibers connected with subsequent fibrosis in the biopsies obtained from the patients suffering from congenital muscular dystrophy. Also observed was a thickening of the capsule of the muscle spindle. Cazzato and Walton (1968) figured out that these changes were more severe than in the progressive muscular dystrophy of Duchené (Figs. 69 and 70) and limb-girdle (Fig. 68) types. They also observed the degeneration of the intrafusal fibers in the cases of dystrophia myotonica.

Dystrophia myotonica is the inherited disease, which in addition to weakness and wasting, is characterized by an inability of muscles to relax after voluntary effort. Muscles show also this abnormal response to mechanical stimulation. Histopathological changes in the extrafusal muscle fibers are characteristic and consist of an increase in fibrous tissue and replacement of muscle fibers by fat. The nuclei are small and darkly stained, and sometimes necrotic muscle fibers may be observed. The role of the muscle spindles in dystrophia myotonica has been discussed in detail by Denny-Brown and Nevin (1941).

Slauck (1921) and Coërs (1952) pointed out that pathological changes might appear in the spindles of the dystrophic muscles. Daniel and Strich (1964) studied the muscle spindles in the intrinsic muscles of

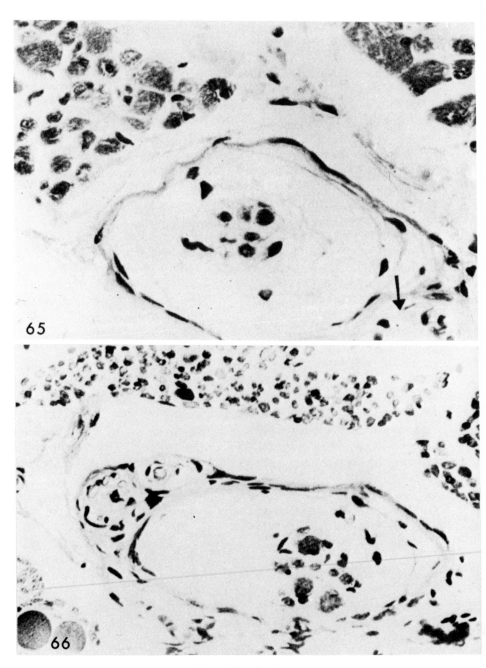

Figs. 65 and 66.

the hand and feet, limb, and trunk muscles, as well as muscles of the tongue and sternomastoids in five cases of dystrophia myotonica, and found abnormal muscle spindles in all studied samples; however, they were most common in the small muscles of the hands. The changes most often observed include an increase in the number of the intrafusal fibers and a decrease in their diameter. Daniel and Strich (1964) described up to sixty intrafusal fibers in the muscle spindle from the lumbrical muscle in the case of dystrophia myotonica (Figs. 71 and 72). The nuclei in the intrafusal fibers were small and darkly stained, and sometimes pyknosis has been observed. The capsules of the spindles were often thickened and filled up by collagen. Daniel and Strich (1964) have also found large inflammatory mononuclear cells inside the muscle spindles which were never observed in any other nerve endings in the case of the dystrophy. Dystrophic muscle spindles were thick, and very fine nerve fibers sprouted on the increased number of intrafusal fibers (Fig. 73). This increase in the number of intrafusal fibers and their altered nerve supply could lead to abnormal discharges from the spindles. Daniel and Strich concluded that similar morphological changes observed in the spindles, as well as in the extrafusal fibers, strongly suggest that both kinds of muscle fibers are affected by the same pathological process.

Kennedy (1969) studied the innervation of spindles obtained during diagnostic muscle biopsies from patients suffering from Guillain–Barré syndrome (acute and recurrent) and Charcot–Marie–Tooth disease associated with alcoholism. He observed a general reduction of the axons entering the spindle. In chronic neuropathies, however, the intrafusal ramification of the persisting axons was greatly enhanced. The most commonly observed change in the motor axons was related to the formation of the terminal coil, surrounding the nuclear bag as well as the nuclear chain intrafusal fibers. The fibers, forming the coil, were less

---

Fig. 65. Transverse section of a muscle spindle from a case of Werdnig–Hoffman disease containing five extremely atrophic intrafusal muscle fibers, the type of which is no longer recognizable; the extrafusal muscle fibers are extremely atrophic, too. A small nerve trunk (arrow) supplying the spindle does not contain nerve fibers. Picro-Mallory, ×960. Figures 65–70 reproduced from Cazzato and Walton (1968).

Fig. 66. Transverse section at about the equatorial region of a muscle spindle from a case of Werdnig–Hoffman disease containing four almost normal sized nuclear bag fibers and several nuclear chain fibers that are markedly atrophic (compare their size with that of the extremely atrophic extrafusal fibers which surround the spindle); in the lower left corner are visible two less atrophic extrafusal fibers. Two small nerve trunks containing few nerve fibers lie just outside the capsule. Picro-Mallory, ×702.

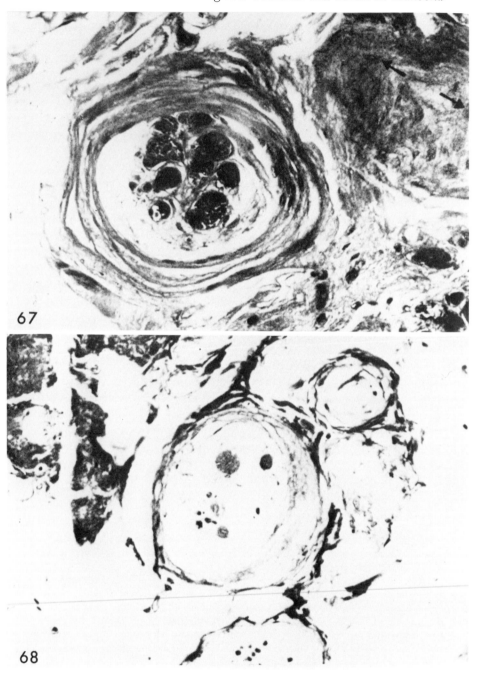

Figs. 67 and 68.

than 0.5 μ in diameter (Kennedy, 1969). Another difference was the appearance of the terminations of some motor axons in the shape of spheres instead motor end-plates. Spheres were observed in the polar and myotube regions of the muscle spindles. Kennedy (1969) observed the absence of primary sensory axons and their ending in the spindles from patients with neuropathy occurring simultaneously with a severe weakness. The surviving primary endings appeared normal. The secondary endings were present in only a few spindles, and Kennedy (1969) concluded therefrom that they are more susceptible to degeneration than the primary endings. These investigations suggested that innervation of the spindles might be disturbed in the muscles of patients suffering from polyneuropathies.

In the case of benign congenital hypotonia, it is only the diameter of the intrafusal fibers that is changed (Fig. 74). In the case of polymyositis, the atrophy of the nuclear chain fibers and the thickening of the capsule and of the connective tissue within the sheath has been observed. Cazzato and Walton (1968) did not observe any significant abnormalities in spindles belonging to cases showing thyreotoxic myopathy and syndrome of myophosphorylase deficiency. Spindles studied in the case of myasthenia gravis showed (Fig. 75) slight atrophy of certain intrafusal fibers, whereas the rest of them seemed to be normal.

VII. Some Aspects of the Function of the Muscle Spindle

In the light of morphological, ultrastructural, cytochemical, and neurophysiological studies, it is evident that the role of the muscle spindles is to form the junction points for information carried from the central nervous system by fusimotor axons as well as information pertaining to the length of the skeletal muscles. The information comes in the

---

Fig. 67. Round structure surrounded by connective tissue and degenerating muscle fibers hardly recognizable as a muscle spindle; the capsule is infiltrated by red cells and does not contain intrafusal fibers. It is possible to identify a small nerve trunk sectioned obliquely and lying partly inside and partly outside the spindle sheath. Congenital dystrophy in advanced stage of evolution. Picro-Mallory $\times 492$.

Fig. 68. Transverse section of two spindles from a case of advanced limb-girdle muscular dystrophy. One has a thickened capsule and shows an evident increase in thickness of the septa, while the intrafusal muscle fibers seem, on the whole, well preserved; the other is changed into a rounded scar, and among the strands of proliferating collagen are still present some sarcolemmal nuclei (arrows). Masson, $\times 492$.

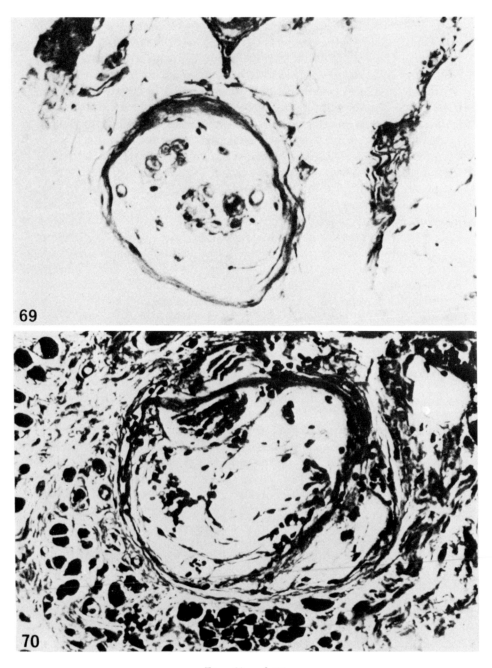

Figs. 69 and 70.

form of an impulse discharge from the sensory endings located on the intrafusal fibers. Two systems of afferent endings supply the muscle spindles. The nuclear bag fibers receive sensory innervation mainly from primary afferent axons, and the nuclear chain fibers receive sensory innervation chiefly from secondary afferent axons. The classification of primary and secondary axons is based on the axonal diameter and conduction velocity. Some intrafusal fibers received both secondary and primary axons. Two kinds of fusimotor or γ-nerve endings appear on the intrafusal fibers: one is the end-plate, and the second is often called the trail ending. There is some disagreement as to whether or not the endings terminate on particular specific type of intrafusal fiber. Boyd (1962) believed that plate endings terminate only on the nuclear bag fibers, whereas trail endings terminate only on the nuclear chain fibers. On the contrary, Barker (1962) maintained that both plate and trail endings are terminating on nuclear bag as well as on nuclear chain fibers. The debate on the problem concerning the types of afferent as well as efferent innervation of the muscle spindle was a prominent topic of discussion at the Symposium in Hong Kong in 1962. At the Nobel Symposium in Stockholm in 1966, the controversy regarding the type of innervation of the muscle spindle continued. Boyd as well as Barker believed that "static" fusimotor fibers terminate as end-plates and that "dynamic" fibers end as trail endings (Barker *et al.*, 1966).

Jansen and Mathews (1962), Bessou *et al.* (1966a,b), and Jansen (1967) showed that the stimulation of dynamic fibers increases the spindle discharge when the muscle is suddenly stretched, and stimulation of static fibers increases the discharge of a resting spindle at the constant length of the muscle but tends to reduce the discharge elicited by sudden muscle stretch. Appelberg *et al.* (1965, 1966) found that static fusimotor fibers show excitatory action on the secondary endings, whereas the dynamic fusimotor fibers do not influence the response of secondaries to stretch. Based on studies carried out by stimulating well identified fusimotor fibers, Bessou *et al.* (1966a) discovered that there are differences in contraction elicited by dynamic and static axons. The results

---

Fig. 69. Transverse section of a muscle spindle from a case of advanced Duchenne-type muscular dystrophy containing some intrafusal muscle fibers which are atrophic and several pyknotic and shrunken nuclei are scattered throughout the lymphatic space. There is discrete thickening of the capsule. Picro-Mallory, ×492.

Fig. 70. Transverse section of a muscle spindle from a case of advanced Duchenne-type muscular dystrophy continuing only one surviving intrafusal muscle fiber which is degenerating inside its thickened reticulin sheath. To be also seen are several empty sarcolemmal tubes and some pyknotic nuclei scattered throughout the periaxial space. Picro-Mallory, ×612.

TABLE X

Muscle Spindle in Polymyositis, Dermatomyositis, Sarcoidosis, McArdle's Disease, Thyrotoxic Myopathy, and Myasthenia Gravis[a]

Case	Age	Muscle	Disease	Capsule	Periaxial space	Septa	Blood vessels
1	45 years	Quadriceps	Polymyositis	Slightly thickened	Normal	Slightly thickened	Thickening of the walls
				Thickened	Normal	Normal	Thickening of the walls
				Thickened	Normal	Normal	—
				Thickened	Normal	Normal	—
				Slightly thickened	Normal	Slightly thickened	Thickening of the walls
				Normal	Reduced	Thickened	—
				Slightly thickened	Reduced	Replacement fibrosis	—
2	50 years	Deltoid	Polymyositis	Edematous swelling	Enlarged	Thickened	—
				Normal	Normal	Normal	Normal
3	3 years, 3 months	Quadriceps	Polymyositis	Normal	Normal	Normal	—
4	70 years	Deltoid	Dermatomyositis	Slightly thickened	Normal	Slightly thickened	—
				Normal	Enlarged	Normal	Normal
				Edematous swelling	Enlarged	Stretched	—
5	45 years	Deltoid	Sarcoid myopathy	Slightly thickened	Normal	Slightly thickened	—
				Slightly thickened	Absent	Thickened	—
6	33 years	Gastrocnemius	McArdle's disease	Slightly thickened	Normal	Normal	Normal
7	23 years	Gastrocnemius	McArdle's disease	Normal	Normal	Normal	Normal
8	27 years	Gastrocnemius	McArdle's disease	Normal	Normal	Normal	Normal
				Normal	Normal	Normal	Normal
9	35 years	Deltoid	Thyrotoxic myopathy	Normal	Normal	Slightly thickened	Normal
				Normal	Normal	Normal	—
10	41 years	Quadriceps	Thyrotoxic myopathy	Normal	Normal	Normal	Normal
11	70 years	Psoas biceps brachii	Myasthenia gravis	Normal	Normal	Normal	—
				Normal	Enlarged	Normal	—
				Normal	Normal	Normal	Normal
				Normal	Normal	Normal	—
				Normal	Normal	Normal	—
				Thickened	Enlarged	Replacement fibrosis	—
				Normal	Normal	Normal	Normal
				Normal	Normal	Normal	Normal

[a] Reproduced from Cazzato and Walton (1968).
[b] Fibers not recognizable as nuclear bag or nuclear chain.

Intrafusal muscle fibers	Sarcolemmal nuclei (nuclear bag fibers)	Overall diameter of spindle (μ)	Number intrafusal muscle fibers	Number nuclear bag fibers	Number nuclear chain fibers	Mean diameter nuclear bag fibers (μ)	Mean diameter nuclear chain fibers (μ)
Normal	Normal	140	3	3	—	16.8	—
Three eosinophilic	Some shrunken	154	7	4	3	19.8	7
Normal	Normal	105	6	3	3	19	5.6
Nuclear chain degenerating	Some pyknotic	126	5	2	3	11.2	4.2
Normal	Some pyknotic	84	3	3	—	18.4	—
Normal	Normal	126	2	1	1	20	14
Fragments of fibers; two relatively normal	Pyknotic and shrunken	126	2	2	—	19.2	—
Atrophied	Pyknotic	455	3	—	—	—	14[b]
Atrophied	Normal	140	5	—	—	—	12.6[b]
Normal	Normal	112	6	2	4	21	8.4
Nuclear chain fibers atrophic	Normal	126	3	2	1	19.6	4.2
Nuclear chain fibers atrophic	Normal	112	3	1	2	19.6	4.2
Signs of degeneration	Some pyknotic	230	11	2	9	20.4	7
Normal	Normal	112	2	1	1	19.6	14
Normal	Normal	126	9	4	5	23.8	9.6
Normal	Normal	120	6	2	4	24	12
Normal	Normal	126	6	4	2	21	11.2
Normal	Normal	140	6	4	2	20	10
Normal	Normal	80	6	3	3	21	11.2
Normal	Two pyknotic	182	6	4	2	22.4	11.2
Normal	Normal	190	4	1	3	23.8	14
Normal	Normal	140	5	2	3	22.4	11.2
Slight atrophy of the nuclear chain	Normal	98	5	2	3	19.4	8.4
Nuclear bag and nuclear chain atrophic	Normal	91	6	4	2	14	5
Slight atrophy of the nuclear chain	Normal	126	5	3	2	19	8.4
Nuclear chain atrophic	One pyknotic	98	3	2	1	16.8	7
Nuclear chain atrophic	Normal	98	9	3	6	19.6	6.8
Only fragments of fibers	Pyknotic	196	—	—	—	—	—
Normal	Normal	70	4	1	3	19	9.8
Normal	Normal	105	7	3	4	21	9.8

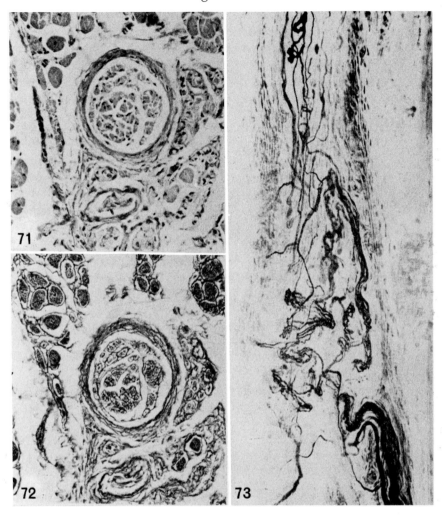

Fig. 71. Transverse section through muscle spindle in a lumbrical muscle in the case of dystrophia myotonica. There are more than forty small intrafusal muscle fibers within this spindle. The capsule is thickened. Hematoxylin and van Gieson, ×200. Figures 71–73 reproduced from Daniel and Strich (1964).

Fig. 72. Adjacent section stained to show reticulin fibers surrounding each intrafusal muscle fiber. Wilder, ×200.

Fig. 73. Longitudinal section through lumbrical muscle in the case of dystrophia myotonica. The pattern of innervation is abnormal. Schofield silver method, ×230.

of such studies on the discharge of spindle primary afferents from m. tenuissimus elicited by stimulation of well defined fusimotor fiber enabled Bessou *et al.* (1966a,b) to conclude that static fusimotor fibers

TABLE XI
PATHOLOGY OF THE MUSCLE SPINDLE: A STATISTICAL ANALYSIS OF THE
VARIATIONS IN INTRAFUSAL FIBER DIAMETER THAT OCCUR IN
DIFFERENT CATEGORIES OF DISEASE[a]

Disease group	Measurement	No. of patients	Mean (μ)	S.D. (μ)
Denervation atrophy	Nuclear bag mean diameter	14	17.29	7.27
	Nuclear chain mean diameter	13	9.04	3.78
	Apparently normal extra-fusal fiber mean diameter	11	36.6	8.77
Muscular dystrophy	Nuclear bag mean diameter	11	17.89	8.34
	Nuclear chain mean diameter	8	8.50	1.35
Congenital muscular dystrophy	Nuclear bag mean diameter	6	16.13	4.24
	Nuclear chain mean diameter	6	8.18	2.08
Polymyositis	Nuclear bag mean diameter	3	19.57	1.63
	Nuclear chain mean diameter	3	7.03	1.93
McArdle's disease	Nuclear bag mean diameter	3	21.83	1.89
	Nuclear chain mean diameter	3	11.27	0.70
Benign congenital hypotonia	Nuclear bag mean diameter	6	16.47	3.79
	Nuclear chain mean diameter	6	8.92	2.79

[a] Reproduced from Cazzato and Walton (1968).

elicit fast contractions and dynamic fusimotor fibers elicit slower ones. R. S. Smith (1966) proposed that nuclear chain fibers are responsible for static effects and nuclear bag fibers for dynamic effects.

Since the different motor endings may coexist on the same intrafusal fibers, it is very interesting to know whether these fibers may contract in different ways according to the particular kind of motor ending actually stimulated. Studies on rabbit hindlimb spindles, which consist only of nuclear bag fibers (Barker and Hunt, 1964), show that they are innervated by large as well as by small γ-nerve fibers (Adal and Barker,

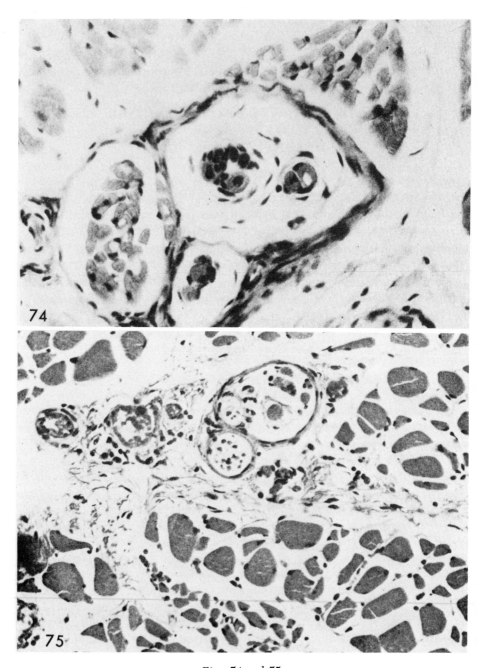

Figs. 74 and 75.

1965a,b), static and dynamic fusimotor fibers (Emonet-Denand *et al.*, 1964, 1966), and two kinds of motor endings. This is suggestive of the fact that the same intrafusal fibers are capable of both twitch and local contraction; however, the existence of two types of ATPase reaction in these intrafusal fibers, as shown by Spiro and Beilin (1969b), indicates that these facts should be reconsidered. Muscle spindle discharges are evoked by the mechanical stretch of extrafusal fibers being in contact with a spindle or by γ innervation of the intrafusal fibers (Boyd, 1962). In both situations, the middle portion of the nuclear bag as well as nuclear chain fibers become stretched. This is probably due to sensory endings located on this portion of the intrafusal fibers. The relative threshold of secondary endings is usually higher as compared to the primary ones (Jansen and Mathews, 1962). The stretch reflex affects the primary and secondary endings and produces differences in muscle patterning reactions. The discharge from the primary endings from nuclear bag as well as from nuclear chain fibers activates α-motoneurons supplying homonymous and synergistic muscles and inhibits antagonistic ones (Cooper, 1960; Eldred, 1965). According to Hunt and Perl (1960), the afferent discharges from the higher threshold secondaries of the muscle spindles are excitatory to the motoneurons of flexor muscles and inhibitory to extensors. There are different stimulations coming from the supraspinal portions of the central nervous system that influence the γ motor activity of the spindles. One such center influencing γ efferent activity is reticular formation. Granit (1952) have shown that the stimulation of particular parts of the upper reticular formation increases the discharge from γ-motoneurons, causing in effect the increase of the excitability of the spindles, whereas the stimulation of the lower portions of the bulbo-reticular area gives just the reverse effect. The reticular formation is responsible for integrating, adding, and combining the influences of all the centers from spinal cord, the cerebral cortex, basal ganglia, and cerebellum.

Fig. 74. Transverse section of a multiloculated muscle spindle from a case of benign congenital hypotonia containing very numerous intrafusal muscle fibers, the type of which is not easily recognizable. Note also the marked smallness of the extrafusal muscle fibers. Hematoxlin–eosin, ×614. From Cazzato and Walton (1968).

Fig. 75. Transverse section at the equator of a muscle spindle from a case of myasthenia gravis; at least two of the intrafusal muscle fibers are atrophic and contain pyknotic nuclei; two small nerve trunks are related to the spindle, one lying inside the capsule and the other just outside. There are also several markedly atrophic extrafusal muscle fibers grouped in small collections or scattered throughout normal or less atrophic fibers. Hematoxylin-eosin, ×384. From Cazzato and Walton (1968).

The relation of muscle spindle to the maintenance of posture was recognized as early as 1895 by Sherrington. The primary endings of the muscle spindle show the characteristics of fast as well as slow receptors (Eldred, 1965). As mentioned earlier, the phasic component of primary endings discharge arise from the nuclear bag fibers, whereas the tonic component arises from nuclear-chain fibers. So the two types of stretch reflex responses evoked from the primary endings afferent discharges can be distinguished. Tendon jerks and Hoffman-wave reactions seem to be phasic stretch responses (Rushworth, 1964a,b). According to Rushworth (1964a,b), slow phasic or static stretches to the primary endings of the spindle probably cause tonic or postural types of reflex reaction. Buller (1963) explained the resistance to postural stretches by somatic muscles by the influence of the phasic and tonic characteristics of the extrafusal fibers and by supraspinal control which will suppress or activate γ-neuron discharges to the muscle spindles. As Rushworth (1964a) put it, the supraspinal sites during postural activity probably transmit tonic facilatory impulses to the γ-motoneurons of antigravity muscles. The increase in γ efferent discharge increases the ability of muscle spindles to respond to gravitational stretches of the extrafusal fibers (Granit, 1955).

Eldred considered the primary endings as a "mainspring" behind reflex posture, since they should be activated by gravitational forces acting on antigravity muscles. Moreover, sensory discharges should be maintained during external stimulation and central connections of the sensory fibers should excite anterior horn cells of the antigravity muscles (Eldred, 1960). It is apparent that indirect γ loop through the muscle spindle plays an important role in the nervous control of the muscle function. This system probably acts during the performance of purposeful movement, although the neurophysiological data are insufficient to prove it. The theory dealing with the significance of indirect γ loop was introduced by Merton (1951). According to his hypothesis, the sensory endings of the muscle spindles respond to length differences between intrafusal and extrafusal fibers (Merton, 1953) with a subsequent action of a servo loop in order to reduce this discrepancy. The data from the experimental studies of Eldred *et al.* (1953) on decerebrated cats supported this hypothesis and led to the suggestion that both reflex and volitional movements might be initiated through indirect γ loop. Rushworth (1964a) has supported this hypothesis and stated that, "The stretch reflex connections from the muscle spindle provide a mechanism for inhibiting the antagonists and facilitating the synergists at the very moment when this synergistic activity and antagonistic inhibition is necessary. It is a mechanism which is there at the segmental level, but it is not available if the α-motor neuron is excited by any route other than the afferent influx along IA fibers. In other words, this is an inbuilt coordinating mechanism at the segmental level." Based on present evidence, it becomes very suggestive that the γ motor system may not only maintain posture, but according to Rushworth (1963), may also "superimpose and merge movement with it over the stretch reflex arc."

For further details on the neurophysiology of the muscle spindle, the reader is referred to the excellent review of Eccles and Lundberg (1958). Kuffler and Hunt (1952) described the fusimotor innervation, and Laporte (1962) discussed in detail muscle spindle physiology and function. The monograph of Buchwald *et al.* (1963) deals with propriocepters in motor learning, and Eyzaguirre (1964) discussed the discharge behavior. Granit (1964) gave an excellent description of the physiology and reflex activity of the muscle spindles, and Eldred (1965) has given a detailed review of the afferent role of the muscle spindles.

VIII. Addendum

Since the manuscript was completed, a few additional reports were published which are not only pertinent, but update the available information on the morphology, ultrastructural anatomy, cytochemistry, pathology, and functional physiology of the muscle spindle. Maier and Eldred (1971) made a detailed comparative study on the structure of avian muscle spindles, especially the m. sartorius and m. adductor profundus in chicken, quail, pigeon, and canary, with a view to comparing it with the mammals. They paid special attention to whether or not avian intrafusal fibers are separable into nuclear bag and nuclear chain fiber types, as observed in mammals. Not only are the intrafusal fibers in these birds not separable, these authors failed to observe prominent bags of nuclei characteristic of mammalian nuclear bag fibers. The authors also noticed the absence of muscle spindles in the extraocular muscles from quail, pigeon, English sparrow, and canary (Maier *et al.*, 1971). Thompson (1970) studied the parallel spindle system in the small muscles of the rat tail and compared it with those of frog (Barker and Cope, 1962). Whereas the frog spindles are long (17–22 mm), having a single type of sensory innervation that is observed in series of two to four regions along the unit spindle, the rat parallel spindles are shorter (1–4 mm) with a single sensory zone in each. Marchand *et al.* (1971) investigated the association of tendon organs with spindles in the muscles of the cat, especially the m. extensor digitorum brevis, medial gastrocnemius, and soleus. According to these authors, an intimate relationship between the tendon organ and the muscle spindle exists in 50% of m. extensor digitorum brevis tendon organ and 20–25% of the tendon organ of other muscles. The samples of "tendon organ-spindle dyads" in the m. extensor digitorum brevis showed large cross-sectional areas at the equatorial region, and the dyad tendon organs received the attachment from more extrafusal fibers than did the solitary tendon organs. Marchand et al. (1971) concluded that the association of tendon organs with muscle spindle is a happenstance of development. They also discussed the possibility that the dyads may serve a specific function in the muscles, especially in the regulation of contraction among the component motor units or sectors of the muscle.

Detailed electron microscopic studies on the motor and sensory innervation of dog muscle spindles (Banker and Girvin, 1971), sensory nerve fibers and nerve fibers and nerve endings of cat lumbrical muscle spindles (Corvaja et al., 1971), and satellite cells of frog muscle spindles (Karlsson and Andersson-Cedergren, 1971) have been made which have brought in new information on the ultrastructural anatomy of the muscle spindles. Corvaja et al. (1971) observed glycogen particles, 200–400 Å in size, in the primary and secondary sensory nerve endings innervating nuclear bag as well as nuclear chain fibers. Describing the difference in the primary and secondary sensory systems, Banker and Girvin (1971) showed that the secondary sensory system lacked the multiple relays of sensory axons from one chain fiber to the other. Instead, the small branches of the sensory nerves entered different nuclear chain fibers directly and separately. Banker and Girvin observed in dog muscle spindles two kinds of vesicles, granular and aggranular, in addition to the neurotubules and mitochondria in the axoplasm. The agranular vesicles are 250–600 Å in diameter and do not show any electron-dense material. The granular vesicles are either small (250–750 Å) with a electron-dense core, or large (750–1500 Å) with diffusely electron-dense material. Several studies have shown that the granular vesicles contain norepinephrine. Banker and Girvin conclude that these small, nonmyelinated nerves, which are conspicuous in the myotubular zones, belong to sympathetic nervous system. Another characteristic feature of the dog muscle spindle is the sheetlike arrangement of myofilaments instead of a separate identity into nuclear bag and nuclear chain fibers. Detailed studies of the satellite cells of the frog revealed that their cytoplasmic components were similar to those of Schwann cells, and occasionally, the intrafusal satellite cells displayed large fenestrations in the cytoplasm, with filamentous ma-

parallel to the axis of the intrafusal muscle fiber. There are two types of motor endings; one of them is the trail ending, distributed in the form of fine diffuse terminals near the secondary sensory endings on the nuclear chain fibers, and the second is the plate ending.

There is no positive evidence in the literature of the existence of sympathetic innervation of the intrafusal fibers. However, some controversial studies have been discussed. Neurotendinous organs are compared to specialized encapsulated fascicles of collagen, which are offshoots from the primary tendon of origin or insertion of the muscle. An electron microscopic study of the tendon organ from the white rat lumbrical muscles shows that the capsule has the same fine structure as that of the muscle spindle, extends into the organ, and divides it into several compartments containing small bundles of collagen lying among the fibroblasts and Schwann cells. The outer part of the capsule is covered by the zone of collagen, and myelinated as well as unmyelinated nerve endings are identified in the tendon organ.

Cytochemical studies on the muscle spindle, regarding the distribution of polysaccharides, lipids, nucleic acids, proteins, and myoglobin and enzymatic proteins including oxidative and hydrolytic enzymes have been discussed in detail. β-Metachromasia is observed at the poles of the intrafusal fibers, and their cytoplasm showed α-metachromasia after toluidine blue staining, but no glycogen or acid mucopolysaccharides. Spiro and Beilin (1969a) investigated the human muscle spindle and did not observe any difference in the PAS reaction between nuclear bag and nuclear chain fibers. Special mention may be made of the excellent histochemical studies of Nakajima *et al.* (1968) on the muscle spindle of the hand and foot interosseous muscles of the Rhesus and squirrel monkeys. The intrafusal muscle fibers show two types of enzyme reaction for ATPase and for the enzymes concerning the carbohydrate metabolism, except glucose-6-phosphate dehydrogenase. The extrafusal fibers show three types of reaction for the enzymes of Embden–Meyerhoff pathway and the tricarboxylic acid (TCA) cycle. Yellin (1969b) observed significant enzymic differences between the polar contractile region of the intrafusal fibers and the equatorial area of the fibers. Significant cytochemical differences between the nuclear bag and nuclear chain fibers are also evident. Nakajima *et al.* believe that the small intrafusal muscle fibers are more capable of energy production through the TCA cycle than the large intra- or extrafusal muscle fibers. From the positive cholinesterase activity around the polar region of the intrafusal fibers, and other histochemical reactions, they believe that the pericapsular epithelial cells of the muscle spindle appear to be metabolically similar to the perineural epithelial cells.

Experimentally induced neuromuscular disorders and naturally occurring neuromuscular diseases have been discussed in the pathology section. Special attention has been paid to the morphology of the muscle spindle in experimentally induced muscular dystrophy and demyelination. Meier (1969) studied in detail the muscle spindle of mice with hereditary neuromuscular diseases. The intrafusal and extrafusal fibers show significant abnormalities, whereas the muscle spindles show comparatively minor changes. The reader is referred to the excellent monographs of Cazzato and Walton (1968) and Patel et al. (1968) on the pathology of muscle spindle. However, briefly discussed are the muscle spindle changes in denervation atrophy, muscular dystrophy, congenital muscular dystrophy and myotonic dystrophy, benign congenital hypotonia and nemaline myopathy, polymyositis, dermatomyositis, sarcoidosis, McArdle's disease, thyreotoxic myopathy, and myasthenia gravis (Tables IV–X). Kennedy (1969) has described in an excellent manner the innervation of muscle spindle after acute and recurrent Guillain–Barré syndrome and Charcot–Marie–Tooth disease associated with alcoholism and observed a general reduction of the axons entering the spindle. Kennedy observed the absence of primary sensory axons and their endings in the spindle from patients with neuropathy occurring simultaneously with a severe weakness.

Finally, in the light of morophological, ultrastructural, pathological, and cytochemical studies outlined above, a few brief remarks on some aspects of the function of the muscle spindle have been made.

ACKNOWLEDGMENTS

This work was supported by Grant RR-00165 from the Animal Resources Branch, National Institutes of Health to Yerkes Primate Research Center and by grant of the Kosciuszko Foundation, New York to Dr. Z. L. Olkowski. We are most grateful to Dr. G. H. Bourne for useful suggestions, informal discussions, and valuable advice on the style of the article and constant encouragement during the preparation of the manuscript. We are thankful to Mrs. Nancy Hiller for technical and general help during the preparation of the manuscript, to Miss Mary Sheldon for typing, to Mrs. Nellie Johns and Mrs. Swaran Manocha for their assistance in the library work, and to Mr. Frank Kiernan for making photographs. We are also thankful to Dr. M. N. Golarz for a very useful and critical appraisal of the article and to Dr. Marilyn Yeagers for editing. We are very thankful to the various reseachers and publishers quoted along with the figures who have kindly given the permission to reproduce their figures, thereby enhancing the value of this review article.

REFERENCES

Adal, M. N. (1969). *J. Ultrastruct. Res.* **26**, 332.
Adal, M. N., and Barker, D. (1965a). *J. Anat.* **99**, 918.

Adal, M. N., and Barker, D. (1965b). *J. Physiol. (London)* **177**, 288.

Adal, M. N., and Barker, D. (1967). *J. Physiol. (London)* **192**, 50P.

Adams, R. D., Denny-Brown, D., and Pearson, C. M. (1962). "Diseases of Muscles," pp. 545, 549, and 558. Harper (Hoeber), New York.

Alexandrowicz, J. S. (1951). *Quart. J. Microsc. Sci.* **92**, 163.

Allen, W. F. (1917). *J. Comp. Neurol.* **28**, 137–213.

Amersbach, K. (1911). *Beitr. Pathol. Anat. Allg. Pathol.* **51**, 56.

Anda, G., and Rebollo, A. (1967). *Acta Anat.* **67**, 437.

Anda, G., and Rebollo, M. A. (1968). *Acta Histochem.* **31**, 287.

Andersson-Cedergren, E., and Karlsson, U. (1967). *J. Ultrastruct. Res.* **19**, 909.

Appelberg, B., Bessou, P., and Laporte, Y. (1965). *J. Physiol. (London)* **177**, 29P.

Appelberg, B., Bessou, P., and Laporte, Y. (1966). *J. Physiol. (London)* **185**, 160.

Babinski, J. (1886). *C. R. Soc. Biol.* [3] **8**, 629.

Bach-y-Rita, P. (1959). *Acta Neurol. Latinoamer.* **5**, 17.

Banker, B. Q., and Girvin, J. P. (1971). *J. Neuropath, Exptl. Neurol.* **30**, 136.

Banker, B. Q., and Girvin, J. P. (1971). *J. Neuropath. Exptl. Neurol.* **30**, 155–195.

Barets, A. (1961). *Arch. Anat. Microsc. Morphol. Exp.* **50**, 31.

Barker, D. (1948). *Quart. J. Microsc. Sci.* **89**, 143.

Barker, D. (1959). *J. Physiol. (London)* **149**, 7P.

Barker, D., ed. (1962). "Symposium on Muscle Receptors." Hong Kong Univ. Press, Hong Kong.

Barker, D. (1966a). *J. Physiol. (London)* **186**, 27P.

Barker, D. (1966b). *In* "Muscular Afferents and Motor Control" (R. Grant, ed.), p. 51. Wiley, New York.

Barker, D. (1967). *Myostatic, Kinesthetic Vestibular Mech., Ciba Found. Symp.* p. 3.

Barker, D. (1968a). *J. Physiol. (London)* **196**, 51P.

Barker, D. (1968b). *Actual. Neurophysiol.* **8**, 23–71.

Barker, D., and Chin, N. K. (1960). *J. Anat.* **94**, 473.

Barker, D., and Cope, M. (1962). *In* "Symposium on Muscle Receptors" (D. Barker, ed.), p. 263. Hong Kong Univ. Press, Hong Kong.

Barker, D., and Gidumal, J. L. (1961). *J. Physiol. (London)* **157**, 513.

Barker, D., and Hunt, J. P. (1964). *Nature (London)* **203**, 1193.

Barker, D., and Ip, M. C. (1961). *Proc. Roy. Soc., Ser. B* **154**, 377.

Barker, D., and Ip, M. C. (1963). *J. Physiol. (London)* **169**, 73P–74P.

Barker, D., and Ip, M. C. (1965). *J. Physiol. (London)* **177**, 27P.

Barker, D., and Stacey, M. J. (1970). *J. Physiol. (London)* **210**, 70P–72P.

Barker, D., Ip, M. C., and Adal, M. N. (1962). *In* "Symposium on Muscle Receptors" (D. Barker, ed.), pp. 257–261. Hong Kong Univ. Press, Hong Kong.

Barker, D., Boyd, I. A., and Granit, R. (1966). *In* "Muscular Afferents and Motor Control" (discussion) (R. Granit, ed.), Wiley, New York.

Barker, D., Stacey, M. J., and Adal, M. N. (1970). *Philos. Trans. Royal Soc. (London)* **B285**, 315–346.

Barrios, P., Haase, J., Heinrich, U., and Schlegel, H. (1966). *Pfluegers Arch. Gesamte Physiol. Menschen Tiere* **290**, 101.

Barrios, P., Haase, J., and Heinrich, W. (1967). *Pfluegers Arch. Gesamte Physiol. Menschen Tiere* **296**, 49.

Batten, F. E. (1897). *Brain* **20**, 138.

Baum, J. (1900). *Anat. Hefte.* **13**, 250–305.

Bennett, H. S. (1956). *J. Biophys. Biochem. Cytol.* **2**, Suppl. 4, 99.

Bennett, H. S. (1960). *In* "The Structure and Function of the Muscle" (G. H. Bourne, ed.), 1st ed., Vol. 1, p. 137. Academic Press, New York.

Bessou, P., Emonet-Denand F., and Laporte, Y. (1963a). *C. R. Acad. Sci.* **256**, 5625.

Bessou, P., Emonet-Denand, F., and Laporte, Y. (1963b). *Life Sci.* **1**, 948.

Bessou, P., Emonet-Denand, F., and Laporte, Y. (1963c). *Nature* (*London*) **198**, 594.

Bessou, P., Emonet-Denand, F., and Laporte, Y. (1965). *J. Physiol.* (*London*) **180**, 649.

Bessou, P., Laporte, Y., and Pages, B. (1966a). *J. Physiol.* (*Paris*) **58**, 31.

Bessou, P., Laporte, Y., and Pages, B. (1966b). *J. Physiol.* (*Paris*) **58**, 467.

Blocq, P., and Marinesco, G. (1893). *Arch. Neurol.* (*Chicago*) **1**, 189.

Boeke, J. (1921). *Brain* **44**, 1.

Bowman, J. P. (1968). *Anat. Rec.* **161**, 483.

Boyd, I. A. (1954). *J. Physiol.* (*London*) **124**, 476.

Boyd, I. A. (1958a). *J. Physiol.* (*London*) **144**, 11P.

Boyd, I. A. (1958b). *J. Physiol.* (*London*) **140**, 14P.

Boyd, I. A. (1959). *J. Physiol.* (*London*) **145**, 55P.

Boyd, I. A. (1960). *J. Physiol.* (*London*) **153**, 23P.

Boyd, I. A. (1962). *In* "Symposium on Muscle Receptors" (D. Barker, ed.), p. 185. Hong Kong Univ. Press, Hong-Kong.

Boyd, I. A. (1966). *J. Physiol.* (*London*) **186**, 109P.

Boyd, I. A., and Davey, M. R. (1962). *In* "Symposium on Muscle Receptors" (D. Barker, ed.), p. 191. Hong Kong Univ. Press, Hong Kong.

Bravo-Rey, M. C., Yamasaki, J. N., Eldred, E., and Maier, E. (1969). *Exp. Neurol.* **25**, 595–602.

Bridgman, C. (1968). *Anat. Rec.* **162**, 209.

Bremer, L. (1883). *Arch. Mikroskop. Anat.* **22**, 318–356.

Broccoli, F. (1970). *Boll. Soc. Ital. Biol. Sper.* **46**, 252.

Brown, M. C., Crowe, A., and Mathews, P. B. C. (1965). *J. Physiol.* (*London*) **177**, 140.

Buchwald, J. S., Standish, M. Eldred, E., Gamble, S., and Halas, E. (1963). *Biol. Inst. Estud. Med. Biol. [Univ. Nac. Auton. Mex.]* **21**, 235.

Buller, A. J. (1963). *In* "Recent Advances in Physiology" (R. Creese, ed.), p. 122. Churchill, London.

Carli, G., Diete-Spiff, K., and Pompeiano, O. (1966). *Experientia* **22**, 583.

Cattaneo, A. (1888). *Arch. Ital. Biol.* **14**, 31.

Cazzato, G., and Walton, J. N. (1968). *J. Neurol. Sci.* **7**, 15.

Chin, N. K., Cope, M., and Pang, M. (1962). *In* "Symposium on Muscle Receptors" (D. Barker, ed.), p. 241. Hong Kong Univ. Press, Hong Kong.

Chor, H. Dolkart, R. E., and Davenport, H. A. (1937). *Amer. J. Physiol.* **118**, 580.

Christensen, E. (1959). *Amer. J. Physiol. Med.* **38**, 65.

Christensen, L. V. (1967). *Arch. Oral. Biol.* **12**, 1203.

Christomanos, A. A., and Strossmer, E. (1891). *Sitzungsber. Kaiserl. Akad. Wiss. Wien, Math.-Naturwiss. Kl., Abt. 1* **100**, 417.

Ciaccio, G. W. (1891). *Arch. Ital. Biol.* **14**, 31.

Climbaris, P. A. (1910). *Arch. Microsk. Anat.* **75**, 692.

Coërs, C. (1952). *Acta Clin. Belg.* **7**, 407.

Coërs, C. (1959). *Bull. Acad. Roy. Med. Belg., Cl. Sci.* **40**, 1000.

Coërs, C. (1961). *Bibl. Anat.* **2**, 139.

Coërs, C. (1962). *In* "Symposium on Muscle Receptors" (D. Barker, ed.), p. 221. Hong Kong Univ. Press, Hong Kong.

Coërs, C., and Durand, J. (1956). *Arch. Biol.* **67**, 685.

Coërs, C., and Woolf, A. L. (1959). *In* "The Innervation of Muscle. A Biopsy Study," p. 39. Blackwell, Oxford.

Cohen, M. J. (1964). *In* "Neural Theory and Modeling" (R. F. Reiss, ed.), p. 273. Stanford Univ. Press, Palo Alto, Caifornia.

Cohen, M. J. (1965). *Cold Spring Harbor Symp. Quant. Biol.* **30**, 587.

Cooper, S. (1953). *J. Physiol. (London)* **122**, 193.

Cooper, S. (1960). *In* "The Structure and Function of Muscle" (G. H. Bourne, ed.), 1st ed., p. 381. Academic Press., New York.

Cooper, S., and Daniel, P. M. (1949). *Brain* **72**, 1.

Cooper, S., and Daniel, P. M. (1956). *J. Physiol. (London)* **133**, 1P.

Cooper, S., and Daniel, P. M. (1963). *Brain* **86**, 563.

Corvaja, N., and Pompeiano, O. (1970). *Pflügers Arch.* **317**, 187–197.

Corvaja, N., Marinozzi, V., and Pompeiano, O. (1967). *Pflügers Arch. Gesamte Physiol. Menschen Tiere* **296**, 337.

Corvaja, N., Marinozzi, V., and Pompeiano, O. (1969). *Arch. Ital. Biol.* **107**, 341.

Corvaja, N., Magherini, P. C., and Pompeiano, O. (1971). *Z. Zellforsch,* **121**, 199–217.

Cuajunco, F. (1940). *Carnegie Inst. Wash. Publ.* **518**.

Daniel, P. M., and Strich, S. J. (1964). *Neurology* **14**, 310.

Dastur, D. K. (1967). *In* "Symposium on Leprosy" (N. H. Antia and D. K. Dastur, eds.), p. 27. Bombay Univ. Press, Bombay.

Denny-Brown, D., and Nevin, S. (1941). *Brain* **64**, 1.

Dogiel, A. S. (1902). *Arch. Mikrosk. Anat.* **59**, 1.

Dudel, J. (1963). *Pflüegers Arch. Gesamte Physiol. Menschen Tiere* **277**, 537.

Dudel, J. (1965a). *Pflüegers Arch. Gesamte Physiol. Menschen Tiere* **282**, 323.

Dudel, J. (1965b). *Pflüegers Arch. Gesamte Physiol. Menschen Tiere* **283**, 104.

Dudel, J. (1965c). *Pflüegers Arch. Gesamte Physiol. Menschen Tiere* **284**, 81.

Dudel, J., and Kuffler, S. W. (1961). *J. Physiol. (London)* **155**, 514.

Düring, M., and Andres, K. H. (1969). *Anat. Anz.* **124**, 566.

Eccles, R. M., and Lundberg, A. (1958). *Experientia* **14**, 197.

Eisenlohr, C. (1876). Cited from Fraenkel (1878).

Eldred, E. (1960). *In* "Handbook of Physiology" (Amer. J. Physiol., J. Field, ed.), Sect. 1, Vol. II, p. 1067. Williams & Wilkins, Baltimore, Maryland.

Eldred, E. (1965). *J. Amer. Phys. Ther. Ass.* **45**, 290.

Eldred, E., Granit, R., and Merton, P. A. (1953). *J. Physiol. (London)* **122**, 498.

Eldred, E., Bridgman, C. F., Swett, J. E., and Eldred, B. (1962). *In* "Symposium on Muscle Receptors" (D. Barker, ed.), p. 207. Hong Kong Univ. Press, Hong Kong.

Eldred, E., Yellin, H., Gadbois, L., and Sweeney, S. (1964). *Exp. Neurol. Suppl.* **3**, 1.

Emonet-Dénand, F., and Laporte, Y. (1966). *C. R. Acad. Sci.* **263**, 1405.

Emonet-Dénand, F., Laporte, Y., and Pages, B. (1964). *C. R. Acad. Sci.* **259**, 2690.

Emonet-Dénand, F., Laporte, Y., and Pages, B. (1966). *Arch. Ital. Biol.* **109**, 195.

Engel, W. K. (1965). *Int. Congr. Neurol., 8th,* Vol. 2, p. 67.

Engel, W. K. (1965). *Int. Congr. Neurol. 8th, 1965,* Vol. 2, p. 67.

Engel, W. K., and Irwin, R. L. (1967). *Amer. J. Physiol.* **213,** 511.

Esaki, K. (1966). *Nagoya Med. J.* **12,** 185.

Eyzaguirre, C. (1964). *In* "The Role of the Gamma System in Movement and Posture" (C. A. Swinyard, ed.), p. 17. Ass. Aid Crippled Children, New York.

Eyzaguirre, C., and Kuffler, S. W. (1961). *J. Gen. Physiol.* **39,** 87.

Farquahr, M. G., and Pallade, G. E. (1963). *J. Cell. Biol.* **17,** 375–412.

Feinstein, B., Lindegard, B., Nyman, E., and Wohlfart, G. (1955). *Acta Anat.* **23,** 127.

Felix, W. (1888). *Anat. Anz.* **3,** 719.

Ferrari, E., and Sonna, G. (1965). *Acta Neurol.* **20,** 250.

Florey, E., and Florey, E. (1955). *J. Gen. Physiol.* **39,** 69.

Forster, L. (1894). *Arch. Pathol. Anat. Physiol. Klin. Med.* **137,** 121.

Forster, L. (1902). *J. Physiol. (London)* **28,** 201.

Fraenkel, E. (1878). *Arch. Pathol. Anat. Physiol. Klin. Med.* **73,** 380.

Franks, A. S. T. (1964). *J. Dent. Res.* **43,** Suppl., 947.

Franzini-Armstrong, C., and Porter, K. R. (1964). *J. Cell Biol.* **22,** 675.

Freimann, R. (1954). *Anat. Anz.* **100,** 258.

Fukami, Y., and Hunt, C. C. (1970). *J. Neurophysiol.* **33,** 9.

Germino, N. I., and D'Albora, H. (1965). *Experientia* **21,** 45.

Gill, H. I. (1971). *J. Anat. (London)* **109,** 157–167.

Gladden, M. H. (1969). *Experientia* **25,** 604.

Golgi, C. (1880). *In* "Opera Omnia," 2nd ed., Vol. 1, p. 171.

Granit, R. (1952). *Acta Physiol. Scand.* **27,** 130.

Granit, R. (1955). "Receptors and Sensory Perception." Yale Univ. Press, New Haven, Connecticut.

Granit, R. (1957). *Ber. Phys.-Med. Ges. Wurzburg* **68,** 81.

Granit, R. (1962a). *In* "Symposium on Muscle Receptors" (D. Barker, ed.), p. 1. Hong Kong Univ. Press, Hong Kong.

Granit, R. (1962b). *In* "Muscle as a Tissue" (K. Rodahl and S. M. Horvath, eds.), p. 190–210 McGraw-Hill, New York.

Granit, R. (1962c). "Receptors and Sensory Perception," 2nd ed. Yale Univ. Press, New Haven, Connecticut.

Granit, R. (1964). *Clin. Pharmacol. Ther.* **5,** 837.

Granit, R., Pompeiano, O., and Waltman, B. (1959a). *J. Physiol. (London)* **147,** 385.

Granit, R., Pompeiano, O., and Waltman, B. (1959b). *J. Physiol. (London)* **147,** 399.

Gray, E. G. (1957). *Proc. Roy. Soc., Ser. B* **147,** 416.

Gray, J. A. B. (1959). *Progr. Biophys. Biophys. Chem.* **9,** 286.

Greenfeld, J. G., Shy, G. M., Alvord, E. C., and Berg, L. E. (1957). "An Atlas of Muscle Pathology in Neuromuscular Diseases." Livingstone, Edinburgh.

Gruner, J. E. (1961). *Rev. Neurol.* **104,** 490.

Gudden, H. (1896). *Arch. Psychiat. Nervenkr.* **28,** 643.

Gutman, E., and Zelena, Y. (1962). *In* "The Denervated Muscles" (E. Gutman, ed.), p. 57. Publ. House Czech. Acad. Sci., Prague.

Haase, G., and Schlegel, H. Y. (1966). *Pflüegers Arch. Gesamte Physiol. Menschen Tiere* **287,** 163.

Häggquist, G. (1960). *Z. Biol.* **112,** 11.

Hagbarth, K. E., and Wolfhart, G. (1952). *Acta Anat.* **15,** 85.

Hagopian, M., and Spiro, D. (1967). *J. Cell Biol.* **32**, 535.
Hariga, J. (1969). *C. R. Soc. Biol.* **158**, p. 21.
Henneman, E., and Olson, C. B. (1965). *J. Neurophysiol.* **28**, 581.
Hennig, A. (1969). *Z. Zellforsch. Mikrosk. Anat.* **96**, 275.
Hess, A. (1961a). *Anat. Rec.* **139**, 173.
Hess, A. (1961b). *J. Physiol. (London)* **157**, 221.
Hess, A. (1962). *Rev. Can. Biol.* **21**, 291.
Hines, M. (1932). *J. Comp. Neurol.* **56**, 105.
Hines, M., and Tower, S. S. (1928). *Bull. Johns Hopkins Hosp.* **12**, 269.
Hinsey, J. C. (1928). *J. Comp. Neurol.* **44**, 87.
Homma, S. (1963). *Chiba Igakkai Zasshi* **39**, 246.
Honee, G. L. J. M. (1966). *Ned. Dent. J.* **73**, 43.
Horsley, V. (1897). *Brain* **20**, 375.
Huber, G. C., and DeWitt, L. (1897). *Science* **5**, 908.
Huber, G. C., and DeWitt, L. (1900). *J. Comp. Neurol.* **10**, 159.
Hunt, C. C. (1959). *J. Gen. Physiol.* **38**, 117.
Hunt, C. C., Kuffler, S. W. (1951). *J. Physiol. (London)* **113**, 283.
Hunt, C. C., and Perl, E. R. (1960). *Physiol. Rev.* **40**, 538.
Hunt, C. C., and Wylie, R. M. (1970). *J. Neurophysiol.* **33**, 1.
Huxley, H. E. (1964). *Nature (London)* **202**, 1907.
Inman, D. R. (1962). *Symp. Soc. Exp. Biol.* **16**, 317.
Ito, F. (1969). *Jap. J. Physiol.* **19**, 641.
James, D. W., and Tresman, R. L. (1968). *Nature (London)* **220**, 184.
James, N. T. (1968). *Nature (London)* **219**, 1174.
James, N. T. (1971). *Histochem. J.* **3**, 457–402.
Jansen, J. K. S. (1967). *Myostatic, Kinesthetic Vestibular Mech., Ciba Found. Symp.* p. 20.
Jansen, J. K. S., and Mathews, P. B. C. (1962). *J. Physiol. (London)* **161**, 357.
Jones, E. G. (1966). *J. Anat.* **100**, 733.
Kadanoff, D. (1956). *Z. Mikroskop. Anat. Forsch.* **62**, 1–15.
Karlsen, K. (1965). *Acta Odontol. Scand.* **23**, 521.
Karlsen, K. (1969). *Arch. Oral Biol.* **14**, 1111.
Karlsson, U., and Andersson-Cedergren, E. (1968). *J. Ultrastruct. Res.* **23**, 417.
Karlsson, U., and Andersson-Cedergren, E. (1971). *J. Ultrastruct. Res.* **34**, 426–438.
Karlsson, U., Andersson-Cedergren, E., and Ottoson, D. (1966). *J. Ultrastruct. Res.* **14**, 1.
Karlsson, U. L., Hooker, W. M., and Bendeich, E. G. (1971). *J. Ultrastruct. Res.* **36**, 743–756.
Karpati, G., and Engel, W. K. (1967). *Nature (London)* **215**, 1509.
Katz, B. (1960). *J. Physiol. (London)* **152**, 13.
Katz, B. (1961). *Phil. Trans. Roy. Soc. London, Ser. B* **243**, 221.
Keene, M. F. L. (1961). *J. Anat. (London)* **95**, 25–29.
Kennedy, W. R. (1969). *Trans. Amer. Neurol. Ass.* **94**, 59.
Kennedy, W. R. (1971). *Mayo Clinic Proc.* **46**, 245–257.
Kerschner, L. (1888). *Anat. Anz.* **3**, 126.
Kidd, G. L. (1966). *In* "Control and Innervation of Skeletal Muscles" (B. L. Andrew, ed.), p. 83. Dundee.
Kidd, G. L. (1967). *J. Physiol. (London)* **191**, 84.
Kidd, G. L. (1969). *Nature (London)* **203**, 1248.
Kölliker, A. (1862). *Proc. Roy. Soc.* **12**, 65.

Kölliker, A. (1863). Z. Wiss. Zool. 12, 149.

Kölliker, A. (1889). "Handbuch der Gewebelehre des Menschen." Engelmann, Leipzig.

Kosaka, K. (1969). Jap. J. Physiol. 19, 160.

Kuffler, S. W. (1954). J. Neurophysiol. 17, 558.

Kuffler, S. W., and Hunt, C. C. (1952). Res. Publ., Ass. Res. Nerv. Ment. Dis. 30, 27.

Kuhne, W. (1863). Arch. Pathol. Anat. Physiol. Klin. Med. 28, 528.

Kupfer, C. (1960). J. Physiol. (London) 153, 522.

Landon, D. N. (1966). In "Control and Innervation of Skeletal Muscle" (B. L. Andrew, ed.), p. 96. Dundee.

Laporte, Y. (1962). Int. Congr. Physiol. Sci. [Lect. Symp.], 22nd, 1962 Vol.. 1, p. 70.

Lapresle, J., and Milhaud, M. (1969). Rev. Neurol. 110, 97.

Latyshev, V. A. (1958). Usp. Lap. Krajnoolgrsk. Gos. Ped. Inst. 14, 47.

Leksell, L. (1945). Acta. Physiol. Scand. 10, 1.

McDonald, W. I., and Gilman, S. (1968). Arch. Neurol. 18, 508.

Maier, A., and Eldred, E. (1971). J. Comp. Neurol. 143, 25–40.

Maier, A., De Santis, M., and Eldred, E. (1971). Exp. Eye Res. 12, 251–253.

Marchand, E. R., and Eldred, E. (1969). Exp. Neurol. 25, 655.

Marchand, E. R., Shumpert, E., Eldred, E., and Bridgman, C. E. (1966). Physiologist 9, 238.

Marchand, R., Bridgman, C. F., Shumpert, E., and Eldred, E. (1971). Anat. Rec. 169, 23–32.

Mather, V., and Hines, M. (1934). Amer. J. Anat. 54, 177.

Mathews, P. B. C. (1960). Physiol. Rev. 89, 219.

Mathews, P. B. C. (1971). Sci. Basis Med. Ann. Res. 1, 99–128.

Mauro, A. (1961). J. Biophys. Biochem. Cytol. 9, 493.

Mavrinskaya, L. F. (1967). Arch. Anat. Histol. Embryol. 53, 42.

Meier, H. (1969). Experientia 25, 965.

Mellstrom, A., and Skoglund, S. (1965). Acta Morphol. Neer.-Scand. 6, 135–145.

Merrillees, N. C. R. (1960). J. Biophys. Biochem. Cytol. 7, 725.

Merrillees, N. C. R. (1962). In "Symposium on Muscle Receptors" (D. Barker, ed.), p. 199. Hong Kong Univ. Press, Hong Kong.

Merton, P. A. (1951). J. Physiol. (London) 114, 193.

Merton, P. A. (1953). In "The Spinal Cord" (G. E. W. Wolstenholme, ed.), p. 244. Churchill, London.

Muir, A. R., Kanji, A. H. M., and Allbrook, D. (1965). J. Anat. 99, 435.

Mummenthaller, M., and Engel, W. K. (1961). Acta Anat. 47, 274.

Nachmias, V. T., and Padykula, H. A. (1958). J. Biophys. Biochem. Cytol. 4, 47.

Nakajima, Y., Shantha, T. R., and Bourne, G. H. (1968). Histochemie 16, 1.

Norberg, K. A. (1967). Brain Res. 5, 125.

Nyström, B. (1967). Science 155, 1424.

Ogata, T., and Mori, M. (1962). Acta Med. Okayama 16, 347.

Onanoff, L. (1890). C. R. Soc. Biol. 2, 432.

Pallot, D. J., and Ridge, R. M. A. P. (1971). J. Physiol. (London) 218, 17P–18P.

Patel, A. N., Lalitha, V. S., and Dastur, D. K. (1968). Brain 91, 636.

Peachey, L. D., and Huxley, A. F. (1962). J. Cell Biol. 13, 177.

Poloumordvinoff, D. (1899). C. R. Acad. Sci. 128, 845.

Porayko, O., and Smith, R. S. (1968). *Experientia* **25**, 588
Ramon Cajal, S. (1897). *Riv. Trim. Microgr.* **2**, 181.
Rebollo, A. M., and Anda, G. (1967). *Acta Anat.* **67**, 595.
Research Group on Neuromuscular Diseases of the World. (1967). Draft Report of a Subcommittee on Quantitation of Muscle Biopsy Findings.
Rexed, B., and Therman, P. (1948). *J. Neurophysiol.* **11**, 133.
Robertson, J. D. (1957a). *Electron Microsco., Proc. Stockholm Conf., 1956* p. 197.
Robertson, J. D. (1957b). *J. Physiol. (London)* **137**, 6.
Robertson, J. D. (1960). *Amer. J. Phys. Med.* **39**, 1.
Romanul, F. C. A., and Van der Meulen, J. P. (1967). *Arch. Neurol.* **17**, 387.
Ruffini, A. (1892). *Atti Reale Accad. Lincei, Rend., El. Sci. Fis., Mat. Natur.* **1**, 31.
Ruffini, A. (1893). *Arch. Ital. Biol.* **18**, 106.
Ruffini, A. (1898). *J. Physiol. (London)* **23**, 190.
Rumpelt, J. F., and Schmalbruch, J. (1969). *Z. Zellforsch. Mikrosk. Anat.* **102**, 601.
Rushworth, G. (1963). *In* "Modern Trends in Neurology" (B. Williams, ed.), p. 36–56. Macmillan, New York.
Rushworth, G. (1964a). *In* "The Role of the Gamma-system in Movement and Posture" (C. A. Swinyard, ed.), p. 45. Ass. Aid Crippled Children, New York.
Rushworth, G. (1964b). *Clin. Pharmacol. Ther.* **5**, 828.
Ruska, H., and Edwards, G. A. (1957). *Growth* **21**, 63.
Sabussow, G. H., Maslow, A. P., and Burnaschewa, D. W. (1964). *Anat. Anz.* **114**, 27.
Schulze, M. L. (1955). *Anat. Anz.* **102**, 290.
Shantha, T. R., Golarz, M. N., and Bourne, G. H. (1968). *Acta Anat.* **69**, 632.
Shanthaveerappa, T. R., and Bourne, G. H. (1962). *J. Cell Biol.* **14**, 343.
Sherrington, C. S. (1895). *J. Physiol. (London)* **17**, 237.
Siemerling, F. (1889). *Charite-Ann.* **11**, 463.
Slauck, A. (1921). *Z. Gesamte Neurol. Psychiat.* **67**, 276.
Smith, C. M. (1963). *Ann. Rev. Pharmacol.* **3**, 223.
Smith, D. S. (1966). *J. Cell Biol.* **28**, 109.
Smith, R. S. (1966). *In* "Muscular Afferents and Motor Control" (R. Granit, ed.), p. 69. Wiley, New York.
Spiro, A. J., and Beilin, R. L. (1969a). *Arch. Neurol. (Chicago)* **20**, 271.
Spiro, A. J., and Beilin, R. L. (1969b). *J. Histochem. Cytochem.* **17**, 348.
Steg, G. (1962). *Acta Neurol. Scand.* **38**, Suppl. 3, 53.
Stein, J. M., and Padykula, H. A. (1962). *Amer. J. Anat.* **110**, 103.
Stewart, D. G., Eldred, E., Hemingway, A., and Kawamura, Y. (1963). *Temp. Meas. Contr. Sci. Ind. Proc. Symp., 4th, 1961* Vol. 3, p. 545.
Stilwell, D. L., Jr. (1957). *Amer. J. Anat.* **100**, 289.
Sunderland, S., and Ray, L. J. (1950). *J. Neurol., Neurosurg. Psychiat.* **13**, 348.
Swash, M., and Fox, K. P. (1972). *J. Neurol. Sci.* **15**, 291–302.
Swett, J. E., and Eldred, E (1960). *Anat. Rec.* **137**, 453.
Szepsenwol, J. (1960). *Cellule* **61**, 19.
Teravainen, H. (1968). *Z. Zellforsch.* **90**, 372–388.
Tello, J. F. (1922). *Z. Anat. Entwicklungsgesch.* **64**, 348.
Thompson, J. (1970). *J. Physiol. (London)* **211**, 781–799.
Tower, S. S. (1932). *Brain* **55**, 77.
Toyama, K. (1966). *Jap. J. Physiol.* **16**, 113.
Uehara, Y., and Hama, K. (1965). *J. Electronmicrosc.* **14**, 34.

Vihvelin, H. (1932). Z. *Zellforsch. Mikrosk. Anat.* **16**, 597.

Viragh, S. (1968). *Electron Microsc. 1968, Proc. Eur. Reg. Conf., 4th, 1968* Vol. 2, p. 295.

von Brzezinski, D. K. (1961a). *Acta Histochem.* **12**, 75.

von Brzezinski, D. K. (1961b). *Acta Histochem.* **12**, 277.

von Brzezinski, D. K. (1961c). *Verh. Anal. Ges.* (*Jens*) **57**, 151.

von Brzezinski, D. K. (1965). *Acta Histochem.* **20**, 45.

Voss, H. (1937). *Z. Mikrosk.-Anat. Forsch.* **42**, 509–524.

Voss, H. (1956). *Anat. Anz.* **103**, 85–88.

Voss, H. (1959). *Anat. Anz.* **107**, 190–197.

Walker, L. B., and Rajagopal, M. D. (1959). *Anat. Rec.* **133**, 438.

Walker, M., and Schrodt, G. R. (1965). *Fed. Proc., Fed. Amer. Soc. Exp. Biol.* **24**, 648.

Weismann, A. (1861a). *Z. Rationelle Med.* **10**, 263.

Weismann, A. (1861b). *Z. Rationelle Med.* **12**, 126.

Wiersma, C. A., Furshpan, E., and Florey, E. (1953). *J. Exp. Biol.* **30**, 136.

Winckler, G. (1961). *Rev. Oto-Neuro-Opthtalmol.* **33**, 1.

Wirsen, C. (1964). *J. Histochem. Cytochem.* **12**, 308.

Wirsen, C., and Larsson, K. S. (1964). *J. Embryol. Exp. Morphol.* **12**, 759.

Wohlfart, G. (1949). *Arch. Neurol. Psychiat.* **61**, 599.

Wohlfart, G., and Henriksson, K. G. (1960). *Acta. Anat.* **41**, 192.

Yellin, H. (1967). *Exp. Neurol.* **19**, 92.

Yellin, H. (1969a). *Exp. Neurol.* **25**, 153.

Yellin, H. (1969b). *Amer. J. Anat.* **125**, 31.

Zelena, J. (1963). *Physiol. Bohemoslov.* **12**, 30.

Zelena, J., and Hník, P. (1963). *Physiol. Behemoslov.* **12**, 277.

⑧

MOTOR END PLATE STRUCTURE

R. COUTEAUX

REVISED BY G. H. BOURNE

I. Introduction

Over a long period during which silver impregnations were almost exclusively used in investigations on the fine structure of motor end

plates, the preparations, at times impressive, obtained either with reduced silver nitrate methods or with ammoniacal silver methods, have
often resulted only in fruitless arguments.

While the silver impregnations have been, and continue to be, of
great value for an overall survey of muscular innervation, as well as
for a general view of nerve terminal arborization, they have proved
inadequate for a cytological study of the end plate. Although they have
in skilled hands produced striking pictures, there are serious drawbacks
to many of these methods. For example, a mediocre, and even sometimes
really unsatisfactory, fixation often has to be used; or they may stain
deeply filamentous structures other than the neurofibrils. This lack of
specificity, which has not always been sufficiently considered, may give
rise to certain misinterpretations.

Another serious shortcoming of the silver methods used in studying
motor end plates is their inability to show clearly the cytoplasmic boundaries. No reliable information can be obtained concerning the nature
of the connection joining the motor nerve fiber and muscle fiber when
using techniques that leave the surface membranes practically unstained.

Rapid progress in our knowledge of the structure of motor end plates
has been made with methods other than silver impregnations; these
methods allow staining of surface membranes or at least one of their
components. By means of postvital stainings, it has been possible to
delimit the sarcoplasm at the neuromuscular junction. Localization with
histochemical techniques of cholinesterase activity, located at the level
of the sarcoplasm surface membrane, has facilitated observation of this
membrane and confirmed conclusions drawn from the results of postvital
stainings. In addition, the use of the electron microscope has filled numerous gaps that previously existed. This new technique, allowing the
best fixations now available to cytologists, enables observations to be
made with a resolution about one hundred times higher than before
and offers motor end plate images on which plasma membranes are
clearly seen.

Although the application of these methods is of recent date, they have
already settled some long-standing debates raised by the study of the
vertebrate motor end plates. It has been demonstrated that the plasma
membranes, which limit, the nerve fiber and the striated muscle fiber,
are present at their junction, as well as at all other points of their surface,
and that they offer everywhere the same general structural patterns.
Thus, the existence of a structural discontinuity between the nerve and
muscle fibers is firmly established, confirming an important point of
the neuron theory.

The electron micrographs have, on the other hand, shown that at the motor end plate the surface membranes of the axoplasm and sarcoplasm are in immediate contact. Neither a collagenous membrane nor an endothelial membrane of the type forming the lamellar sheath of the nerve is found between the nerve ending and the sarcoplasm, and even the teloglia associated with each of the nerve ending branches leave the synaptic side of the branch exposed. In grown animals, the motor end plates are the only points of the muscle at which the axoplasm of the motor nerve fiber and the sarcoplasm are in such proximity to each other. At all other points, a more or less thick layer of collagenous fibrils and sheaths surrounding the axon are interposed between the axoplasm and the sarcoplasm.

If the motor neuromuscular junction consisted only in the apposition of two plasma membranes, its morphological study would be reduced to specifying the site of the junction and the form and dimensions of the apposition surface of the pre- and postsynaptic membranes.

In fact, the motor nerve fiber and the muscle fiber each show, in the immediate vicinity of their junction, several distinguishing characteristics which they exhibit nowhere else, and which confer upon the motor junctional region a structural originality in its pre- and postsynaptic components.

Following the development of the striated muscle, one may observe the motor nerve ending and junctional sarcoplasm progressively acquiring their distinguishing characteristics, some of which probably result from local interactions between the axoplasm and sarcoplasm, interpretable as late inductions of the organogenesis. In regard to the sarcoplasm, the morphological features of the subneural area and the appearance of its level of a high acetylcholinesterase activity are typical signs of this local, and obviously epigenetic, differentiation. Further study of motor end plates, more particularly by new cytochemical techniques, will doubtless soon reveal other signs of specialization of the junctional sarcoplasm.

In this paper, the structure of the motor end plate will be considered from a very general point of view, without discussing in detail the particularities it exhibits in each type of striated muscle and in the different zoological divisions. Information concerning the comparative morphology of motor end plates may be found in several reviews (Hines, 1927; Hinséy, 1934; Tiegs, 1953) for all classes of vertebrates, and in a very recent monograph in which Hoyle (1957) deals with "nervous control of the muscular contraction" in both vertebrates and invertebrates. Regarding the pathological changes that may affect the structure of the motor end plates, a paper by Cöers (1955) sets forth the principal known data.

II. Early Observations

Before the year 1840, histologists agreed with Valentin (1836) and Emmert (1836) that nerves possessed bow-shaped endings in striated muscles, the ultimate nerve branches joining up and continuing in one another. This opinion denied, in effect, the existence of true nerve endings with direct connections between the motor nerve and each muscle fiber.

From Doyère (1840), we have our first notion of a nerve ending that ensures the close union of nerve and muscle fiber. But his conclusion dealt with the motor endings of an invertebrate, *Milnesium tardigradum,* endowed as all tardigrades with smooth muscle fibers only; the importance of this finding, confirmed by de Quatrefages (1843) on the muscles of a gastropod mollusk, *Eolidina paradoxa,* was not immediately appreciated by those histologists studying striated muscles in vertebrates.

It was not until 1847 that Wagner's observations on the striated muscles of the frog shed doubt for the first time on the existence of the terminal bows in striated muscles of vertebrates, described by Valentin and Emmert. Wagner was the first to reveal two fundamental features of the neuromuscular junction of vertebrates; he concluded from his studies of the hyoidean muscles of the frog that the motor nerve fiber, after branching, loses its myelin sheath and closely connects with the muscle fiber.

Two other important characteristics, although less constant throughout the various types of neuromuscular junctions found in striated muscles, were subsequently discovered. The first was described by Kühne (1862), who showed, also on frog muscles, that, after piercing the sarcolemma, the motor nerve fiber branches again and provides a terminal arborization; the second was described by Rouget (1862), who reported the presence, in the muscles of reptiles, birds, and mammals, of a flattened heap of granular nucleated substance at the level of the neuromuscular junction, which he interpreted as the spreading of the axis cylinder substance at the surface of the myofibrils and which he called the end plate (*plaque terminale*).

At the time of his discovery of the end plate, Rouget did not perceive the nerve ending itself. It was Krause (1863) who the following year described the terminal branching at the level of the end plate on the retractor muscle of a cat's eye, which he called motor end plate (*motorische Endplatte*). This term is often applied today, conveniently but improperly, to all neuromuscular junctions, whether or not they present a plate of granular nucleated protoplasm.

The first really detailed observations of the structure of the motor end plate were made by Ranvier (1878) and Kühne (1883, 1887, 1892). All the later morphological research is based on the work of these two authors, especially on the very impressive paper by Kühne, which appeared in 1887.

According to Cöers (1967), "the modern era of knowledge of the morphology of the myoneural junction began with Couteaux's work." In 1947, Couteaux identified and described the subneural apparatus from pictures obtained by supravital staining with Janus green B. He observed that this pallisadelike structure, which outlined the motor arborization, represented the postsynaptic surface of the myoneural junction and was the site of the cholinesterase activity involved in neuromuscular transmission. This hypothesis was proved by the histochemical method for cholinesterase, which clearly showed the selective localization of the enzyme in this subneural apparatus (Couteaux and Taxi, 1952; Cöers, 1953a).

Subsequently, electron microscope studies by a number of authors confirmed the presence of membranes between the axoplasm and the sarcoplasm and that at that point the sarcoplasmic membrane was complexly folded.

III. General View of the Motor End Plate

For a general description of the vertebrate motor end plate, we shall refer mainly to end plates of reptiles and mammals. The accumulation of nuclei and granular cytoplasm of the junctional area forms a rounded heap at the surface of the myofibrillar bundle. It is this platelike heap of protoplasm lying against the myofibrillar bundle and at the level of the ramification of the nerve ending (Fig. 1) that Kühne called the sole.

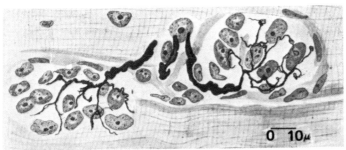

Fig. 1. Drawing of two motor end plates of guinea pig (m. gastrocnemius internus). Silver preparation.

Some of the various nuclei observed in the vicinity of the nerve ending are closely connected with the terminal nerve branches. These are the nuclei described by Ranvier (1878) as "arborization nuclei." Then there are the "fundamental nuclei" of Ranvier (Kühne's "sole nuclei," 1864), which are larger and much less stainable than the former, and which contain much larger nucleoli. In addition to these two categories of nuclei, there are the endothelial nuclei of the blood capillaries, sometimes very closely attached to the end plate, and last, the "vaginal nuclei" of Ranvier, which are not always present, and which are linked to the endothelial and collagenous coating, prolonging Henle's sheath of the nerve fiber over the end plate (see below).

Of these four kinds of nuclei observable at junction level, the only ones that are frequently easy to identify are the fundamental nuclei; these are generally distinguished by their size and especially their structure. But this is not the case for the arborization nuclei, the endothelial nuclei of the blood capillaries, and the vaginal nuclei, at least in adult muscles. This doubtless explains the long, inconclusive debate between Kühne and Ranvier on the end plate nuclei, and the numerous controversies on this subject which have arisen constantly since the end of the last century.

Data obtained from a study of the development of the end plate and from an analysis of its structure in the adult, both with light and electron microscopes, enable us to give a coherent interpretation of the highly complex neuromuscular junction, summarized in Fig. 2.

After the loss of its myelin sheath, the motor axon branches and con-

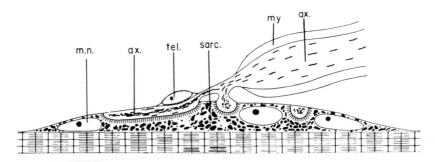

Fig. 2. Schematic drawing of a motor end plate seen in a longitudinal section of the muscle fiber. ax. = axoplasm with its mitochondria; my. = myelin sheath; tel. = teloglia; sarc. = sarcoplasm with its mitochondria; m.n. = muscle nuclei (fundamental nuclei). The terminal nerve branches lie in synaptic gutters, or troughs. Immediately under the interface axoplasm–sarcoplasm, the ribbon-shaped subneural lamellae, cut transversely, may be seen as rodlets. The bell mouth by which the Henle's sheath ends above the motor end plate has not been represented.

nects with the sarcoplasm. The nerve branches are accompanied by teloglia (terminal Schwann cells), to which the arborization nuclei belong; the branches occupy hollows of widely varying depth, shape, and size according to the animal species considered. These "synaptic gutters" or "synaptic troughs" are hollowed out into the sole, a flattened-out heap of sarcoplasm rich in nuclei and mitochondria. In spite of its close connections with each terminal nerve twig, the teloglia does not penetrate the interior of the synaptic gutters and does not come between the axoplasm and sarcoplasm.

By means of selective stainings, it is possible with a light microscope to distinguish the interface, at the level of which axoplasm and sarcoplasm are joined. On a section cut through a motor end plate, this interface appears as a thin line, to which are attached by one of their edges on the sarcoplasm side the elongated ribbon-shaped lamellae of the subneural apparatus. When they are cut transversally and observed with a light microscope, these subneural lamellae appear as rodlets measuring generally about 1 μ in length. They are always perpendicular to the axoplasm and sarcoplasm apposition surface. The juxtaposition of these rodlets, situated on the deeper face of the nerve terminals, results in a picture resembling that of a palisade or picket fence.

On a front view of a motor end plate, the lamellae, which run under the nerve terminals, appear as a fingerprint. Their orientation differs widely according to their position in the subneural apparatus; they are of unequal lengths and generally more close set near the edges of gutters.

After this brief description of the main morphological features of the motor end plate, we shall proceed to a more detailed analysis of its diverse components.

IV. Terminal Axoplasm

As a result of extensive work, Kühne (1887) using a gold method, Dogiel (1890), and Retzius (1892), with methylene blue, had already described the general form of motor nerve endings in each class of vertebrates. Their description was supported by numerous excellent plates.

Subsequently, many other investigators using the same methods, as well as silver impregnations, confirmed and completed previous observations; but, although these methods permit a general survey of various types of motor nerve endings, they unfortunately provide very little reliable information concerning their fine structure. They are indeed quite often used either in fresh tissues or after poor fixations, and in both cases it is difficult to know which artifacts are included in the stained structures.

These staining methods leave, moreover, the border line separating the terminal axoplasm from the teloglia practically unstained. In addition, because of their small size, most of the axoplasm components cannot be clearly seen, if at all, with the light microscope.

The nerve ending, as described by Kühne, is made up of two parts, differentially stained by his gold method. The first part, stained deep violet, is axial; this is the "axialbaum." The second part, the "stroma," stained red, encloses the first. When the structure described by Kühne as the stroma is located at the level of the still myelinated part of the axon preceding the branching, it must be admitted that this structure, found inside the myelin sheath, is part of the axoplasm. The nature of what he designated under the name of stroma at the level of the nerve ending itself is much more uncertain. As we shall see in connection with the teloglia, it seems that Kühne's "stroma" brings together quite different structures.

Kühne observed that the stroma is not quite a homogeneous structure, being more granular in its peripheral part. Contrary to a rather widely held opinion, he never referred to the stroma as a striated layer. There is an obvious confusion here with a superficial fringe (*Borstensaum*), also described by Kühne (1883), the significance of which will be discussed later. According to Kühne, this fringe belongs to the membrane coating the end plate. In his paper of 1887 (p. 21) and in particular, in the legends accompanying the drawings of the fringe, he invariably and carefully distinguished it from the stroma.

Most of the investigators who have stained the nerve arborization of motor end plates with methylene blue by intravascular injection of the staining solution or who have placed the excised muscle tissue in direct contact with the staining solution have often observed much more deeply stained filament in the axis of the arborization branches (Dogiel, 1890; Feist, 1890; Kulchitsky, 1924; Tiegs, 1932; Tello, 1944; Couteaux, 1947). It may happen that the axial filament is the only stained part of the nerve ending, but this picture is rarely obtained; in general, the axial filament stands out from the rest of the arborization, which is pale blue, by its deep blue color. A quite similar picture may sometimes be observed at the level of the axon myelinated part; a deep blue filament occupying the axis of the nerve fiber is separated from the myelin sheath by a poorly stained or unstained space.

Of all the methods used to selectively stain the motor nerve ending, those using silver gave the most clear-cut pictures, especially in the hands of Ramón y Cajal (1904, 1925), Tello (1905, 1944), Boeke (1909, 1911, 1927), and Lawrentjew (1928). Silver impregnation discloses a neurofibrillar framework in the nerve arborization, prolonging the neurofibrillar bundle of the nerve fiber at the level of the branches of this ending (Figs. 1, 3, and 4a–c,e–g).

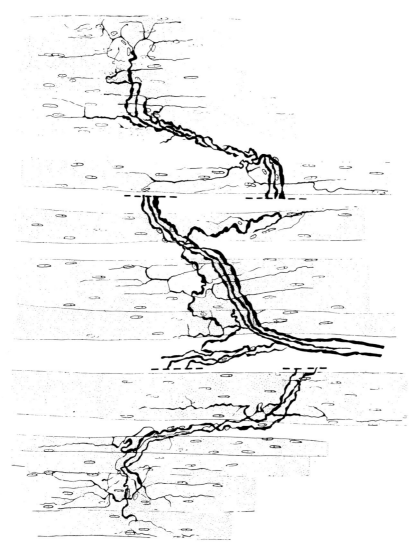

Fig. 3. Drawing of frog terminal arborizations (abdominal muscle). Silver preparation. The three median muscle fibers have been displaced in relation to the other fibers in order to show entirely their motor innervation. (×190.)

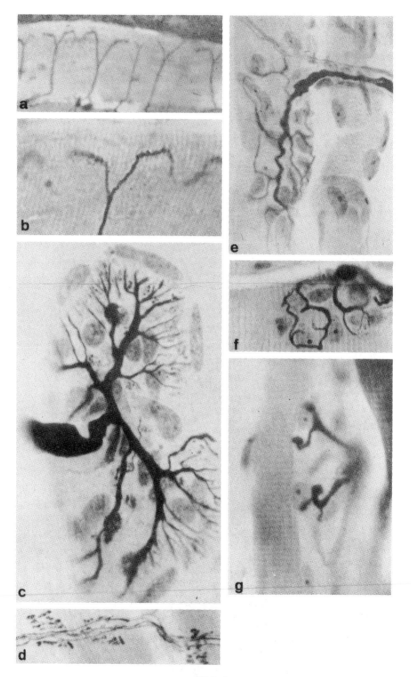

Fig. 4.

Many investigators who have studied the neurofibrillar framework of nerve arborization have pointed out that the neurofibrillar bundles of the terminal branches often end in an olive-shaped enlargement, in a small racketlike net, in loops, or in rings. The neurofibrillar bundle, often very slender, which is found inside each of the nerve terminal branches, is surrounded by a clear space similar to the one observed, following staining by the methylene blue method, around a deeply stained axial filament. It is quite probable that this clear space is, to a certain degree at least, a fixation artifact, resulting from either a shrinkage of the entire axoplasm or the condensation of a fibrillar component. It is also likely that the teloglia, whose relationship with the nerve branches will be described later, is part of the perifibrillar layer.

The assumption put forward by several authors, principally Ramón y Cajal (1925), to the effect that a peripheral layer of axoplasm lacking the fibrillar component always exists at the level of the terminal nerve branches, cannot be entirely rejected today; the presence of such a layer would also explain the prsence of a clear space surrounding the neurofibrillar bundle.

The only inclusions of the terminal axoplasm observable with the light microscope are very small granules, which can be stained either by Regaud's technique or by Altmann's aniline fuchsin with fixation by osmic mixtures. From their special stainability, it may be concluded that, in all probability, these granules are mitochondria. They appear more numerous at the level of the nerve ending than in the other regions of the motor axon. When examined with the electron microscope (Fig. 5), these tiny granules, observable with the light microscope inside the nerve branches, exhibit the typical internal structure of mitochondria (Robertson, 1956).

Fig. 4. Neuromuscular junctions of animal species belonging to different zoological divisions.   (a–b) Motor nerve endings at the surface of an insect muscle fiber (*Musca domestica,* thoracic muscle). Silver method; (a) ×520, (b) ×1600. Original photomicrographs from Auber.   (c) Motor end plate of a Selachian muscle fiber (Raja clavata, pelvic fin muscle). Silver method; ×800. Original photomicrograph from Barets.   (d) Motor nerve endings at the surface of slow muscle fibers of the frog (m. rectus abdominis). Methylene blue method; ×300. Original photomicrograph from Barets.   (e) Motor nerve ending (twitch system) of the frog (m. rectus abdominis). The neuromuscular junction represented here, more concentrated than usually is the case in the frog muscle, effects the transition between the end-bush (see Fig. 3) and the true end plate. (×480.) From Couteaux (1947). (f) Motor end plate of the guinea pig (m. gastrocnemius internus). Silver method; ×750.   (g) Motor end plate of man (larynx muscle). Silver method; ×750. Original photomicrograph from Barets. As frequently seen in human muscles, the nerve terminal branches are connected with the muscle fibers only by button-shaped extremities.

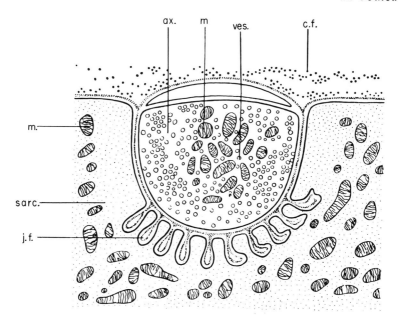

Fig. 5. Schematic drawing of a synaptic gutter, or trough, seen in cross section; ax. = axoplasm; m. = mitochondria; ves. = vesicles; sarc. = sarcoplasm; j.f. = junctional folds; c.f. = collagen fibrils. After Robertson (1956), slightly modified.

Electron microscopy has further shown that, in addition to the mitochondria, the terminal axoplasm contains some inclusions which are much more numerous and much smaller (300–500 Å). These "vesicles" have been described in the motor nerve endings of the striated muscle by Palade and Palay (1954) and by Robertson (1956); they appear to be homologous to the presynaptic "vesicles" described for some interneuronal synapses. These synaptic vesicles are most numerous at the edge of the ending of the nerve close to the sarcoplasm and may even be in contact with the plasma membrane. De Robertis (1955) has brought forward evidence on the morphological parts of this cell that contain acetylcholine. They open through the axolemma and in this way empty small "quanta" of acetylcholine into the sarcolemmal gutters. Thus, an end plate potential would be produced and this could lead to depolarization of the sarcolemma.

In spite of these recent findings, there still exist some important gaps in our knowledge of the fine structure of the terminal axoplasm. Studies made with the electron microscope have not yet allowed a satisfactory interpretation of the different pictures obtained with gold, silver, and methylene blue methods. In particular, neurofibrils of a nerve ending stained in the usual manner continue to pose a difficult problem. It seems probable today that the neurofibrils are a result of the artificial agglutination of submicroscopic filaments called neurofilaments; observations made on nerve fibers of the leech (Couteaux, 1957a) suggest that this agglutination depends above all on the manner the fixation is used. Up to now, however, the presence of neurofilaments in the motor endings of striated muscles has not been clearly determined, and it might be asked if thse filaments are not completely absent from the terminal nerve branches. This conclusion appears, nevertheless, quite unlikely, since preparations in which the neurofibrils are stained clearly show the existence of terminal neurofibrils extending the motor axon neurofibrils beyond the preterminal constriction. The presence at the same time of two kinds of submicroscopic neurofilaments of different size and nature (Palay, 1958) in the axoplasm of the nerve fibers may explain this apparent discrepancy between the data furnished by light and electron microscopy. On one hand, it is possible that the neurofilaments of larger caliber are not prolonged in the terminal axoplasm, and on the other hand, the neurofibrils stained by the silver methods inside the terminal nerve branches may be the result of the agglutination of the other submicroscopic neurofilaments, whose existence in the nerve fibers is not always easy to demonstrate with the present methods of electron microscopy.

V. Teloglia

The cells accompanying the terminal branches of the motor axon, and which collectively form the teloglia, are visible with a light microscope only at the level of their nuclei (arborization nuclei), each of which is generally surrounded by a small heap of cytoplasm.

A. Arborization Nuclei

Ranvier (1878, 1888) was the first author to distinguish clearly the arborization nuclei from, (1) those nuclei belonging to the endothelial sheath, which continues the Henle's sheath above the motor end plate (see below), and (2) from the fundamental nuclei (sole nuclei of Kühne). Kühne (1886, 1887, 1892) always disagreed with this distinction and would admit only the presence of sole nuclei and of nuclei attached to the telolemma, i.e., to the membrane coating the motor end plate.

Kühne was not the only one who denied the existence of a special category of nuclei bound to the terminal nerve branches. Ramón y Cajal (1899, 1909), Tello (1907, 1917), Iwanaga (1925), Stöhr (1928), Cuajunco (1942), and many others, although generally admitting that the relationship of certain nuclei with the nerve end-ramifications is particularly intimate, do not consider them as quite different from the other sole nuclei. As for the origin of arborization nuclei, it was long debated. It seems that Ranvier himself, uncertain as to the nature of the fundamental nuclei, did not wish to commit himself fully concerning the nature of the arborization nuclei. Del Rio-Hortega (1925), while he notes that the significance of thse nuclei is "not very clear," regards their muscular origin as probable.

As we shall see further, the existence of the arborization nuclei and their homology with the Schwann cell nuclei may be very clearly established by studying the development of the motor end plate (Couteaux, 1938a,b, 1941). The teloglia is most easily distinguished from the rest of the end plate when it forms, together with the nerve branches, a rounded corpuscle which depresses the neighboring nuclei.

The embryonic muscle preparations, however, are conclusive in this respect only if the end plates are seen as a whole and have been stained by a method that brings out the differences in the structure of the end plate nuclei, especially those concerning their nucleoli.

A study of motor end plate development led Boeke also (1942, 1944, 1949) to distinguish nuclei of lemmoblastic origin, which become arborization nuclei, from those of muscular origin found at the level of the neuromuscular junction. But, although the silver methods he used show the neurofibrils very clearly and stain the nuclei chromatin quite deeply, they do not always stain the nucleoli of the muscle nuclei with sufficient strength to enable them to be distinguished from other nuclei. According to Boeke, the cells to which the arborization nuclei belong are homologous to the "interstitial cells" described at the level of the terminal plexus of the autonomous nervous system; the cytoplasm of these cells and the neighboring sarcoplasm would "merge" into the sole protoplasm.

Further research on motor end plates by Tello (1944), using grown animals, showed that a nucleated sheath, probably prolonging the Schwann sheath at the level of the end plate, accompanies the terminal nerve branches along their entire length.

In spite of important differences in the general interpretation of the motor end plate, two points are agreed upon by the three above-mentioned authors: (1) cells having the same origin as Schwann cells are found in the end plate and (2) their nuclei are those described by Ranvier as arborization nuclei.

B. *Relationship between Teloglia and Terminal Axoplasm*

Referring to the structure of the terminal axoplasm, it has already been pointed out that complex problems are raised by the interpretation of pictures resulting from methods that use impregnation by gold, silver, or postvital stainings. The same difficulties arise when these procedures are used to define the relationship between teloglia and nerve terminal branches.

Tello (1944) concluded that each branch of the end arborization is completely enclosed in a nucleated sheath. The axoplasm would be, as a consequence, entirely separated from the junctional sarcoplasm by this sheath, even inside the synaptic gutters.

Through comparison of aspects of the nerve ending obtained by postvital staining with methylene blue or Janus green B and by silver methods, Couteaux (1945c, 1947) described two distinct parts in the end-bush of a frog: one, axial, extending the axon, and the other surrounding the first, which he regarded as belonging to the teloglia. This peripheral part forms a continuous sheath around each nerve branch. Couteaux compared it to the "stroma" described by Kühne (1887) by means of gold methods, but which the latter had considered as a component of the nerve end arborization.

This view of Kühne's stroma as a sheath, made of nonnervous protoplasm was not new. Tiegs (1932) showed on reptilian muscles that the stroma (which he called perilemma in order to emphasize its independence from the axon) survives degeneration of the nerve after section and remains in direct continuity with the cord of Büngner, i.e., Schwann cells of the degenerated nerve. This observation appeared to give an interesting evidence in favor of the glial nature of the Kühne's stroma (Tiegs, 1953).

Examination of sections made through the motor end-plate after fixation by means of good osmic fixatives (Bensley, osmiochromic fluid of Laguesse) and staining by mitochondrial methods shows that there is no room in the synaptic gutters for a periaxoplasmic layer as thick as Kühne's stroma.

The conclusions drawn from electron microscope observations have been in this respect much more categorical, since all the researchers who have hitherto studied the structure of the end plates with the electron microscope (Palade and Palay, 1954; Reger, 1954, 1955, 1957; Robertson, 1954, 1956) deny that any teloglial cytoplasmic layer, even of the thinnest dimensions, is present between the axoplasm and the sarcoplasm.

Due to its high resolution and the thinness of its sections, electron microscopy is much more suitable than light microscopy for solving such a problem, and consequently, it may be considered as established that there is no interposition at all of teloglia between the axoplasm and sarcoplasm at the level of the motor endplate. Although the teloglia is closely tied to the nerve ending, beginning with the first stage of the formation of the motor end plate (Couteaux, 1938a,b, 1941), it does not completely enclose the nerve ending and does not penetrate to the interior of the synaptic gutters (Fig. 5). It is possible, particularly in the frog, that slender teloglial processes sometimes form a loop around the terminal nerve branches. It seems, however, that such an interposition between the pre- and postsynaptic membranes would be practically negligible as far as the functional relationships of these two membranes are concerned.

A similar problem has been posed by interneuronal synapses. Hitherto, in every case where an examination with the electron microscope has been made, the conclusion was the same, i.e., there is no neuroglial interposition between the pre- and postsynaptic membranes.

Concerning the general arrangement of the teloglia, it seems, however, to have been previously demonstrated with the light microscope that the teloglia follows each of the branches of the nerve ending along its entire length. Several electron micrographs by Reger (1955) show the intimate connections of the teloglia with the nerve terminals.

Robertson (1956) established, in a chameleon lizard, that above the nerve twig a thin layer of cytoplasm exists which differs decidedly in appearance from the underlying axoplasm, from which it is separated by a double membrane. Though the nature of this superficial layer of cytoplasm has not been established beyond doubt, it seems highly probable that it may be considered an expansion of the teloglial cell. According to this hypothesis, the teloglia closes the synaptic gutter as a lid or an operculum and does not, therefore, lie between the axoplasm and the sarcoplasm, but between the axoplasm and the extracellular medium. Further research will show whether this firm closing of the synaptic gutter by the teloglia is the rule, and a constant feature of the end plate organization.

Now that it has been demonstrated that the axon terminal branches have no teloglia lining inside the synaptic gutters, the stroma described by Kühne as completely enclosing each branch of the "axialbaum" and as being more granular in its peripheral part can no longer be considered as formed uniquely by the teloglia. It also includes, without any doubt, an axoplasmic part, since Kühne describes it as existing not only around the nerve terminal branches, but also at the level of the last myelinated segment of the motor axon, inside the myelin sheath (1887, Fig. 47).

On the other hand, the structures described as stroma on the end plates, seen in front view and on cross sections, are doubtless not the same. Several of the cross sections of the motor end plates drawn by Kühne show beyond doubt that, at the level of the space located between the fibrillar axis deeply stained by a gold method and the sarcoplasm, the stroma is formed, at least in its peripheral part, by the subneural apparatus. It seems probable, therefore, that the stroma described with a gold method includes some very different structures belonging to the three tissues which enter into the make-up of the motor end plate.

As the most recent data show that the teloglia accompanies the terminal nerve branches, although not interposed between the axoplasm and the sarcoplasm, the only part of the stroma that can be regarded as teloglia at the level of the junction is the one which is found above the nerve branches, outside the synaptic gutters.

VI. Junctional Sarcoplasm

In mammals, the junctional sarcoplasm, together with the fundamental nuclei and its granules, generally form only a slight protuberance on the surface of the muscle fiber when the latter is not contracted. As Kühne (1887) had already noted, however, this mass of sarcoplam may become quite prominent if the muscle fiber is contracted.

In the frog, whose nerve branches are nearly always quite long, the junctional sarcoplasm does not form a proturberance that can be seen, even when the muscle fiber is tightly contracted. This difference between the neuromuscular junction of batrachians and those of higher vertebrates was emphasized by Rouget (1864) in a paper on motor endings, in which he pointed out also that the prominence of the "terminal cone or eminence" in the Tardigrada, described by Doyère (1840), depends above all on the "mechanical tension" of the motor nerve, and varies widely as a result with the movements of the muscle fibers.

Whether a protuberance of sarcoplasm exists or not at the neuromuscular junction, the terminal branches of the motor nerve ending are invariably found in the surface depressions of the junctional sarcoplasm. In vertebrates, these depressions often take the form of gutters, modeled on the nerve branches.

This surface position of the motor nerve ending was observed by Gutmann and Young (1944), using silver impregnation, in lateral views of the motor end-plate of the rabbit, and by Couteaux (1944, 1945a), while studying the subneural apparatus, stained with Janus green B, in cross sections of musle fibers of the mouse. The nervous and muscular substances enter directly into contact at the bottom of these "synaptic gutters."

In those vertebrates whose motor end plates have been studied in detail, it has been remarked that the sarcoplasm presents at the level of the synaptic gutters structural or cytochemical particularities that it exhibits nowhere else. On account of these particularities, the layer of sarcoplasm located immediately below the terminal nerve branches appears as a differentiated zone of the muscle fiber described as the subneural apparatus.

A. Subneural Apparatus

In the layer of subneural sarcoplasm, selective stainings show a collection of elongated ribbon-shaped lamellae attached by one edge along their entire length to the sarcoplasm membrane. These lamellae are attached along fairly equidistant lines, giving a periodic structure to the subneural apparatus.

The subneural lamellae may be stained by several methods. The first procedures employed were postvital stainings, with substances ordinarily used to distinguish the mitochondria vitally (e.g., Janus green B, methyl violet, and dahlia violet), the stained muscles being observed fresh or after fixation with ammonium molybdate (Couteaux, 1944, 1947). Using these methods, it was possible not only to stain the subneural lamellae but also to demonstrate that the membrane to which they are attached is prolonged with the one that delimits the sarcoplasm outside the synaptic gutters.

Thanks to the development of enzymic histochemistry, and in particular to the discovery of methods for localizing cholinesterases, these postvital methods have been replaced by far more specific ones, which are at the same time more reproducible, and which, applied for formaldehyde-fixed muscles, give quite conclusive pictures (Fig. 6).

Lastly, the usual cytological methods, such as staining with ferric hematoxylin or Altmann's aniline fuchsin, associated with other stains, provide images of the subneural apparatus on muscles fixed with osmic mixtures—images that are less selective than those obtained with histochemical methods, but which nevertheless are quite clear.

Studies of the motor end plates using these methods show that the general form of the subneural apparatus varies considerably from one animal species to another. It naturally depends, above all, on the shape of the nerve endings, and also, apparently, on the thickness of the junctional sarcoplasm. In cases where the junctional sarcoplasm presents only a slight thickness, for example, in the frog, the depression limited by the subneural apparatus is very shallow. It is deeper when the junctional sarcoplasm accumulates in a true plate, as is generally the case in the mammals, and to a greater degree, in the lizard, where the plate is particularly thick.

On a section cut perpendicularly to one of the nerve terminal branches of the end-plate of the lizard or mammals, the subneural apparatus appears on the deep face of the nerve branch as a very thin layer, curved into an arc (Fig. 6b–c). This characteristic aspect can be observed only on sections cut exactly perpendicularly to the plane of the nerve arborization through an uncontracted muscle fiber. More or less oblique sections, or those cut through a motor end-plate projecting considerably, due to the contraction of the myofibrils, display the subneural apparatus in very different aspects and do not always show its edges reaching the surface of the end plate.

At the level of each synaptic gutter, the lamellae are oriented in different directions in relation to the gutter axis—transversally, obliquely, or longitudinally. In following the curve of the gutters, the subneural lamellae crop out here and there at the surface of the end plate; they are prolonged at this level by threads, which are sometimes quite long. If only the superficial part of these lamellae is stained, we obtain, from the front view of an end plate, a ciliated fringe appearance similar to the one described in reptiles by Kühne (1883) at the surface of the end-plate (*Borstensaum*), and which he interpreted as a superficial structure situated "in or on" the membrane covering the end plate. Although Kühne was unable to find the equivalent in mammals, there seems to be no doubt that it corresponds to the outcroppings of the subneural lamellae at the surface of the sole. Studies made on the motor end plates of lizards, using postvital staining, are quite conclusive on this point. It does not appear possible, on the other hand, to interpret so easily the fringe, described by Kühne, surrounding the terminal nerve branches of the frog, since the most recent findings do not agree with such an interpretation. Kühne considered these fringes to be formed only by the soldering lines (*Lothlinien*) of the sarcolemma and the perineural sheath, i.e., Henle's sheath.

Light microscopy, with the aid of selective stainings, makes it possible to describe the lamellae of the subneural apparatus. Using only this means, however, it would be difficult to define the nature of the lamellae. Since they are placed at the boundary between axoplasm and sarcoplasm and penetrate into the junctional sarcoplasm, the subneural lamellae might be considered as a dependence of the membrane limiting sarco-

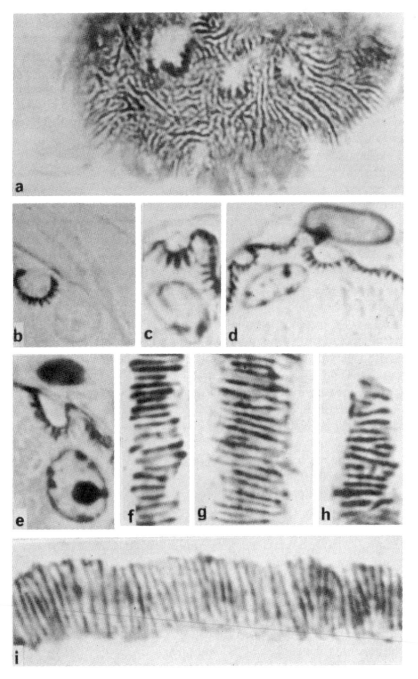

Fig. 6.

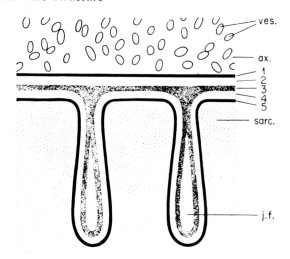

Fig. 7. Schematic drawing of two junctional folds, showing how the sarcoplasm surface membrane joins with the axoplasm surface membrane and forms the synaptic compound membrane; ax. = axoplasm; ves. = vesicles; sarc. = sarcoplasm; j.f. = junctional fold. After Robertson (1956), slightly modified.

plasm; new methods of research, however, were necessary to verify this point.

Using the electron microscope, Palade and Palay (1954), Robertson (1954, 1956), and Reger (1957) showed that the subneural lamellae are formed by narrow infoldings of the sarcoplasm surface membrane (Figs. 5 and 7). By combining the electron microscopy data with those furnished by light microscopy, one may conclude that the junctional infoldings are not tubular recesses but long folds orientated in a different manner in relation to the gutter axis (Fig. 8).

In the frog, the folds are transversal (Fig. 6f–i) and stretch from

---

Fig. 6. Subneural apparatus. Acetylthiocholine method after fixation by formaldehyde. (a–e) Hedgehog motor end plate (intercostal muscles). (a) Front view of the end plate with the focus on the deeper part of the subneural apparatus (×2700). (b) Synaptic gutter seen in cross section, without counterstaining (contrast of the unstained structures slightly increased by diaphragm adjustment); the subneural lamellae (folds of the electron micrographs) appear as rodlets if they are cut perpendicular to their axis; on the right, a fundamental nucleus, unstained, is hardly visible (×2700). (c–e) Cross sections through end plates, with staining of the nuclei by hematoxylin; above the subneural apparatus, teloglial nuclei, and below it, fundamental nuclei (from Couteaux, 1955). (c) ×4000; (d) ×3300; (e) ×3700. (f–i) End-bushes of the frog (gastrocnemius); front view of the subneural apparatus at different levels of end-bushes. (f) ×3850. (g) ×3500. (h) Terminal segment belonging to the same end-bush branch as the preceding one (×3500). (i) ×3400.

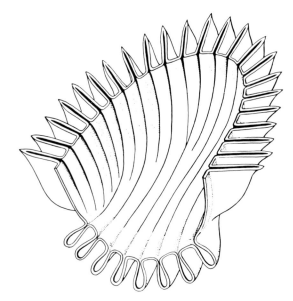

Fig. 8. Schematic three-dimensional presentation of the subneural apparatus at the extremity of a synaptic gutter. This interpretation is based upon data hitherto obtained with both the light and the electron microscope (cf. Fig. 5).

one edge of the gutter to the other. This arrangement may be found in reptiles and in mammals, but it is much less frequent, and is observed in general only along short segments of the gutter. In reptiles and mammals, the orientation often varies considerably, depending on whether we are dealing with the deep part of the gutter or the areas adjacent to the edges.

In the deep part of the gutter, the folds may be oriented in different patterns, often longitudinally. In the latter case, the folds may be as long as the gutter itself, and the appearance of the synaptic gutter, seen on a section cut perpendicular to its axis, resembles part of a cog wheel (Fig. 6b–e), the cogs corresponding to the folds seen in cross section. Bordering the edges of each gutter, the subneural folds have, on the contrary, a fixed orientation, perpendicular to each edge. The arrangement of folds in these areas explains why the sections cut transversally to the gutter axis have no cogs near the edges, except at the gutter extremity.

It is possible that the unstained zone existing in the deepest part of each infolding corresponds to the section of a canaliculus. Observations with the light microscope, moreover, invariable show, on the sides of the synaptic gutter, outcroppings of the subneural lamellae at the surface of the sole. If, therefore, the unstained zone visible inside the folds is really a "free" space, it is probable that the canaliculi of the subneural folds open at the end plate surface by means of slot-shaped openings placed on either side of each nerve branch (Couteaux, 1957b).

In the animal species where numerous anastomoses unite the subneural folds, the canaliculi would form a kind of network below each terminal nerve branch. It is obvious that such a system of canaliculi may play an important part in the normal functioning of the normal motor end plate and may allow, moreover, the pharmacodynamic agents to reach more easily both the synapse and the junctional sarcoplasm. It may be assumed that the canaliculi bring into relationship the diverse regions of the postsynaptic membrane, including even those regions located in the deepest part of the gutters and the extracellular medium. The folds are often prolonged at the surface of the end-plate by other canaliculi, which correspond to the superficial threads, sometimes quite long, observed with the light microscope.

The study of the submicroscopic structure of the subneural apparatus has only just begun, but it is already certain that this layer of sarcoplasm, having a thickness of about 1 μ, and limiting the synaptic gutters, presents morphological features of a highly specialized zone.

Let us now consider to what degree the sarcoplasmic membrane in the region of the subneural infoldings differs from that of the remaining muscle fiber. Postvital methods allow the selective staining, on the surface of the striated muscle fiber, of a component of the sarcolemma that appears under the light microscope as a continuous and homogeneous membrane (Couteaux, 1947). This membrane extends under the terminal nerve branches at the motor end plate and is prolonged by the folded membrane of the subneural apparatus. It is probable that the membrane stained by the periodic acid–Schiff method (PAS reaction) at the level of the sarcolemma by MacManus (1948) is identical with the one stained by postvital methods. These staining methods reveal no important difference between the membrane coating the sarcoplasm within the synaptic area and the one enveloping the rest of the muscle fiber.

Differences are apparent between the two zones of the sarcoplasm surface membrane when other methods are used, such as Heidenhain's Azan method, which stains the collagen fibrils of the endomysium, or Laidlaw's silver technique, which stains the argyrophilic fibrillar network very deeply at the surface of the muscle fiber. None of these procedures bring to light the existence of any fibrillar structure at the surface of the subneural sarcoplasm. Nor has it been possible to detect collagen fibrils at this level with the electron microscope.

Further data have resulted from observations made with the electron microscope after osmic fixation. There is a continuity between, on one hand, the three layers of the subneural sarcoplasmic membrane (made up of a very dense inner layer, an outer layer less dense and thicker than the former, and, between these two layers, a light layer), and on the other hand, the three corresponding layers of the sarcolemma (Figs. 5 and 7). Robertson (1956), however, remarked that the total thickness of these three layers is slightly less at the level of the synaptic gutters, due mainly to the thinning out of the light middle layer. An interpretation of this fact would at present be premature.

Histochemical methods that reveal cholinesterase activity have proved, up to the present moment, to be the most effective ones for clearly defining the differences between the sarcoplasm membrane located within the synaptic area and the remainder of this membrane. These procedures disclose at the motor end plate an important concentration of cholinesterase, mainly located at the postsynaptic membrane.

The location of cholinesterase at this point is quite typical of the subneural apparatus. In vertebrates, this cytochemical characteristic seems to be a more constant attribute to the subneural differentiation of the sarcoplasm than the presence of folds, which appear to be wanting at some neuromuscular junctions, for example, on frog slow muscle fibers and teleost muscle fibers. The question of enzymic location will be discussed later in more detail.

Further investigation will be necessary not only to complete the study of the sarcoplasmic membrane at the level of the synaptic gutters, but also to examine the particularities of the thin sarcoplasm layers located between the folds.

Since it appears probable that the deep part of the subneural apparatus rather far distant from the axoplasmic membrane does not play exactly the same role as the properly synaptic part of this apparatus in the transmission of the excitation from the motor nerve to the muscle, a more thorough morphological and cytochemical investigation will perhaps reveal significant differences between the two parts.

B. *Relationship between Axoplasmic and Sarcoplasmic Membranes*

Due to the existence of the subneural infoldings, the sarcoplasm surface membrane at the end plate presents two parts having very different relationships with the axoplasm surface membrane. One is closely joined to the axoplasm surface membrane, whereas the other corresponds to the infoldings at the level of which the membrane limiting the junctional sarcoplasm leaves the one limiting the terminal axoplasm.

Between the infoldings, the surface membranes of the nerve fiber and muscle fiber are closely united to each other, and their connections are, in this respect, comparable with those existing between the pre- and postsynaptic membranes of a central synapse (Palade and Palay, 1954; Palay, 1956), and of a ganglionic synapse (De Robertis and Bennett, 1954, 1955; Taxi, 1957, 1958), where there is also an apposition of the two membranes. Besides the junctional folds, however—nothing comparable to them has hitherto been noted at interneuronal synapses— the pre- and postsynaptic membranes present at the end plate a somewhat different structure from that of the membrane forming the synaptolemma of interneuronal synapses.

At the point where the two membranes are apposed and form a synaptic compound membrane (Fig. 5), it is difficult to determine exactly what belongs to each of the two membranes. The junctional fold, however, formed solely by the sarcoplasm surface membrane, offers particularly favorable conditions for the analysis of the latter. In the chameleon lizard, Robertson (1956) describes three layers: (1) a very dense inner layer less than 100 Å thick; (2) an outer layer about 200 Å thick, which is also dense, but less dense than the former; and (3) between these two layers a light layer about 100–200 Å thick.

As one proceeds from the junctional folds, their three layers may be seen extending into the synaptic compound membrane that results from the apposition of the axoplasmic and sarcoplasmic surface membranes. The surface membrane of the axoplasm, like that of the sarcoplasm, contains a very dense inner layer, less than 100 Å thick. It seems probable that the axoplasm surface membrane also includes a light layer and a dense outer layer.

The synaptic membrane complex, about 500–700 Å thick, resulting from apposition, comprises five layers (Fig. 7). The two very dense inner membranes, limiting the axoplasm and sarcoplasm, respectively, are separated from each other by three other layers: a dense middle layer and two light layers. One of the light layers obviously corresponds to the light layer of the sarcoplasm surface membrane, and the dense middle layer probably results from a fusion of the two dense outer layers of the axoplasmic and sarcoplasmic membranes.*

* This description given by Robertson of the synaptic membrane complex was based for the most part upon osmium tetroxide-fixed phosphotungstic acid-stained preparations. Later investigations led Robertson (1957, 1958) to consider as plasma membranes only the very dense layers next to axoplasm and sarcoplasm. Each of these plasma membranes would be made up of two dense layers 25 Å thick separated by a light layer of about the same width. The rest of the synaptic membrane complex would be extracellular, perhaps comparable to a basement membrane, and would probably represent a highly hydrated gel.

Although the axoplasmic and sarcoplasmic membranes are only apposed, it is quite difficult to separate them from each other. If a strong traction is applied after fixation on the motor axon segment immediately adjacent to the end plate, it is possible to detach the entire nerve ending from the muscle fiber, but at the same time, the subneural apparatus is carried with it (Couteaux, 1952, Plate II,B). Consequently, the sarcolemma breaks off around the end plate without the axoplasmic membrane separating from the sarcoplasmic membrane. This phenomenon can doubtless be explained in large part by the ties, in particular fibrillar, uniting the coating of the motor end plate to that of the motor axon; in all likelihood, it is also a result of the strong adherence, at least after fixation, of the axoplasmic and sarcoplasmic membranes at their junction.

C. Fundamental Nuclei and Inclusions of the Junctional Sarcoplasm

As the study of the development of motor end plates and the delimitation of the sarcoplasm in adult end-plates (Fig. 6c–e) have established, the sole nuclei of fundamental nuclei belong to the muscle fiber. For quite a long time, they have been described as large, clear, rounded nuclei, containing one or two rather bulky nucleoli.

The shape of the fundamental nuclei often contrasts with that of ordinary muscle nuclei, the latter often being elongated, with their axes disposed parallel to the direction of the myofibrils. This difference serves as one of the premises of those authors who do not consider the fundamental nuclei and the surrounding granular cytoplasm as muscular, attributing to them the same origin as the Schwann cells (Iwanaga, 1925; Noël, 1927, 1950).

In fact, only the sarcoplasm-poor fiber nuclei exhibit a marked difference; these fibers make up the greater part of the white muscles, where the ordinary muscle nuclei are surrounded only by a very small quantity of sarcoplasm. The difference in shape of nuclei occasionally disappears completely when it is a question of sarcoplasm-rich fibers having an accumulation on their surface of an important quantity of peripheral sarcoplasm distributed mainly around their nuclei.

It appears that the muscle nuclei—the ordinary muscle nuclei as weil as the fundamental nuclei—invariably present a rounded form when they are located inside a considerable mass of the sarcoplasm, as if they were no longer subject to the direct mechanical action of the myofibrils, with the resulting deformations. As a consequence, the form of the fundamental nuclei would depend above all on the quantity of sarcoplasm surrounding them and would not be an intrinsic feature distinguishing them from ordinary muscle nuclei. Moreover, because the form and size of the nuclei are quite variable, one cannot distinguish the fundamental nuclei from the other muscle nuclei on the basis of the structural differences of the nucleoli.

In the case of frog muscle fibers, on whose surface are found the motor axons terminating in end-bushes, Ranvier (1878) concluded that the fundamental nuclei were absent. In reality, there exist here also, along the end-bush twigs, some muscle nuclei, satellites of the nerve branches like the fundamental nuclei of other kinds of neuromuscular junctions (Couteaux, 1945b). At the level of the end-bush, however, the junctional sarcoplasm is not accumulated in a single plate; it is scattered along the terminal nerve branches, which are frequently quite long. Each of the muscle nuclei accompanying the nerve ending is surrounded by only a very small quantity of sarcoplasm. One may deduce that in the frog the nuclei of the junctional sarcoplasm generally take the form of ordinary muscle nuclei of the sarcoplasm-poor fibers and, like the latter, are nuclei stretched out parallel to the myofibrils.

Whether it be scattered as in the frog or accumulated in a sole, the junctional sarcoplasm invariably contains very numerous granules, observed for the first time without staining by Rouget (1862) and by Held (1897), using a poorly selective staining. These granules were later stained by Boeke and Noël (1925) and by Noël (1925, 1950), using Regaud's methods, and described as mitochondria. All the other usual methods of staining mitochondria, employed after fixation or vitally, stain them as deeply. Because of their characteristic stainability and occasionally their form, almost all of these granules may be considered mitochondria. Studies made with the electron microscope have recently confirmed this conclusion.

Granules apparently identical with those surrounding the fundamental nuclei of the end plate may be observed around the ordinary muscle nuclei on the surface of the sarcoplasm-rich fibers. In addition to the granules, quite a number of investigators have described special fibrillar structures at the level of the junctional sarcoplasm, which would establish a direct connection between the nerve ending and the myofibrils.

At the end of the last century, several authors, among them Foettiger (1880) on observation of "Doyère's cones" in insects, stated that the nerve endings are prolonged by slender fibrils as far as the Z bands. In the lizard, Tiegs (1932) reported a similar connection at the motor end plate. Boeke (1911, 1932) also recognized a connection between the nerve ending and the myofibrils. According to him, however, this connection is effected by a network of polygonal meshes, which he called the periterminal network. Some authors confirmed Boeke's findings and adopted his conclusions; others questioned their validity. Ramón y Cajal (1925), Tower (1931), and Tello (1944) all noted in the junctional sarcoplasm a network (called the *réseau intraplaculaire* by Cajal) comparable to Boeke's periterminal network; none of them, however, could observe that it was continuous with the terminal branches of the nerve fiber.

It is still difficult today to interpret the different pictures obtained by silver impregnantions, described as fibrillar structures with or without reticulations, located between the motor nerve ending and the myofibrils. In any event, whatever may be the nature of the structures observed, it is very unlikely that they can be continuous, at the same time, with axoplasmic components and the myofibrils, since it is now established that the nerve fiber and muscle are each limited by a plasma membrane at their junction as at any other points of their surfaces.

VII. Coating of the Motor End Plate

Investigations with the light microscope have shown that the terminal nerve branches are accompanied by teloglia along their whole length. With the aid of electron microscopy, it has been established that the teloglial cells do not penetrate into the synaptic gutters and are not found between the axoplasm and the sarcoplasm at the motor end plate. It is possible that these cells, arranged like a lid over the synaptic gutters, separate entirely the terminal axoplasm from the extracellular medium.

Outside of its teloglial component, the coating of the motor end plate includes the terminal bell mouth of Henle's sheath. This sheath, which prolongs the system of the lamellar nerve sheath as far as the level of the isolated nerve fiber, has at its end the form of an inverted funnel whose edge is attached to the muscle fiber and covers the motor end plate. It is formed from an endothelium. Renaut (1899) was able to stain with a mixture containing silver nitrate the boundaries of the uninucleated flat cells of this endothelium on a rabbit tongue muscle. In the preparations he thus obtained, the boundaries of the endothelial cells appear less clearly and broken along the line of the junction of the Henle's sheath and muscle fiber.

On the chameleon lizard, Robertson (1956) showed that the Henle sheath cells are limited, on the inner and outer surfaces, by a plasma membrane of a type similar to those of the Schwann cells and of the muscle fiber.

For a long time, it has been known that collagen fibrils are found adherent to the endothelium and strengthening it. Many collagen fibrils are also present in the space between this sheath and the surface of the end plate.

An extensive investigation of the motor end plate by Shanthaveerappa and Bourne (1968) has indicated that the terminal bell mouth of Henle's sheath is actually an extension of the perineural epithelium, and they suggested that the term "bell mouth of Henle" should be discontinued, since it indicated an inaccurate origin of the cells composing this structure. They suggest that the cells composing this bell mouth should be called perisynaptic cells and the motor end plate be referred to as the perisynaptic membrane. Figures 18–25 in the paper by Robertson (1960), which show some electron micrographs of these cells, and L. M. Brown's (1961, Figs. 5 and 6) paper, which deals with the localization of cholinesterase in the motor end plate, show the bell mouth cells very clearly, and they are obviously identical with the cells described by Shanthaveerappa and Bourne in a series of publications as perineural epithelial cells. Illustrations of these cells by many other workers demonstrate the same thing. Most of the latter actually refer to the membrane as being composed of Schwann cell cytoplasm interposed between the nerve terminal in the synaptic trough and the extracellular fluid. It is also of interest that the cells found in the cuff surrounding the space between the electroplaques in the electric organ of the torpedo are identical with the perineural epithelial cells (Sheridan, 1965).

Most workers who have studied the motor end plate with the light microscope have described two types of sheath covering the motor end plate. One is said to be a prolongation of the Schwann cell and is described as the terminal teloglial sheath. The second sheath covering the motor end plate on the top of the teloglial cell is what has been described as the bell mouth of Henle, which is said to be derived from the sheath of Henle; however, the collagenous sheath of Henle is much too scanty to form a sheath over the motor end plate, and the Schwann cell terminates by sequential peeling before the end plate is reached and does not form any kind of a sheath over it.

Various workers with the electron microscope have shown only one type of flat cell covering the motor end plate. These cells, although called various names by the different authors, are undoubtedly perineural epithelial cells. They attach to the sarcolemma and therefore, continue to its termination, the isolation from the body fluids that the nerve enjoys from its origin in the central nervous system to its termination on the muscle by the sheath which the perineural epithelium forms.

When a nerve is cut or destroyed, Wallerian degeneration sets in,

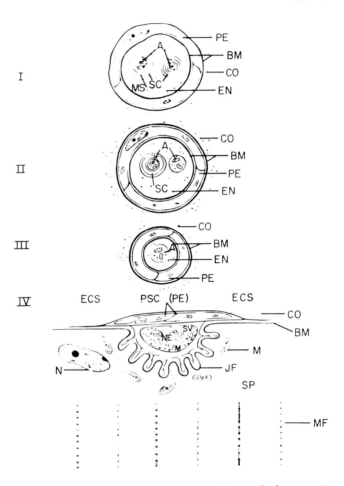

Fig. 9. A diagrammatic representation based on light and electron microscopic studies of the nerve fasciculus before its termination in the motor end plate (I, II, III) and of the motor end plate (IV). All these diagrams represent transverse sections.   (I) The nerve fasciculus is composed of two axons (arrow A) in which one is completely surrounded by myelin sheath and Schwann cell (arrows MS, SC), and the other only by Schwann cell cytoplasm. Both axons are surrounded by scanty collagen fibers which form the endoneurium (arrow EN). The axons, Schwann cells, and endoneurium are surrounded by a layer of flat perineural epithelial cells (arrow PE), having a basement membrane on both sides. Outside this cell covering are found collagen fibers forming the perineural connective tissue (arrow CO).   (II) This diagram is similar to I, but in this case one axon is covered by a Schwann cell and the other has lost its Schwann cell. (III) The structure of the nerve fiber just before termination at the motor end plate. Note the single naked axon (arrow A) without any myelin sheath or Schwann cell covering, and the very scanty endoneural collagen (arrow EN) surrounded by the perineural epithelial cells having the same character as in I and II and which covers the axon completely.

and eventually only the neurilemma remains. Similarly, it has been shown by Shanthaveerappa and Bourne (1964) that the perineural epithelium also remains intact under these conditions. The so-called bell mouth of Henle—the perisynaptic membrane—also remains intact, but collapses onto the motor end plate only to resume its former position when that particular motor end plate is reinnervated.

The relationship between the sarcolemma and the nerve terminal branches has been the subject of lively controversy since the first investigations on the motor end plate. The final conclusion was that the nerve terminal arborization is located below the sarcolemma, i.e., is hypolemmal. (See Fig. 10)

The above description of the synaptic gutters—where the axoplasm of the terminal nerve branches is separated from the sarcoplasm not only by the axoplasmic membrane, but also by a membrane prolonging the sarcolemma—showed that at present it is not possible to consider the nerve ending as hypolemmal in the usual sense of the word, i.e., as located below the sarcolemma. Whether or not the nerve branches are found deep inside the sarcoplasm, the sarcoplasm surface membrane always passes below them and presents the same general structural pattern at this level as at all other points of the sarcoplasm surface. One might conclude from this fact that the nerve ending is located above the sarcolemma and that therefore it is epilemmal, but this term is not exactly appropriate. The sarcoplasm surface membrane seen below the nerve terminal twigs does not, in fact, seem to possess a fibrillar lining, and may not, therefore, be considered as quite the same as ordinary sarcolemma.

The discarding of the terms "hypolemmal" and "epilemmal" used

---

This protoplasmic epithelial cell is in turn surrounded by scanty collagen fibers which represent the perineurial connective tissue. Actually, the collagen is so scanty it can be demonstrated only electron microscopically. (IV) Transverse section of the motor end plate as seen electron microscopically, showing the myofibrils (MF), nucleus (N), mitochondria (M), sarcoplasm (SP), and junctional fold (JF), along with diffuse amorphous substance (BM) of the muscle at the nerve terminal. Note that the axon (NE) in the synaptic gutter of the muscle contains numerous mitochondria and synaptic vesicles and is completely naked, being devoid of any Schwann cell covering. This nerve terminal in the synaptic gutter is surrounded by the perineural epithelial cells (PE), which completely separate this end organ from the surrounding tissue fluids of the extracellular space (ECS). These cells, because they completely cover the end organ, we call the presynaptic cell layer (PSC). They are flat squamous cells like the lamellar cells of the Pacinian corpuscles and the perineural epithelial cells. Above this perisynaptic cell layer, the end plate is surrounded by scanty collagen fibrils (CO). From Shanthaveerappa and Bourne (1968).

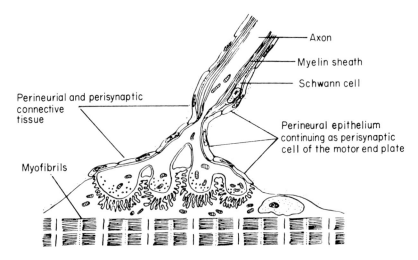

Fig. 10. Diagram based on light and electron microscopic studies of the myoneural junction showing the relationship of various membranes of the terminal nerve fiber to the motor end plate. The Schwann cell along with the myelin sheath is shown to terminate by sequential peeling before the axon comes in contact with muscle fiber, leaving the axon naked. The endoneurium at this point is virtually absent. The perineural epithelial cells covering the terminal nerve fiber continue as the perisynaptic cells of the motor end plate cover the whole of the motor end plate and finally attach to the sarcolemma. There are only a few collagen fibers covering the terminal nerve fiber and the motor end plate, representing the continuation of the epi- and perineurial connective tissue. This diagram shows that the Henle sheath, which is said to continue as the bell mouth of Henle's sheath, or the Schwann cells, which are said to continue as the teloglial cells of the motor end plate, have no role in the formation of the coverings of the motor end plate. From T. R. Shanthaveerappa and G. H. Bourne (1967).

hitherto has already been proposed (Couteaux, 1947). These terms are today even more out of date, since they refer chiefly to a morphological distinction between motor and sensory endings, a distinction not confirmed by electron microscope investigations.

VIII. Localization of the Junctional Cholinesterase

A study of the distribution of cholinesterase activity at the neuromuscular junction was undertaken subsequent to the work of Dale and Feldberg (1934), Dale *et al.* (1936), and G. L. Brown *et al.* (1936), who had shown that the transmission of excitation from motor nerve to striated muscle is accompanied by a release of acetylcholine. Its main object was to determine whether the cholinesterase, which exists in a weak concentration in nearly all animal tissues, has at the end plate a concentration high enough to ensure that the acetylcholine released at each nerve impulse is destroyed during the refractory periods, i.e., in a few milliseconds.

The first research on the distribution of activity in the striated muscle was carried out on the sartorius of the frog (Marnay and Nachmansohn, 1937) and toad (Feng and Ting, 1938). Manometrically, this research established that hydrolysis is more rapid in that region of the muscle where motor nerve endings are located than in other regions of the muscle, and more rapid than at the level of the motor nerve itself. Concerning mammalian muscles, a few manometric determinations on the gastrocnemius of the dog corroborated results obtained with batrachians, but the first really significant data were obtained only later, on the m. gastrocnemius internus of the guinea pig (Couteaux and Nachmansohn, 1940, 1942).

As the motor end plates represent only an extremely small fraction of the total volume of the muscle, it was inferred from the differences in enzymic activity observed between muscle regions rich in or deprived of motor end plates, that enzymic concentration is probably extremely high at the level of the motor end plate itself. This conclusion was subsequently verified histochemically for the first time by Koelle and Friedenwald (1949) and Koelle (1950) with the thiocholine method on the muscles of the rat.

Once it was established that the cholinesterase is more highly concentrated at the level of the neuromuscular junctions than at the level of the motor nerve and the aneural part of the muscle, it still had to be determined in exactly what part of the junctional area this large amount of enzyme was located. One opinion already put forward was that the junctional cholinesterase was located in the nerve endings. A study of the variations in the cholinesterase content of the striated muscle after section of the motor nerve and the consequent disappearance of the axoplasmic component of the end plate has, on the contrary, shown that the greater part of the junctional cholinesterase long survives degeneration of the axon, and therefore is located not in the interior of the nerve endings, but outside them (Couteaux and Nachmansohn, 1938, 1940, 1942).

Histochemical methods later confirmed this interpretation. The remaining problem was to find out which of the various motor end plate components is the site of this cholinesterase activity located outside of the nerve endings. By comparing several experimental and cytochemical data, the assumption was made that the junctional cholinesterase is mainly located at the level of the subneural apparatus (Couteaux, 1947).

The histochemical pictures obtained by most authors (Couteaux and Taxi, 1951, 1952; Couteaux, 1951; Holt and Withers, 1952; Cöers, 1953a,b, 1955; Denz, 1953; Gerebtzoff, 1953; Gerebtzoff *et al.* 1954; Harris, 1954; Holt, 1954; Woolf and Till, 1955; Kovac *et al.*, 1955; Crevier and Bélanger, 1955) showed that the structures selectively stained are those that have been described as the subneural apparatus (Fig. 6).

With thiocholine methods, using acetylthiocholine and butyrylthiocholine as substrates, and when the pH and buffers have been appropriately chosen, the subneural apparatus appears as the main site of the cholinesterase activity of the end plate. The appearance of pH values of important diffusion artifacts, entailing the staining of other structures, challenges the validity of this localization. It has been proved that enzyme diffusion may be practically disregarded and that only the thiocholine diffusions should be taken into account.

By mechanical separation of the subneural apparatus from the other juxtasynaptic components before incubation (Couteaux, 1958) and by a histochemical study of the cholinesterase activity of the denervated end plates, it has been possible to establish that selective staining of the subneural apparatus is not the result of a thiocholine diffusion from a neighboring structure.

Histochemical observations have already been made repeatedly on the cholinesterase activity of denervated end plates. Sawyer and associates (1950), in a preliminary report, and Kupfer (1951) indicate for the rat, the guinea pig, and the rabbit, persistence of cholinesterase concentration at the level of motor end plates after section of motor nerve and degeneration of nerve endings. Subsequently, after a more thorough histochemical study, Cöers (1953c), Snell and MacIntyre (1955, 1956), and Savay and Csillik (1956) arrive at the same result. These authors have shown that the cholinesterase activity which persists in the denervated end plate of the rat and the guinea pig is in this case also located at the level of the subneural apparatus. The whole appearance of this apparatus generally altered 2 weeks after the section of the motor nerve, but at this stage, the lamellae may still be seen and stained by histochemical methods.

Appreciable differences are shown between figures hitherto published regarding the time limit for persistence of cholinesterase activity in the denervated end plates of the rat and the guinea pig. Three months after the section of a motor nerve, Cöers still found, in the gastrocnemius of the rat some end plates that showed a deformed, faintly stained, but easily recognizable subneural apparatus. In the same species, and on the same muscle, Savay and Csillik noted a residual activity after 4 months, and even after 6 months; while with the guinea pig, Snell and MacIntyre found no activity after 45 days.

It seems probable that these differences are mainly due to minor differences in the technique used. With Koelle's technique, and checking for the absence of nerve regeneration, Carric (unpublished results) observed, on certain greatly altered motor end plates of the m. gastrocnemius internus of the rat and guinea pig, the persistence of subneural lamellae faintly stained but still visible 6 weeks after section of the motor nerve.

In any case, conclusions reached today from histochemical research on the cholinesterase activity of the denervated end plates prove that the staining of the subneural apparatus by means of the thiocholine method is not the result of a diffusion from the terminal nerve branches.

Histochemical methods for localizing cholinesterases using substrates other than acetylthiocholine and butyrylthiocholine, such as the long-chain fatty acid esters [e.g., myristoylcholine (Denz, 1953) and α-naphthyl acetate (Denz, 1953)], indoxyl derivatives (Holt, 1954), and thiolacetic acid (Crevier and Bélanger, 1955) also selectively stain the subneural apparatus.

After the investigations made by Denz (1953, 1954) and Holmstedt (1957a,b), with the aid of several substrates and a number of selective inhibitors of cholinesterases, it seems that the end plate contains two cholinesterases, an acetylcholinesterase, and an unspecific cholinesterase, the second of these being present in a much smaller quantity than the first.

Before electron microscopy enabled us to conceive of the subneural apparatus in ultrastructural terms, the possibility that a thin teloglial layer was incorporated in the subneural lamellae and survived degeneration of the nerve endings after section of the motor nerve could not be excluded. Neither, therefore, did it seem possible to exclude completely the possibility that a cholinesterase attached to the subneural apparatus might be located in teloglia. As a result of electron microscope examination, the subneural apparatus may now be considered as formed exclusively by the sarcoplasm surface membrane. We must therefore conclude that the cholinesterase activity observed at the level of the subneural apparatus, after denervation of the end plate, is exclusively located at the level of the sarcoplasm surface membrane.

As, at normal end plates, the terminal axoplasm membrane, i.e., the presynaptic membrane, is intimately apposed to the postsynaptic membrane, it is impossible to distinguish one from the other by the use of the light microscope alone when localizing the cholinesterase. Further

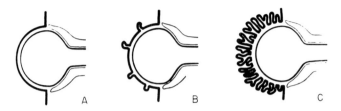

Fig. 11. Evolution of the subneural apparatus. (A) No secondary synaptic clefts; (B) A few shallow synaptic clefts; (C) Profuse folding of the sacroplasmic membrane. From C. Cöers, *Int. Rev. Cytol.* (1967). **22**, 239.

histochemical research with the electron microscope will therefore be necessary before deciding whether or not a cholinesterase site exists at the level of the presynaptic membrane. In any case, results obtained from the above-mentioned biochemical investigations on denervated muscle show that the cholinesterase activity at a possible presynaptic site is incomparably smaller than the activity located in the subneural part of the postsynaptic membrane (see fig. 11).

Based upon biochemical data, it may already be admitted that outside the subneural lamellae there exists a certain amount of cholinesterase, as the aneural part includes a noticeable part of the total quantity of enzyme contained by the muscle. The failure to find the site of this moiety of the enzyme by the acetylthiocholine method is not surprising, since this method is not very sensitive.

Zacks and Welsh (1951) showed that unspecific and specific cholinesterases are present in rat liver mitochondria. We have consequently to admit as possible a similar location at the level of the muscle mitochondria, and particularly of the numerous mitochondria of end plates.

IX. Morphogenetic Significance of the Motor End Plate

Insofar as the terminal axoplasm and the teloglia are concerned, the preceding description of the different components of the motor end plate does not bring to light any characteristics essentially different from those observed at the level of the central and ganglionic synapses. Such is not the case for the structure of the junctional sarcoplasm, i.e., the postsynaptic component of the motor end plate, and its relationship with the terminal axoplasm. In general, at an interneuronal synapse, the postsynaptic membrane is simply apposed to the presynaptic membrane, forming a double membrane; the postsynaptic axoplasm, at least up to now, seems not to present any really significant particularity in the juxaterminal area. The existence of the subneural apparatus at the motor end plate clearly shows, on the contrary, the existence of a quite special organization of the postsynaptic sarcoplasm at the neuromuscular junction. The study of the motor end plate development allows the observation of the genesis of this differentiation, and suggests assumptions regarding its mechanism. Two main phases may be distinguished in the evolution of neuromuscular relations during embryonic development in mammals (Fig. 12).

In the first phase, the muscle fibers are still in myotube form, with axial nuclei and peripheral myofibrils. The motor nerve fibers, still few in number, proceed through the muscle, thus clearing the first motor intramuscular pathways. These are the "exploring" nerve fibers, whose extremities have as yet developed no lasting connections with the muscle fibers. Even when they are closely apposed to the myotubes, no significant reaction thus far discernible from the sarcoplasm and muscle nuclei is provoked by these "free" motor endings situated at their level.

It is in the second phase that the final junction of the motor nerve fibers and muscle fibers is established, when the muscle nuclei, originally axial, have completed their migration to the surface of the myofibrillar bundle. Once the margination of the muscle nuclei is finished, the nerve fibers send out either terminal or lateral sprouts, generally accompanied by Schwann cell nuclei, and each of these sprouts enters into close relation with a muscle fiber at the level of one of its nuclei, now peripheral. At the contact of the nerve sprout, the muscle nucleus divides repeatedly, while at this level the sarcoplasm becomes more abundant (Fig. 9). These divisions of muscle nuclei are generally preceded by nucleolar bipartition, and appear in all respects comparable to the amitoses described by Naville (1922), in the course of muscular development of frogs (Fig 13).

On the preparations of muscles stained by means of the Bielschowsky–Gros method, the nucleoli of the muscle nuclei appear very clearly, and it is not difficult to find, in the vicinity of the nerve sprouts, the characteristic aspects of nucleolar bipartition. What is most frequently found in the muscle nuclei during the process of division are enlarged nucleoli, exceeding in volume those of all the neighboring muscle nuclei. This enlargement of the nucleoli, doubtless a prelude to their division, may be so obvious that it sometimes allows the distinction of the motor innervation areas from other regions of an embryonic muscle.

The accumulation of muscle nuclei and sarcoplasm at the junction, and the growth of the neuroteloglial expansion, whose nuclei also become more numerous, results firstly in the formation of the primitive

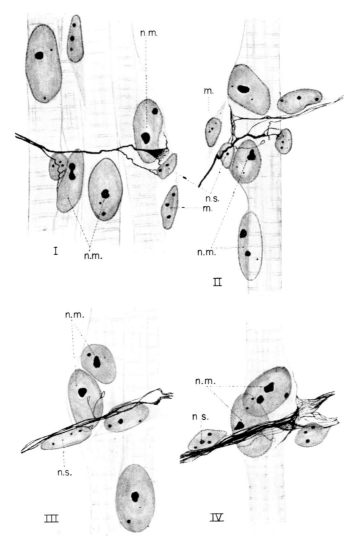

Fig. 12. Drawing of neuromuscular junctions of a 22 mm mouse embryo (m. cutaneous maximus); n.m. = muscle nuclei; n.s. = Schwann cell nuclei; m. = mesenchymatous cells. Bielschowsky method; ×1000. Couteaux (1938b.)

eminence, which is prominent at the surface of a myofibrillar bundle, as yet very slender. At this stage of its development, the embryonic end plate appears to be formed of two distinct parts, one consisting of the nerve ending to which are closely attached one, two, or three nuclei, which will later turn into arborization nuclei; and the other con-

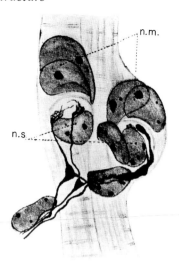

Fig. 13. Drawing of two neuromuscular junctions of an 80 mm guinea pig embryo (m. latissimus dorsi); n.m. = muscle nuclei; n.s. = Schwann cell nuclei (teloglia). Bielschowsky–Gros method; ×1200.

sisting of a cluster of large nuclei, accompanied by the granular substance of the sarcoplasm, which will become fundamental nuclei (Fig. 14).

The distinction between these two parts of the primitive eminence is made particularly clear by the fact that the first part, the neuroteloglial expansion, depresses the second and often deforms the fundamental nuclei, as if it were exerting pressure on them (Fig. 10). Nevertheless, it should be pointed out that, whereas these aspects (particularly significant in guinea pigs) are easily observed on whole muscle fibers, they are much more difficult to observe on the sections. This is a possible source of error that may perhaps explain certain differences of opinion.

Not only does the study of the setting up of neuromuscular connections enable the main components of the motor end plate to be defined, but it also brings to light some of the interactions that occur when a junction is being formed (Fig. 16).

From an analysis of the first stages of the process leading to the formation of the motor end plate, one fact stands out clearly: to effect the junction of a muscle fiber and nerve fiber from which a motor end plate will result, it is not sufficient that these fibers should come into direct contact with each other; it is also necessary for one, or perhaps both of these fibers, to have reached a certain stage of their development. It is, indeed, at a relatively late period of the muscle development, when the nerve fibers have already been in contact with the embryo

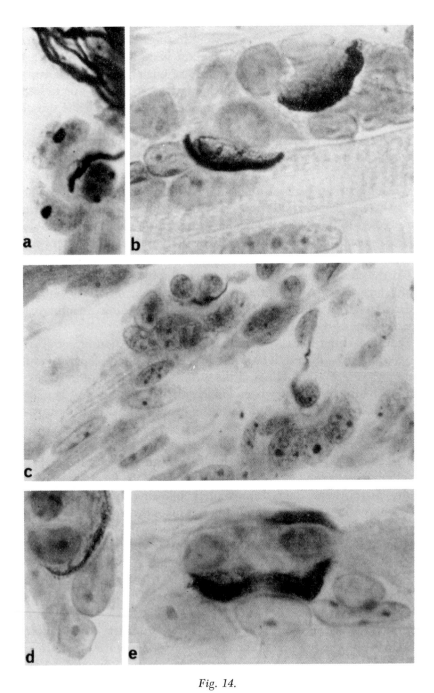

Fig. 14.

muscle fibers for some time, that the first outline of the final connections appears.

These connections are established only after the margination of the muscle nuclei, and always at their level, at least in mammals (London and Pesker, 1906; Tello, 1971; Couteaux, 1941). It seems probable that the muscle nuclei, or the sarcoplasm nucleated zones, intervene directly in the junctional process by exerting an attraction on the neighboring neurites. At the moment of this attraction, the axon that crosses a muscle fiber is never very far from one of the nuclei that protrude at the myofibrillar bundle surface; thus, the attraction always comes into play at a short distance. It determines the formation of a short nerve sprout, and the nearer the axon from which it originates to a muscle nucleus, the shorter the nerve sprout.

The mechanism of this attraction exerted by muscle fiber nucleated zones on the motor axons is still not well understood. Its intervention marks the beginning of a new phase in neuromuscular relations and ends the "free" growth phase of the motor nerve fibers, during which the growth orientation appears to be governed by the "contact guidance," as defined by Weiss (1941).

The study of the motor end plate development shows that the accumulation of muscle nuclei and sarcoplasm observed at the end plate is not, as believed in the past by several authors, antecedent to the junction, but results from this junction.

The multiplication of the muscle nuclei at the point of contact of the motor nerve sprout and the muscle fiber differs, in at least one respect, from the nuclear multiplication observed over the whole length

Fig. 14. Neuromuscular junctions of 80–85 mm guinea pig embryos (m. latissimus dorsi).   (a) Three nuclei may be observed at the level of this embryonic neuromuscular junction; one teloglial nucleus accompanying the nerve ending and two muscle nuclei. Silver preparation; ×1550. From Couteaux (1941).   (b) Lateral view of two neuromuscular junctions. At this stage of the development, the subneural apparatus assumes the form of a shallow cupule in which lie the nerve ending and the teloglia. Acetylthiocholine method after fixation by formaldehyde, with staining of the nuclei by hematoxylin; ×1620.   (c) The muscle nuclei, already numerous, form an arc of a circle around the nerve ending and the teloglia. Silver preparation; ×1070. From Couteaux (1941).   (d–e) Acetylthiocholine method after fixation by formaldehyde, with staining of the nuclei by hematoxylin.   (d) Front view of a neuromuscular junction. The subneural apparatus, seen in optical section, appears as a curved line separating the neuroteloglial expansion from the muscle nuclei cluster. (×1680.)   (e) Lateral view of a neuromuscular junction. Three kinds of nuclei are visible: below the subneural apparatus, several muscle nuclei, above it two teloglial nuclei, surmounted by one nucleus belonging to the endothelial coating of the junction. (×1850.)

of the muscle fiber during its growth. As a rule, during muscle development, muscle nuclei resulting from a nuclear division tend to draw away from one another as if by a mutual repulsion; in the case of divisions which take place at junction level, they remain juxtaposed. Thus, there is formed at each junction a cluster of muscle nuclei, and correlatively, an accumulation of sarcoplasm that can be regarded as due to local effects caused by contact of the motor nerve sprout with the muscle fiber.

The formation of the subneural apparatus also appears to be a local effect due to the influence of the embryonic motor nerve ending on the muscle fiber. When the striated muscle fibers of a guinea pig, removed shortly before its birth (e.g., from an 80 mm embryo) are postvitally stained with Janus green B, the sarcoplasmic membrane at the level of the embryonic motor end-plate is depressed by the nerve ending, the latter being surmounted by one or two teloglial nuclei.

The thiocholine method shows that a cholinesterase activity is present at the level of this subneural membrane, which, observed with the light microscope, appears thicker than the sarcoplasmic membrane of other regions of the muscle fiber (Fig. 14b,d,e). This thickening is easily explained by the formation of subneural lamellae, which observations with the electron microscope on adult mammals reveal as infoldings of the sarcoplasmic membrane. On the muscles of a newborn guinea pig, these subneural lamellae—very close-set and not yet penetrating deeply into the sarcoplasm—are clearly preceptible with the light microscope, after staining either with Janus green B or by the thiocholine method.

The configuration of the subneural apparatus at each stage of the developing motor end plate depends directly on the form of the nerve ending at this same stage. Immediately following the junction of the motor nerve fiber and the muscle fiber, it takes the form, in the guinea pig, of a shallow cupule, whose edges are level with the surface of the junctional protuberance. The concavity, however, of this primary subneural apparatus is sometimes extremely slight. The subneural cupule subsequently widens, correlatively with the development of the nerve ending, which extends to the surface of the muscle fiber and branches out.

Under the pressure of the terminal nerve branches, whose dimensions increase, the subneural apparatus that heretofore was of simple form and presented an almost regular curve, gradually acquires a lobed outline and complicated relief. Gutters appear under the nerve ending branches, constituting the first outlines of adult synaptic gutters. From this moment, the subneural lamellae reach the surface of the junctional protuberance on both sides of each nerve branch. During the course of postnatal growth, these gutters, which are molded on the nerve branches, widen and lengthen at the same time as the branches. In the guinea pig, they become much deeper, but, as has already been mentioned in connection with the description of the subneural apparatus, the final depth of the synaptic gutters is quite variable, depending on the animal species. Many other specific morphological differences accompany these variations in depth.

The formation of the subneural apparatus characterized by special folds and acetylcholinesterase activity, clearly results from an inductive action exerted directly by the nerve ending on the subneural sarcoplasm.

If one compares the local influences at work on the sarcoplasm exerted by the motor nerve endings and the sensory nerve endings, respectively, during the formation of the motor end plates and the neuromuscular spindles, it will be seen that they both entail the formation of a mass of muscle nuclei in their immediate vicinity. However, one may observe two important differences between them: (1) the sensory nerve endings of the neuromuscular spindles do not determine, as do the motor nerve endings, the formation of a subneural apparatus at the level of the junctional sarcoplasm; and (2) they generally exercise on the myofibrillogenesis a local inhibitory action, which the motor nerve endings never seem to exercise at the neuromuscular junctions (Fig. 15).

Since the presynaptic nerve endings of the ganglionic and central synapses are in several respects homologous to the terminals of the motor axons innervating the striated muscle, one may ask if at the level of these synapses the nerve endings do not exert on the subsynaptic zones of the nerve cell a morphogenetic influence comparable in some degree to that which determines the formation of the subneural apparatus at the level of the junctional sarcoplasm of the muscle fiber. The morphological and cytochemical methods have not up to the present moment provided data that allow one to consider the subsynaptic zones of the nerve cell as specialized areas; however, in the present state of these methods, the negative results they furnish cannot be held as conclusive.

Cöers (1967) has drawn attention to the absence of "secondary synaptic clefts in the motor ending of some somitic-primitive muscles, of some 'slow' muscles, and of intra fusal muscle fibers." He goes on to point out that the degree of folding of the sarcolemma is a morphological indication of the degree of differentiation of a motor end plate. The object of the folding of the sarcolemma in the plate is presumably to provide a greater surface for the localization of cholinesterase. This has been supported by the work of Couteaux (1956) and Csillik (1961) on frog muscle in which they found that those fibers that had a tonic

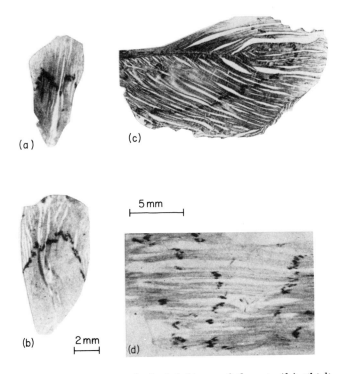

Fig. 15. Modified thiocholine method: (a) biceps of the rat; (b) tibialis anticus of the rat; (c) human palmaris longus (3-month-old infant); (d) sartorius of the cat, proximal part.   From C. Coërs (1967). *Int. Rev. Cytol.* **22**, 239.

function had much less cholinesterase activity in their motor end plates than those associated with the fast fibers. Correlated with this is the fact that the end plates of slow fibers show no or little sarcolemmal folding, whereas those of the fast fibers show considerable folding. Cöers points out that Buller *et al.* (1960) have shown that if the nerves serving fast and slow muscles in mammals are reversed, the fast muscle then acts as a slow muscle and vice versa, and in similar studies with the frog, Miledi and Orkand (1966) showed that if the nerve of the sartorius muscle (fast) is joined to the ileofibularis muscle (slow), the length of time of the contraction of the muscle due to the prolonged action of acetylcholine from the neural element of the end plate is reduced, suggesting increased cholinesterase activity. Since this change is brought about without any obvious ultrastructural change in the muscle from a slow to a fast fiber, perhaps the change is primarily in the motor end-plate. Cöers suggests it would be interesting to note whether such a nerve cross would cause a more complex folding of the sarcolemma in the end plate leading to the accumulation of more cholinesterase.

Cöers (1967) has described a variety of motor innervation patterns in various animals. In one pattern, the nerve endings are scattered along the length of the muscle fiber. In another pattern, the endings of the nerve form a basketlike structure over the end of the fiber. In yet another pattern, the endings are confined mainly to the equator of a muscle. The nerve endings may be of two types, the standard motor end-plates and the scattering of endings known as *en grappe*. In amphibia, only a primitive motor end plate is present; in reptiles, characteristic and defined motor end-plates are found distributed over the middle of the muscle. Multiple innervation occurs in some muscles of birds, and in mammals the major innervation is by defined motor end plates. Beckett and Bourne (1957) have shown the presence of cholinesterase-positive basket structures over the ends of muscle fibers in human muscles and also have recorded and figured a curious "cake frill"-type cholinesterase positive ending in human muscle which appears to completely encircle the individual muscle fiber. Cöers (1967) has published photographs of a variety of nerve endings in muscle. The variety of endings met with in human muscle are also shown by Beckett and Bourne (1957).

REFERENCES

Beckett, E. C., and Bourne, G. H. (1957). *J. Neurol., Neurosurg. Psychiat.* **20**, 191.
Boeke, J. (1909). *Anat. Anz.* **35**, 192.
Boeke, J. (1911). *Int. Monatschr. Anat. Physiol.* **28**, 376.
Boeke, J. (1927). *Z. Mikrosk.-Anat. Forsch.* **8**, 561.
Boeke, J. (1932). *In* "Cytology and Cellular Pathology of the Nervous System" (W. Penfield, ed.), Vol. 1, p. 243. Harper (Hoeber), New York.
Boeke, J. (1942). *Proc., Kon. Ned. Akad. Wetensch.* **45**, 444.
Boeke, J. (1944). *Acta Neer. Morphol.* **5**, 189.
Boeke, J. (1949). *Acta Anat.* **8**, 18.
Boeke, J., and Nöel, R. (1925). *C. R. Soc. Biol.* **92**, 263.
Brown, G. L., Dale, H. H., and Feldberg, W. (1936). *J. Physiol. (London)* **87**, 394.
Brown, L. M. (1961). *Bibl. Anat.* **2**, 21.
Buller, A. T., Eccles, J. C., and Eccles, R. M. (1960). *J. Physiol. (London)* **150**, 417.
Cöers, C. (1953a). *Arch. Biol.* **64**, 133.
Cöers, C. (1953b). *J. Belge Pathol. Med. Exp.* **22**, 306.
Cöers, C. (1953c). *Acad. Roy. Belg., Cl. Sci., Mem.* **39**, 447.
Cöers, C. (1955). *Acta Neurol. Psychiat. Belg.* **55**, 741.
Cöers, C. (1967). *Int. Rev. Cytol.* **22**, 239.

Couteaux, R. (1938a). *C. R. Soc. Biol.* **127**, 218.
Couteaux, R. (1938b). *C. R. Soc. Biol.* **127**, 571.
Couteaux, R. (1941). *Bull. Biol. Fr. Belg.* **75**, 101.
Couteaux, R. (1944). *C. R. Soc. Biol.* **138**, 976.
Couteaux, R. (1945a). *C. R. Acad. Sci.* **220**, 567.
Couteaux, R. (1945b). *C. R. Soc. Biol.* **139**, 376.
Couteaux, R. (1945c). *C. R. Soc. Biol.* **139**, 641.
Couteaux, R. (1947). *Rev. Can. Biol.* **6**, 563.
Couteaux, R. (1951). *Arch. Int. Physiol.* **59**, 52.
Couteaux, R. (1952). In "Le Muscle, Etude de Biologie et de Pathologie" (C. C. I. C. M. S., ed.), p. 173. L'Expansion Scientifique, Paris.
Couteaux, R. (1955). *Int. Rev. Cytol.* **4**, 335.
Couteaux, R. (1956). *C. R. Acad. Sci.* **242**, 820.
Couteaux, R. (1957a). *Proc. Reg. Conf. (Eur.) Electron Microsc., 1956* p. 188.
Couteaux, R. (1957b). In "Microphysiologie comparée des éléments excitables" (C. N. R. S., ed.) p. 255. C. N. R. S., Paris.
Couteaux, R. (1958). *Exp. Cell Res., Suppl.* **5**, 294.
Couteaux, R., and Nachmansohn, D. (1938). *Nature* **142**, 1481.
Couteaux, R., and Nachmansohn, D. (1940). *Proc. Soc. Exp. Biol. Med.* **43**, 177.
Couteaux, R., and Nachmansohn, D. (1942). *Bull. Biol. Fr. Belg.* **76**, 14.
Couteaux, R., and Taxi, J. (1951). *C. R. Ass. Anat.* **70**, 1030.
Couteaux, R., and Taxi, J. (1952). *Arch. Anat. Microsc. Morphol. Exp.* **41**, 352.
Crevier, M., and Bélanger, L. F. (1955). *Science* **122**, 316.
Csillik, B. (1961). *Bibl. Anat.* **2**, 161.
Cuajunco, F. (1942). *Contrib. Embryol. Carnegie Inst.* **30**, 127.
Dale, H. H., and Feldberg, W. (1934). *J. Physiol. (London)* **81**, 320.
Dale, H. H., Feldberg, W., and Vogt, M. (1936). *J. Physiol. (London)* **86**, 353.
del Rio-Hortega, P. (1925). *C. R. Soc. Biol.* **92**, Suppl. 3.
Denz, F. A. (1953). *Brit. J. Exp. Pathol.* **34**, 329.
Denz, F. A. (1954). *Brit. J. Exp. Pathol.* **35**, 459.
de Quatrefages, A. (1843). *Ann. Sci. Nat. Zool.* [2] **19**, 299.
De Robertis, E. D. P. (1955). *Acta Neurol. Latinoamer.* **1**, 1.
De Robertis, E. D. P., and Bennett, H. S. (1954). *Fed. Proc., Fed. Amer. Soc. Exp. Biol.* **13**, 35.
DeRobertis, E. D. P., and Bennett, H. S. (1955). *J. Biophys. Biochem. Cytol.* **1**, 47.
Dogiel, A. S. (1890). *Arch. Mikrosk. Anat. Entwicklungsmech.* **35**, 305.
Doyère, L. (1840). *Ann. Sci. Nat. Zool.* [2] **14**, 269.
Emmert, F. C. (1836). "Ueber die Endigungsweise der Nerven in den Muskeln nach eigenen Untersuchungen." Jenni, Bern.
Feist, B. (1890). *Arch. Anat. Physiol., Anat. Abt.* p. 116.
Feng, T. P., and Ting, Y. C. (1938). *Chin. J. Physiol.* **13**, 141.
Foettiger, A. (1880). *Arch. Biol.* **1**, 279.
Gerebtzoff, M. A. (1953). *Acta Anat.* **19**, 366.
Gerebtzoff, M. A., Philippot, E., and Dallemagne, M. J. (1954). *Acta Anat.* **20**, 234.
Gutmann, E., and Young, J. Z. (1944). *J. Anat.* **78**, 15.
Harris, C. (1954). *Amer. J. Pathol.* **30**, 501.
Held, H. (1897). *Arch. Anat. Physiol., Anat. Abt.* p. 204.
Hines, M. (1927). *Quart. Rev. Biol.* **2**, 149.

Hinséy, J. C. (1934). *Physiol. Rev.* 14, 514.
Holmstedt, B. (1957a). *Acta Physiol. Scand.* 40, 322.
Holmstedt, B. (1957b). *Acta Physiol. Scand.* 40, 331.
Holt, S. J. (1954). *Proc. Roy. Soc., Ser. B* 142, 160.
Holt, S. J., and Withers, R. F. J. (1952). *Nature (London)* 170, 1012.
Hoyle, G. (1957). "Comparative Physiology of the Nervous Control of Muscular Contraction." Cambridge Univ. Press, London and New York.
Iwanaga, I. (1925). *Mitt. Allg. Pathol. Pathol. Anat.* 2, 257.
Koelle, G. B. (1950). *J. Pharmacol. Exp. Ther.* 100, 158.
Koelle, G. B., and Friedenwald, J. S. (1949). *Proc. Soc. Exp. Biol. Med.* 70, 617.
Kovac, M., Kraupp, O., and Lassmann, G. (1955). *Acta Neuroveg.* 12, 329.
Krause, W. (1863). *Z. Rationelle Med.* 18, 136.
Kühne, W. (1862). "Ueber die peripherischen Endorgane der motorischen Nerven." Engelmann, Leipzig.
Kühne, W. (1864). *Arch. Pathol. Anat. Physiol. Klin. Med.* 29, 433.
Kühne, W. (1883). *Z. Biol.* 19, 501.
Kühne, W. (1886). *Verh. Naturh.-Med. Ver. Heidelberg* 3, 277.
Kühne, W. (1887). *Z. Biol.* 23, 1.
Kühne, W. (1892). *Verh. Naturh.-Med. Ver. Heidelberg* 4, 1.
Kulchitsky, N. (1924). *J. Anat.* 59, 1.
Kupfer, C. (1951). *J. Cell. Comp. Physiol.* 38, 469.
Lawrentjew, B. J. (1928). *Z. Mikrosk-Anat. Forsch.* 13, 388.
London, E. S., and Pesker, D. J. (1906). *Arch. Mikrosk. Anat. Entwicklungsmech.* 67, 303.
MacManus, J. F. A. (1948). *Stain Technol.* 23, 99.
Marnay, A., and Nachmansohn, D. (1937). *C. R. Soc. Biol.* 125, 41.
Miledi, G. E., and Orkand, P. (1966). *Nature (London)* 209, 717.
Naville, A. (1922). *Arch. Biol.* 32, 37.
Noël, R. (1925). *Bull. Histol. Appl. Tech. Microsc.* 2, 124.
Noël, R. (1927). *Bull. Histol. Appl. Tech. Microsc.* 4, 382.
Noël, R. (1950). *Biol. Med. (Paris)* 39, 273.
Palade, G. E., and Palay, S. L. (1954). *Anat. Rec.* 118, 335.
Palay, S. L. (1956). *J. Biophys. Biochem. Cytol.* 2, Suppl. 192.
Palay, S. L. (1958). Personal communication.
Ramón y Cajal, S. (1899). "Textura del sistema nervioso del Hombre y de los Vertebratos." N. Moya, Madrid.
Ramón y Cajal, S. (1904). *Trab. Lab. Invest. Biol. Madrid* 3, 97.
Ramón y Cajal, S. (1909). "Histologie du système nerveux de l'Homme et des Vertébres," Vol. 1. Maloine, Paris.
Ramón y Cajal, S. (1925). *Trab. Lab. Invest. Biol. Madrid* 23, 245.
Ranvier, L. (1878). "Leçons sur l'histologie du systéme nerveux." Savy, Paris.
Ranvier, L. (1888). "Traité technique d'histologie," 2nd ed. Savy, Paris.
Reger, J. F. (1954). *Anat. Rec.* 118, 344.
Reger, J. F. (1955). *Anat. Rec.* 122, 1.
Reger, J. F. (1957). *Exp. Cell Res.* 12, 662.
Renaut, J. (1899). "Traité d'histologie pratique," Vol. 2, Part 2, p. 972. Rueff, Paris.
Retzius, G. (1892). *Biol. Untersuch* [N. S.] 3, 41.
Robertson, J. D. (1954). *Anat. Rec.* 118, 346.

Robertson, J. D. (1956). *J. Biophys. Biochem. Cytol.* **2**, 369.

Robertson, J. D. (1957). *J. Physiol. (London)* **40**, 58P.

Robertson, J. D. (1958). *J. Biophys. Biochem. Cytol.* **4**, 349.

Robertson, J. D. (1960). *Am. J. Physiol. Med.* **39**, 1.

Rouget, C. (1862). *C. R. Acad. Sci.* **55**, 548.

Rouget, C. (1864). *C. R. Acad. Sci.* **59**, 851.

Savay, G., and Csillik, B. (1956). *Acta Morphol. Acad. Sci. Hung.* **6**, 289.

Sawyer, C. H., Davenport, C., and Alexander, L. M. (1950). *Anat. Rec.* **106**, 287.

Shanthaveerappa, T. R., and Bourne, G. H. (1962). *J. Anat.* **96**, 527.

Shanthaveerappa, T. R., and Bourne, G. H. (1964). *Anat. Rec.* **150**, 35.

Shanthaveerappa, T. R., and Bourne, G. H. (1967). *Int. Rev. Opt.* **21**, 353.

Shanthaveerappa, T. R., and Bourne, G. H. (1968). *In* "The Structure and Function of Nervous Tissue" (G. H. Bourne, ed.), Vol. 1, p. 379. Academic Press, New York.

Sheridan, M. (1965). *J. Cell Biol.* **24**, 129.

Snell, R. S., and MacIntyre, N. (1955). *Nature (London)* **176**, 884.

Snell, R. S., and MacIntyre, N. (1956). *Brit. J. Exp. Pathol.* **37**, 44.

Stöhr, P. (1928). *In* "Handbuch der mikroskopischen Anatomie des Menschen" (W. von Möllendorff, ed.), Vol. 4, Part 1. Springer-Verlag, Berlin and New York.

Taxi, J. (1957). *C. R. Acad. Sci.* **245**, 564.

Taxi, J. (1958). *C. R. Acad. Sci.* **246**, 1922.

Tello, J. F. (1905). *Trab. Lab. Invest. Biol. Madrid* **4**, 105.

Tello, J. F. (1907). *Trab. Lab. Invest. Biol. Madrid* **5**, 117.

Tello, J. F. (1917). *Trab. Lab. Invest. Biol. Madrid* **15**, 101.

Tello, J. F. (1944). *Trab. Inst. Cajal Invest. Biol.* **36**, 1.

Tiegs, O. W. (1932). *J. Anat.* **66**, 300.

Tiegs, O. W. (1953). *Physiol. Rev.* **33**, 90.

Tower, S. S. (1931). *J. Comp. Neurol.* **53**, 177.

Valentin, G. (1836). *Nova Acta Leopold. Carol.* **18**, 51.

Wagner, R. (1847). "Neue Untersuchungen uber den Bau and die Endigungen der Nerven." Leipzig (cited by Ranvier, 1878).

Weiss, P. (1941). *Growth* **5**, Suppl., 163.

Woolf, A. L., and Till, K. (1955). *Proc. Roy. Soc. Med.* **48**, 189.

Zacks, S. I., and Welsh, J. H. (1951). *Amer. J. Physiol.* **165**, 620.

MEMBRANOUS SYSTEMS IN MUSCLE FIBERS

CLARA FRANZINI-ARMSTRONG

I. Introduction

At the time the first edition of this treaties appeared, ten years ago, electron microscopy had already provided some basic information on the structure and disposition of the membranes that pervade striated muscle fibers (Bennett, 1960). It was evident from those excellent initial studies (Bennett and Porter, 1953; Porter, 1956; Porter and Palade, 1957; Andersson-Cedergren, 1959; Peachey and Porter, 1959) that knowledge of morphology was essential to understanding control of muscle fiber activity (cf. Porter, 1961). Progress since has been enormous and particularly exciting in the strict interplay that has developed between knowledge of the physiology, biochemistry, and morphology of muscle. These are so closely correlated that understanding of one aspect is no longer possible without at least some knowledge of the other.

Ten years ago, for example, it was known that an action potential triggers a contractile response of the fiber, but the mechanisms of this process were largely unknown. Now, most of the steps involved in excitation–contraction coupling and in relaxation as well have been recognized at least in their broad outlines. One of the systems of membranes present in muscle fibers, the tubular or T system, which is connected with the sarcolemma, is responsible for bringing the electrical impulse to the fiber's interior. There the signal is transmitted to the sarcoplasmic reticulum (SR), which releases an activator substance (calcium ions) stored within its interior. Relaxation occurs when elements of the SR again segregate calcium from the fibrils.

Some of the major contributions of morphology to this knowledge of the processes involved in muscle activity have been the discovery of the T system and its definition as a sarcolemmal derivative and the identification of the SR as the component of the fraction of muscle homogenate that acts as a calcium accumulating and relaxing agent. Indispensable to any inquiry into the electrical properties of the fiber surface is also a knowledge of the relative size and distribution of the T system elements, which can only be provided by electron microscopy (see also Chapter 1 in volume III). SR and T system elements are described in Sections II,B,C,D, and their functions are briefly explored in Sections III,A,B. Sections II,E,F are dedicated to other membrane-bound structures—mitochondria and nuclei. One important question is left both for morphologists and physiologists: What is the exact nature of the junction between SR and T system elements (at dyads and triads) and what is the nature of the signal that is transmitted across such junction? Section III,C gives an account of what is so far known on the triad.

A comparative approach to the study of muscle structure has shown how structure and distribution of SR and T system as well as of the fibrils are modulated and combined with variations in electrical and biochemical properties to produce a whole spectrum of fibers with distinctive functional characteristics adapted to special requirements. The distribution of mitochondria also correlates well with what is known or expected of the basic type of metabolism (either preferentially or only partially aerobic) with which different fiber types meet the energy requirements of the contractile system. In sections IV,A–D and V,A–D, the specific distribution of SR, T system and mitochondria of fibers with known physiological properties are described both for vertebrates and invertebrates.

II. Structure of the Surface and Internal Membranes

A. *Plasmalemma and Sarcolemma*

The surface of the fiber is covered by a unit membrane (Robertson, 1956) 7.0 nm thick (Figs. 1 and 2), which is taut in stretched fibers and forms loose folds in shortened fibers. The surface membrane is called plasmalemma or plasma membrane and it forms a permeability barrier.

The plasmalemma is decorated by numerous caveolae. These are small, membrane-limited vesicles of fairly uniform size, which open to the outside via a narrow neck (Figs. 1 and 2). The number of caveolae varies considerably in different fibers, even within the same preparation and it is not clear whether this variability has any significance. Often caveolae are multiple, i.e., two or three of them are confluent and open to the outside via a common neck (Fig. 2).

In some fibers, caveolae are not uniformily distributed; on the contrary, they are mostly distributed in bands that run over the fiber's surface in correspondence to the level of the Z line and I bands region of the underlying fibrils (Figs. 3 and 4). This is the region where T system openings are also present (cf. Section II,C).

Two important questions may be raised: (1) At any given moment are the caveolae all opened to the outside? When the surface of the fiber is exposed by fracture following quick freezing (Rayns *et al.*, 1968), numerous inpocketings mark the surface (Fig. 3), indicating that most of the caveolae are opened. Similar indication is given by the fact that when the fiber is exposed to a solution containing an electron dense tracer (see Section II,C) all caveolae are filled with the tracer. Thus

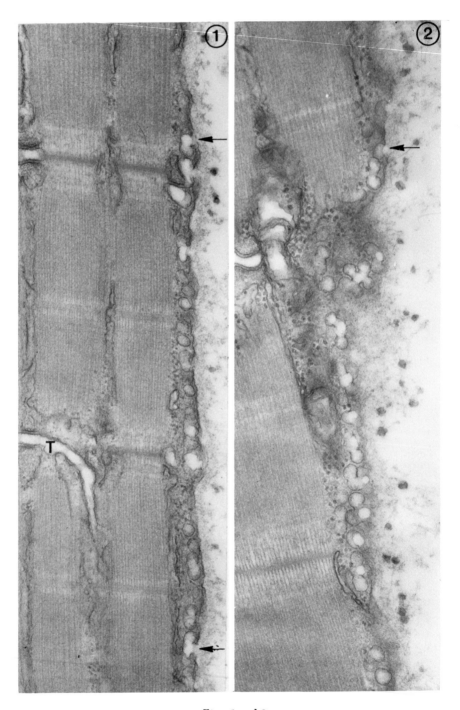

Figs. 1 and 2.

all caveolae seem to be open. (2) May the caveolae be considered functionally analogous to the pinocytotic vesicles of other cell types? In adult fibers, caveolae are exclusively located under the sarcolemma and thus cannot be considered pinocytotic. It is possible that the membrane forming the caveolae possesses properties different from the rest of the plasma membrane (e.g., it may be the site of entrance of specific metabolites).

The contribution of caveolae to the total surface area is probably significant, but it has not so far been precisely calculated. It certainly should be kept in mind when considering the electrical properties of the fiber surface (Section III,A; see also Chapter 1 in Volume III).

The outer surface of the plasmalemma is coated by a basement membrane, analogous to that covering epithelial cells, which is about 100 nm thick (Figs. 1 and 2). A loose network of collagen fibrils is also associated with the fiber surface. The network varies in size in different animals (cf. Figs. 2 and 4) and it is usually thicker in older specimens. In tightly packed muscles, fibers sometimes share a common basement membrane; this is partly what makes dissection of single fibers from most mammalian muscles rather difficult. Plasma membrane, basement membrane, and the collagen fibrils network are collectively called sarcolemma, a term initially employed by light microscopists to designate the line visibly marking the outer edge of the fiber. Occasionally, the term sarcolemma has been incorrectly employed in the literature to indicate the plasmalemma only. The sarcolemma is responsible for some of the mechanical properties of the muscle fibers, notably for its parallel elastic component.

B. The SR—Definition and Structure

The contractile material that almost completely fills striated muscle fibers is not a continuous mass, but it is separated into columns, the fibrils, by thin layers of sarcoplasm (Fig. 5). The sarcoplasm is pervaded by a fine reticular network of membrane limited elements belonging

---

Figs. 1 and 2. Longitudinal section near the edge of two fibers from the tail of bullfrog tadpoles. In Fig. 1, the plasmalemma and basement membrane are clearly visible, because they are transversely cut. The small dense dots at the outside of the fibers are cross-sectioned collagen fibrils. Numerous caveolae are visible and their openings indicated by arrows. In Fig. 2, most caveolae are multiple. Unless specified, all fixations are in glutaraldehyde followed by osmium tetroxide. ($\times 50,000$).

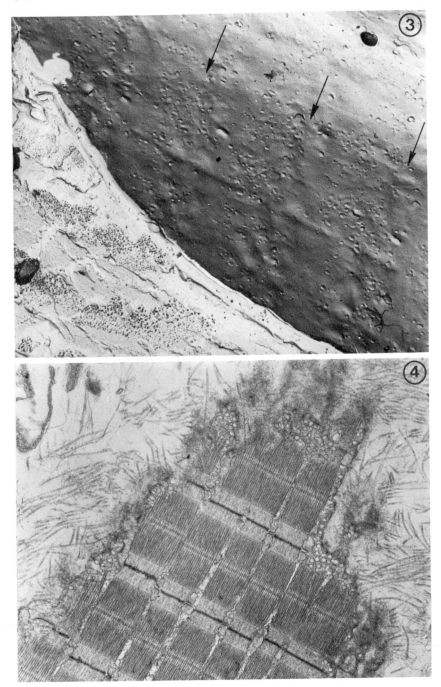

Figs. 3 and 4.

to the sarcoplasmic reticulum or SR (Fig. 6). The names sarcotubular system (Andersson-Cedergren, 1959) and, occasionally, longitudinal system have also been employed for the same structure.

Appropriately, the first extensive electron microscope study of the sarcoplasmic reticulum and of its distribution in different muscle fibers (Porter and Palade, 1957) was the continuation and extension of a series of studies on the endoplasmic reticulum (ER), then known to be a more or less abundant, but constantly present, organelle of all cell types. The SR has several properties in common with the ER. Most important, the interior of both systems of membranes is a compartment separated from the cytoplasm or, in muscle, the sarcoplasm. The SR segregates into its cavity calcium ions and, possibly, enzymes. Like the ER, the lumen of the SR is continuous over large portions of the cell. Implied in the original term reticulum (Porter, 1954) is that the components of the system are not separated from one another, so that their content is probably equilibrated throughout the entire cell, or at least selected portions of it (Porter, 1961). Both SR and ER have a large surface to volume ratio, thus emphasizing surface-associated functions. Finally, both are continuous with and/or homologous to the nuclear envelope (Porter and Palade, 1957; D. S. Smith, 1961a; Franzini-Armstrong and Porter, 1964a; Peachey, 1965a). In the case of muscle fibers, the functional significance of this continuity has not been explored.

All muscle fibers contain elements of the SR, but in different amounts. The other major membrane system, the T system, on the other hand is absent in a few fiber types. For clarity of presentation, the specific distribution of SR and T system elements will first be considered in several types of twitch fibers—e.g., frog twitch fibers (Birks, 1964; Page, 1965; Peachey, 1965a; Franzini-Armstrong, 1970a) and some fish fibers (Franzini-Armstrong and Porter, 1964a; Kilarski, 1965; Y. Nakajima, 1969). The description can then be easily extended to include all the numerous variations that have been described in other fiber types.

Just as the fibrils are divided into transverse bands (see Chapter 7, Volume I), which form a repeating pattern and confer the characteristic

---

Fig. 3. Freeze-fracture preparation of guinea pig leg muscle. In the upper right part of the image, the outer surface of the plasmalemma is exposed. The numerous inpocketings on the fiber surface are caveolae, perferentially located in bands over the I band of the underlying fibrils (arrows indicate the approximate position of the Z line). T system in these fibers is at the A–I junction. (×9800.) Reprinted with permission from Rayns *et al.* (1968).

Fig. 4. Grazing section through the periphery of a frog twitch fiber to show disposition of caveolae. Most of the caveolae are located in the neighborhood of the Z line where the T system opening is located. Notice numerous collagen fibrils. (×15,000.)

Fig. 5. Low-power view of a frog twitch fiber in longitudinal section. The long axis of the fiber is oblique, from lower left to upper right corner. Striation is at right angles. An elongated nucleus runs between the fibrils. (×11,000.)

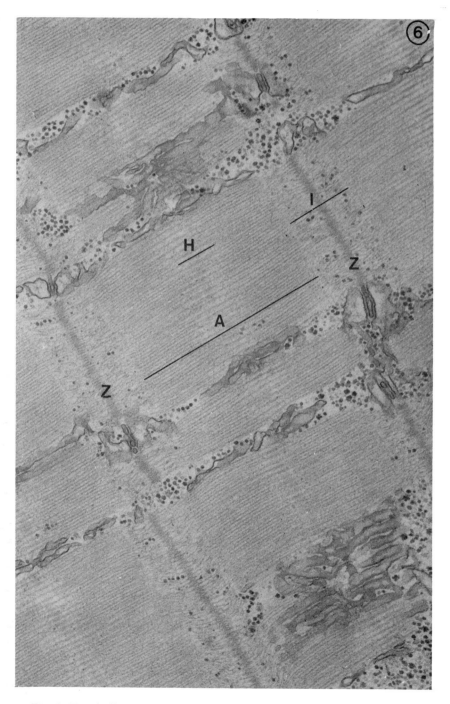

Fig. 6. Twitch fiber from the body myotomes of a small fish, the guppy, fixed in osmium tetroxide. The Z lines (Z) mark the limits of the sarcomere. A, I, and H bands are indicated. The SR continuous structure is prominent where the section cuts tangentially to a fibril (×36,000.)

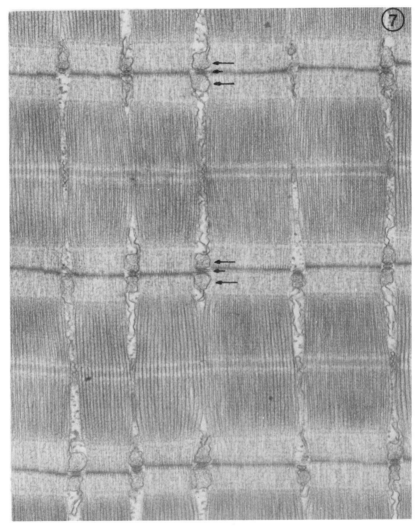

Fig. 7. Frog twitch fiber, showing the regular occurrence of triads at the Z line level. (×22,500.) Reprinted from Franzini-Armstrong (1970a).

cross striation to the fiber, the SR is divided into segments, which repeat in register with the striations. This structural parallelism is an indication that the major role of the SR is related to the contractile activity. The cross striations offer a useful landmark for the description of the SR elements distribution. The major bands and lines used here as a reference are the A, I, Z, and M (Figs. 6 and 7). The repeating unit, the sarcomere, is by convention taken to be the portion of fibril between two consecutive Z lines.

Two planes of section, both parallel to the fiber long axis, are advantageous in providing a general view of the SR. One cuts across the center of several adjoining fibrils and across the intervening sarcoplasmic layer (Fig. 7), the other cuts for some distance tangentially to a fibril and it includes in its thickness a complete layer of sarcoplasm, so that the SR and T system elements are displayed as they are distributed over the surface of the fibrils (Figs. 6, 8, and 9).

At the level of the I band, the SR forms dilated sacs or cisternae which are variously called. The name "terminal sacs" is used in consideration of the fact that they are at the end of an SR segment. The name "lateral sacs of the triad" is used to indicate the same portion of the SR, because opposite the I bands and Z lines of every sarcomere there always are two SR terminal sacs that face each other across a narrow space where a T system tubule runs. The complex formed by two SR sacs and one T system element is named triad (Porter and Palade, 1957) (Fig. 7). In order to emphasize the importance of the triad, which will become readily evident when considering the likely series of events in e–c coupling (see Section III,C), I shall use here the term lateral sacs of the triad to indicate the SR portions close to the T system. The lateral sacs of the triad run transversely in the fiber (Figs. 8 and 9), meander between the fibrils, and are continuous across a large part or perhaps all of the fiber, as can be seen in cross sections that happen to cut at the appropriate level (Fig. 10).

At the approximate level of the A–I junction, there is a dramatic change in the SR configuration. Longitudinally oriented slender sacs run along the fibril, forming a palisade around the A band, which is continuous at either end with the lateral sacs of the triad (Figs. 8 and 9).

In frog fibers, an intermediate, more flattened SR cisterna is interposed between the lateral sacs of the triad and the longitudinally oriented tubules of the SR (Peachey, 1965a) (Fig. 7). This intermediate cisterna is more evident in fibers fixed at long sarcomere lengths, and it is virtually absent in some types of fibers (e.g., see Fig. 12). Its function is not known. Its diameter is certainly too large to preclude diffusion of solutes between the lateral sacs of the triad and the rest of the SR.

In the center of the sarcomere, opposite the H zone, the longitudinal SR elements fuse to form a flattened cisterna which is oriented crosswise, similarly to the lateral sacs of the triad (Figs. 8, 9, 12, and 19). The central cisterna's distinctive trait is the presence of small (150–200 Å) pores or fenestrae (Porter and Palade, 1957). These, which were wrongly interpreted as actual interruptions in the continuity of the SR membrane

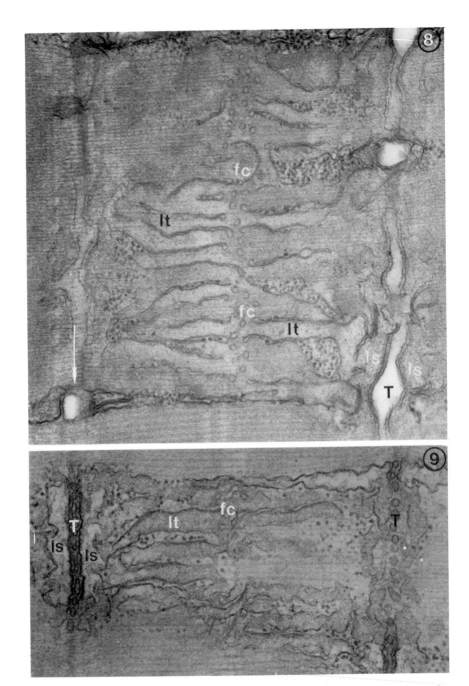

Figs. 8 and 9. Fibers from tail of frog tadpoles. The three major portions of the SR: fenestrated collar (fc), longitudinal tubules (lt) and lateral sacs of the triad (ls) are indicated. An arrow points to a bypass of the triadic interruption. The specimen in Fig. 9 is fixed in osmium tetroxide only; notice the difference in T system shape. (×50,000.)

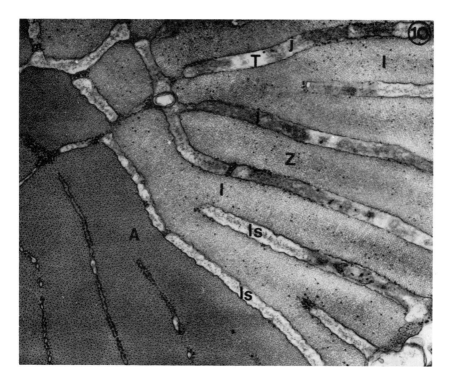

Fig. 10. Cross section through a white twitch fiber from a small fish, the black mollie. The A, I, and Z bands are shown. The transversely continuous sacs are either part of the T system (T) or the lateral sacs of the triad (ls). At j, the triad junction gap is included in the plane of sectioning. (×23,000.)

(Franzini-Armstrong, 1963), are small canals of unknown function running from one side to another of the SR (Page, 1965; Peachey, 1965a). The name fenestrated collar (Peachey, 1965a) is appropriately descriptive of the orientation of the central SR cisterna (which forms a collar around the fibrils) and of the presence of the pores or fenestrae.

By means of the A band longitudinal elements and of the fenestrated collar, continuity is provided, within the limits of a sarcomere, between all composing elements of each SR segment. At the level of the Z line, on the other hand, the SR is longitudinally interrupted by the presence of the intervening T system tubule. However, the interruption at the Z line level is not complete; one can occasionally observe a sarcoplasmic reticulum bridge which crosses the level of the Z line, thus joining lateral sacs belonging to adjacent sarcomeres (Fig. 8) (Franzini-Armstrong and Porter, 1964a; Peachey, 1965a; Kilarski, 1965). Such connections are but rarely observed, so that it is difficult to assess the frequency of their occurrence. They may be in sufficient number to provide effective longitudinal continuity of the SR content along the whole fiber length. In frog twitch fibers, it has been calculated (Peachey, 1965a) that the SR occupies approximately 13% of the total fiber volume.

In most vertebrate fibers, the lateral sacs of the triad are filled by a dense matrix (Porter, 1956; Revel, 1962; Franzini-Armstrong and Porter, 1964a; Peachey, 1965a; Page, 1965; Walker and Schrodt, 1966a,b) (see Figs. 7, 8, and 10), and this is particularly prominent when ruthenium red is added to the fixing solution (Luft, 1966; Goldstein, 1969a; D. E. Kelly, 1969). Not much information is available on the structure of this dense matrix, particularly since it varies considerably in appearance in fixed preparations (Walker and Schrodt, 1965, 1966a; Bertaud *et al.*, 1970; Franzini-Armstrong, 1970a). The rest of the SR is apparently empty in conventionally fixed preparations, but following freeze-etching some structure may be identified within the longitudinal sacs (Bertaud *et al.*, 1970). The presence of a dense content in the lateral sacs of the SR has been taken to indicate a specific function of this SR portion, possibly in binding calcium ions. There are some indications that this may be the case (see Section III,B).

C. The T System

The small tubule located in the center of the triad is part of a system of tubules, which in vertebrate twitch fibers run around the fibrils, lying in a plane transverse to the fiber long axis (Fig. 8 and 9). The name transverse tubular (or T) system has been given to the whole tubular complex (Andersson-Cedergren, 1959), and it has been maintained in the literature, despite the more recent demonstrations that elements of the T system may also run longitudinally (see Section II,E). It has been suggested that excitatory (E) system is a more universally applicable term than T system (Hoyle, 1965).

It has now been convincingly demonstrated that the T system tubules arise at the periphery of the fiber as invaginations of the plasma membrane (Figs. 11, 12, 14, and 15) and that the content of the tubules is in direct continuity with the fluid bathing the fiber. The structure of the most peripheral portion of the T system and its behavior upon fixation vary considerably in different fiber types. In Insecta and Crustacea, for example, the T system is open to the outside even following osmium tetroxide fixation (D. S. Smith, 1961a,b; Peterson and Pepe, 1961). In heart muscle fibers, where present, the T system has a large mouth, and both the mouth and the rest of the tubule are large enough that the basement membrane penetrates into it (Simpson and Oertelis, 1962) (see Chapter 2). In most vertebrate skeletal muscle fibers, on the other hand, osmium tetroxide fixation produces an interruption of the continuity of the T system tubule, particularly at the periphery of the fiber where the T system is not accompanied by the lateral sacs of the triad (Andersson-Cedergren, 1959; Porter and Palade, 1957; Franzini-Armstrong and Porter, 1964a). Thus, although it has long been clear, for comparative reasons, that the T system of all fibers must be a sarcolemmal derivative (see Porter, 1961), it was not until the introduction of a new fixative (glutaraldehyde) and of appropriate experimental techniques (such as the use of tracer molecules) that it was possible to demonstrate the real nature of the T system tubule in most vertebrate fibers. Following fixation in glutaraldehyde, the T system of some fish muscles, which is accompanied by the lateral sacs of the triad to the very edge of the fiber, is obviously opened, with a funnel-shaped mouth (Fig. 11) (Franzini-Armstrong and Porter, 1964a; Kilarski, 1966; Goldstein, 1969b). The fairly large size of the actual opening has been confirmed by freeze-etch studies (Bertaud *et al.*, 1970). The same is true for *Rana* tadpoles (Figs. 14 and 15). The striking feature of the openings in fish muscle is their frequent occurrence, an opening being visible wherever the plane of section exactly coincides with a sarcoplasmic layer between two fibrils at the edge of the fiber (Fig. 11).

Even following fixation in glutaraldehyde, however, the T system mouth is very difficult to visualize in most vertebrate fibers. Usually one may follow a triad almost to the edge of the fiber, the T system may be seen to come close to and even to touch the sarcolemma, but the actual opening is not visible in the section (Fig. 13). It has taken long hours of fiber-edge watching by a number of authors currently engaged in morphological research on muscle to provide few examples of T system openings (Hoyle *et al.*, 1966, for a snake; Peachey, 1965a; Eisenberg, 1967; Page, 1965; Franzini-Armstrong, 1970a, for the frog; Ashurst, 1969, for the pigeon; Walker and Schrodt, 1965, for the rat; Rayns *et al.*, 1968, for the guinea pig).

A common characteristic can be found in these few scattered examples, and this is illustrated in a frog T system opening in Fig. 12. The T system tubule is wide and has fairly straight wall and a transverse orientation as long as it is within a triad. Between the triad and the edge of the fiber, however, the tubule changes abruptly in size, becoming much narrower, and it follows a winding path. The actual opening is very small, about the same size as the opening of one of the caveolae.

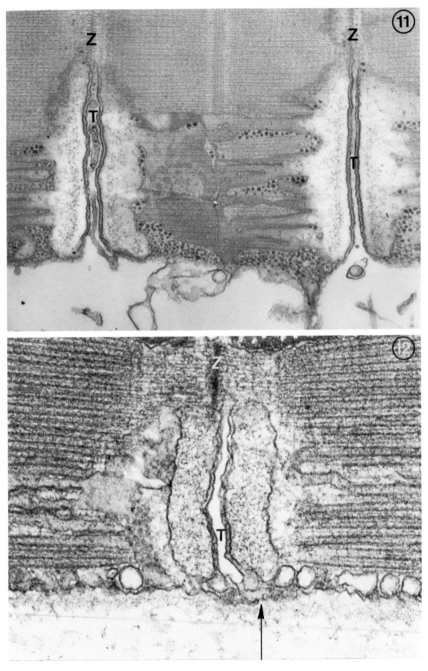

Figs. 11 and 12.

It is easy to see that only an occasional section may happen to follow such a tubule without interruption from the periphery of the fiber into a recognizeable triad. The obliquity of the most peripheral portion of the T system is sometimes very striking (Hoyle *et al.*, 1966) and it is frequent in developing fibers (Fig. 15) (see Chapter 3, Volume I).

The similarity in size of T system and caveolae openings in most fibers, which has been so well demonstrated in freeze fracture (Rayns *et al.*, 1968) (see Section II,A), is not mere coincidence. It has been shown that during development of some fibers the T system is produced as a proliferation and fusion of vesicles identical to the caveolae (Ezerman and Ishikawa, 1967; Schiaffino and Margreth, 1968) (see Chapter 3, Volume I). As proposed by Rayns *et al.* (1968), it is possible that in most adult fibers the T system opens in the back of one of the caveolae and also that it is capable of shifting its position slightly to open into any nearby caveola. In rat muscles, a different mechanism has been proposed for the T system development (A. M. Kelly, 1969). This would occur by initial attachment of SR sacs to the plasma membrane and their subsequent movement to the interior of the fiber, pulling in the T system. This type of development may have occurred in the case illustrated in Fig. 14. In this case also, the peripheral portion of the T system is more narrow and winding than the part of it that forms a junction with the SR.

Unfortunately, due to the geometrical difficulties, morphology cannot provide exact information on the number of T system openings around the circumference of the fiber, except in the somewhat exceptional case of some fish muscles, where the openings are wider. Indirectly, the local stimulation experiments of A. F. Huxley and Taylor (1958) (see Section III,A) indicate that in frog fibers the T system is not open at every interfibrillar space.

A significant breakthrough in providing definite proof that in all types of fibers no barrier exists between the T system content and the outer fluid came when muscle fibers were exposed to a solution containing molecules of relatively large size, which are visible either directly or indirectly in the light or electron microscopes. Endo (1964) used for the purpose single frog fibers and a fluorescent dye, whose penetration

---

Fig. 11. Periphery of a white twitch fiber from the black mollie. One sarcomere only is illustrated. At the Z line (Z) the plasmalemma penetrates into the fiber to form the T system. At left, a thin strand coming from a neighbor fiber penetrates into the T system. This is a somewhat unusual finding. (×50,000.)

Fig. 12. Periphery of a frog twitch fiber at the Z line level. The T system, upon leaving the triad, becomes narrow and winding. The opening is with a small mouth (arrow). Reprinted from Franzini-Armstrong (1970a). (×64,000.)

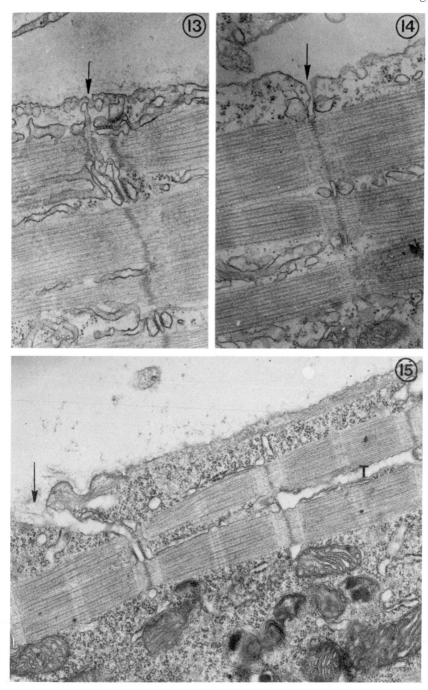

Figs. 13, 14 and 15.

into the fiber he could follow. The dye penetrated into a compartment which could be identified with the T system on the basis of its relative volume and location. More direct proof that large molecules, i.e. ferritin, are able to penetrate into the T system of frog fibers was obtained by using tracers visible in the electron microscope (Fig. 16) (H. E. Huxley, 1964; Page, 1964, 1965). Ferritin also freely penetrates into the T system of toad fibers (Peachey and Schild, 1968). Lanthanum (Rayns *et al.*, 1968; Zacks and Saito, 1970), horseradish peroxidase (Karnovsky, 1965; Eisenberg and Eisenberg, 1968, Fig. 25), ruthenium red (Luft, 1966; Goldstein 1969a), and Imferon have similarly been traced inside the T system of a variety of vertebrate fibers. Since the T system is filled by a fluid in probable equilibrium with that bathing the fiber, it is not surprising that insoluble sodium pyroanthiomonate precipitates are found within the T system of fibers exposed to a solution of potassium pyroanthiomonate (Zadunaisky, 1966; Tice and Engel, 1966).

Once it penetrates into the fiber, the T system runs mostly transversely among the fibrils surrounding each of them with a ring. Each ring is shared by adjacent fibrils, and ideally, one can consider all rings to be intercommunicating, to form a complete network across the fiber in whose openings run the fibrils (Fig. 17). The exact configuration of the network depends on the shape of the fibrils' cross section. Since a transverse section of a fiber never includes the T system over a long distance, so far it has not been possible to directly ascertain how complete the T system network is. Eisenberg and Eisenberg (1968) notice that most, but not all (i.e., about 82%), of the triads encountered in a random series of longitudinal sections contain a profile of the T system. This would indicate that interruptions in the T system exist, but are few and that the network may be considered as continuous. The transverse continuity of the T system is more easily demonstrated in arthropod fibers, where, for geometrical reasons the tubules can be followed for a considerable distance in cross sections of the fiber (Sections IV,A,C).

In longitudinal sections of the fiber, the T system tubule often appears cut exactly crosswise (Figs. 7 and 17). Following fixation in glutaraldehyde and either conventional embedding and sectioning, or freeze frac-

Figs. 13, 14, and 15. Three fibers from the tail of young frog tadpoles in the course of differentiation. In Fig. 13, the T system approaches a caveola, but the actual communication between the two, if existing, is lost from the plane of the section. In Fig. 14, an arrow indicates the obviously open T system. In Fig. 15, the T system penetrates obliquely into the fiber. A longitudinal portion of the T system is indicated (T). Figs. 13, 14, ×32,000; Fig. 15, ×22,000.

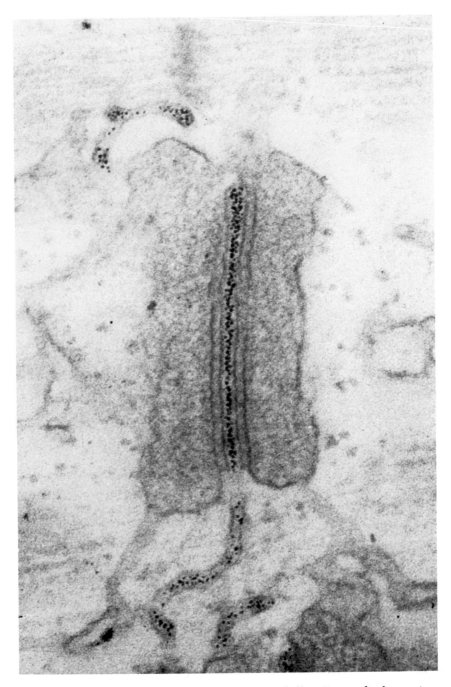

Fig. 16. Ferritin-infiltrated T system in a frog twitch fiber. Longitudinal extensions of the T system are evident. ($\times 96,000$.) Reprinted with permission from H. E. Huxley (1964).

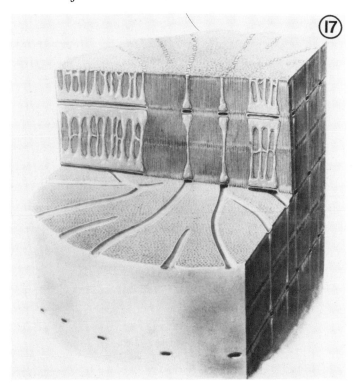

Fig. 17. Tridimensional reconstruction of SR and T system in a fish white fiber. Notice array of T system openings on the fiber surface. Other fibers (e.g., frog twitch fibers) may have less numerous openings. See Porter and Franzini-Armstrong (1965).

ture, the T system cross section has a flattened oval shape, the short axis being longitudinally oriented and 30–50 nm, the long axis being crosswise and variable between 500 and 120 nm (Page, 1965; Peachey, 1965a, for frog fibers). Most of the T system surface thus faces toward the lateral sacs of the triad, from which it is separated by a narrow, approximately 10 nm gap (Section III,C). Fixation in osmium tetraoxide produces a distinctly different-looking T system, i.e., a small (about 30 nm) circular diameter tubule, which goes meandering in a tight zigzag in the space between the fibrils (Figs. 6 and 9) (Porter and Palade, 1957; Andersson-Cedergren, 1959; Franzini-Armstrong and Porter, 1964a; Bübenzer, 1966). Occasionally the shape is less regular and the small tubule presents blebs and deformations. The reason for this striking difference in shape and surface-to-volume ratio in the T system tubules when exposed to the action of different fixatives is not known. Available evidence is not sufficient to state with certainty which of the two shapes more faithfully represents the *in vivo* situation, but on the basis of the generally superior preservation of all structural details which is obtained following glutaraldehyde fixation, it is tacitly assumed that the image provided by this fixative is closer to the *in vivo* structure. Thus all recent qualitative and quantitative information on the T system has been deduced from glutaraldehyde-fixed material. Peachey (1965a) and Page (1965) agree on the estimate of the T system volume relative to that of the whole fiber, i.e., 0.2–0.3%. Notice that T system volume is considerably smaller than SR volume. The relative surface area of the T system depends on the fiber size. In a 100 μ diameter fiber, it is calculated (Peachey, 1965a) that the T system surface area is seven times that of the sarcolemma (not considering the extra sarcolemmal area due to the caveolae). Above surface and volume values are for frog twitch fibers, the only vertebrate fiber type for which precise estimates have been made, for use in the understanding of physiological parameters (see Chapter 1, Volume III). In other fiber types, the T system is either more or less well developed (see Sections II,d; IV,A–D; and V,C).

D. SR and T System—Variations on a Theme

The relative amount and the shape and distribution of SR and T system elements vary considerably in different fibers, to the point that few micrographs are often sufficient, to the expert eye, to determine the source of the particular fiber being examined.

For example, the triads may be located at the approximate level of the edges of the A band (usually referred to as A–I junction level), rather than at the Z lines (cf. Figs. 18–21). In this case, two T system networks rather than one pervade the fiber at each sarcomere. In the search for the possible functional significance of the different location of the triads, several hypothesis have been formulated, all of which are only partially true. It was noticed that faster acting fibers usually have two triads per sarcomere, while fibers that contract more slowly have only one triad per sarcomere, located at the Z line level. This is the case when one compares, for example, most mammal leg muscles (contraction time of less than 30 msec) with frog twitch fibers that take up to 70 msec to contract. On the other hand, several exceptions have been described. Notable among these is the case of tortoise muscles, which are known to be several times slower than frog twitch fibers and in which the triads are at the A–I junction (Page, 1968a). Also, different fiber types in rat leg muscles, with up to twofold differences in contraction speed all have two triads per sarcomere.

Fibers with two triads per sarcomere are more common in mammals than in lower vertebrates. However, they also occur in some chicken muscles (Page, 1969) (Fig. 22) and in fast-acting fish muscles (Fawcett and Revel, 1961; Reger, 1961; Kilarski and Bigai, 1969) (Figs. 44 and 45). Two fiber types, with either one or two triads per sarcomere, coexist in a fish eye muscle (Kilarski, 1965). Location of the triad thus cannot be attributed to a species difference.

Page (1968b) provides a clue of almost universal applicability. The triads are two per sarcomere in those fibers that have long thin filaments and which are thus capable of large changes in sarcomere lengths. Interestingly, the only exceptions to this rule are some very fast acting fibers (see Section IV,D) in which short thin filaments coincide with two triads per sarcomere. In these fibers the T system distribution probably reflects the need for fast e–c coupling (see Section III,A).

One disposition of the triads, so far only observed in a fast twitch muscle from a fish (Goldstein, 1969b), is worth mentioning here as a curiosity. The triads are located at the A–I junction, but they are not, as usual, present at every interfibrillar space. Rather, triads occur at regularly alternating interfibrillar spaces, due to the fact that the T system tubule surrounds group of two fibrils, whereas the SR forms a complete sleeve around each individual fibril.

In some fibers, pentads have been described (Revel, 1962; Gori *et al.*, 1965). These occur where a T tubule branches into two parallel running tubules which are separated from one another and accompanied on either side by sacs of the sarcoplasmic reticulum, thus forming a group of five elements (Fig. 47). All sides of the reticulum facing a T tubule in the pentads show the characteristic surface differentiations that occur at the triads (see Section II,C). Since T system surface and junctional area with the SR are thus increased, it is not surprising that pentads have first been described in a very fast acting type of fiber (Revel, 1962) (see Fig. 47). However, they are not exclusive of fast fibers, since they are occasionally encountered in frog twitch fibers (Peachey, 1965a), in rat muscles (Gori *et al.*, 1965), and, strangely, in the course of atrophic processes (see Pellegrino and Franzini-Armstrong, 1969).

The opposite situation, that is, reduction of the area of junction between SR and T system, occur in those fibers where the lateral sacs of the triad are not continuous all across, but interrupted into segments, so that the T system is alternatively running between two SR sacs and free in the sarcoplasm (Figs. 22 and 23) (Porter and Palade, 1957;

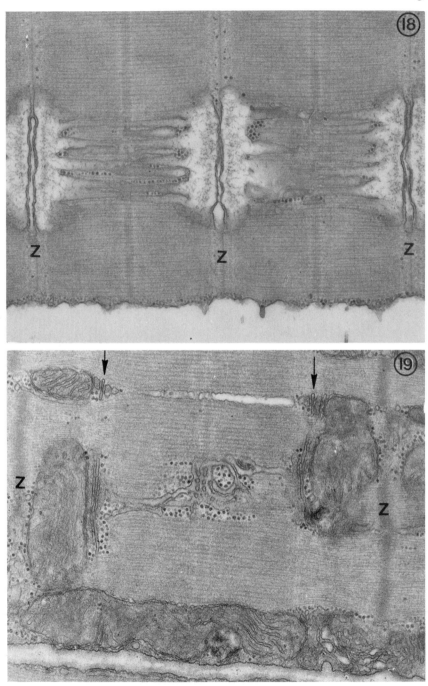

Figs. 18 and 19.

Page, 1968a, 1969). Fibers where this occurs are among fairly slow acting fibers although of the twitch type.

An incomplete coverage of T system elements by SR sacs is a characteristic of all slow (nontwitch) type fibers so far described (see Section IV,A,B). In most of these, elements of the T system run at random in the fiber, i.e., as much longitudinally and obliquely as transversely (Page, 1965, 1969; Hoyle *et al.*, 1966), and only rarely come to close proximity with elements of the SR. The junctions that occur at these points of proximity have been identified as equivalent to triads, because the content of the junctional SR and the structure of the junctional gap are the same as in the triads (Page, 1965; Franzini-Armstrong, 1970b), (see Section II,C). Often one sac only of the SR participates in the junction, which is then called dyad (Fig. 37), using a term initially introduced for some insect muscles (D. S. Smith, 1961a) where only two components, one SR sac and a T tubule form the junction (see Section V,C). The junctional area of triads and dyads of slow fibers is irregular in dimensions, usually being small and round or oval (Figs. 36–38). The axis of the junction is either transverse, as in the previously described triads, or longitudinal and oblique in the fiber. Most of the T system surface is free (Fig. 37).

In most arthropod fibers, the junctions between SR and T system are also oriented longitudinally; i.e., SR sacs are interposed laterally between the T system and the fibrils. This occurs usually because the long axis of the flattened T tubule cross section is oriented longitudinally, rather than transversely (Figs. 51, 54, and 55). In addition to triads and dyads, or in place of them, where the T system is absent, some fibers have peripheral junctions between SR and the plasma membrane. Such is the case, for example, in cardiac muscle (Johnson and Sommer, 1967) and in oblique striated fibers (see Section V,D).

The shape of the SR is largely dependent on the shape and disposition of the triadic or dyadic junctions. Where the triads are two per sarcomere, two or three almost completely independent SR segments, rather than a continuous one cover each sarcomere. The segment covering the A band is usually identical to that covering the whole sarcomere, as described in Section II,C, except that the longitudinal tubules are shorter (Figs. 18 and 19). Opposite the Z line–I band regions, the SR

Fig. 18. In this white fiber from a fish, the T system is at the Z line (Z). Two SR segments are included in the micrograph. (×30,000.) Reprinted from Franzini-Armstrong and Porter (1964a).

Fig. 19. Twitch fiber from the rat soleus muscle. The T system (arrows) and triads are at the A–I junction, so that there are two triads within the limits of one sarcomere. Z = Z lines. (×37,000.)

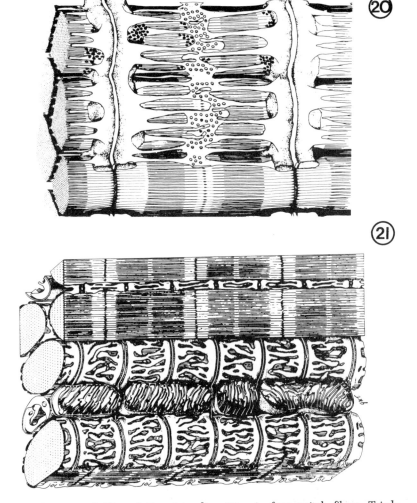

Fig. 20. Drawing of SR and T system disposition in frog twitch fibers. Triads are at the Z line. Reprinted with permission from Peachey (1965a).

Fig. 21. SR, T system, and mitochondria in the very fast fibers of the bat crycothyroid muscle. Triads are at the A–I junction, the SR is well developed. Mitochondria are large and numerous. Reprinted with permission from Revel (1962).

is either continuous—[with a structure similar to the fenestrated collar (Figs. 21, 22, 43, and 46)—or it is interrupted almost completely at the Z line level (Figs. 44 and 45), being in the form of longitudinal tubules which meet but do not fuse at the level of the Z line. Often two layers of SR are interposed between adjacent fibrils at the level of the I bands, whereas a single layer divides the fibrils at the level

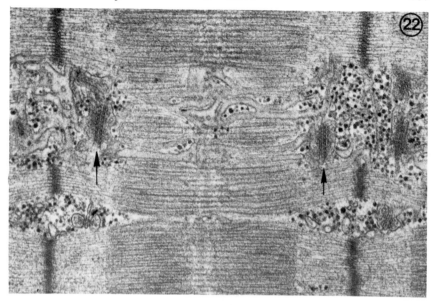

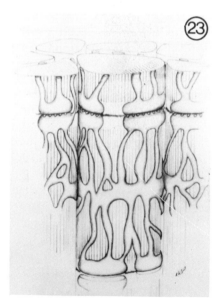

Fig. 22. Twitch fiber from the chicken posterior latissimus dorsi. Triads (arrows) are at the A–I junction level, but the lateral sacs of the triad are not continuous across the fiber (cf. Figs. 11 and 12). (×35,000.) Reprinted with permission from Page (1969).

Fig. 23. Drawing of the structure of Amblystoma tail myotome fibers. As in Fig. 22, the lateral sacs of the triad are interrupted into short segments. Reprinted with permission from Porter and Palade (1957).

of the A band (Fig. 39). Where the triads are oriented either at random or longitudinally in the fiber, the SR is not interrupted longitudinally. This occurs in slow fibers and in most arthropod fibers. The significance of the extensive longitudinal continuity of the SR has not been explored. Most cardiac muscle also has a longitudinally uninterrupted SR, although the triads are transversely oriented (Fig. 29).

In the fenestrated collar of vertebrate twitch fibers, the fenestrae are strikingly uniform in size. It is conceivable that the small diameter of the opening is determined by the minimum ray of the curvature that the SR membrane may assume. Where the triads are at the A–I junction, the SR segment opposite the I bands and Z line is also perforated by small openings, which are essentially similar to the fenestrae, except for the fact that they are less numerous, more variable in size, and sometimes elongated (Figs. 43 and 46). In arthropod fibers numerous, mostly round fenestrae perforate both H region and Z line–I band region SR, but they are usually larger than in vertebrates (Figs. 51 and 54). One may finally consider analogous to fenestrae the much wider, round openings that perforate the whole, continuous SR of chicken slow fibers (Fig. 37).

The T system does not exclusively lie in transverse planes even in those fibers where it has been classically thought to do so, i.e., vertebrate twitch fibers. Occasional short longitudinal excursions of T tubules were noticed by Andersson-Cedergren (1959) in her accurate study of rat leg muscles. Introduction of tracers inside the T system allowing a more clear and immediate visualization of the T tubules and differentiation of them from similar-looking SR tubules has brought to the recognition that T system longitudinal extensions are fairly numerous (H. E. Huxley, 1964; Page, 1965). In toad fibers, Peachey and Schild (1968) calculate that longitudinal T tubule extensions along the I bands may increase by an additional 30% of the total surface area. Eisenberg and Eisenberg (1968) suggest that this is an overestimate and, for frog fibers, use a more conservative 10% figure. In addition, in adult frog fibers, approximately 3% is contributed by longitudinal extensions of the T system that run for the length of a whole sarcomere, connecting two T system transverse networks (Fig. 24) (Eisenberg, 1972). These longer longitudinal tubules are fairly rare and often coincide with areas of the fiber where the cross striations are out of register.

On the other hand, in young tadpole tail fibers, where originally noticed (Franzini-Armstrong and Porter, 1964b), longitudinal T tubules are far more numerous (Figs. 1, 15, and 25). The reason for their abundance is clear when one considers that in developing muscles T system and triads are randomly oriented before assuming their final position (Page, 1969; Ezermann and Ishikawa, 1967; Schiaffino and Margreth, 1968; A. M. Kelly, 1969; Walker and Schrodt, 1967, 1968; Edge, 1970) (see Chapter 3, Volume I).

An interesting example of the way in which longitudinal extensions of the T system are employed in the regular architecture of an adult fiber is that encountered in turtle muscles (Page, 1968a). The triads are located at the A–I junction and the T system, rather than running in two distinct cross planes at that level, courses back and forth along the I band to form triadic junctions with the SR elements on either side of the Z line. The I bands of adjacent sarcomeres thus are synchronously excited, so that the division of the fibril in sarcomeres starting at the Z line is artificial; the e–c coupling unit is more likely to be from the center of one A band to the next.

A more thorough appreciation of how much variability exists in the distribution of T system elements in different fiber types and at different stages of development may be obtained from reexamining the now famous early work by Veratti (1902) on striated muscle fibers. The advantage of the light microscope over the electron microscope in providing larger fields of view is evident in this type of problem. The extremely fine and complete visualization by the author of such small elements as the T system tubules, achieved by the combination of a special silver staining method, much patience, a keen eye, and an artistic hand is impressive. All possible types of T system distribution slowly "rediscovered" with the use of the electron microscope are already described in Veratti's work—from the one or two transverse networks per sarcomere, only occasionally connected by longitudinal elements, which characterize adult vertebrate fibers, to the irregular orientation of the T system found in young fibers, to the two transverse networks at the A–I junctions, plus another set of tubules at the Z line, which is met in Crustacea (see Section V,A).

Perhaps the most curious and fascinating example among those described by Veratti (1902) is that of the dorsal fin fibers of the sea horse (Figs. 26 and 27). These fibers have an unusual, disposition of the contractile material; the fibrils are located in horseshoe-shaped segments near the center of the fiber (Fig. 26).

Mitochondria, SR, and T system are abundant and regularly distributed around the fibrils. The periphery of the fiber is occupied by apparently structureless sarcoplasm. Veratti, however, describes a fine network of tubules running across this space from the periphery of the fiber to the fibrillar core. Careful examination of electron micrographs reveals the presence of few transversely oriented tubules which run in the empty sarcoplasmic spaces (Fig. 27). Veratti's work allows the

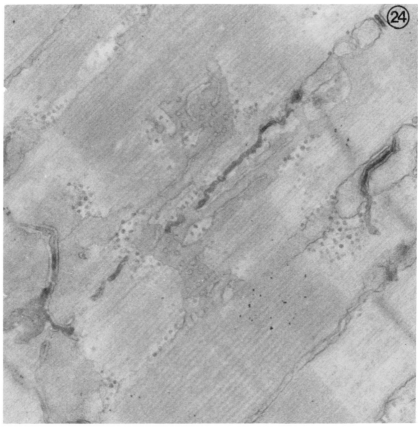

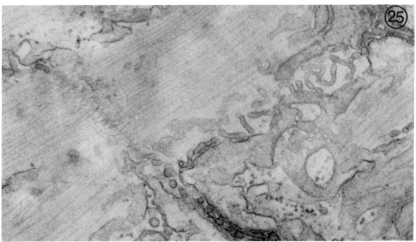

Figs. 24 and 25.

conclusion to be reached that these are parts of the T tubule connecting to the plasma membrane, and thus allowing e–c coupling to follow the familiar path even in these unusual fibers (Bergman, 1964).

Only one type of fiber has so far been described in vertebrates completely lacking T system elements; these are fibers composing the body muscles of amphioxus (Peachey, 1961). Each fiber is composed of a single fibril, that is one continuous mass of contractile material. The fibril is a flat lamella, with a minimum diameter slightly less than 1 μ; i.e., of the same order of magnitude as the diameter of an ordinary fibril within a multifibrillar fiber. The plasma membrane closely envelopes each lamella, so that any component of the contractile material is at a short distance from it (Fig. 28). If one considers the principal role of the T system as that of speeding conduction of excitation from the periphery to the interior of a large fiber (see Section III,A), then it is apparent that these fibers do not need a T system component. The few sacs present between the periphery of the single fibril composing the amphioxus fibers and the sarcolemma may be identifiable as a reduced expression of the SR. Similar subsarcolemmal SR sacs characterize several types of fibers such as those of oblique striated muscles (see Section V,D).

Perhaps amphioxus fibers should be considered as a primitive type of fiber, from which larger fibers have evolved by fusion with preservation of some of the interfibers space to constitute the T system. Indeed, amphioxus myotomes have another primitive characteristic; the muscle fiber sends long processes back to the spinal cord, so that neuromuscular junctions occur centrally, at the surface of the cord (Flood, 1966). A similar situation has only been described in some worm muscles (Rosenbluth, 1965b; Reger, 1965).

E. The Mitochondria

The fine structure of skeletal muscle mitochondria, or sarcosomes, does not differ in its basic details from that of mitochondria from other cell types; i.e., mitochondria are bags delimited by two membranes,

Fig. 24. Peroxidase-infiltrated T system in a frog twitch fiber. A T tubule extends longitudinally for the length of a sarcomere. Notice that this occurs where striations are out of register. ($\times 45{,}000$.) Reprinted with permission from Eisenberg and Eisenberg (1968).

Fig. 25. Longitudinal extension of the T system in a fiber from the tail myotomes of frog tadpole. The T system is distinguishable from the longitudinal SR sacs because it has a smaller diameter in this osmium tetroxide fixed preparation. ($\times 50{,}000$.)

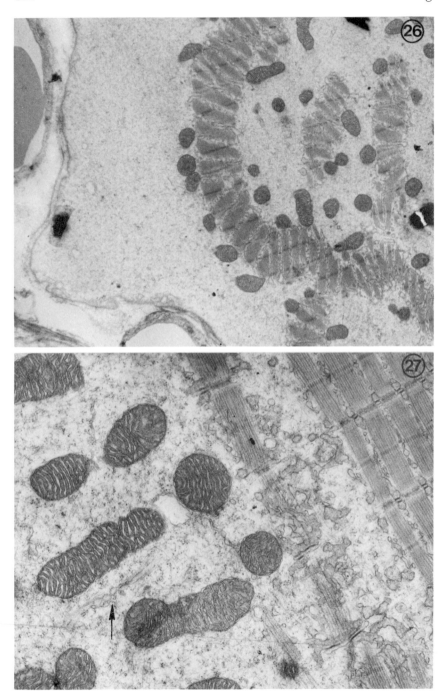

Figs. 26 and 27.

Fig. 28. Structure of the body myotomes of *Amphioxus*. Each fiber is composed of a single ribbonlike fibril, closely surrounded by the sarcolemma. No T system is present. Reprinted with permission from Peachey (1961).

the inner one of which infolds to form the cristae mitochondriales (Andersson-Cedergren, 1959) (Fig. 29). Most sarcosomes are elongated, probably to adapt to the shape of the interfibrillar spaces where they are located. Mitochondria in subsarcolemmal and paranuclear positions are more rounded.

Mitochondria vary greatly in their size, distribution, and internal structure and, in contrast to SR and T system, they only rarely take a precise relationship relative to the cross striation. In most fibers they are located in longitudinal columns irregularly scattered among the fibrils (Figs. 5, 21, and 37). Where more numerous, large groups of them may be present under the sarcolemma (Fig. 40 and 41) (see also Section IV,C), in addition to those dispersed in the middle of the contractile material. More unusual is the aggregation of all mitochondria in a central core of the fiber and under the sarcolemma, while none are between the fibrils (Section V,D). Mammalian skeletal muscle fibers are among the few examples where mitochondria are located at precise levels of the sarcomere (Fig. 42) (Section IV,C).

Fig. 26. Cross section of a fiber from the dorsal fin muscle of the sea horse. The fibrils are grouped in the center of the fiber and surrounded by SR, T system, and mitochondria. The peripheral sarcoplasm is apparently devoid of any organized structure. (×5250.)

Fig. 27. Longitudinal section of a sea horse dorsal fin muscle. Structure of SR and T system is similar to that of other fish muscles (cf. Fig. 18). Radially oriented string of vesicles (arrow) is probably part of the T system reaching the periphery of the fiber; for further explanation, see text. (×15,000.)

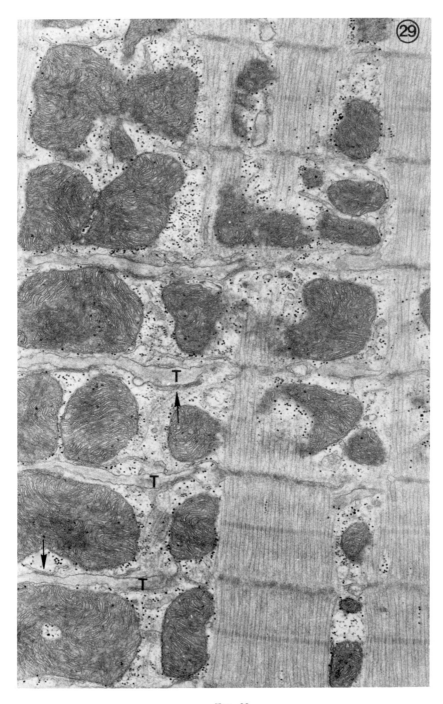

Fig. 29.

The distribution of mitochondria is regulated by the predominant type of metabolism fibers use to maintain their ATP stores. Each fiber is capable of anaerobic glycolysis (the enzymes for which are in the sarcoplasm, or possibly partially segregated within the SR, see Section III,B) and of oxidative phosphorylation, which is localized in the mitochondria (see Chapter 9, Volume III). Since the latter is a slower but more efficient way of utilizing the energy sources (glycogen and saturated fatty acids), it is not surprising that mitochondria are numerous in fibers capable of sustained periods of activity (see also Sections IV,C and V,C). Large mitochondria, evenly dispersed between the fibrils characterize fibers that contract continuously over long periods of time. Typical examples are the heart muscle fibers of most animals (Fig. 29) (e.g., see Slautterback, 1965; DiDio, 1967; Fawcett and McNutt, 1969) (see also Chapter 2), the flight muscle of insects (Figs. 53, 54, and 56) (D. S. Smith, 1961a, 1963, 1966a; Ashurst, 1967), and those of actively flying birds (Grinyer and George, 1969a,b; Ashurst, 1969). The relationship between continuous activity and mitochondria content is well exemplified by the observation that in hummingbirds, where it has an important role in flight, the supracoracoideus muscle is rich in mitochondria, whereas it has fewer mitochondria in other birds, where its role in flight movements is reduced (Lasiewski et al., 1965). In some flight muscles, the mitochondria are giant, that is unusually large (D. S. Smith, 1963; Rockstein and Bhatnager, 1965). Mitochondria up to 1.5 μ wide and 10.0 μ long have been described in the hummingbird pectoralis muscle (Lasiewski, et al., 1965). Particularly well packed and also peculiarly shaped cristae are present in mitochondria of some active fibers (Leeson and Leeson, 1969; D. S. Smith, 1963; Revel et al., 1963; Fawcett and McNutt, 1969). The cristae either have angular configuration or they are in the form of fenestrated sheets, the fenestrae in adjacent sheet coinciding to constitute tubular channels running for the length of the mitochondrion (D. S. Smith, 1963). Abundance of the cristae is taken as an indication of functional hypertrophy.

Lipid droplets are present in close association with the mitochondria, particularly in fibers that mostly rely on oxidative phosphorylation (Fig. 43). In pigeon's breast muscles, for example, there are two types of fibers—one with rare mitochondria, abundant glycogen and supplied with glycolytic enzymes; the other type is rich in mitochondria and

Fig. 29. Longitudinal section of the papillary muscle from cat heart. The mitochondria occupy approximately as much volume as the fibrils. Their cristae are tightly packed. The T system (T) is wide and filled by the basement membrane. The SR forms a longitudinally continuous network and flattened sacs forms dyads (arrows) with T system. ($\times 25,000$.)

lipid droplets and relies mostly on oxidative phosphorylation (George and Naik, 1959; Ashurst, 1969).

Abundance of mitochondria in the fibers is accompanied by a rich vascularization of the muscle (Romanul, 1964) and a higher concentration of the oxygen-carrying molecule myoglobin within the fibers themselves (e.g., see Morita *et al.*, 1969; Leeson and Leeson, 1969). Both factors, it has long been known, contribute to the deeper red color which characterizes muscles rich in mitochondria. Since most muscles are not homogenous in their fiber content (see Section IV,C), the term red and white are now restricted to indicate single fibers (see Gauthier, 1969, for a recent review). A pinkish color also distinguishes bundles of mitochondria-rich fibers in crustacea (Hoyle, 1968; Hoyle and McNeill, 1968). As a rule, red fibers have a smaller diameter than white fibers, probably because oxygen supply to mitochondria deep in the fiber is restricted by the diffusion of myoglobin from the surface. Further examples of the variability in the distribution of mitochondria and its relationship to the overall pattern of fiber activity are discussed in Section IV,C.

F. The Nucleus and Other Components

The nucleus is surrounded by an envelope which is formed by two membranes separated by a narrow nuclear cisterna. As noted previously (Section II,B) the nuclear envelope membrane is continuous with the SR, so that the nuclear cisterna content is in communication with the SR content. As in all other cells, numerous pores perforate the nuclear envelope, chromatin is usually clumped at the periphery (Fig. 5) and the nucleolus varies in size depending on the state of the fiber (e.g., development versus maturity).

Being derived from the fusion of numerous cells (see Chapter 3, Volume I), muscle fibers contain numerous nuclei. These are usually elongated longitudinally and their surface is marked by furrows, which are deeper the more shortened the fiber is at the moment of fixation. These furrows, or incisures, have commonly been considered to be secondary to some pressure exercised by the contractile material, particularly because they are exactly localized at the Z line level. However, the envelope folds are a permanent feature, as indicated by the fact that they persist in isolated nuclei when separation is done carefully (Franke and Schinko, 1969).

The location of the nucleus is most often, but not necessarily, at the periphery of the fiber. In some muscles, a specific location of the nucleus (i.e., either peripheral or more or less central) is typical of different fiber types and can thus be employed as a useful, quick identifying landmark. Such is the case for frog slow and twitch fibers, where the nucleus is peripheral and central, respectively (W. K. Engel and Irwin, 1967).

In addition to the fiber nuclei and not distinguishable from them in shape and structure are the nuclei of the satellite cells (Mauro, 1961; Muir *et al.*, 1965). These are very small cells nested in depressions of the fiber surface under the basement membrane. They are almost completely filled by a nucleus, and their cytoplasm is isolated from that of the muscle cell (Muir *et al.*, 1965). The role and significance of satellite cells in development and regeneration of muscle fibers is discussed in Chapter 3, Volume I.

The Golgi complex of adult muscle fibers is small and usually located at the nuclear poles (Andersson-Cedergren, 1959). Rough-surfaced endoplasmic reticulum is rarely encountered. Lysosomes are occasionally visible, but they are in significant quantities only in conditions of degeneration (cf. Pellegrino and Franzini-Armstrong, 1969).

III. Functions of SR and T System

A. *The T System and Excitation–Contraction Coupling*

The presence of the T system is sufficient to account for most of the major differences in electrical behavior of muscle relative to nerve membrane, which are listed below. The subject is more thoroughly discussed in Chapter 1, Volume III.

It was noticed (Katz, 1948; Fatt and Katz, 1951) that the membrane capacitance of frog muscle fibers, when referred to the then estimated total surface area, has a value of 5 $\mu F/cm^2$, which is considerable higher than the analogue value for squid nerve fibers (1 $\mu F/cm^2$) (Hodgkin *et al.*, 1952). On the other hand, when the total capacitance is referred to the plasmalemma plus T system surface area, then the values for nerve and muscle are not significantly different (Falk and Fatt, 1964; see also Peachey, 1965a,b).

When a fiber is exposed to a solution containing a high concentration of potassium ions, it depolarizes very rapidly, but repolarization following quick removal of potassium ions from the bathing solution has a slower time course (Hodgkin and Horowicz, 1960). Hodgkin and Horowicz calculate that the effect can be accounted for by retention of potassium ions in a space delimited by a membrane electrically continuous with the plasmalemma, that takes 0.2–0.5% of the fiber volume and from which diffusion is limited. There is little doubt that the T system elements fit into this category.

One last phenomenon which is attributed to the presence of the T system is the late after potential; i.e., following a train of action potentials, the membrane tends to recover more slowly from depolarization than following a single action potential (Freygang *et al.*, 1964).

A recently discovered and immediately very popular method of selectively destroying the T system tubules without greatly affecting the other membranes has been of considerable importance as a means of directly assessing the T system contribution to the fiber physiological properties. The method was developed by Howell and Jenden (1967) following some observations by Fujino *et al.* (1961), and it consists in soaking the muscle in a Ringer solution containing a low concentration of glyercol (400 mM). When the muscle is returned to the regular Ringer solution, some of the fibers give action potential, but they do not contract. Electron microscopy has shown that in fibers treated this way, the T system tubules are interrupted, but a slight disagreement exists on the extent of the damage. Eisenberg and Eisenberg (1968) and Howell (1969) find the T system tubules almost completely absent in the interior of the fiber, whereas S. Nakajima *et al.* (1969) report a much smaller degree of disruption. All authors are in good agreement on one essential point: tracers do not penetrate into the remnants of the T system, which may still be present in the glycerol-treated fibers. This is taken to indicate that the T system continuity is certainly interrupted at or near its point of entry into the fiber. Elements of the SR are somewhat affected by the procedure (e.g., swollen and partly separated into vesicles), but in general this is not considered significant. In glycerol-treated, detubulated fibers, the surface capacitance is similar to that of nerve fibers (Gage and Eisenberg, 1967), repolarization following removal of a high potassium solution is rapid (S. Nakajima *et al.*, 1969), and finally, the late after potential is abolished (Gage and Eisenberg, 1969).

Studies of electrical parameter thus are in agreement with the morphological observation that the T system interior is in communication with the outside and its membrane continuous with the plasmalemma and ideally suited for conducting an impulse to the interior of a fiber. The T system must be considered a necessary link in the process of excitation–contraction coupling, i.e., the series of events that leads from the initial depolarization of the sarcolemma to contraction of the fibrils. Hill (1948, 1949) initially calculated that in frog twitch fibers the release of an activator from the surface of a fiber cannot account for the known rapidity of the spread of activation throughout the fiber cross section. Diffusion of a substance from the periphery would take much longer than the observed e–c delay, which in most twitch fibers is of the order of a few milliseconds. The presence of the T system tubule, capable of transmitting an electrical signal to the interior of the fiber to within a few microns at most from any part of the contractile material, solves this apparent paradox (Peachey and Porter, 1959).

The best direct evidence that the T system has a role in e–c coupling comes from the classic experiments of A. F. Huxley and co-workers (see A. F. Huxley, 1964). The experiments consisted in producing a localized area of depolarization on the sarcolemma of a single fiber, which was examined under a microscope. The diameter of the tip of the electrode was small enough that its position could be accurately determined relative to the bands of the sarcomere. In frog fibers, the depolarization produced a visible, local, graded contraction when depolarization was applied at the level of the Z line. Depolarization at other levels of the sarcomere did not have any effect; a clear indication that to produce contraction of even the superficial fibrils the depolarization has to be mediated via the T system tubule. Similar experiments when carried out on lizard and crab muscle fibers, where the T system is located at the level of the A–I junction, were also positive only when depolarization coincided with the zone of entry of the T system in these fibers (see also Section V,B).

Due to the small size of a tubule, the electrical parameters of the T system can only be determined indirectly, by observing, for example, the contractile response of the fiber when the surface membrane is depolarized of known amounts (Gonzales-Serratos, 1966; Adrian *et al.*, 1969). The most exciting question in the minds of most researchers interested in the spread of activation along the T system tubule is whether this is an active or passive process. Recent results by Costantin indicate that the depolarization may spread actively along the tubules (Costantin, 1970). These and other questions are dealt with in more detail in Chapter 2, Volume III.

B. The SR and Relaxation

Following depolarization of the T system, the next step in e–c coupling is the release from the SR of a substance (calcium ions) that activates the contractile material (or better, removes inhibition to interact) (see Chapter 7, Volume III). In order to explain this step, it is convenient to consider first the process of relaxation, and the involvement of the SR in this part of the fiber activity cycle.

A crude muscle extract, the Marsh factor (Marsh, 1952) produces a series of effects on muscle systems *in vitro,* which can be regarded as mimicking the *in vivo* phenomenon of relaxation (for reviews, see Sandow, 1965; Hasselbach, 1964; Weber, 1966; Ebashi and Endo, 1968). The effective component of the Marsh factor, that is, the component which has a maximum relaxing activity (relaxing factor), may be separated by differential centrifugation from the rest of the muscle extract, and it is a suspension of smooth-surfaced vesicles (Fig. 30) derived from the breakdown of SR and T system elements (Muscatello *et al.,* 1961; Ebashi and Lipmann, 1962). By analogy with similar fractions obtained from other cell types, the relaxing factor is also called microsomal fraction. Only occasionally whole triads can be identified in the suspension and thus the two components of the microsomal fraction can be distinguished (Ebashi and Lipmann, 1962; Hasselbach and Elfvin-Lars, 1967). Most of the time, it is not possible to tell how much T system elements contribute to the composition of the relaxing factor, but since *in vivo* the SR volume largely exceeds that of the T system, it is reasonable to assume that most of the relaxing factor vesicles derive from SR elements and that the main functional characteristics of the relaxing factor are attributable to the SR.

In negatively stained preparations whole vesicles, rather than profiles can be seen. Following this procedure, the relaxing factor vesicles are in the shape of tadpoles (Martonosi, 1968), that is, the vesicles have a rounded portion to which an elongated tail is attached. However, identification of these peculiarly shaped vesicles with specific portions of the SR is not possible, particularly since other types of membrane in suspension (e.g., red blood cell ghosts) tend to assume very similar appearances. The extent to which this is due to the preparative procedure is not clear in either case. Analogously, the approximately 40 Å subunits which several authors have described on the surface of the relaxing factor vesicles (Martonosi, 1968; Deamer and Baskin, 1969; Greaser *et al.,* 1969) are not specific to these types of membranes and are not involved in the major function of the vesicles (Ikemoto *et al.,* 1970).

The relaxing factor particles accumulate calcium from the medium using ATP as a source of energy (Nagai *et al.,* 1960; Ebashi, 1961, 1962), and their capability of reducing the calcium concentration to less than 10^{-7} M is now thought to be the sole mode of action in producing relaxation (Ebashi, 1962; Weber *et al.,* 1964, see Chapter 7, Volume III).

When the relaxing factor is allowed to accumulate calcium in the presence of oxalate, its calcium-accumulating ability is enhanced, and calcium oxalate precipitates within the vesicles and can be directly visualized as a dense content in the electron microscope of sectioned or negatively stained relaxing factor preparations (Hasselbach and Makinose, 1961; Deamer and Baskin, 1969; Greaser *et al.*, 1969). It is interesting that even when the microsomal fraction has been purified of mitochondrial contamination, not all vesicles are equally capable of accumulating calcium ions. Indeed, one may obtain from skeletal muscle two microsomal fractions, the light and heavy fractions, differing by their sedimenting characteristics and their calcium-binding or accumulating ability, the heavy fraction being more effective (cf. Weber, 1966).

Studies of the microsomal fraction, that is of the SR isolated from the fiber, leave one major question unanswered: Is the division of the SR into different structural components (i.e., lateral sacs of the triad, longitudinal tubules, and fenestrated collar) of any significance in relation to the major function of the SR as a relaxing factor? The answer to this question must be sought by exploring, if possible, the SR function in the whole fiber. Three different approaches have been used:

1. The SR ATPase activity has been probed by cytochemical techniques, which allow direct visualization of ATP splitting sites. The dense final products of the reaction have been traced by several authors within the SR lumen, either throughout it or sometimes specifically in the lateral sacs of the triad. However, identification of these regions as those involved in calcium uptake is uncertain because the SR membrane possesses several ATPases, one only of which is involved in calcium uptake (cf. Hasselbach, 1964) and these are not so far distinguishable by cytochemical methods (A. G. Engel and Tice, 1966). In addition, calcium-binding-associated ATPase is particularly sensitive to glutaraldehyde, the fixative most commonly employed (Sommer and Hasselbach, 1967).

2. Calcium movements within the fiber at different moments of the activity cycle have been followed by autoradiographic techniques (Winegrad, 1965, 1970). These indicate that calcium may be taken up mostly by the A band region SR (longitudinal sacs) during the relaxation phase and subsequently accumulated in the lateral sacs of the triad.

3. Sites of calcium accumulation may be identified *in vivo* as they had previously been *in vitro*, by precipitating the calcium with oxalate and then observing the dense calcium oxalate precipitates in the electron microscope. When applied to the whole fiber, however, this technique encounters one major difficulty—the sarcolemma is not

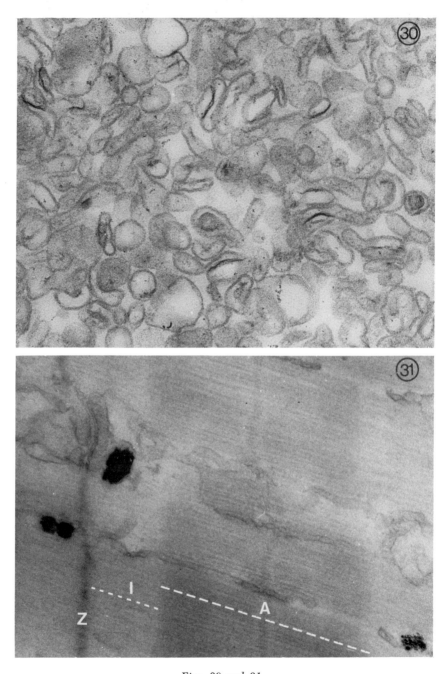

Figs. 30 and 31.

permeable to oxalate. The problem has been overcome by first exposing the fibers to glycerol in fairly high concentration for a short time, thus altering the sarcolemmal permeability, but only slightly damaging the internal structure. By this technique, calcium oxalate deposits have been observed in the SR of vertebrate twitch fibers (Hasselbach, 1964; Pease *et al.,* 1965) and of chicken slow fibers (Page, 1969) without a specific location. Similar results have also been obtained by exposing the fiber to a fixing solution containing oxalate (Komnick, 1969), with the difference, however, that calcium oxalate deposits are in this case restricted to the lateral sacs of the triad. More precisely controlled experiments, performed on "skinned" fibers, also come to the important conclusion that the regions of highest calcium concentration at rest are the lateral sacs of the triad (Costantin *et al.,* 1965). The experiments consist in mechanically interrupting the continuity of the sarcolemma using a sharp needle while the fiber is immersed in neutral oil and its surface is free of calcium ions (Natori, 1954). The skinned portion of the fiber is an intact array of fibrils, mitochondria, T system, and SR, the latter being capable of normally functioning as a calcium sink (Podolsky and Costantin, 1964). When a small drop of a solution containing oxalate is added to this preparation, few dense deposits form in the SR, specifically in the lateral sacs of the triad (Fig. 31). Previous addition of small amounts of calcium, which cause a localized area of contraction followed by relaxation, produce a larger number of deposits in the same region of the SR. Positive identification of these as calcium oxalate has been achieved by the use of a microprobe (Podolsky *et al.,* 1970).

4. Binding of ruthenium red and thorium dioxide to the SR (Goldstein, 1969a,b; Philpott and Goldstein, 1967) using techniques designed to identify the presence of polyanions is interpreted as indicating calcium binding sites. The whole SR seems to react uniformly in this case.

To summarize, there are strong but not conclusive indications that the SR is functionally differentiated into at least two parts—one is active in uptake of calcium from the fibrils and the other serves as a storage area where calcium is accumulated at rest. It is easy to speculate that the presence of a fairly high concentration of calcium ions in the lateral sacs of the triad seems reasonable if one considers that the signal for

Fig. 30. Microsomal fraction (relaxing factor) from a white muscle from rabbit. Most of the vesicles are the product of breakage of SR elements. ($\times 75,000$.)

Fig. 31. Section through a "skinned" frog twitch fiber treated with oxalate. The dense deposits within the SR at the I band level are calcium oxalate. Z, I, and A, bands are indicated. ($\times 38,000$.) Reprinted from Costantin *et al.* (1965) copyright 1965 by the American Association for the Advancement of Science.

the release of calcium must come from the adjacent T system tubules. The dense content of the lateral sacs of the triad may serve as a calcium binding material, thus permitting calcium to be accumulated in that particular portion of the SR.

The SR may have other functions besides binding calcium ions during relaxation. The possibility has been explored that some enzymes used in glycolysis are either segregated within the SR or associated with its membrane. There is, however, one major difficulty inherent to biochemical investigations of the enzyme content of the microsomal fraction: the enzymes that are only weakly bound to the membrane and those that are simply contained within the SR are lost during preparative procedures. In view of this consideration, the recovery of 15% of the total phosphofruktokinase activity in the microsomal fraction is taken as an indication that this enzyme is within the SR *in vivo* (Margreth *et al.*, 1963). More precise is the cytochemical localization of lactic dehydrogenase within the SR (Fahimi and Amarasingham, 1964; Fahimi and Karnovsky, 1966). Both studies indicate some involvement of the SR in glycolytic activity.

C. The Triad—Structure and Function

Since excitation is conducted to the fiber interior along the T system and calcium is accumulated in the SR preferentially in the lateral sacs of the triad, it is likely that some interaction between T system and SR at the triadic junction causes the release of calcium from the SR. The mechanisms of this step in e–c coupling are still obscure, and morphology, despite the recent progress toward a clarification of the structural details of the triad, has not been able to provide an unequivocal answer.

The fine structure of the junctional area of triads, dyads, and peripheral junctions of most fiber types is basically the same. In all cases, flattened portions of SR and T system (or sarcolemma) run parallel at a distance of 10–13 nm (e.g., see Porter and Palade, 1957; Revel, 1962; Franzini-Armstrong and Porter, 1964a; Page, 1965; Peachey, 1965a; Walker and Schrodt, 1965; D. E. Kelly, 1969; Franzini-Armstrong, 1970a,b, for vertebrate skeletal muscle; Johnson and Sommer, 1967; Forssman and Girardier, 1970, for heart muscle; Hoyle, 1965, for Crustacea; D. S. Smith, 1966a; Hagopian and Spiro, 1967; D. S. Smith and Sacktor, 1970, for insects; Rosenbluth, 1969a, for annelids). The narrow space separating SR and T system, here referred to as the junctional gap, is crossed at periodical intervals by small dense spots (or feet), which join SR and T system membranes (Figs. 32, 33, 36, and 55). In vertebrates, the feet are made up to two components:

1. Scallops, which project from the SR membrane, reducing the junctional gap to less than 10 nm (Figs. 33 and 36). It is debated whether the SR scallops protrude far enough to form localized areas of tight junction with T system (cf. D. E. Kelly, 1969; Franzini-Armstrong, 1970a).

2. In most cases, the scallops do not reach the T system and the rest of the junctional gap is filled with small lumps of amorphous material. In invertebrates, the SR membrane is a flat surface without protrusions and the whole 12 nm gap is crossed by the amorphous material lumps (Fig. 55) (D. S. Smith, 1966a; Hagopian and Spiro, 1967; Rosenbluth, 1969a).

The connecting feet occur at regular intervals of approximately 30 nm and are separated by apparently empty spaces. In the triads of most vertebrate twitch fibers, the feet are disposed in parallel rows, which run along the flattened faces of the T system, usually two on either side (Franzini-Armstrong, 1970a,b). The feet are in register in the two rows, forming a tetragonal pattern. In one type of fiber, four, rather than two, parallel rows of feet in register have been described (D. E. Kelly and Cahill, 1969). In dyads, such as those of frog slow fibers (Page, 1965), where the junctional area covers small, round patches of the T system surface, the feet are also disposed in a tetragonal pattern, which covers most of the junction (Franzini-Armstrong, 1970b). The bidimensional disposition of the feet in invertebrates has not been explored.

Regarding the nature of the signal that causes ions release from the SR, two major possibilities exist: (i) Ionic current flows directly between T system and SR lumens at the moment of excitation, thus producing a change in polarization of the SR membrane, which in turn causes a calcium permeability increase. If this were true, the feet should be regarded as channels through which ion movement may occur. (2) the properties of the SR membrane may be altered by a trigger substance released from the T system walls (e.g., see Bianchi and Bolton, 1967). If this latter hypothesis were true, then the function of the feet would be purely mechanical, to maintain a small separation between the two membranes.

There is some indication that when exposed to the appropriate solution, the SR of "skinned" fibers is capable of a regenerative calcium release (Ford and Podolsky, 1970). Studies in this direction may very well solve the last obscure problem in e–c coupling.

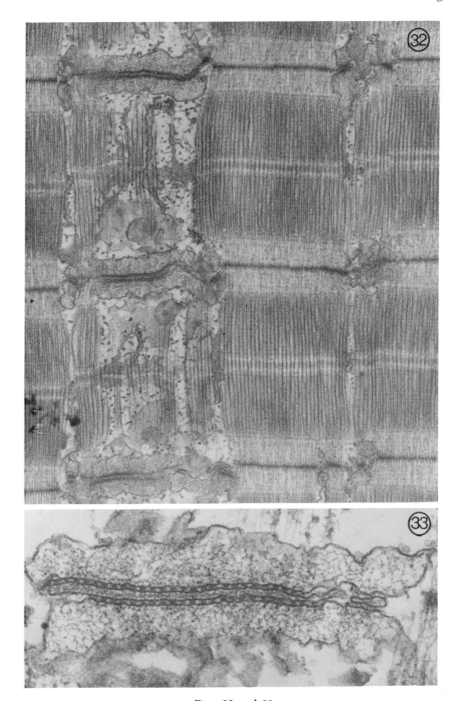

Figs. 32 and 33.

IV. Correlation between Physiology and Morphology
of Fiber Types in Vertebrates

Vertebrate muscles are capable of various types of performances: some are used in postural or tonic (relatively slow and long lasting) movements, others are used in phasic (faster, usually shorter in duration) movements. Still others are specialized for extremely fast motions or slow, holding contractions. These differences are due to the functional properties of the single fibers composing the muscles and are in turn reflected in the structural composition of the fiber. Most muscles have a mixed content of fibers, and in these a meaningful correlation between morphology and function can only be achieved when either single fibers or small identifiable and uniform bundles are examined (see Peachey and Huxley, 1962; Lännergren and Smith, 1966). Properties of different fiber types are to a great extent under the influence of the innervation (see Miledi *et al.*, 1968), and even completely differentiated fibers will at least partially alter some of their properties when innervated with a nerve previously going to a different type of fiber (Buller *et al.*, 1960; Eccles, 1963; Bücher and Pette, 1965; Romanul and Van der Menlen, 1966; Guth and Watson, 1967). When fibers have very different functional characteristics (e.g., twitch and slow fibers, see Sections 14,A,B,C), cross innervation experiments are not always successful in changing the fiber properties (Hník *et al.*, 1967; Close and Hoh, 1968). However, Elul *et al.* (1970) and Miledi *et al.* (1971) have been able to report that some characteristics of slow fibers, i.e., incapability of producing an action potential and the presence of prolonged contractures, are dependent on the type of nerve supply. Within a motor unit (i.e., the group of muscle fibers innervated by a single axon), the properties of the fibers are essentially the same (Edström and Kugelberg, 1968), although it cannot be excluded that even there some minor variations may occur (R. S. Smith and Länngren, 1968).

Until recently, it was thought that all fibers could be classified into a few clearly distinct categories, each endowed with a set of electrical, functional, and morphological properties. However, in recent years it has become evident that the distinctions between the major fiber types

Fig. 32. Frog twitch fiber. At left an SR layer is included in the plane of section. In the three triads thus cut tangentially, a small gap separates SR and T system membranes. ($\times 30{,}000$.)

Fig. 33. Detail of a triad from the same muscle as Fig. 32. The SR feet periodically crossing the junctional gap are clearly visible. The SR facing the T system is slightly scalloped. ($\times 100{,}000$.) Reprinted from Franzini-Armstrong (1970a).

are blurred by the existence of fibers with either intermediate character-istics or a mixture of properties such that classification is difficult. In the following four sections, specific examples are described that illustrate to what extent morphology of fiber types may be correlated to their function.

A. Frog Slow Fibers

Slow fibers from some frog muscles are chosen here to be described as the slow fiber prototype, because most of their properties have been more thoroughly investigated than for other slow fibers. The major physiological parameters of slow fibers are listed below (Kuffler and Vaughan Williams, 1953). For comparison, see the corresponding list for twitch fibers (section IV,C).

1. The sarcolemma is not capable of propagated action potential, but it responds with a graded depolarization to both direct and indirect (via the nerve) excitation.

2. The contraction is also graded (i.e., stronger the larger the de-polarization) and slow (or the order of 1 second). The contracture induced by potassium depolarization develops slowly and is maintained over long period of time.

3. The whole surface of the fiber is uniformly sensitive to acetyl-choline (ACh), and a slow contraction follows application of the drug to the fiber surface.

Morphologically, slow fibers are characterized by multiple nerve end-ings of the *en grappe* type (Hess, 1960), i.e., each junctional area is small and not raised from the fiber surface (see Chapter 8). Post junc-tional folds are fewer and less deep than in twitch fibers (Page, 1965).

The fibrils are large, irregular in shape, and mostly confluent at the A band level (Peachey and Huxley, 1962). This produces a characteristic appearance in cross section (fig. 34), which is visible in the light micro-scope and is named Felderstruktur pattern, as suggested by German authors (see Krüger, 1962). The Z line is considerably wider than in twitch fibers, and in cross sections it has a fine, punctuated appearance, rather than the regular cross-hatched pattern (Page, 1965). In fixed mate-rial, the Z line and the edges of the A band are wavy and not well defined; this is likely to be due to irregular contractions of the sarcomere during fixation. The M line is absent. The structure of the fibril, as shown in Fig. 35, should be compared with that of a typical twitch fiber (Fig. 7), which displays a narrow Z line with a zigzag pattern, straight and sharp A band edges, and an evident M line. Histochemically

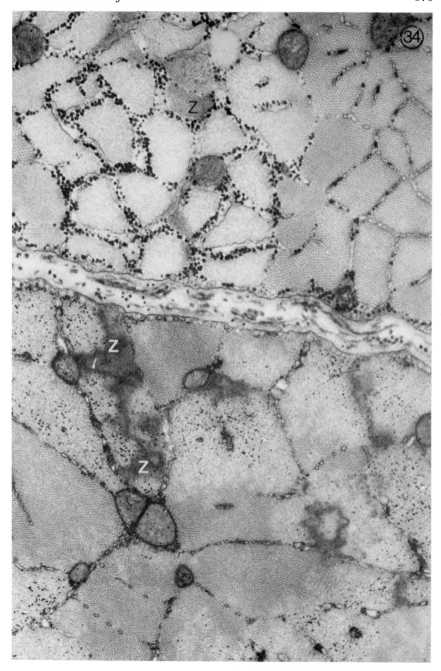

Fig. 34. Two fibers from a frog muscle. (*Top*) Fibrillen structure, twitch-type fiber. (*Bottom*) Felder structure, slow-type fiber. Notice difference in fibrils size and in the structure of the Z line (Z). (×20,000.)

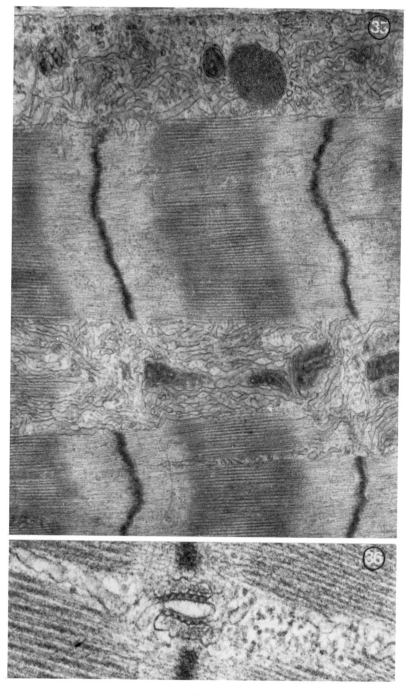

Figs. 35 and 36.

demonstrated myosin ATPase activity is much lower in slow than in twitch fibers (W. K. Engel and Irwin, 1967).

Elements of the T system in slow fibers do not enter the fiber at regular places relative to the cross striation (Page, 1965). Hence the results of the local stimulation experiments of Peachey and Huxley (1960) indicating that in slow fibers local contractions can be obtained when the stimulating electrode is placed at random on the fiber surface. From the large size of the fibrils and the random orientation of the T system (see Section II,D), it is calculated that the T system surface area is only approximately equal to the plasmalemmal area. For comparison, the T system of an average twitch frog fiber has a surface area seven times larger.

Large fibril diameter also results in an SR-to-fiber volume ratio considerably lower than in twitch fibers, although, where present, the SR forms a rich network (Fig. 35). Probably most significant is the fact that the dyadic and triadic junctions (Fig. 36) (see Section III,C) are few, irregularly scattered in the fiber, and cover only a minor portion of the T system surface (Page, 1965). As in twitch fibers, calcium is accumulated in the SR, probably in the lateral sacs of the triad (Costantin *et al.*, 1967).

Several factors probably jointly contribute to produce the slow activity cycle of these fibers: (1) the sites of calcium release, presumably the lateral sacs of the triad, are few; (2) once released, calcium has to travel relatively large distances to produce full activation of the fibrils; (3) as indicated by the low ATPase activity, contraction itself may be intrinsically slower; indeed, when calcium ions are directly applied to the fibrils of a "skinned" slow fiber, the response is slow (Costantin *et al.*, 1967); (4) finally, calcium uptake during relaxation is also slower, due to the relatively small SR volume. The relative importance of the above factors is still a matter of speculation. The significance of the structure of the Z line and of the lack of M line are also not known.

B. Other Slow Fiber Systems

Fibers that share with frog slow fibers at least some functional and morphological characteristics have been described in a variety of

Fig. 35. Frog slow-type fiber in longitudinal section. Notice thick, wavy Z line, lack of M line, uneven A band edge. The SR is mostly in the form of longitudinal sacs and it is not interrupted in segments. (×21,000.) Reprinted with permission from Page (1965).

Fig. 36. Detail of a longitudinally oriented small triadic junction in a frog slow fiber. The SR is scalloped and the disposition of the junctional feet is analogous to that of twitch fibers (cf. Fig. 33). (×70,000.)

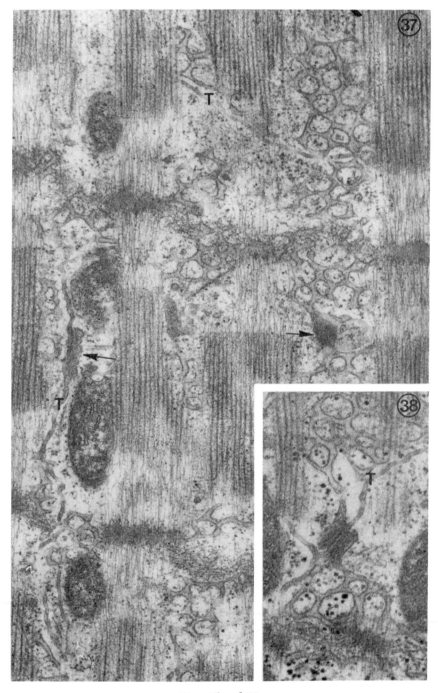

Figs. 37 and 38.

vertebrates. These range from fibers that are obviously to be classified as slow, to others whose location either in the slow or the twitch category is a matter of choice.

On the basis of their response to ACh, slow contraction cycle and multiple, *en grappe* innervation, slow fibers have been identified in the garter snake (Hess, 1963, 1965). These also have in common with frog slow fibers the *Felderstruktur* disposition of the fibrils and a wide Z line (Hess, 1963), but differ by having an M line (Hess, 1965). The T system, initially thought to be totally absent (Hess, 1965), is irregularly distributed and forms oddly oriented triadic and even pentadic junctions with the SR (Hoyle *et al.*, 1966).

The ability to produce a propagated action potential, at least under experimental conditions, has not excluded some fibers from being classified as slow. Such is the case of some chicken muscles, where two types of fibers, one capable of a twitch and the other giving a slow response are conveniently located in two separate muscles; the posterior and anterior latissimus dorsi, respectively. The fibers from the latter muscle, besides contracting slowly, have overall sensitivity to ACh (Ginsborg, 1960; Fedde, 1969) and, analogously to other slow fibers, they have multiple, *en grappe* innervation and *Felderstruktur* fibrils (Hess, 1961). The Z line is broad and wavy, but the M line is evident (Fig. 37). The irregular course of the T system (Page and Slater, 1965; Hess, 1967) and particularly the paucity and often longitudinal orientation of dyads and triads (Fig. 37 and 38) are strikingly similar to that of frog slow fibers (Page, 1969). The SR is a curious lacy network, perforated by wide fenestrae and longitudinally continuous. Calcium accumulation occurs within the elements of the SR (Page, 1969). Interestingly, embryonic chick skeletal fibers, differentiated *in vitro*, develop an SR identical in structure to that of the slow fibers (Ezerman and Ishikawa, 1967; Shimada *et al.*, 1967).

Basically, the structure of chicken slow fibers is the same as that of the frog slow fiber and, correspondingly, the structure of twitch fibers from the posterior latissimus dorsi is that of typical twitch fibers, with triad located at the A–I junction (Fig. 22) (Page, 1969).

Extraocular and ear muscles are the only muscles in mammals where slow fibers have been demonstrated (Pilar and Hess, 1966; Fernand

Fig. 37. Slow fiber from the chicken anterior latissimus dorsi. Notice wide Z line and lack of M line. The SR is a continuous network, perforated by large fenestrae. The T system (T) is oblique and longitudinal, so are dyads and triads (arrows). (×30,000.) Reprinted with permission from Page (1969).

Fig. 38. Obliquely oriented triad in the same muscle. T = T system. (×approx. 40,000). Reprinted with permission from Page (1969).

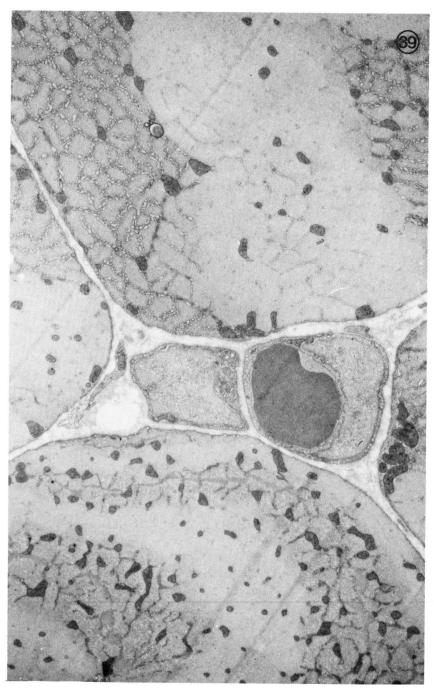

Fig. 39.

and Hess, 1969; Hess, 1970). Taking clever advantage of a difference in the pattern of innervation, Hess and Pilar (1963) first showed that smaller diameter fibers in one of the cat's extraocular muscles have an overall sensitivity to ACh, of the type encountered in slow fibers and that they respond with a slow, sustained contraction to application of the drug. These slow contracting fibers have multiple, *en grappe* innervation (Pilar and Hess, 1966), a wide Z line, and a barely visible M line. The fibrils are separated by a single layer of reticulum at the level of the I bands, but are almost completely fused at the level of the A band, the SR there being very scarce (Fig. 39). This results in a Felderstruktur cross section, similar to that of frog slow fibers. One distinctive feature of eye muscle slow fibers is that the T system and the triads are predominantly oriented crosswise and are at the A–I junction level (Peachey, 1965c). The electrical behavior of slow contracting, multi-innervated eye muscle fibers is still controversial, since it has been alternately reported that (Hess and Pilar, 1963; Bach-y-Rita and Ito, 1966) they are not and that they are capable of propagated action potential. The problem is not easily solved because of the small diameter of the fibers being examined and the difficulty in dissecting them away from the muscle. In addition, histochemical and electron microscopic studies of eye muscles from a variety of animals have shown that they possess at least three (Cheng and Breinin, 1966; Peachey, 1965c) or possibly more (Miller, 1967; Mayr, 1971) types of fibers, distinguishable on the basis of their mitochondria and SR content, size of the fibrils, and myosin ATPase activity (see also Matyushkin, 1963). It is likely that there are two types of multiply innervated slow contracting fibers, one is a real slow fiber, sensitive to ACh and with graded depolarization and contraction, whereas the other, also contracts slowly, but it is capable of a propagated action potential. The latter perhaps should be classified, with other fibers, in the slow twitch category (see Peachey, 1968, 1972, for reviews on the subject).

Slow fiber systems in fishes may be really intermediate between slow and twitch fibers in the sense that they have properties in common with both types. In the body muscles of the snake fish, for example, there are fibers that do not seem capable of a propagated action potential and that contract slowly (Takeuchi, 1969). Unfortunately, no information is available on ACh sensitivity and response to potassium depolariza-

---

Fig. 39. Cross section through the superior oblique (extraocular) muscle of the cat. At top is a fast twitch fiber; fibrils are small and separated by double SR layers at I band level. At bottom is a slow fiber; the fibrils are almost completely fused at the A band level and separated by a single SR layer at the I band. Mitochondria are more numerous in the slow fibers. (×10,500.)

Fig. 40.

1965; McPhedron *et al.,* 1965; Buchtal and Schmalbruch, 1969) as well as in amphibia (Lännergren and Smith, 1966; Smith and Lännergren, 1968) and in fishes (Hidaka and Toida, 1969). Even in animals as far away on the zoological scale as the crustacea, a large mitochondria content accompanies low fatiguability (Hoyle, 1968). Where red fibers are encountered, it is permitted to speculate that they may be used for some sustained motion (e.g., see Ashurst, 1969).

In mammalian legs, the motor units of phasic muscles contract about twice as fast as those of tonic muscles (Close, 1965; Buller *et al.,* 1960; Schmalbruch, 1967). Triads are at the A–I junction in all these fibers, but the A band level SR is slightly less abundant in tonic muscles (Schiaffino *et al.,* 1970). In addition, tonic motor units have a larger number of red fibers, in which mitochondria occupy a larger proportion of the interfibrillar spaces (Pellegrino and Franzini, 1963; Bübenzer, 1966; Tice and Engel, 1967). However, the difference in SR content is not very large, and it is difficult to assess its significance in relation to differences in the speed of the activity cycle. Indeed, there is sufficient evidence that this is mostly determined by intrinsic differences in the contractile material; the myosin ATPase activity of slow twitch muscles is lower than that of fast twitch muscles (Bárány *et al.,* 1965, 1967). When ATPase determination is done histochemically, more active fibers, mostly white fibers, appear dark, whereas less active fibers appear lighter. Hence, some white fibers have been classified as dark, whereas some red fibers are called light (W. K. Engel *et al.,* 1966).

Red, mitochondria-rich fibers are not necessarily slower than white fibers. Notable is the case of rat extensor digitorum longus, a physiologically homogeneous fast twitch muscle, which is composed of red as well as white fibers (Schiaffino *et al.,* 1970). The red fibers, the authors comment, are likely to be less easily fatiguable, although as fast in contraction as the white fibers. Also, in rat soleus, which on the whole is a red tinted tonic muscle, the red fibers contraction is faster (18 msec) than that of intermediate fibers (38 msec) and their myosin ATPase activity is correspondingly higher (Edgerton and Simpson, 1969). Analogously, in some rabbit laryngeal muscles, fast-contracting fibers (65 msec) have a high mitochondrial content as well as high myosin ATPase activity (Hall-Craggs, 1968). The same is true of the bat crycothyroid muscle (following section).

Fig. 42. Three fibers from the rat diaphragm. At left, a red fiber, rich in mitochondria between the fibrils and under the sarcolemma. At lower left, a presumably intermediate fiber, with less mitochondria, particularly under the sarcolemma. At right, a white fiber, with few mitochondria. All fibers are Fibrillen structure and twitch type. L = lipid droplets. (×7500.)

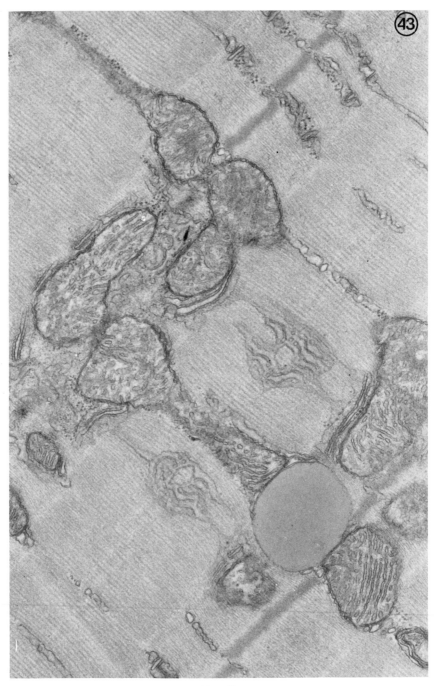

Fig. 43.

A demonstrative example of how mitochondria content is coupled to the requirements of continuous activity but not necessarily to the speed of contraction of mammalian fibers is illustrated by the work of Gauthier and Padykula (1966) on mammalian diaphragm. It was found that smaller, faster breathing animals have a red diaphragm, whereas, larger, slower-breathing animals have an almost completely white diaphragm. This is not unexpected when one considers that red fibers are usually less easily fatiguable, but it is certainly not in accord with the preconceived idea that red fibers are necessarily slower. Judgment on the properties of a type of fiber should be based only on a complete histochemical profile accompanied by at least some functional parameters.

D. Some Very Fast Acting Fibers

Few twitch fibers are known to be capable of extremely fast contractions, the contraction times being only a few milliseconds. Two of these form muscles that produce vibrations in sound-emitting organs–the bat crycothyroid muscle (Revel, 1962) (Fig. 21) and the toadfish swimbladder muscle (Fawcett and Revel, 1961) (Figs. 44 and 45). Fast twitch fibers are also present in the extraocular muscles of mammals (Figs. 46 and 47) and of fishes (Kilarski, 1967a,b; Reger, 1961; Kilarski and Bigai, 1969; Goldstein, 1969b). All these fibers have one striking feature in common; the elements of the SR are very abundant, forming a double layer either all along the sarcomere (Fig. 44) or at least opposite the I band (Figs. 39 and 46). The fibrils have a small diameter (Fig. 39), and where elongated, as in the case of fish muscles, they are very thin (Fig. 44). Thus, the distance calcium ions must travel to produce full activation is very short, and SR-to-fibril volume ratio is high. The SR shape is also strikingly convoluted (Figs. 45–47), so that SR surface is large. All the above factors may contribute to produce a fast contraction–relaxation cycle. Unfortunately, no information is available on the intrinsic properties of the fibrils, such as myosin ATPase activity.

The only discordant note to the above generalizations is brought by the observation that the swimbladder muscle in the burbot (*Lota lota*),

---

Fig. 43. Red fiber from the rat soleus (a tonic leg muscle). Mitochondria have branches encircling the fibrils at the I band and longitudinal branches. A large lipid droplet is closely associated with them. T system is at A–I junction level, SR is well developed. (×30,000.) Reprinted from Pellegrino and Franzini-Armstrong (1969).

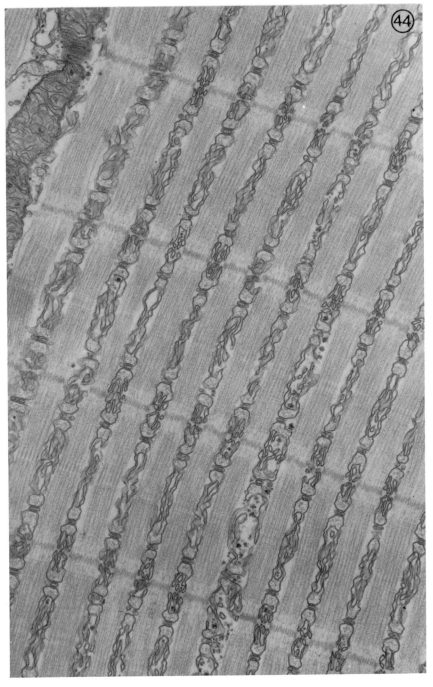

Fig. 44.

a fish related to the toadfish but which does not emit sounds, also has a highly developed SR (Kilarski, 1964). The author comments, however, that possibly in far-off times the burbot was capable of producing sound and it has now "lost this faculty without losing the organ."

Mitochondria vary from few, centrally placed in the toadfish swimbladder, to numerous, more evenly distributed in the bat crycothyroid (Fig. 21) and the extraocular fast-twitch fibers (Fig. 39). This reflects the extent to which the fibers are used either for discontinuous bursts of activity or for more sustained contractions.

V. Correlation between Physiology and Morphology of Fiber Types in Invertebrates

Although considerable structural similarities exist between fibers belonging to arthropods, it is convenient here to describe separately the structure of fibers from Insecta and Crustacea. This section also includes a very brief account of SR distribution in fibers from some lower invertebrates.

A. *Crustacea*

In crustacean muscle fibers, structural and functional variabilities are most strikingly represented (see Chapter 9, Volume I). Even within a single muscle, one sometimes encounters fibers that go from typical twitch to slow fibers through a continuous spectrum of intermediates (Dorai-Ray and Cohen, 1964; Atwood, 1967; Hoyle, 1968). Also, the concept of motor unit (see Section IV,C) does not apply, since fibers varying in their properties are sometimes innervated by a common axon. In these more primitive animals, control of movement is peripheral.

Apart from some highly specialized fibers, to be described in the following section, even large variations in function are associated with modulations of one basic type of structure, which is described below. Fibers conforming to the following general description have been encountered, for example, in crayfish leg muscles (Brandt *et al.*, 1965), in the fast and slow abdominal muscles of the same animal (Jahromi

---

Fig. 44. Very fast fiber from the toadfish swimbladder. The flat thin fibrils are cut across their short diameter. The SR occupies almost as much volume as the contractile material. Triads are at the A–I junction, but the SR is also interrupted at the Z line level. (×25,000.)

Fig. 45.

and Atwood, 1967), in the leg muscles of a variety of crabs (Peachey, 1967; Selverston, 1967; Reger, 1967b; Franzini-Armstrong, 1970c) and of the lobster (Franzini-Armstrong, 1967), in the giant barnacle scutal depressor (Hoyle *et al.*, 1965; Selverston, 1967), and finally, in the eyestalk muscle of a crab (Hoyle and McNeill, 1968). Most of these fibers are very large, even compared with the largest from vertebrates (1–4 mm versus 0.1–0.2 mm). Such large size is evidently inefficient for the problems it creates in the diffusion of metabolites and in the process of spreading of excitation along the T system tubules (cf. Selverston, 1967). Thus, the fiber is subdivided into smaller portions by numerous large clefts or folds of the sarcolemma which penetrate deep into the fiber (Fig. 49) and run longitudinally for the length of several sarcomeres. The basement membrane participates with the plasmalemma in the formation of the clefts. Two distinct categories of tubules, having their walls continuous with the plasmalemma, invaginate into the fiber from its periphery and from the clefts (Peachey, 1967). Both are flattened, and run transversely into the fiber. The cross sectional long axis of both tubules is oriented longitudinally, rather than transversely as is the case for vertebrates. Hence the possibility of following the tubules for a long distance in cross sections of the fiber (Figs. 49 and 50).

One system of tubules pervades the fiber at the level of the Z line; it has a dense content continuous with the basement membrane, and it presents thickenings on the sarcoplasmic side, with which it is thought to be attached to the Z line material (Fig. 48). It is conceivable that these tubules serve to anchor the fibrils so that excessive changes in position of the internal structures relative to the sarcolemma are not possible during contraction.

Tubules of the second group are located opposite the A band near both edges, and they are apparently empty (Fig. 49). Upon coming in proximity to elements of the SR, these tubules form dyads with enlarged SR cisternae, which are located on alternate sides of the tubules between them and the fibrils (Figs. 49, 50). Each junctional spot is round or oval (Fig. 51), similar to those in chicken and frog slow fibers. In the stomach muscle of the lobster, the second set of tubules is unusually located at the center of the sorcomere (Atwood, 1971).

Fig. 45. Grazing section through an SR layer, same fiber as Fig. 44. Notice abundance of SR tubules, giving SR a large surface area. The lateral sacs of the triad, distinguishable due to their dense content, occupy a large volume and even extend longitudinally (arrow). A by-pass of the triad is indicated by white arrow. (×25,000.) From "The Sarcoplasmic Reticulum," K. R. Porter and C. Franzini-Armstrong. Copyright 1965 by *Scientific American Inc.* All rights reserved.

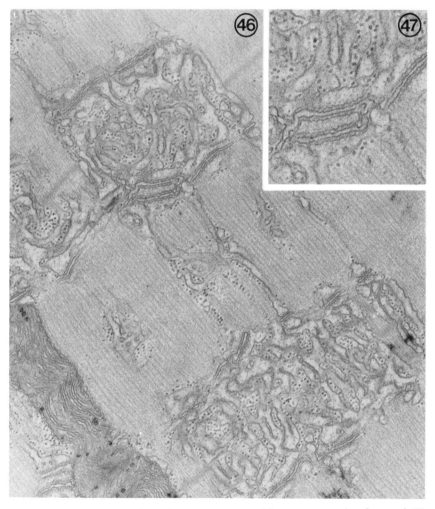

Fig. 46. A fast-twitch fiber from cat superior oblique. Notice abundance of SR, particularly at I bands level. Triads at A–I junction. (×25,000.)

Fig. 47. Detail of a pentad produced by a bifurcation of T system, enlarged from Fig. 46. (×40,000.)

Elements of the SR envelope the fibrils with either a single or a double layer (Figs. 50, 51). Their distinctive characteristics are the longitudinal continuity and the large number of fenestrations both at the A band and at the I band levels (Fig. 51). The fenestrae are slightly larger and more irregular than in vertebrates.

A local depolarization applied to the surface of crab fibers, in experiments similar to those described in Section III,A for vertebrates, produces a localized contraction of the peripheral fibrils only when applied at the level of the edges of the A band, in correspondence with the level of entry into the fiber of the second set of tubules (Peachey and Huxley, 1964; Peachey, 1967). The obvious conclusion is that only those tubules that form dyadic contact with the SR are capable of producing release of an activator from the SR. The morphological pathway followed by e–c coupling in crab fibers is thus equivalent to that of vertebrate fibers. It must be noted however, that occasionally longitudinal extensions arise from the Z line level tubules and run longitudinally to the level of the A band edge, where they form dyadic junctions from the SR. Why then have local stimulation experiments failed to obtain a response when depolarization was applied at the Z line? The answer is probably to be found either in the fact that these secondary extensions are very rare and thus did not happen to be detected, or that they only arise deeply into the fiber and thus were not affected by the relatively small depolarization applied in the experiments. Less probable is the fact that fibers explored in the local stimulation experiments did not happen to have such secondary extensions.

The dyadic junctions in crustacean fibers are, as mentioned earlier, morphologically equivalent to the triads of vertebrates, and there is reason to believe that they serve the same general function of coupling events in the T system and SR. Indeed, it is known that SR from crustacean fibers is capable of accumulating calcium ions (Portzehl *et al.*, 1964; Van der Kloot, 1966) and that contraction of the fibrils is produced by the application of calcium (Gillis, 1969).

B. Variations in the Structure of Crustacean Fibers

For purposes of description crustacean, fibers may be divided into two categories: (1) fibers that although varying considerably in function have an overall structure similar to that described in the previous section; (2) fibers, mostly very fast, in which SR and/or T system distribution is unusual.

In fibers of group one, differences in contraction speed are accompanied by modulations in the structure of the fibrils. Variations occur in four major components. (1) The number of thin filaments surrounding each thick filament is larger the slower the fiber. The thin to thick filaments ratio is 3:1 in the fastest fibers and 6:1 in the slowest. All possible intermediate ratios exist (Fahrenbach, 1967; Reger, 1967b; Hoyle and McNeill, 1968; Franzini-Armstrong, 1970c). (2) The length

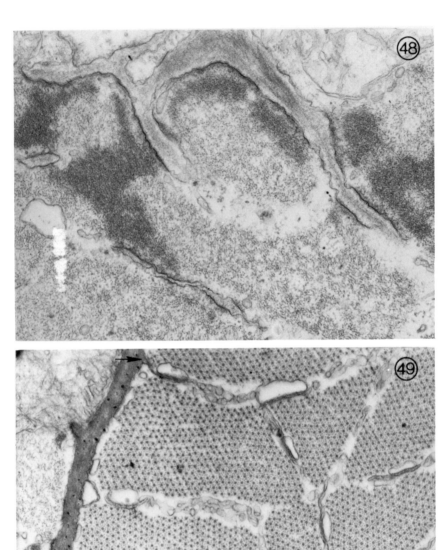

Figs. 48 and 49.

of the A band is also related to the speed of contraction and continuously variable (Dorai-Ray and Cohen, 1964). Lengths anywhere between 2 and 7 μ exist. The larger the A band, the farther away are the dyads from one another. The length of the thin filaments is proportional to that of the thick filaments. (3) The Z line is wider in slower fibers (Franzini-Armstrong, 1970c). (4) The fibrils are large and partially fused in the slow fibers (Fahrenbach, 1967; Cohen and Hess, 1967), thus producing a Felderstruktur pattern very similar to that of vertebrate slow fibers. However, smaller variations in contraction speed (and A band length) are not accompanied by variations in the size of the fibrils (Selverston, 1967; Franzini-Armstrong, 1970c). The SR is more abundant in the faster fibers (Cohen and Hess, 1967; Hoyle and McNeill, 1968).

Variations in SR content, size of the fibrils, and Z line thickness are very much similar to those described for vertebrate fibers. However, the variability in length of thin and thick filaments and in number of thin filaments is unique to arthropod fibers.

To the second group belong fibers in which either SR or T system or both are more abundantly distributed. In all cases, a fast speed of contraction is either demonstrated or presumably present. In the small copepod *Macrocyclops albidus*, for example, the muscle fibers responsible for the fast swimming motions are able to contract in no more than 40 msec and probably less, a relatively fast speed for crustacean fibers. The small diameter fibers (20–25 μ) are pervaded by an extensive network of SR and T system elements, as demonstrated by the artistic micrographs of Fahrenbach (1963). The T system tubules not only surround the fibrils at the Z line level, but also penetrate within them, between small bundles of filaments, and run longitudinally. The SR is also within the fibrils, so that in most cases about 0.2 μ separate the contractile material from the nearest SR tubule. Dyads occur at several not precisely located levels of the sarcomere.

In another small crustacean, an ostracod, the small-diameter fibers are filled by a mass of contractile material not divided into fibrils (Fahrenbach, 1964). SR and T system elements, however, pervade this mass with such abundance that no more than 0.2 μ distance exist between them and any point of the contractile material. Although no precise information exists on the speed of contraction of these muscles, it is

---

Fig. 48. Cross section at Z line–I band level of a walking-leg muscle fiber from the lobster. Z line level invaginations of plasmalemma and basement membrane pervade the fiber. The sarcoplasmic surface of the membrane presents thickenings. (×30,000.)

Fig. 49. Walking-leg muscle fiber from a crab. At left, one of the clefts that pervade the fiber. From it the T system (arrows) penetrates into the fibers. (×50,000.)

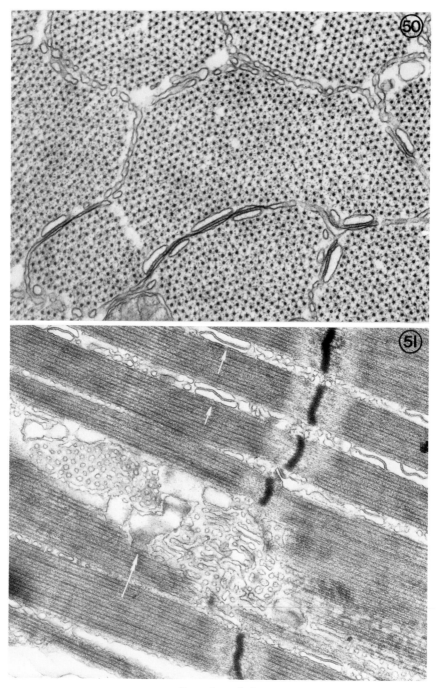

Figs. 50 and 51.

likely that they are fast, as indicated also by the 3:1 ratio of thin and thick filaments.

Interestingly, the fastest fibers so far described in crustacea are, analogously to those of vertebrates, employed in sound production. In the lobster, the second antenna vibrates when muscles having a fusion frequency of more than 125 per second (Mendelson, 1969) contract. In these fibers, the SR fills the fiber, whereas small diameter, short A band fibrils are scattered in small groups of two to four (Rosenbluth, 1969b) (Fig. 52). As calculated (Van der Kloot, 1969), the large quantity of SR is capable of reducing the calcium concentration fast enough to allow for the quick relaxation necessary to attain such high fusion frequency. It is noticeable that the general size and conformation of the whole fiber and the location of T system elements and dyads reflect those described for not particularly fast fibers in the previous section. Apparently, this structural conformation is capable of e–c coupling times fast enough even for these exceptional fibers.

C. Insecta

The structure of insect muscle fibers has many points in common with that of Crustacea. Namely, the structure of the fibril is very similar, having a 3:1 ratio of thin to thick filaments in fast-contracting fibers (e.g., flight muscles), a 6:1 ratio in the slowest fibers (e.g., visceral and body segment muscles), and intermediate values of the ratio in other fibers (Auber, 1966, 1967a,b; Reger, 1967a; Reger and Cooper, 1967; D. S. Smith, 1966c; D. S. Smith *et al.*, 1966; Toselli and Pepe 1968). Incidentally, the same is true for muscle fibers of scorpions (Auber, 1963a,b). The A band length is also variable, being longer in slow muscles. The fibrils are either almost perfectly round, as in some flight muscles, or flat, ribbonlike. In slow fibers, they are large and fused together.

As for Crustacea, there is one basic type of distribution of SR and T system elements to which most fibers conform with only slight variations. Such is that encountered in synchronous flight and leg muscles. The T system tubules are transversely oriented and flattened, with the

---

Fig. 50. Section through the A band of a crab leg muscle fiber. The T system surrounds the fibrils and forms dyads with enlarged SR sacs. The SR is often in double layer. (×110,000.)

Fig. 51. Fiber from a crab, showing continuous, fenestrated SR. Dyads are opposite the A band near its edges (arrows). Double arrow indicates a grazing view of dyadic junction. (×25,000.)

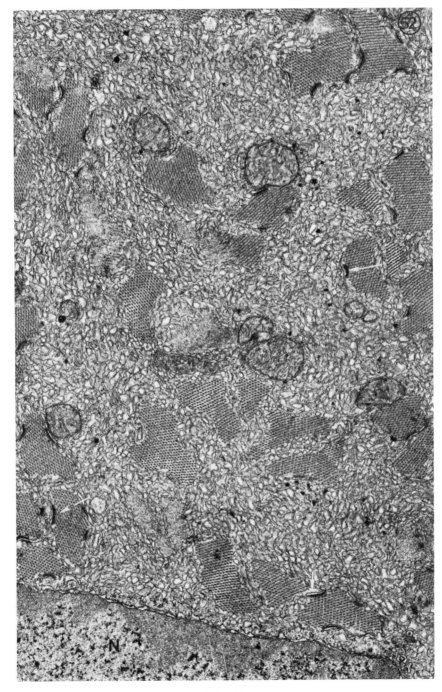

Fig. 52.

flat faces oriented longitudinally, so that they are best followed in sections transverse to the fibers, where the peripheral openings are easily detectable (Fig. 53) (D. S. Smith, 1961b, 1965, 1966a; Hagopian and Spiro, 1967). It is not surprising that ferritin penetrates into the T system of these fibers as easily as it does in those of vertebrates (D. S. Smith, 1965). The T system elements are confluent (Fig. 53) and probably form a continuous network across the fiber (D. S. Smith, 1966a; Pasquali-Ronchetti, 1969). Two T system networks cross the fiber at every sarcomere and are located opposite the A band, approximately midway between the Z line and the center of the sarcomere (D. S. Smith, 1966a) (Fig. 54). In some flight muscles where measured, the T system relative volume is considerably larger than in vertebrate twitch fibers—2% of the fiber volume (D. S. Smith, 1966b). The SR is in the form of either a single or a double fenestrated sheet, and it is not interrupted at the level of the longitudinally oriented dyadic junctions with the T system. Discrete dense spots protruding from the SR membrane cross the dyadic junction gap (Fig. 55) (D. S. Smith, 1966a; D. S. Smith and Sacktor, 1970; Hagopian and Spiro, 1967), thus making the structure of the junction very similar to that of vertebrates (cf. Figs. 55 and 33) (see Section III,C).

The mitochondria are particularly abundant in the flight muscle, where they occupy approximately as much volume as the contractile material (Figs. 53, 54, and 56), and are often very regularly arranged relative to the fibrils (Fig. 54). This is not surprising in view of the tremendous work load that these muscles are capable of producing during flight and of the known high metabolic rates accompanying their activity (D. S. Smith, 1961a; D. S. Smith and Sacktor, 1970). Except in Odonata, folds of the plasma membrane invade the fiber carrying to within short distance of all mitochondria the oxygen-carrying tracheolae. These folds are probably homologous to those of Crustacea (D. S. Smith, 1966a).

Some interesting exceptions to the above scheme have been reported. In the basalar muscle of the wing of a lepidopteran, for example, the SR penetrates within the fibril at the level of the M line, there forming a fine meshwork between the filaments (Reger and Cooper, 1967). Pasquali-Ronchetti (1969) describes an unusual arrangement of the T system in the femoral muscle of the household fly; longitudinally oriented outpocketings project out of the T tubules at regular intervals. It is with these outpocketings, rather than with the rest of the T system,

---

Fig. 52. A very fast fiber from the lobster in cross section. The fibrils, in small groups, are separated by large spaces filled by elements of the SR. The T system and dyads (arrows) are essentially the same as in other lobster muscles (and as in Fig. 50). (×15,000.) Reprinted with permission from Rosenbluth (1969b).

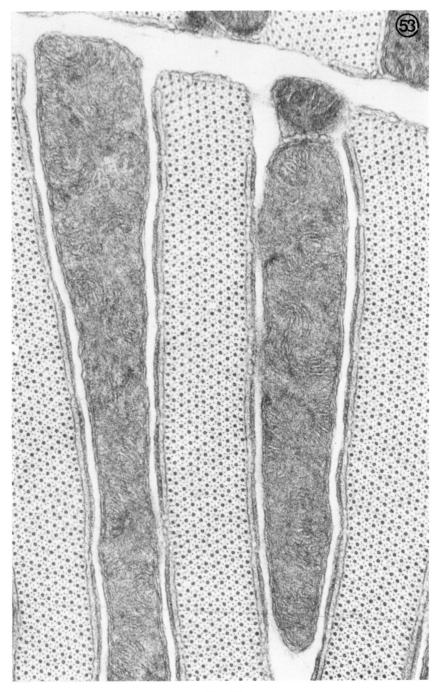

Fig. 53.

that the SR forms dyadic junctions. The structure is most striking when caught in a section tangential to the fibril, where one can visualize several of the regularly repeating round T system outpocketings and dyadic contacts. In these fibers, the SR is not continuous, but segmented, the interruptions occurring regularly at each sarcomere.

In visceral muscles, which probably contract very slowly, the T system is also oriented transversely, and it forms dyadic contact with the SR (Franzini-Armstrong, 1964). However, the T tubules are not regularly distributed, and due to the large confluent fibrils, the SR elements are scarce (D. S. Smith *et al.*, 1966).

In order to attain the high wing beat frequencies necessary for their flight, insects of the orders Diptera, Hymenoptera, Coleoptera, and most Hemiptera have developed muscle fibers that have unique properties (see Pringle, 1965, for a review). For details of the physiology of these fibers, the reader is referred to Chapter 10, Volume I. It is sufficient here to say that the fibers respond with a burst of rapid contractions to each nerve impulse and that the possibility of producing these rapid small changes in length are thought to reside in the contractile material, whereas the process of e–c coupling and of calcium removal from the fibrils are relatively slow. Since nerve impulses and contraction proceed at different frequencies, these fibers are called asynchronous (while the rest of the insect fibers are synchronous, i.e., they behave more conventionally).

Asynchronous fibers are fairly large (100–200 μ), but adequate supply of oxygen to the large, numerous mitochondria is provided by the system of tracheolae-carrying clefts which penetrate into the fiber at 20–30 μ intervals (D. S. Smith, 1961a; D. S. Smith and Sacktor, 1970) (Fig. 57). The mitochondria separate the fibrils from one another, and this may be responsible for the fact that these fibers separate easily into fibrils. Hence the name fibrillar muscle used in early literature (D. S. Smith, 1961a). The fibrils are regularly round, with perfectly arranged thin filaments in a $3:1$ ratio. As opposed to other insect fibers there is an evident M line, which is made of some amorphous material, apparently holding the thick filaments together (Auber, 1967b,c). The I band is extremely short, and it does not vary much during the contractions, which change the fiber length only of a few percent. The structure of the Z line is also distinctive of these fibers (Auber and Couteaux, 1963).

Fig. 53. Cross section of a synchronous flight muscle fiber from a dragonfly (Insecta) at the level of the T system, showing T system openings. Flattened SR cisternae are interposed between T system and fibril, forming dyadic junctions (cf. Fig. 54). Notice confluence of T system tubules at lower right. Ratio of thin to thick filaments is $3:1$ ($\times 40,000$.) Courtesy of D. S. Smith.

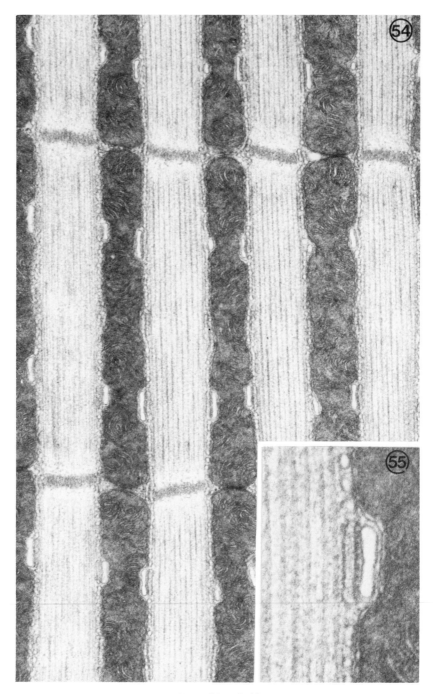

Figs. 54 and 55.

The T tubules, similar in structure to those of synchronous fibers, run in some muscles regularly either at the M line (Fig. 56) or midway between H and Z lines (Shafiq, 1964; D. S. Smith, 1966). In most fibers, however, irregular orientation of the T system prevails (Fig. 57). Continuity with the sarcolemma has been demonstrated either directly (D. S. Smith, 1961a; Ashurst, 1967; D. S. Smith and Sacktor, 1970) or indirectly by infiltration with ferritin (Ashurst, 1967; D. S. Smith and Sacktor, 1970). The SR is singularly reduced; only few, scattered, and mostly isolated vesicles without continuity with one another represent the SR (Fig. 57). Most of these vesicles are close to the T system, with which they form dyadic associations. The portion of SR equivalent to longitudinal tubules and fenestrated collar of vertebrates is virtually absent. Exceptionally, the SR may form a more continuous network, at least at some level of the sarcomere (Ashurst, 1967).

In one of the asynchronous flight muscles (from Phormia, a dipteran) it is calculated (D. S. Smith and Sacktor, 1970) that the T system occupies approximately 0.5–1% of the fiber volume, a value slightly higher than the correspondent for a vertebrate twitch fiber (see Section II,C). The SR, on the other hand, is considerably reduced; the volume of the SR participating in dyads is 0.2%, i.e., much less than the corresponding compartment in frog twitch fibers, which is approximately 5% of the fiber volume. It is interesting to note that if SR of these muscles behaves, analogously to that of vertebrates, as a calcium sink, then the lateral sacs of the dyads must have calcium-binding ability. As suggested by Pringle (1965), the paucity of SR may very well be responsible for the slow onset of activity. Also, oscillations occur in the presence of calcium, so that the fibers do not need an abundant "relaxing factor."

D. Oblique Striated Fibers

Fibers from different groups of lower invertebrates (mollusks and worms) are grouped in this section because they share one major structural characteristic—their fibrils are obliquely rather than transversely striated. The reader is referred to Chapter 8, Volume I for a complete description of the structure and function of oblique striated muscles.

Fig. 54. Longitudinal section of same muscle as Fig. 53. Dyads are located midway between Z line and the center of the sarcomere. The T system is longitudinally flattened. Large mitochondria packed with cristae and small fibril diameter are typical of flight muscles. (×36,000.) Courtesy of D. S. Smith.

Fig. 55. Detail of a dyad from the same muscle as Fig. 54. Small densities periodically cross the junctional gap (cf. Fig. 33). (×110,000.) Reprinted with permission from D. S. Smith (1966a).

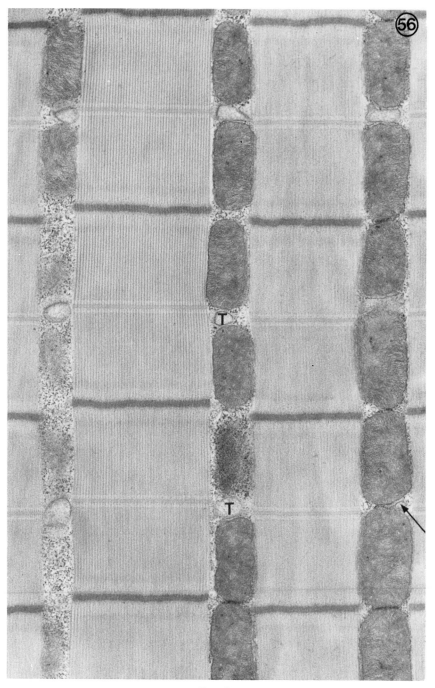

Fig. 56.

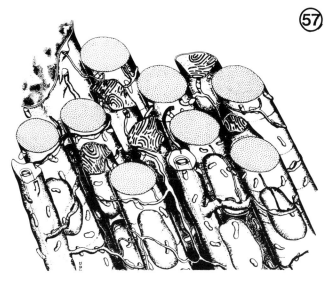

Fig. 57. Drawing illustrating distribution of membranes in fibrillar (asynchronous) flight muscle of the beetle. Two tracheolae pervade the fiber, within a fold of the plasma membrane. T system runs irregularly, SR is represented by small isolated vesicles. Nuclear envelope is analogous to SR. N = nucleus. Reprinted with permission from D. S. Smith (1961a).

Here it is only important to note that the striations, although oblique, are quite regular in most of these fibers and that there are sarcomeres formed by bands and lines essentially equivalent to those of cross-striated fibers (Rosenbluth, 1967). Most of the oblique striated fibers contract slowly, but variations in the contraction speed exist.

In some muscles, the contractile material fills the whole fiber, and the fibrils appear in cross sections as thin ribbons, cursing obliquely through the fiber. Such is the case for the worms *Glycera dibranchiata* (the bloodworm) (Rosenbluth, 1968), *Lumbricus terrestris* (the earthworm) (Lanzavecchia, 1968; Franzini-Armstrong, 1964) and for the oyster (a mollusk) (Hanson and Lowy, 1961). The few mitochondria and presumably the nuclei are located at the periphery of the fiber under the sarcolemma.

In other fibers contractile material occupies the periphery of the fiber, while the center is occupied by sarcoplasm containing mitochondria

Fig. 56. Asynchronous flight muscle of a wasp. Notice short I band and noticeable M line. The T system (T) is at the level of the M line. SR elements (arrows) are scarce and not continuous. Mitochondria are numerous, fibrils round and small. (×17,000.) Reprinted with permission from D. S. Smith (1966a).

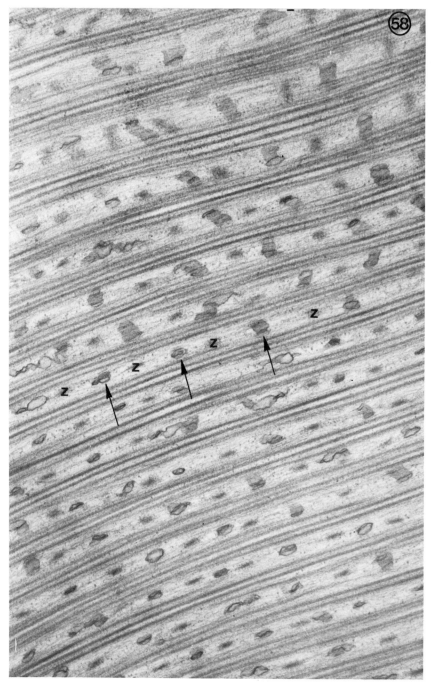

Fig. 58.

and nuclei. The fibrils are radially arranged. Fibers of this type have been described in the leech (Pucci and Afzelius, 1962), in *Ascaris lumbricoides* (Reger, 1964; Rosenbluth, 1965a, 1967), and in a snail (*Helix aspersa*) (Rogers, 1969).

As in the cross striated fibers, the fibrils are separated by elements of the sarcoplasmic reticulum which in most cases take a precise disposition relative to the bands of the fibrils (Fig. 58) (see Chapter 8, Volume I). Also, the SR is more abundant and regular in fibers which, judging from the movements of the animal, are faster acting (Rosenbluth, 1968). The SR has two components—wide sacs, usually in subsarcolemmal position, and elongated tubules, which run obliquely across the fiber, sometimes precisely following the I bands of the sarcomere (Hanson and Lowy, 1961; Pucci and Afzelius, 1962; Franzini-Armstrong, 1964; Rosenbluth, 1968). In *Glycera* musculature (Rosenbluth, 1968), the SR forms a network that extends over the I band, but in this as in all other oblique muscles, no longitudinal SR tubules exist opposite the A band. At the periphery of the fiber, the oblique tubules open into the peripheral sacs. These in turn form a peripheral junction with the sarcolemma, which is probably equivalent to the SR junction with the T system of cross-striated fibers.

So far, T system elements have been described only in the oblique striated fibers of the worm *Ascaris* (Reger, 1964; Rosenbluth, 1965a). In these fibers, the T system has a dense content and forms dyads with elements of the SR. The structure of the dyadic junction is essentially similar to that of vertebrate and arthropod fibers described in the previous sections (Rosenbluth, 1969a). It is likely that the SR and T systems (the latter where present) perform similar roles in oblique striated fibers as in the cross-striated ones.

Acknowledgment

Written under the support of USPHS Grant No. 5R01–NS08893.

REFERENCES

Adrian, R. H., Costantin, L. L., and Peachey, L. D. (1969). *J. Physiol.* (*London*) **204**, 231.

Albuquerque, E. X., and Thesleff, S. (1968). *Acta Physiol. Scand.* **73**, 471.

Fig. 58. Oblique striated fibers from earthworm. Transversely oriented SR tubules (arrows) run at regular intervals, alternated with dense bodies, equivalent to the Z line (Z). No T system is present, but the SR makes peripheral junctions with the sarcolemma, see Chapter 8, Volume I. ($\times 30,000$.)

Andersson-Cedergren, E. (1959). *J. Ultrastruct. Res., Suppl.* **1**.
Ashurst, D. E. (1967). *J. Cell Sci.* **2**, 435.
Ashurst, D. E. (1969). *Tissue Cell* **1**, 485.
Atwood, H. L. (1967). *Amer. Zool.* **7**, 527.
Atwood, H. L. (1971). *J. Cell Biol.* **50**, 264.
Auber, J. (1963a). *C. R. Acad. Sci.* **256**, 2022.
Auber, J. (1963b). *J. Microsc. (Paris)* **2**, 233.
Auber, J. (1966). *J. Microsc. (Paris)* **5**, 28a.
Auber, J. (1967a). *Amer. Zool.* **7**, 451.
Auber, J. (1967b). *C. R. Acad. Sci.* **264**, 621.
Auber, J. (1967c). *C. R. Acad. Sci.* **264**, 2916.
Auber, J., and Couteaux, R. (1963). *J. Microsc. (Paris)* **2**, 309.
Back-y-Rita, P., and Ito, F. (1966). *J. Gen. Physiol.* **49**, 1177.
Bárány, M., Bárány, K., Reckard, T., and Volpe, A. (1965). *Arch. Biochem. Biophys.* **109**, 185.
Bárány, M., Conover, T. E. Bárány, K., and Goffart, M. (1967). *Fed. Proc., Fed. Amer. Soc. Exp. Biol.* **26**, 728 (abstr.).
Bennett, H. S. (1960). *In* "The Structure and Function of Muscle" (G. H. Bourne, ed.), Vol. 1, pp. 137–150. Academic Press, New York.
Bennett, H. S., and Porter, K. R. (1953). *Amer. J. Anat.* **93**, 61.
Bergman, R. A. (1964). *Bull. Johns Hopkins Hosp.* **114**, 325
Bergman, R. A. (1967). *J. Cell Biol.* **32**, 751.
Bertaud, W. S., Rayns, D. G., and Simpson, F. O. (1970). *J. Cell Sci.* **6**, 537.
Bianchi, C. P., and Bolton, T. C. (1967). *J. Pharmacol. Exp. Ther.* **157**, 388.
Birks, R. I. (1964). *In* "Muscle" (W. M. Paul *et al.*, eds.), pp. 199–216. Pergamon, Oxford.
Brandt, P. W., Reuben, J. P., Girardier, L., and Grundfest, H. (1965). *J. Cell Biol.* **25**, 233.
Bübenzer, H. J. (1966). *Z. Zellforsch. Mikrosk. Anat.* **69**, 520.
Bücher, T., and Pette, D. (1965). *Verh. Deut. Ges. Inn. Med.* **71**, 104.
Buchtal, F., and Schmalbruch, H. (1969). *Nature (London)* **222**, 89.
Buller, A. J., Eccles, J. C., and Eccles, R. M. (1960). *J. Physiol. (London)* **150**, 417.
Cheng, K., and Breinin, G. M. (1966). *Invest. Ophthalmol.* **5**, 535.
Close, R. (1965). *Nature (London)* **206**, 831.
Close, R., and Hoh, J. J. (1968). *J. Physiol. (London)* **198**, 103.
Cohen, M. J., and Hess, A. (1967). *Amer. J. Anat.* **121**, 285.
Costantin, L. L. (1970). *J. Gen. Physiol.* **55**, 703.
Costantin, L. L., Franzini-Armstrong, C., and Podolsky, R. J. (1965). *Science* **147**, 158.
Costantin, L. L., Podolsky, R. J., and Tice, L. W. (1967). *J. Physiol. (London)* **188**, 261.
Deamer, D. W., and Baskin, R. J. (1969). *J. Cell Biol.* **42**, 296.
DiDio, L. A. J. (1967). *Anat. Rec.* **159**, 335.
Dorai-Ray, B. S., and Cohen, M. J. (1964). *Naturwissenschaften* **51**, 224.
Ebashi, S. (1961). *Progr. Theor. Phys.* **17**, 35.
Ebashi, S. (1962). *Nature (London)* **194**, 378.
Ebashi, S., and Endo, M. (1968). *Progr. Biophys. Mol. Biol.* **18**, 123.
Ebashi, S., and Lipmann, F. (1962). *J. Cell Biol.* **14**, 389.
Eccles, J. C. (1963). *In* "The Effect of Use and Disuse on Neuromuscular Functions" (E. Gutmann and P. Hník, eds.), p. 111. Publ. House Czech. Acad. Sci., Prague.
Edge, M. B. (1970). *Develop. Biol.* **23**, 634.

Edgerton, V. R., and Simpson, D. R. (1969). *J. Histochem. Cytochem.* **17**, 828.
Edström, L., and Kugelberg, E. (1968). *J. Neurol., Neurosurg. Psychiat.* **31**, 424.
Eisenberg, B. R. (1972). *J. Cell. Biol.* **55**, 68a.
Eisenberg, B. R., and Eisenberg, R. S. (1968). *J. Cell Biol.* **39**, 451.
Eisenberg, B. (1969). Personal communication.
Elul, R., Miledi, R., and Stefani, E. (1968). *Nature* (*London*) **217**, 1274.
Endo, M. (1964). *Nature* (*London*) **202**, 1115.
Engel, A. G., and Tice, L. W. (1966). *J. Cell Biol.* **31**, 473.
Engel, W. K., and Irwin, R. L. (1967). *Amer. J. Physiol.* **213**, 511.
Engel, W. K., Brooke, M. H., and Nelson, P. G. (1966). *Ann. N.Y. Acad. Sci.* **138**, 160.
Ezerman, E. B., and Ishikawa, H. (1967). *J. Cell Biol.* **35**, 405.
Fahimi, H. D., and Amarasingham, C. R. (1964). *J. Cell Biol.* **22**, 29.
Fahimi, H. D., and Karnovsky, M. J. (1966). *J. Cell Biol.* **29**, 113.
Fahrenbach, W. H. (1963). *J. Cell Biol.* **17**, 629.
Fahrenbach, W. H. (1964). *J. Cell Biol.* **22**, 477.
Fahrenbach, W. H. (1967). *J. Cell Biol.* **35**, 69.
Falk, G., and Fatt, P. (1964). *Proc. Roy. Soc., Ser. B* **160**, 69.
Fatt, P., and Katz, B. (1951). *J. Physiol.* (*London*) **115**, 320.
Fawcett, D. W., and Revel, J. P. (1961). *J. Biophys. Biochem. Cytol.* **10**, Suppl. 4, 89.
Fawcett, D. W., and McNutt, N. S. (1969). *J. Cell Biol.* **42**, 1.
Fedde, M. R. (1969). *J. Gen. Physiol.* **53**, 624.
Fernand, V. S. V., and Hess, A. (1969). *J. Physiol.* (*London*) **200**, 547.
Flood, P. R. (1966). *J. Comp. Neurol.* **126**, 58.
Ford, L. E., and Podolsky, R. J. (1970). *Science* **167**, 58.
Forssman, W. C., and Girardier, L. (1970). *J. Cell Biol.* **44**, 1.
Franke, W. W., and Schinko, W. (1969). *J. Cell Biol.* **42**, 326.
Franzini-Armstrong, C. (1963). *J. Cell Biol.* **19**, 637.
Franzini-Armstrong, C. (1964). *Fed. Proc., Fed. Amer. Soc. Exp. Biol.* **23**, 887.
Franzini-Armstrong, C. (1970a). *J. Cell Biol.* **47**, 488.
Franzini-Armstrong, C. (1970b). *J. Cell Biol.* **47**, 64a.
Franzini-Armstrong, C. (1970c). *J. Cell Sci.* **6**, 559.
Franzini-Armstrong, C. (1964). Unpublished observations.
Franzini-Armstrong C. (1967). Unpublished observations.
Franzini-Armstrong, C., and Porter, K. R. (1964a). *J. Cell Biol.* **22**, 675.
Franzini-Armstrong, C., and Porter, K. R. (1964b). *In* "From Molecule to Cell" (P. Buffa, ed.), pp. 331–346. Consiglio Nazionale delle Ricerche, Roma.
Freygang, W. H., Goldstein, D. A., Hellam, D. C., and Peachey, L. D. (1964). *J. Gen. Physiol.* **48**, 235.
Fujino, M., Yamaguchi, T., and Suzuli, K. (1961). *Nature* (*London*) **192**, 1159.
Gage, P. W., and Eisenberg, R. S. (1967). *Science* **158**, 1702.
Gage, P. W., and Eisenberg, R. S. (1969). *J. Gen. Physiol.* **53**, 298.
Gauthier, G. F. (1969). *Z. Zellforsch. Mikrosk. Anat.* **95**, 462.
Gauthier, G. F. (1970). *In* "The Physiology and Biochemistry of Muscle as a Food" (E. J. Briskey, R. G. Cassens, and B. B. Marsh, eds.), Vol. 2, pp. 103–130. Univ. of Wisconsin Press, Madison.
Gauthier, G. F., and Padykula, H. A. (1966). *J. Cell Biol.* **28**, 333.
George, J. C., and Naik, R. M. (1959). *Biol. Bull.* **116**, 239.
Gillis, J. M. (1969). *J. Physiol.* (*London*) **205**, 849.
Ginsborg, B. L. (1960). *J. Physiol.* (*London*) **154**, 581.
Goldstein, M. A. (1969a). *Z. Zellforsch. Mikrosk. Anat.* **102**, 459.
Goldstein, M. A. (1969b). *Z. Zellforsch. Mikrosk. Anat.* **102**, 31.

Gonzales-Serratos, H. (1966). *J. Physiol. (London)* **185**, 20P.

Gori, Z., Pellegrino, C., and Pollera, M. (1965). *Atti Congr. Ital. M. E. Bologna* **5**, 203.

Greaser, M. L., Cassens, R. G., Hoekstra, W. G., and Briskey, E. J. (1969). *J. Cell. Physiol.* **74**, 37.

Grinyer, I., and George, J. C. (1969a). *Can. J. Zool.* **47**, 517.

Grinyer, I. and George, J. C. (1969b). *Can. J. Zool.* **47**, 771.

Guth, L., and Watson, P. K. (1967). *Physiol. Bohemoslov.* **16**, 244.

Hagopian, M., and Spiro, D. (1967). *J. Cell Biol.* **32**, 535.

Hall-Craggs, E. G. B. (1968). *J. Anat.* **102**, 241.

Hanson, J., and Lowy, J. (1961). *Proc. Roy. Soc., Ser. B* **154**, 173.

Hasselbach, W. (1964). *Proc. Roy. Soc., Ser. B* **160**, 501.

Hasselbach, W., and Elfvin-Lars, G. (1967). *J. Ultrastruct. Res.* **17**, 598.

Hasselbach, W., and Makinose, M. (1961). *Biochem. Z.* **333**, 518.

Henneman, C., and Olson, C. B. (1965). *J. Neurophysiol.* **28**, 581.

Hess, A. (1960). *Amer. J. Anat.* **107**, 129.

Hess, A. (1961). *J. Physiol. (London)* **157**, 221.

Hess, A. (1963). *Amer. J. Anat.* **113**, 347.

Hess, A. (1965). *J. Cell Biol.* **26**, 467.

Hess, A. (1967). *Invest. Ophthalmol.* **6**, 217.

Hess, A. (1970). *Physiol. Rev.* **50**, 40.

Hess, A., and Pilar, G. (1963). *J. Physiol. (London)* **169**, 780.

Hidaka, T., and Toida, N. (1969). *J. Physiol. (London)* **201**, 49.

Hill, A. V. (1948). *Proc. Roy. Soc., Ser. B* **135**, 446.

Hill, A. V. (1949). *Proc. Roy. Soc., Ser. B* **136**, 399.

Hník, P., Jirmanová, I., Vylický, L., and Zelená, J. (1967). *J. Physiol. (London)* **193**, 309.

Hodgkin, A. L., and Horowicz, P. (1960). *J. Physiol. (London)* **153**, 370.

Hodgkin, A. L., Huxley, A. F., and Katz, B. (1952). *J. Physiol. (London)* **116**, 424.

Howell, J. N. (1969). *J. Physiol. (London)* **201**, 515.

Howell, J. N., and Jenden, D. J. (1967). *Fed. Proc., Fed. Amer. Soc. Exp. Biol.* **26**, 553.

Hoyle, G. (1965). *Science* **149**, 70.

Hoyle, G. (1968). *J. Exp. Zool.* **167**, 471.

Hoyle, G., and McNeill, P. A. (1968). *J. Exp. Zool.* **167**, 487.

Hoyle, G., McAlear, J. H., and Selverston, A. (1965). *J. Cell Biol.* **26**, 621.

Hoyle, G., McNeill, P. A., and Walcott, B. (1966). *J. Cell Biol.* **30**, 197.

Huxley, A. F. (1964). *Ann. N.Y. Acad. Sci.* **81**, 446.

Huxley, A. F., and Taylor, R. E. (1958). *J. Physiol. (London)* **144**, 426.

Huxley, H. E. (1964). *Nature (London)* **202**, 1067.

Ikemoto, N., Sreter, F., and Gergely, J. (1970). *J. Cell Biol.* **47**, 94A.

Jahromi, S. S., and Atwood, H. L. (1967). *Can. J. Zool.* **45**, 601.

Johnson, J. R., and Sommer, J. R. (1967). *J. Cell Biol.* **33**, 103.

Karnovsky, M. J. (1965). *J. Cell Biol.* **27**, 49A.

Katz, B. (1948). *Proc. Roy. Soc., Ser. B* **135**, 506.

Kelly, A. M. (1969). *J. Cell Biol.* **43**, 65A.

Kelly, D. E. (1969). *J. Ultrastruct. Res.* **29**, 37.

Kelly, D. E., and Cahill, M. A. (1969). *J. Cell Biol.* **43**, 66A.

Kilarski, W. (1964). *Acta Biol. Cracov., Ser. Zool.* **7**, 161.

Kilarski, W. (1965). *Acta Biol. Cracov., Ser. Zool.* **8**, 51.

Kilarski, W. (1966). *Bull. Acad. Pol. Sci., Ser. Sci. Biol.* **14**, 575.

Kilarski, W. (1967a). *Z. Zellforsch. Mikrosk. Anat.* **79**, 562.

Kilarski, W. (1967b). *Acta Med. Pol.* **8**, 399.

Kilarski, W., and Bigai, J. (1969). *Z. Zellforsch. Mikrosk. Anat.* **94**, 194.

Komnick, H. (1969). *Histochemie* **18**, 24.

Krüger, P. (1962). "Tetanus und Tonus der quergestreiften Skelettmuskel der Wielbertiere und des Menschen." Akad. Verlag, Leipzig.

Kuffler, S. W., and Vaughan Williams, E. (1953). *J. Physiol. (London)* **121**, 289.

Lännergren, J., and Smith, R. S. (1966). *Acta Physiol. Scand.* **68**, 263.

Lanzavecchia, G. (1968). *Atti Accad. Naz. Lincei, Cl. Sci. Fis., Mat. Natur., Rend.* **44**, 32.

Lasiewski, R. C., Galey, F. R., and Vasquez, C. (1965). *Nature (London)* **205**, 404.

Leeson, C. R., and Leeson, T. S. (1969). *J. Anat.* **105**, 363.

Luft, J. H. (1966). *Electron Microsc., Proc. Int. Congr., 6th, 1966* Vol. 2, pp. 65–66.

McPhedron, A. M., Wuerker, R. B., and Henneman, E. (1965). *J. Neurophysiol.* **28**, 71.

Margreth, A., Muscatello, U., and Andersson-Cedergren, E. (1963). *Exp. Cell Res.* **32**, 484.

Marsh, B. B. (1952). *Biochim. Biophys. Acta* **9**, 247.

Martonosi, A. (1968). *Biochim. Biophys. Acta* **150**, 694.

Matyushkin, D. P. (1963). *Exp. Biol. Med.* **55**, 3.

Mauro, A. (1961). *J. Biophys. Biochem. Cytol.* **9**, 493.

Mayr, R. (1971). *Tissue Cell* **3**, 433.

Mendelson, M. (1969). *J. Cell Biol.* **42**, 548.

Miledi, R., and Zelená, J. (1966). *Nature (London)* **210**, 855.

Miledi, R., Stefani, E., and Zelená, J. (1968). *Nature (London)* **220**, 497.

Mileoli, R., Stefani, E., and Steinbach, A. B. (1971). *J. Physiol. (London)* **217**, 737.

Miller, J. E. (1967). *Invest. Ophthalmol.* **6**, 18.

Morita, S., Cassens, R. G., and Briskey, E. J. (1969). *Stain Technol.* **44**, 283.

Muir, A. R., Kanji, A. H. M., and Allbrook, D. (1965). *J. Anat.* **99**, 435.

Muscatello, U., Andersson-Cedergren, E., Azzone, G. F., and Von der Decken, A. (1961). *J. Biophys. Biochem. Cytol.* **10**, Suppl. 4, 201.

Nagai, T., Makinose, M., and Hasselbach, W. (1960). *Biochim. Biophys. Acta* **43**, 223.

Nakajima, S., Nakajima, Y., and Peachey, L. D. (1969). *J. Physiol. (London)* **200**, 115P.

Nakajima, Y. (1969). *Tissue Cell* **1**, 229.

Natori, R. (1954). *Jikeikai Med. J.* **1**, 119.

Padykula, H. A., and Gauthier, G. F. (1967). *Excerpta Med. Found. Int. Congr. Ser.* **147**.

Padykula, H. A., and Gauthier, G. F. (1970). *J. Cell Biol.* **46**, 27.

Page, S. G. (1964). *J. Physiol. (London)* **175**, 10P.

Page, S. G. (1965). *J. Cell Biol.* **26**, 477.

Page, S. G. (1968a). *J. Physiol. (London)* **197**, 709.

Page, S. G. (1968b). *Brit. Med. Bull.* **24**, 170.

Page, S. G. (1969). *J. Physiol. (London)* **205**, 131.

Page, S. G. (1965). Unpublished observation.

Page, S. G., and Slater, S. R. (1965). *J. Physiol. (London)* **179**, 58P.

Pasquali-Ronchetti, I. (1969). *J. Cell Biol.* **40**, 269.

Peachey, L. D. (1961). *J. Biophys. Biochem. Cytol.* **10**, Suppl. 4, 159.

Peachey, L. D. (1965a). *J. Cell Biol.* **25**, 209.

Peachey, L. D. (1965b). *Fed. Proc., Fed. Amer. Soc. Exp. Biol.* **24**, 1124.

Peachey, L. D. (1965c). *J. Cell Biol.* **31**, 84A.

Peachey, L. D. (1967). *Amer. Zool.* **7**, 505.

Peachey, L. D. (1968). *Annu. Rev. Physiol.* **30**, 401.
Peachey, L. D. (1972). In press.
Peachey, L. D., and Huxley, A. F. (1960). *Fed. Proc., Fed. Amer. Soc. Exp. Biol.* **19**, 257.
Peachey, L. D., and Huxley, A. F. (1962). *J. Cell Biol.* **13**, 177.
Peachey, L. D., and Huxley, A. F. (1964). *J. Cell Biol.* **23**, 70A.
Peachey, L. D., and Porter, K. R. (1959). *Science* **129**, 721.
Peachey, L. D., and Schild, R. F. (1968). *J. Physiol. (London)* **194**, 249.
Pease, D. C., Jenden, D. J., and Howell, J. (1965). *J. Cell. Comp. Physiol.* **65**, 141.
Pellegrino, C., and Franzini, C. (1963). *J. Cell Biol.* **17**, 327.
Pellegrino, C., and Franzini-Armstrong, C. (1969). *Int. Rev. Exp. Pathol.* **7**, 139.
Peterson, R. P., and Pepe, F. A. (1961). *Amer. J. Anat.* **109**, 277.
Philpott, C. W., and Goldstein, M. A. (1967). *Science* **155**, 1019.
Pilar, G., and Hess, A. (1966). *Anat. Rec.* **154**, 243.
Podolsky, R. J., and Costantin, L. L. (1964). *Fed. Proc., Fed. Amer. Soc. Exp. Biol.* **23**, 933.
Podolsky, R. J., Hall, T., and Hatchett, S. L. (1970). *J. Cell Biol.* **44**, 699.
Porter, K. R. (1954). *J. Histochem. Cytochem.* **2**, 346.
Porter, K. R. (1956). *J. Biophys. Biochem. Cytol.* **2**, 163.
Porter, K. R. (1961). *J. Biophys. Biochem. Cytol.* **10**, Suppl. 4, 219.
Porter, K. R., and Franzini-Armstrong, C. (1965). *Sci. Amer.* **212**, (3), 75.
Porter, K. R., and Palade, G. E. (1957). *J. Biophys. Biochem. Cytol.* **3**, 269.
Portzehl, H. P. C., Caldwell, P. C., and Ruegg, J. C. (1964). *Biochim. Biophys. Acta* **79**, 581.
Pringle, J. W. S. (1965). *In* "The Physiology of Insecta" (M. Rockstein, ed.), Vol. 2, p. 283. Academic Press, New York.
Pucci, I., and Afzelius, B. A. (1962). *J. Ultrastruct. Res.* **7**, 210.
Rayns, D. G., Simpson, F. O., and Bertaud, W. S. (1968). *J. Cell Sci.* **3**, 475.
Reger, J. F. (1961). *J. Biophys. Biochem. Cytol.* **10**, Suppl. 4, 111.
Reger, J. F. (1964). *J. Ultrastruct. Res.* **10**, 48.
Reger, J. F. (1965). *Z. Zellforsch. Mikrosk. Anat.* **67**, 196.
Reger, J. F. (1967a). *J. Ultrastruct. Res.* **18**, 595.
Reger, J. F. (1967b). *J. Ultrastruct. Res.* **20**, 72.
Reger, J. F., and Cooper, D. P. (1967). *J. Cell Biol.* **33**, 531.
Revel, J. P. (1962). *J. Cell Biol.* **12**, 571.
Revel, J. P., Fawcett, D. W., and Philpott, C. W. (1963). *J. Cell Biol.* **16**, 187.
Robertson, J. D. (1956). *J. Biophys. Biochem. Cytol.* **2**, 369.
Rockstein, M., and Bhatnager, P. L. (1965). *J. Insect Physiol.* **11**, 481.
Rogers, D. C. (1969). *Z. Zellforsch. Mikrosk. Anat.* **99**, 315.
Romanul, F. C. A. (1964). *Nature (London)* **201**, 307.
Romanul, F. C. A., and Van der Menlen, J. (1966). *Nature (London)* **212**, 1369.
Rosenbluth, J. (1965a). *J. Cell Biol.* **25**, 495.
Rosenbluth, J. (1965b). *J. Cell Biol.* **26**, 579.
Rosenbluth, J. (1967). *J. Cell Biol.* **34**, 15.
Rosenbluth, J. (1968). *J. Cell Biol.* **36**, 245.
Rosenbluth, J. (1969a). *J. Cell Biol.* **42**, 817.
Rosenbluth, J. (1969b). *J. Cell Biol.* **42**, 534.
Sandow, A. (1965). *Pharmacol. Rev.* **17**, 265.
Schiaffino, S., and Margreth, A. (1968). *J. Cell Biol.* **41**, 855.
Schiaffino, S., Hanzlíková, V., and Pierobon, S. (1970). *J. Cell Biol.* **47**, 107.
Schmalbruch, H. (1967). *Z. Zellforsch. Mikrosk. Anat.* **79**, 64.
Selverston, A. (1967). *Amer. Zool.* **7**, 515.

Shafiq, S. A. (1964). *Quart. J. Microsc. Sci.* **105**, 1.
Shimada, Y., Fischman, D. A., and Moscona, A. A. (1967). *J. Cell Biol.* **35**, 445.
Simpson, F. O., and Oertelis, S. J. (1962). *J. Cell Biol.* **12**, 91.
Slautterback, D. B. (1965). *J. Cell Biol.* **24**, 1.
Smith, D. S. (1961a). *J. Biophys. Biochem. Cytol.* **10**, Suppl. 4, 123.
Smith, D. S. (1961b). *J. Biophys. Biochem. Cytol.* **11**, 119.
Smith, D. S. (1963). *J. Cell Biol.* **19**, 115.
Smith, D. S. (1965). *J. Cell Biol.* **27**, 379.
Smith, D. S. (1966a). *Progr. Biophys. Mol. Biol.* **16**, 107.
Smith, D. S. (1966b). *J. Cell Biol.* **28**, 109.
Smith, D. S. (1966c). *J. Cell Biol.* **29**, 449.
Smith, D. S., and Sacktor, B. (1970). *Tissue Cell* **2**, 355.
Smith, D. S., Gupta, B. L., and Smith, U. (1966). *J. Cell Sci.* **1**, 49.
Smith, R. S., and Lännergren, J. (1968). *Nature (London)* **217**, 281.
Sommer, J. R., and Hasselbach, W. (1967). *J. Cell Biol.* **34**, 902.
Stein, J. M., and Padykula, H. A. (1962). *Amer. J. Anat.* **110**, 103.
Takeuchi, A. (1969). *J. Cell. Comp. Physiol.* **54**, 211.
Tice, L. W., and Engel, A. G. (1966). *J. Cell Biol.* **31**, 118A.
Tice, L. W., and Engel, A. G. (1967). *Amer. J. Pathol.* **50**, 311.
Toselli, P. A., and Pepe, F. A. (1968). *J. Cell Biol.* **37**, 445.
Van der Kloot, W. G. (1966). *Comp. Biochem. Physiol.* **17**, 75.
Van der Kloot, W. G. (1969). *J. Cell Biol.* **42**, 562.
Veratti, E. (1902). *Mem. Ist. Lomb., Cl. Sci. Nat.* **19**, 87 [translated in *J. Biophys. Biochem. Cytol.* **10**, Suppl. 4, 3 (1961)].
Walker, S. M., and Schrodt, G. R. (1965). *J. Cell Biol.* **27**, 671.
Walker, S. M., and Schrodt, G. R. (1966a). *Anat. Rec.* **155**, 1.
Walker, S. M., and Schrodt, G. R. (1966b). *Nature (London)* **211**, 935.
Walker, S. M., and Schrodt, G. R. (1967). *Nature (London)* **216**, 985.
Walker, S. M., and Schrodt, G. R. (1968). *J. Cell Biol.* **37**, 564.
Weber, A. (1966). *Curr. Top. Bioenerg.* **1**, 203–254.
Weber, A., Herz, R., and Reiss, I. (1964). *Fed. Proc., Fed. Amer. Soc. Exp. Biol.* **23**, 896.
Winegrad, S. (1965). *Fed. Proc., Fed. Amer. Soc. Exp. Biol.* **24**, 1146.
Winegrad, S. (1970). *J. Gen. Physiol.* **55**, 77.
Zacks, S. I., and Saito, A. (1970). *J. Histochem. Cytochem.* **18**, 302.
Zadunaisky, J. A. (1966). *J. Cell. Biol.* **31**, C11.

AUTHOR INDEX

Numbers in italics refer to the pages on which the complete references are listed.

A

Aagaard, O. C., 219, *237*
Abe, Y., 20, 21, 28, 29, *89*
Abraham, R., 146, *153*
Adal, M. N., 376, 389, 391, 392, 393, 395, 396, 403, 409, 410, 411, 413, 415, 419, 463, 465, *474, 475*
Adams, N. I., Jr., 119, *156*
Adams, R. D., 316, 319, 320, 343, 351, 352, 353, 354, *359, 363*, 367, 440, 449, *475*
Adrian, R. H., 125, *151*, 569, *613*
Afifi, A. K., 355, *359*
Afzelius, B. A., 613, *618*
Aguayo, A. J., 322, 323, *359*
Alanis, J., 102, 143, *155*
Albuquerque, E. X., 589, *613*
Alexander, L. M., 516, *530*
Alexandrowicz, J. S., 373, *475*
Allbrook, D., 407, *480*, 567, *617*
Allbrook, D. B., 344, *360*
Allen, C. R., 299, *309*
Allen, W. F., 373, *475*
Allen, W. M., 35, *88*
Altschul, R., 346, *359*
Alvord, E. C., 440, *449*
Amakawa, T., 146, *157*
Amarasingham, C. R., 574, *615*
Amersbach, K., 377, *475*
Anda, G., 372, 373, 374, 388, 422, 423, 424, 425, 427, 430, 437, *475, 481*
Anderson, A. B. M., 54, *87*
Anderson, G. G., 53, *87*

Andersson, E., 105, *157*
Andersson-Cedergren, E., 99, 105, 143, 144, 145, 149, *151, 157*, 230, *241*, 382, 396, 405, 409, 468, 469, *475, 479*, 532, 537, 544, 545, 551, 558, 563, 567, 570, 574, *614, 617*
Andreasen, E., 345, *359*
Andreoli, T. C., 120, *157*
Andres, K. H., 397, *477*
Andrew, N. V., 345, *359*
Andrew, W., 345, *359*
Anitschkow, W., 356, *359*
Appelberg, B., 459, *475*
Appleton, A. B., 204, 205, 207, *239*
Araya, K., 144, 146, *155*
Arey, L. B., 197, *237*
Arnold, J., 189, *237*
Ashurst, D. E., 545, 565, 566, 591, 609, *614*
Atwood, H. L., 595, 597, *614, 616*
Auber, J., 603, 607, *614*
Aubert, X., 293, *306*
Aufrecht, E., 312, 313, *359*
Azzone, G. F., 570, *617*

B

Babinski, J., 366, 439, *475*
Baccetti, B., 105, 151, *152*
Bach-y-Rita, P., 367, *475*, 585, *614*
Baker, D. D., 205, *241*
Baldwin, K. M., 131, 137, *152*
Banker, B. Q., 468, 469, *475*
Banks, J. B., *239*

SUBJECT INDEX

A

A bands, 214–215
Abortion
  incomplete, prostglandins and, 53, 57
  prostglandins as inducers, 52–63
    contracture development, 56
    placenta role, 57–59
    progesterone role, 58–59
Actin filaments, contraction and, 15
Actinopoda, contractile elements, 160
Actomyosin, uterine
  ATPase activity, 8
  ovariectomy and, 16
Adenine nucleotides, rigor changes,
    271–272
Adenosine triphosphate
  plasticizer, experimental, 244–245
  resynthesis, 245–246, 286
  rigor process and
    level during, 261–264
    nutrition, 263
    phosphocreatine, 283–284
    resynthesis, 261, 303
Adrenaline
  cause of dark beef, 268
  exhaustion-producing agent, 267–268
Amphioxus fibers, 561
Arthropods
  SR fenestrae, 558
  SR-T system junctions, 555
Atrial muscle
  electrical properties, 136–138
    time constant, 136–137
  myoplasmic continuity, 137
  transverse cell boundaries, 137
Atrioventricular bundle, 236

Atrium
  electrical properties, 118
  mammalian, 102
Atrophy
  denervation, 449, 463
  experimental, 443–448
  infantile spinal, see Werdnig–Hoffman
    disease
Axons, motor, 456–457
Axoplasm, terminal, 489–495
  arborizations, 490–493
    neurofibrillar bundle, 493
    neurofibrils, 495
    neurofilaments, 495
    peripheral layer, 493
    vesicles, 494
  inclusions, 493–494
  staining methods, 489–491

B

Basal cells, see Myoepithelial cells
Basal lamina, 209, 223–224
Beef sternomandibularis
  ATP levels and phosphocreatine, 283
  rigor, changes in, 266–267, 270–271
Bell mouth of Henle, 510–511, 513
Bowman's degeneration, 319–320
Bundle of His, 102

C

Capacitance, 117, 119
  membrane, 119
Cardiac cell boundaries
  class I
    classification, 96–97